ADVANCES IN CHEMICAL PHYSICS

VOLUME XCV

EDITORIAL BOARD

Advances in CHEMICAL PHYSICS
Surface Properties

Edited by

I. PRIGOGINE

University of Brussels
Brussels, Belgium
and
University of Texas
Austin, Texas

and

STUART A. RICE

Department of Chemistry
and
The James Franck Institute
The University of Chicago
Chicago, Illinois

VOLUME XCV

AN INTERSCIENCE® PUBLICATION
JOHN WILEY & SONS, INC.
NEW YORK • CHICHESTER • BRISBANE • TORONTO • SINGAPORE

This text is printed on acid-free paper.

An Interscience® Publication

Library of Congress Catalog Number: 58-9935

ISBN 0-471-15430-X

Printed in the United States of America

10 9 8 7 6 5 4 3 2 1

CONTRIBUTORS TO VOLUME XCV

DOUGLAS J. DOREN, Department of Chemistry and Biochemistry, University of Delaware, Newark, Delaware

DOMINIK MARX, Max-Planck-Institut für Festkörperforschung, Stuttgart, Germany

KEIJI MOROKUMA, Cherry L. Emerson Center for Scientific Computation and Department of Chemistry, Emory University, Atlanta, Georgia

DJAMALADDIN G. MUSAEV, Cherry L. Emerson Center for Scientific Computation and Department of Chemistry, Emory University, Atlanta, Georgia

SANFORD A. SAFRON, Department of Chemistry, Florida State University, Tallahassee, Florida

HORST WIECHERT, Institut für Physik, Johannes Gutenberg-Universität, Mainz, Germany

INTRODUCTION

Few of us can any longer keep up with the flood of scientific literature, even in specialized subfields. Any attempt to do more and be broadly educated with respect to a large domain of science has the appearance of tilting at windmills. Yet the synthesis of ideas drawn from different subjects into new, powerful, general concepts is as valuable as ever, and the desire to remain educated persists in all scientists. This series, *Advances in Chemical Physics*, is devoted to helping the reader obtain general information about a wide variety of topics in chemical physics, a field which we interpret very broadly. Our intent is to have experts present comprehensive analyses of subjects of interest and to encourage the expression of individual points of view. We hope that this approach to the presentation of an overview of a subject will both stimulate new research and serve as a personalized learning text for beginners in a field.

I. PRIGOGINE
STUART A. RICE

CONTENTS

ADVANCES IN CHEMICAL PHYSICS

VOLUME XCV

KINETICS AND DYNAMICS OF HYDROGEN ADSORPTION AND DESORPTION ON SILICON SURFACES

DOUGLAS J. DOREN

Department of Chemistry and Biochemistry, University of Delaware, Newark, Delaware

CONTENTS

Advances in Chemical Physics, Volume XCV, Edited by I. Prigogine and Stuart A. Rice.
ISBN 0-471-15430-X

I. INTRODUCTION

Hydrogen adsorbed on silicon may be the simplest semiconductor adsorption system. Over the last decade, this system has become established as a model in surface science. This system has practical importance, because many of the chemical reactions used in silicon device processing leave hydrogen on the surfaces [1]. The strong Si–H bond can passivate the surface and limit further reaction. Adsorbed hydrogen limits the growth rate in low-temperature (below 770 K) chemical vapor deposition, [1–4], and it alters the morphology of epitaxially grown films [5, 6].

In the 1970s and 1980s both the clean and H-covered Si surfaces were characterized by diffraction and spectroscopic methods, but only in the last decade have there been reproducible studies of chemical kinetics and dynamics on well-characterized silicon surfaces. Despite the conceptual simplicity of hydrogen as an adsorbate, this system has turned out to be rich and complex, revealing new principles of surface chemistry that are not typical of reactions on metal surfaces. For example, the desorption of hydrogen, in which two adsorbed H atoms recombine to form H_2, is approximately first order in H coverage on the Si(100) surface. This result is unexpected for an elementary reaction between two atoms, and recombinative desorption on metals is typically second order. The fact that first-order desorption kinetics has now been observed on a number of covalent surfaces demonstrates its broader significance.

Adsorption on Si also provides a meeting ground between theory and experiment. Accurate predictions from first principles are available to complement experimental work. From the theorist's viewpoint, semiconductors allow several technical simplifications. In contrast to metals, the electrons in semiconductors are localized, and a small cluster of silicon atoms has bonding properties similar to those of the bulk. Sophisticated quantum chemistry methods that reliably predict structure and energetics can be applied directly to discrete clusters. This approach has been successful on Si surfaces, though such properties as adsorbate binding energies on metals are difficult to predict using cluster models.

Semiconductor surfaces present new complications as well. Silicon surfaces exhibit complex reconstructions, and the surface structure is altered by adsorbates. Although silicon is available with exceptionally high bulk purity, observed structures have many defects that may affect reactivity. The dynamical behavior of adsorbates on silicon should also have novel features. Due to the low mass of hydrogen, energy transfer to the surface by impact must be small. Likewise, hydrogen vibrational frequencies are much higher than surface vibrational modes, so that vibrational energy transfer must occur via slow, multiple quantum transitions. The band gap of Si is large enough to prevent energy transfer from adsorbed molecules to electron–hole pairs. Since energy transfer to surface vibrations and electronic modes is inefficient, the vibrational lifetimes of H on Si are much longer than those observed on metals.

This chapter focuses on adsorption and desorption of molecular hydrogen, the best understood reactions on Si, though the discussion will also refer to related reactions such as surface diffusion. The discussion is also restricted to results from well-characterized single-crystal surfaces, with special emphasis on the (100) surface. This is the surface on which electronic devices are primarily grown. It reconstructs to a 2×1 surface unit cell, but this remains simple enough for accurate first principles calculations. The Si(111) surface is also important, but because of the complexity of the 7×7 reconstructed surface, the quality of feasible calculations remains limited. Reactions on the (111) surface are discussed here for comparison against the analogous properties of the (100) surface.

This chapter reviews what is known about kinetics and dynamics of hydrogen on Si. Section II is a brief description of the surface structure and basic properties of adsorbed hydrogen. Results from experiments and simple modeling of kinetics and dynamics are described in Section III, followed by a discussion of results from first principles theory in Section IV. The presentation generally follows a chronological organization, with results described in approximately the order that they were published. Many reported results appear mutually contradictory. To provide a guide to the detailed discussion, the experimental section begins with a brief summary of our current understanding. This serves to define the problems for theory, so that readers not interested in the details of the experimental work can turn directly to the discussion of theoretical work after this summary. Finally, Section V is a summary of outstanding issues that remain to be addressed.

II. OVERVIEW AND NOMENCLATURE

The early work of Law [7] showed that H_2 has a very low sticking probability on silicon, though H atoms (typically generated by passing H_2 over a hot tungsten filament) stick readily. Law found that H atoms have a sticking probability close to unity at low coverage ($\theta < 0.15$ ML), though this discreases dramatically at higher coverage. While the surfaces used by Law were known to be roughened by the cleaning procedure, these results have been repeatedly confirmed on various silicon surfaces. This implies that adsorption of H_2 is a rare event and therefore difficult to study directly. Similarly, studies of H_2 desorption typically dose the surface with atomic H and observe the subsequent recombination.

Law also found that the desorption behavior depended on coverage. At high coverage, there are two species that desorb at different temperatures, while at low coverage only the high-temperature state (β_1) is present. He suggested that the low-coverage species is formed from atomic H without activation barrier and corresponds to hydrogen bound to dangling bonds on the clean surface [8]. Formation of the high-coverage species is activated. While Law's methods do not meet current standards of surface analysis, his conclusions as stated here have been confirmed and extended by successive work over the last three and a half decades.

Current understanding of adsorption on Si(100) is summarized below. More detailed reviews are readily available. The structure of the clean surface has been determined experimentally by electron diffraction [9] and x-ray diffraction [10]. Liu and Hoffman [11] have given a concise review of the geometry and electronic structure of the bare surface. The classic review of Appelbaum and Hamann [12] describes the electronic structure of the bare surface and the bonding of H atoms. Accurate diffraction studies of the H-covered surfaces are not available. Infrared spectroscopy of H-covered Si surfaces has been reviewed by Chabal et al. [13], and the tunneling microscopy of these systems has been reviewed by Boland [14].

The bare Si(100)-2 $\times$ 1 reconstructed surface, consists of rows of "dimers" (Figure 1). Each surface atom has a single dangling bond. The dimer bond length is about 2.3 Å, slightly shorter than the bulk Si–Si bond length of 2.35 Å. On a bare surface with symmetric dimers, the dangling bonds can form a weak π bond. However, the surface energy is lowered slightly if dimers are allowed to buckle. This disrupts the π bond and leaves the "down" side of the dimer with a lower electron density than the "up" side. In fact, the ideal surface unit cell is larger than 2 $\times$ 1, since adjacent dimers tend to be tilted in opposite directions at low temperature [15]. This "anticorrelated" buckling relieves strain in the subsurface layers [15, 16]. Although the actual bonding on the buckled surface does not involve a typical π bond, the interaction between the dangling bonds in the dimer is often referred to as a π bond for simplicity.

At low H coverage, a monohydride state is populated, in which each

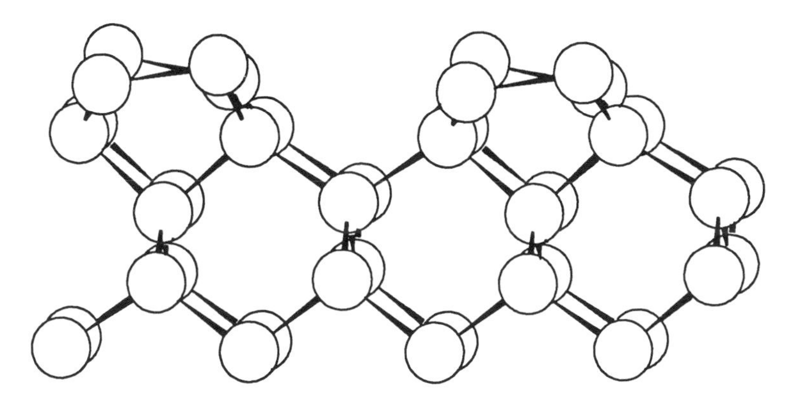

Figure 1. View of the clean Si(100)-2 × 1 surface. Two dimer rows are shown, which extend perpendicular to the plane of the paper. Successive dimers in the same row are buckled in opposite directions.

surface Si is bound to (at most) one H atom (Figure 2*a*). The Si–Si dimer bonds are undisturbed in this state, though theory predicts that the dimer bond length is increased by about 0.1 Å and the dimer buckling is removed. When this state is saturated, at 1 ML coverage, it has the same 2 × 1 periodicity as the bare surface. [Throughout this article, a monolayer (ML) refers to the number of atoms on the ideal surface, as distinguished from the number of dangling bonds or a saturation coverage.] At higher coverage, dihydrides (two H atoms bound to a single surface Si) are formed. This requires breaking the Si–Si dimer bond, but the stronger Si–H bonds compensate for the loss. Depending on the temperature, 3 × 1 and 1 × 1 diffraction patterns are observed. The 3 × 1 pattern is attributed to alternating rows of monohydride and dihydride (Fig. 2*b*); the 1 × 1 pattern is due to disorder or regions of dihydride (Fig. 2*c*). At high dihydride coverage and low temperature, further exposure to atomic H causes disruption of back bonds, leading to formation of trihydride and etching of the surface by loss of SiH_4 [17–19].

Gupta et al. [20] took time-dependent infrared spectra simultaneously with desorption rate measurements on porous silicon. They showed that the low temperature (β_2) desorption peak removes the dihydride, while the high temperature (β_1) peak corresponds to desorption from the monohydride state. This means that a pure monohydride state can be established by exposing the surface to atomic H, then annealing to about 630 K, a temperature high enough to desorb the dihydride but not the monohydride. Porous silicon was

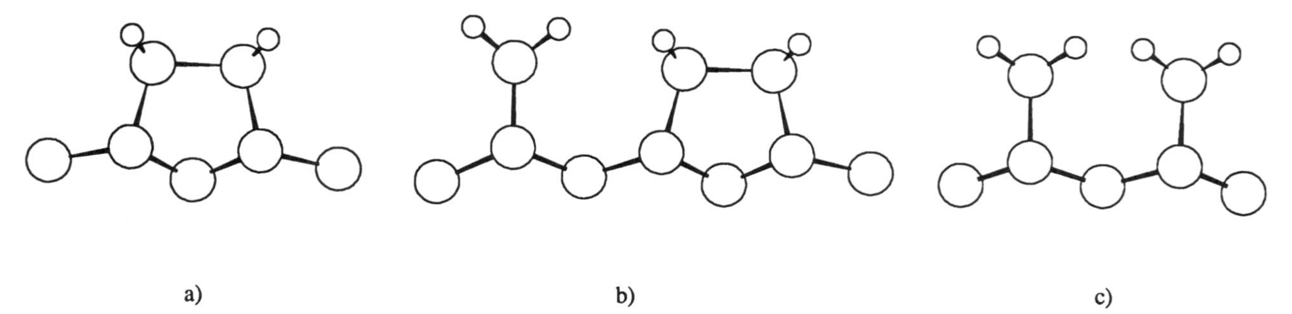

Figure 2. Schematic views of (*a*) the 2×1 monohydride phase, (*b*) the 3×1 mixed monohydride and dihydride phase, and (*c*) a 1×1 dihydride phase.

used because the Fourier transformer infrared (FTIR) measurement required a sample of high surface area, though Gupta et al. [20] argued that most of the surface consisted of (100) planes.

The Si(111)-7 × 7 surface has a much more complex structure, which is described in detail by Takayanagi et al. [21]. A concise description by Zangwill [22] is also useful. Adsorption of H on the Si(111)-7 × 7 surface also begins with a monohydride phase at low coverage, though higher hydrides can also be formed and etching occurs at high exposure. The number of dangling bonds on this surface is only 19/49 of the density of surface atoms, and the minimum distance between dangling bonds is over 4 Å, much greater than on the (100)-2 × 1 surface.

III. EXPERIMENTAL OBSERVATIONS

A. Summary

The low-coverage monohydride state has been most widely studied. Desorption of H_2 from Si(111) follows approximately second-order kinetics, with an activation barrier of 55–60 kcal/mol. At low coverage (below 0.1 ML), the kinetics are noninteger order, which has been modeled by assuming that H may bind to two types of sites, having different binding energies, barriers to diffusion, and barriers to recombination. Measurements of H diffusion on Si(111) show that the barrier is 35 kcal/mol, which is a much larger fraction of the desorption energy than is typical for diffusion on metals.

Surprisingly, desorption of H_2 from Si(100) obeys first-order kinetics, with an activation barrier of about 58 kcal/mol. Deviations from first-order behavior are seen below 0.1 ML. A model has been developed to fit the desorption kinetics at all coverages below 1 ML, assuming that it is energetically favored for two H atoms to bind on the same dimer rather than on two independent dimers. This ''prepairing'' model is supported by scanning tunneling microscope (STM) observations of pairing as well as accurate predictions of desorption kinetics over a wide range of coverage. This mechanism is not universally agreed upon. No measurements of the diffusion rate of H on Si(100) have been published.

Several dynamical measurements provide more insight into the mechanism. The sticking probability for H_2 on each surface is small (on the order of 10^{-6}). This implies that dissociative adsorption is an activated process or that adsorption occurs only at very rare defect sites. Recent measurements of the sticking probability have shown that adsorption is activated, though there is disagreement concerning the size of the barrier. The average energy in desorbing H_2 molecules has been measured. On Si(100) the average rotational energy is colder than expected at equilibrium, while there is excess vibrational excitation. The translational energy is nearly that expected at equilibrium. Similar results have been found for desorption from the monohydride and dihydride phases on the Si(100)-2 $\times$ 1 surface and the monohydride on the Si(111)-7 $\times$ 7 surface. The similarity of the energy distributions also suggests that the different surfaces have similar transition states. Since the internal energy distribution in the products presumably reflects the energy of the transition state, these results seem to imply that the transition state is not appreciably higher in energy than the products. If adsorption is truly activated, then the only explanation for the low energy in desorbing H_2 is that at least part of the excess energy at the transition state for desorption is deposited into the surface.

Proposed mechanisms for adsorption and desorption of H_2 on Si surfaces must address the following issues:

1. The kinetic order, including deviations from integer-order behavior
2. The origin of the low sticking probability
3. The energy distribution of desorbing molecules, including the similar distributions for different phases and surfaces

Despite the effort devoted to these issues over the last few years, particularly on Si(100), a clear concensus remains elusive. The situation on Si(111) is even less clear, due to the greater complexity of the surface structure.

B. Desorption Kinetics

1. *Early Work*

In early measurements of desorption kinetics, Kleint et al. [23] and Belyakov et al. [24] reported that desorption of H_2 from the Si(111) surface follows second-order kinetics with an activation barrier of 42 kcal/mol. Second-order kinetics was also reported on silicon films with the same activation barrier and on Si(100) with an activation barrier of 46 kcal/mol [25]. None of these surfaces were well-characterized, and coverages were unknown. Despite the similarity among these results, they have not been confirmed in more recent work. Several authors have commented on possible problems with this early work, ranging from poor surface preparation to inaccurate measurement methods [26–28].

2. *Si(111)-7 $\times$ 7*

The 1983 results of Schulze and Henzler [26] on Si(111)-7 $\times$ 7 have been repeatedly reproduced. They used temperature-programmed desorption (TPD) to measure desorption kinetics. TPD measurements involve dosing a clean surface with a given coverage of adsorbate at a relatively low temperature, then heating at a constant rate while observing the desorption rate as a function of temperature, typically using a mass spectrometer as detector. The desorption intensity as a function of time, temperature, and coverage can be modeled to determine the order and rate of desorption. At low coverage, Schulze and Henzler found one peak in the TPD curve, which became asymmetric and shifted to lower temperature as the coverage was increased. This indicates second-order kinetics. By fitting the TPD curves at low coverage they determined a desorption energy of 59 kcal/mol, with a prefactor of 136 cm^2/s. They noted that the TPD peak shift decreased at higher coverage, and they suggested that the kinetics might approach first-order near saturation. However, they cautioned that coverage-dependent kinetic parameters might be required for an exact description.

Schulze and Henzler calibrated their dosing exposure and desorption intensity, so that they could determine sticking probabilities and absolute coverage. They found that the saturation coverage was about 1.0 H atoms per surface Si atom. This is greater than the density of dangling bonds on this surface, so that higher hydrides must be formed, implying that some Si–Si back bonds must be broken. Schulze and Henzler observed roughening in the low-energy electron diffraction (LEED) pattern at coverages above roughly 0.5 ML. They also detected SiH_3 and SiH_4 in their mass spectrometer, indicating that the surface was etched by H exposure. Since a coverage of $19/49 = 0.39$ would saturate all the dangling bonds on a clean surface, their calibration of coverage is consistent with the interpretation that coverages over 0.5 cause roughening and even etching.

A few years later, Koehler et al. [27] applied TPD and laser-induced thermal desorption (LITD) to H/Si(111)-7 $\times$ 7. LITD is an isothermal method, in which the surface coverage is monitored as a function of time while the surface is held at constant temperature. This direct measure of the desorption rate at a fixed temperature removes the complications caused by the constantly changing temperature in TPD measurements. Interpretation of isothermal measurements is less model-dependent than interpretation of TPD experiments. Isothermal experiments may also be less susceptible to errors caused by uneven surface heating. The isothermal measurements confirmed that desorption is a second-order process on this surface, with an activation energy of 61 ± 4 kcal/mol and a prefactor of $1.2 \times 10^{1 \pm 1.3}$ cm^2/s. However, when Koehler et al. used the isothermal results to simulate TPD spectra, the predicted shifts of the TPD peak to lower temperature with increasing coverage were much larger than those observed in actual TPD spectra. They also found that standard analysis of the TPD spectra yielded Arrhenius parameters that depended on initial coverage, with apparent activation energies as low as 46 ± 5 kcal/mol.

These conflicts between the Arrhenius parameters derived from TPD and isothermal measurements suggest that the desorption process is not a simple second-order process. Koehler et al. suggested that desorption may be diffusion-influenced, and they tried to use the LITD method to measure the diffusion rate. Diffusion is slow enough that they were only able to determine an upper bound to the diffusion rate, with the diffusion coefficient $D \leq 10^{-9}$ cm^2/s. This fact is striking in itself, since it implies that the rate of diffusion is much slower (relative to the desorption rate) than is typical on metal surfaces. This upper bound to the diffusion rate is consistent with the hypothesis that desorption is limited by diffusion. However, without being able to determine actual diffusion rates, Koehler et al. could not determine an activation energy or prove that the barrier to diffusion is comparable to the desorption barrier. Later work by Reider et al. [29], using optical second-harmonic diffraction to detect diffusion, was more sensitive to diffusion across short distances. They determined a diffusion barrier of 35 kcal/mol, with a ''normal'' prefactor of 10^{-3} cm^2/s. This barrier is small enough that diffusion cannot be the rate-limiting step in desorption on this surface. However, this barrier is more than half the barrier to desorption, while on a metal the diffusion barrier is typically less than 20% of the desorption barrier [30]. The difference reflects the directional, covalent bonding that distinguishes semiconductors from metals.

It is significant that the LITD experiment of Koehler et al. was sensitive only to coverages above 0.1 ML. In principle, this might have prevented detection of deviations from second-order kinetics, since such deviations would be more pronounced at low coverage. However, subsequent work from this group demonstrated that this range of coverage was adequate to distinguish between first- and second-order kinetics [31].

Reider et al. [32] measured desorption of H_2 from Si(111)-7 $\times$ 7 at coverages below 0.14 ML. These measurements also used the isothermal technique, but changes in coverage were detected by an optical method, surface second-harmonic generation (SHG), which measures the density of surface dangling bonds. Koehler et al. [27] used mass spectrometry to detect desorbed H_2. The SHG signal is very sensitive to low coverages of H, allowing measurements with good signal-to-noise ratio at well below 0.01 ML. Reider et al. [32] showed that their data did not obey either first- or second-order kinetics. To obtain an empirical fit to the data they used a more general model,

$$-\frac{d\theta}{dt} = k_{\mathrm{des}}\, \theta^{m} \tag{1}$$

where m may take noninteger values. For each initial coverage (0.04–0.14 ML) and temperature (680–800 K) considered, a good fit was achieved with

$m = 1.5 \pm 0.2$. Non-integer-order kinetics have no direct microscopic interpretation, but they indicate either competing reaction paths or a non-elementary process. The activation energy was found to be 55 ± 2 kcal/mol for all coverages. From the Arrhenius plot Reider et al. [32] determined effective first- and second-order prefactors of $6 \times 10^{13}\ s^{-1}$ and $0.8\ cm^2/s$, respectively [32].

To rationalize the fractional-order kinetics, Reider et al. [32] proposed a model in which hydrogen may bind at two sites with different densities, residence times and desorption rates. These might correspond to the adatom and rest atom dangling bond sites, for example. These sites should have similar, but not identical, binding energies, diffusion rates, and desorption rates. Some extreme cases illustrate the possible behavior from this model. If site energies are sufficiently different that only one type of site is populated, desorption will have second-order kinetics. On the other hand, suppose that both sites are occupied, but only hydrogen at site A is mobile, while desorption occurs only at site B. Then desorption will be first order in the coverge at A. Intermediate cases yield kinetics between first and second order. This family of models is not the only possible mechanism, but Reider et al. [32] showed that they could reproduce the fractional-order kinetics of Eq. (1) with small differences in site energy (0.1–0.2 eV), a modest propensity for desorption at one site (ratio of desorption rates ≈ 20), and desorption occurring at sites representing 5–10% of a monolayer (consistent with the density of rest atom dangling bond sites). With these parameters, the model also predicts nearly second-order behavior at higher coverage, as observed by Koehler et al. [27]. In this sense, the work of Reider et al. [32] confirms the kinetic order (as well as the activation energy) found by Schulze and Henzler [26] and Koehler et al. [27] on this surface.

3. *Si(100)-2 × 1*

Shortly after the work of Koehler et al. on Si(111) appeared, Sinniah et al. [33] reported the unexpected result that desorption from Si(100) followed first-order kinetics. These experiments used an isothermal technique like that of Koehler et al. However, the measurements of Sinniah et al. were sensitive to coverages as low as 0.006 ML, allowing them to use initial coverages as low as 0.06 ML. This result demonstrates that the mechanism of recombinative desorption on Si(100) is qualitatively different from that on metal surfaces, where kinetics are second order [34]. It is easy to rationalize second-order desorption kinetics by a mechanism in which two independently diffusing H atoms must approach one another to recombine. A mechanism that yields first-order kinetics must imply either (a) some interaction between the H atoms (so that their positions are correlated) or (b) a multistep mechanism where the rate-limiting step involves motion of only one H atom.

Sinniah et al. [33] determined an activation barrier of 45 kcal/mol for desorption. They argued that this was inconsistent with a concerted desorption mechanism. If both Si–H bonds were broken and H_2 formed in a single elementary step, the reaction energy for that step would be given approximately by the sum of bond energies,

$$E_{\text{rxn}} \approx 2E_{\text{Si-H}} - E_{\text{H-H}} \tag{2}$$

The desorption activation energy must be at least as large as the (positive) reaction energy. While the energy of a surface Si–H bond was unknown, Sinniah et al. estimated a value of 90 kcal/mol from bond strengths in molecular silanes. This does not account for effects of surface strain or interactions, such as those between adjacent dangling bonds, that are unique to the constrained geometry of the surface. Sinniah et al. also assumed that both Si–H bonds have equal strength. With the H_2 bond energy equal to 104 kcal/mol, these assumptions imply that the activation energy for a concerted mechanism must be greater than $(2 \times 90) - 104$ kcal/mol, or 76 kcal/mol. This clearly conflicts with their measured activation energy of 45 kcal/mol.

However, later work has shown that the energy values used in this argument are likely to be incorrect. Later measurements, described below, have determined an activation energy of 57–58 kcal/mol, which is higher than that determined by Sinniah et al. The estimate of Si–H bond energies used by Sinniah et al. are also too high, as discussed in Section IV.A. This is partly because Sinniah et al. used values from gas-phase experiments with molecules containing only one or two Si atoms. Measurements [35] and calculations [36] on substituted silanes have shown that Si–H bond energies decrease with additional silyl substitution. In addition, the bare surface is stabilized by rebonding between dangling bonds. This both lowers the Si–H bond dissociation energy and provides a thermodynamic driving force for pairing. This work is discussed in more detail below, but the conclusion is that the activation energy is higher and the Si–H bonds are weaker than the estimate used by Sinniah et al. The recent measured values of the activation energy for desorption are indeed greater than the current estimates of the reaction energy. A concerted mechanism cannot be ruled out on energetic grounds alone.

Believing that a concerted mechanism was not possible, Sinniah et al. [33] considered alternatives. They were particularly concerned with a "prepairing" mechanism, in which H atoms are paired up on adjacent sites due to an (unspecified) attractive interaction. This would easily rationalize first-order kinetics because the desorption rate would depend on the density of correlated pairs of H atoms, rather than the probability for two uncorrelated atoms to be on adjacent sites. They performed an isotopic mixing experiment to test the prepairing idea. They dosed the surface with deuterium atoms, annealed it (to overcome any kinetic barrier to prepairing), cooled it, and dosed it with an equal coverage of hydrogen atoms. The LITD experiment with mass spectrometric detection yielded desorbing H_2, HD, and D_2 in the ratios 1:2:1. This is consistent with complete isotopic mixing and was interpreted as an argument against a stable prepaired state. However, since the diffusion rate is faster than the desorption rate, one expects that the prepaired atoms could easily scramble before desorbing. Isotopic mixing in the desorption products may simply be evidence that the prepaired state is isotopically mixed, rather than evidence that prepairing does not occur. Most atoms could be in the prepaired state at any instant, even if a significant amount of diffusion occurs during the course of the desorption measurement. A quantitative argument depends on the actual diffusion rate, which is unknown at present.

As an alternative to the prepairing mechanism, Sinniah et al. proposed a two-step mechanism, in which the rate-limiting step was the excitation of a hydrogen atom into a ‘‘delocalized’’ state, denoted H*. This step is first order; the second step is abstraction of H from an Si–H bond by H* to form H_2. Although the H* state was described as ‘‘band-like,’’ it is sufficient for their argument that it be bound to the surface with nearly free lateral motion. In order to obtain direct evidence that H* exists, Sinniah et al. dosed a deuterium-covered surface with H atoms and measured the coverage of remaining D. They found that the probability for incident H to abstract D from the surface was comparable to the probability for incident H to chemisorb. While this experiment shows that incident H atoms do not equilibrate immediately, it does not show that incident atoms are sampling a state involved in molecular desorption. Indeed, the incident atoms are initially much more energetic than can be expected for the H* state at equilibrium. Taking, for simplicity, the numerical values given by Sinniah et al., H* is bound by 45 kcal/mol, relative to gas-phase atomic H (with no initial kinetic energy). This means that incident H atoms are strongly accelerated to reach the H* state, and they will have much higher average kinetic energy than thermally excited H*. Thus, the experiment is not compelling evidence for the role of H* in desorption.

This abstraction experiment is significant in its own right, since it is apparently one of the first direct observations of a surface reaction via an Eley–Rideal process. Abstraction appears to be temperature-independent down to 123 K, suggesting that abstraction is unactivated or that the incident H atoms do not equilibrate with the surface before abstraction occurs. Both suggestions are probably true. A low activation barrier for abstraction from a Si surface (4.5 kcal/mol) has been measured by Abrefah and Olander [37], and a low barrier for abstraction from disilane (2.4 kcal/mol) has been calculated by Dobbs and Doren [38]. On the other hand, H atoms are presumably accelerated as they approach the surface, and there is no efficient way to dissipate the energy gained into surface vibrations. Collisions with adsorbed H or D atoms are therefore likely to occur before the incident atom is equilibrated.

4. Verification of Structure-Dependent Desorption Mechanism

The work of Sinniah et al. initiated a controversy. The H* mechanism does not appear to be specific to the (100) surface. Sinniah et al. suggested that this mechanism may also operate on the (111) surface and that the earlier work on Si(111) should have been interpreted in terms of first order kinetics. The novelty of the H* mechanism, and its weak foundations, also attracted attention. Since the work of Sinniah et al., further experiments have confirmed that desorption from the monohydride phase (at coverages that are not too low) from Si(100) is approximately first order and that desorption from Si(111) is approximately second order. As an explicit test of the differences in kinetics on the (100) and (111) surfaces, Wise et al. [31] used TPD and isothermal LITD to measure desorption kinetics on both surfaces. By doing experiments in the same apparatus, preparing both crystals by similar methods, and analyzing the data with the same methods, any uncertainties due to variations between laboratories are removed. For example, errors in the measurements of heating rate or surface temperature can be caused by the way the sample is mounted and attached to the thermocouple. Wise et al. confirmed both the approximate first-order kinetics on Si(100)-2×1 and approximate second-order kinetics on Si(111)-7×7 by comparing isothermal LITD measurements at coverages between 1 ML and 0.1 ML to predicted kinetics of each order. Some fractional-order kinetic schemes may also be consistent with their data, though precise determination of a non-integer order would not be possible from their data. The TPD measurements confirmed these findings qualitatively: TPD from Si(100) showed no peak shift with coverage (except for a small shift upward at the lowest coverages), while on the (111) surface the peaks consistently shift to higher temperature with decreasing coverage. Arrhenius parameters from TPD are in rough agreement with the more accurate isothermal measurements.

The Arrhenius parameters determined by Wise et al. [31] have been reproduced by Höfer et al. [39] and Flowers et al. [40] using different techniques (a detailed description of this work is given below). All three groups find activation energies of 57–58 kcal/mol. This raises the issue of why Sinniah et al. [33] found a much lower value. Differences in surface preparation or impurities must be considered, but there is no evidence that the surface of Sinniah et al. was of substantially better or worse quality than the others. An error in temperature measurement, caused for example by poor thermal contact with a thermocouple, could cause an underestimate of the actual temperature, so that the activation barrier appears artificially low. Kolasinski et al. [41] have noted that the rates determined by Sinniah et al. were close to the values found by others and have suggested that the difference in Arrhenius parameters might simply be an error in fitting. This would be surprising, since only a linear fit is required, and the data of Sinniah et al. covered more than an order of magnitude of rates. Indeed, direct comparison of the data of Sinniah et al. [33] to that of Wise et al. [31] shows that they have distinct Arrhenius parameters. The reasons for the differences remains unresolved, but the preponderance of experiments in close agreement has created a consensus that the activation barrier is about 58 kcal/mol, with prefactors of about 10^{15} s^{-1}.

5. *Evidence for Prepairing*

The activation energy on Si(100) determined in the isothermal measurement by Wise et al. [31] was 58 ± 2 kcal/mol, substantially higher than the value of Sinniah et al. [33]. With this new value, the surface Si–H bond energy can be estimated (again ignoring any stabilization of the clean surface) from

$$2E_{\text{Si-H}} \approx E_{\text{rxn}} + E_{\text{H-H}} \tag{3}$$

Since the measured activation energy for desorption, E_a^{des}, must be an upper bound to the overall reaction energy, E_{rxn}, the measurement of Wise et al. implies that the Si–H bond energy must be less than (58 + 104)/2 kcal/mol = 82 kcal/mol. This is similar to some gas-phase Si–H bond strengths [35, 42], so if it is a good estimate of the Si–H bond strength, it eliminates the argument cited by Sinniah et al. against a concerted mechanism. Wise et al. suggested that the prepairing model was responsible for first-order kinetics on this surface. Atoms are forced to form pairs on a saturated surface, and it is possible that in desorption experiments starting from a saturated state (as for the isothermal experiments by Wise et al.), pairing persists due to a kinetic barrier. However, TPD experiments with lower initial coverage also show first-order kinetics, and Wise et al. interpreted this as evidence that pairing is both energetically favored and kinetically feasible during the desorption process.

Additional evidence for prepairing became available in short order. Boland [43] used an STM to observe the distribution of H atoms on Si(100)-2 × 1 at submonolayer coverages. Using tunneling spectroscopy, he assigned features in the STM images as bare dimers, isolated dangling bonds on one side of a dimer (a singly occupied dimer), and doubly occupied dimers that have reacted with a pair of H atoms. Dosing at room temperature yields a random distribution of isolated dangling bonds and bare dimers.

After briefly annealing at 630 K (which causes negligible desorption) or dosing at 630 K, the number of isolated dangling bonds is sharply reduced. Annealing at higher temperature or for longer times leaves more bare dimers. Observations after annealing show primarily (though not exclusively) the bare dimer or the doubly occupied dimer. Thus, the combination of a doubly occupied dimer and a bare dimer is more stable than two singly occupied dimers. The pairing energy, E_{pair}, defined as the energy change for reaction (4),

$$\text{H--Si--Si--H} + \text{Si--Si} \longrightarrow 2\ \text{Si--Si--H}^{\cdot} \tag{4}$$

is positive. There is a kinetic barrier to the pairing process, as evident from the need for annealing to achieve pairing. Furthermore, there is some probability to find singly occupied dimers at finite temperatures. Thus, if desorption can occur in a concerted mechanism from prepaired H atoms, the kinetics of pair formation may influence the desorption kinetics at low coverage.

Boland [43] noted that Sinniah et al. did not account for stabilization of the bare dimer final state in their estimate of the activation energy from the prepaired state. Stabilization of the final state lowers the overall reaction energy and reduces the activation energy. Boland estimated E_{pair} from the differences in tunneling spectra (which reflects changes in electronic energy) between a bare and singly occupied dimer. He determined a value of 18 kcal/mol, though later work has indicated that this estimate of E_{pair} is too high. Boland noted that his estimate neglects differences of surface relaxation and electron correlation between the two states. D'Evelyn et al. [44] showed that when Boland's method is used to estimate the π bond strength in a molecular analogue, $H_2Si{=}SiH_2$, the value agrees well with more traditional estimates based on bond rotation barriers. However, the surface states of the buckled dimer are not similar to a traditional π bond. Even on the symmetric surface, the electronic states corresponding to the π bond have substantial dispersion, indicating delocalizing interactions between adjacent dimers. Since the electronic states probed by Boland are delocalized, while E_{pair} reflects bond energy changes caused by absorption on a single dimer, D'Evelyn et al. [44] suggested that Boland's method should be less accurate on a surface.

D'Evelyn et al. [45] independently recognized that π bonding in the bare dimer could make pairing energetically favored. Their evidence is indirect, but it draws from a variety of sources. Surface vibrational spectroscopy shows coupling between adsorbed H atoms, consistent with doubly occupied dimers, even at low coverage. Theoretical calculations show that the π bond on the bare dimer is broken by adsorption of a single H atom. Two π bonds must be broken to form two singly occupied dimers, but only one must be broken to form a doubly occupied dimer. Thus, E_{pair} is approximately equal to the interaction energy of the dangling bonds in a dimer. Molecular analogues of the surface bond show a strong π bond, but this is weakened by surface strain. D'Evelyn et al. [45] derived a lattice gas model that predicts the density of singly occupied, doubly occupied, and bare dimers for given values of coverage and E_{pair}. For small values of E_{pair} (relative to $k_B T$), hydrogen is randomly distributed on the surface. Higher values of E_{pair} make doubly occupied dimers more probable at a given coverage. At lower coverage, for a given value of E_{pair}, entropy favors distributions with a larger fraction of hydrogen on singly occupied dimers.

D'Evelyn et al. [45] modeled desorption experiments by assuming that the desorption rate was proportional to the density of doubly occupied dimers, taken from the lattice gas model. For large values of E_{pair}, most H is paired on doubly occupied sites and desorption is close to first order. As E_{pair} approaches zero, the predicted kinetics become second order. D'Evelyn et al. [45] reanalyzed the isothermal measurements of Wise et al. using this model and determined slightly different Arrhenius parameters than obtained from a strictly first-order analysis ($E_{\text{des}} = 54.9$ kcal/mol, prefactor $= 5.6 \times 10^{14}$ s^{-1}).

To make an empirical estimate of E_{pair}, D'Evelyn et al. [45] noted that the TPD experiments of Sinniah et al. [33] and Wise et al. [31] exhibited a small peak shift to higher temperature at low coverge. They interpreted this as a deviation from first-order behavior caused by the larger fractional population of singly occupied dimers at low coverage. Using their model to simulate the TPD spectra of Wise et al., D'Evelyn et al. found that good fits were obtained with pairing energies of 5–10 kcal/mol. At typical temperatures of isothermal desorption experiments, the pairing energy corresponds to several times $k_B T$. When the pairing energy is $5 k_B T$, more than 70% of adsorbed H is paired on doubly occupied dimers and deviations from first-order behavior in isothermal desorption are small.

The model of D'Evelyn et al. [45] was used by Höfer et al. [39] to interpret isothermal LITD measurements, using SHG detection. This work primarily concerned low coverages, with initial coverages of 0.15 ML and detection effective for two orders of magnitude below that. Höfer et al. [39] estimated the pairing energy by fitting the observed time dependence of the coverage to the model of D'Evelyn et al. The kinetics are very sensitive to

E_{pair} below 0.01 ML. The derived pairing energy is 5.8 $\pm$ 1.1 kcal/mol. At 700 K, the middle of the temperature range used (670–730 K), the model predicts with this value of E_{pair} that about 70% of H atoms are on doubly occupied dimers at 0.15 ML. This makes the kinetics indistinguishable from first order to almost 0.01 ML. The fact that the prepairing model accurately fits the kinetics over the entire coverage range, including substantial deviations from simple first- or second-order kinetics at low coverage, constitutes strong evidence in favor of the model.

Höfer et al. [39] determined a value for the desorption activation energy of 57.3 $\pm$ 2.3 kcal/mol, with a prefactor of 2×10^{15} s^{-1}, essentially the same as found by Wise et al. [31] but quite different from the values found by Sinniah et al. [33]. The work of Höfer et al. is another example in which kinetics on the (100) [39] and (111) [32] surfaces were determined in the same laboratory and found to have different orders. No doubt can remain that the kinetic order of desorption is different on the two surfaces.

Höfer et al. [39] introduced the pairing energy into the estimate of the overall reaction energy. Since pairing reduces the energy of the surface after desorption, the Si–H bond energies are limited by

$$2E_{\text{Si-H}} \leq E_{\text{des}} + E_{\text{pair}} + E_{\text{H-H}} \tag{5}$$

Using the values determined by Höfer et al., this implies that $E_{\text{Si-H}}$ is less than 83.7 kcal/mol, which is well within the range of literature values for both measured and predicted gas-phase Si–H bond energies [35, 42]. This refinement of the argument introduced by Sinniah et al. shows that a concerted mechanism is indeed feasible.

A final piece of experimental evidence for the prepairing mechanism comes from the TPD experiments of Flowers et al. [40]. They included higher initial coverages in their work, with a range from 0.05 to 1.5 ML. This required a model of desorption from the dihydride state as well as the monohydride, but in the range below 1 ML, where only the monohydride is present, their model is the same as that of D'Evelyn et al. [45] and Höfer et al. [39]. Excellent fits to the TPD spectra are achieved over the whole coverage range, yielding energetic parameters that confirm the earlier results with minor variations.

Boland [43] had observed some singly occupied dimers in high coverage STM images taken at room temperature. He interpreted them as remnants of a nonequilibrium distribution, partly based on his high estimate of E_{pair}. The model of D'Evelyn et al. [45] predicts a low but easily measurable population of singly occupied sites at room temperature for coverages below 1 ML. Annealing at high temperature, as Boland did, should permit rapid equilibration. Using the measured diffusion rate for H on Si(111)-7 $\times$ 7,

D'Evelyn et al. estimated that diffusion should allow H atoms to move over 50 Å in 1 s at 750 K. If this crude estimate is accurate, an equilibrium distribution of dimer occupations should be easily achieved during a desorption measurement.

Boland also observed that doubly occupied dimers cluster together along dimer rows, indicating an attractive interaction. Boland attributed this to the change in Si–Si dimer bond length caused by H adsorption, which causes subsurface relaxation. Since adjacent dimers share subsurface neighbors, there is an advantage to bringing doubly occupied dimers together. The spatial distribution of doubly occupied dimers depends on the desorption rate: Only at slow desorption do the bare dimers cluster together [43]. The original model of D'Evelyn et al. [45] makes no predictions about the spatial distribution of dimer occupation, but a later version by Yang and D'Evelyn [46] adds attractive near-neighbor interactions between doubly occupied dimers to mimic the observed clustering of occupied dimers. Monte Carlo simulation was used to predict the probability distribution of cluster sizes from this model. The cluster size distribution observed by Boland could not be quantitatively reproduced, perhaps because the near-neighbor model of dimer interactions is inadequate, or perhaps because the observed distribution was not at equilibrium. Nevertheless, an approximate fit to the observed frequency of singly and doubly occupied dimers and the distribution of dimer clusters yields a value for E_{pair} of 4–6 kcal/mol. This value is close to those obtained by fitting desorption rates, though the source of the experimental data is entirely independent.

To summarize, the prepairing model can predict both the observed distribution of dimer occupations and the coverage dependence of desorption kinetics with a single set of parameters. This is strong evidence for the effect of pairing on desorption, though further confirmation of the model with STM images of equilibrated distributions at different temperatures would be a useful refinement. As described below, measurements of adsorption and desorption dynamics are qualitatively consistent with pairing. However, first principles calculations disagree about the likelihood of this mechanism, as described in Section IV.

6. *Desorption After Dosing with Silanes*

Several experiments have measured H_2 desorption from surfaces dosed with silane [47–49] (SiH_4) or disilane [50, 51] (Si_2H_6), under various conditions of temperature and pressure. These surfaces may better reflect the structure and hydride species typical of realistic chemical vapor deposition (CVD) growth conditions. It is possible that the surfaces will have different defect densities than the typically prepared Si surfaces. STM experiments [52] have shown that surfaces dosed with disilane are covered with randomly distributed adatoms, but after annealing to 670 K, the adatoms form epitaxial islands in the (100)-2 × 1 structure. Tunneling spectra are similar to those for the monohydride surface. Okada et al. [50] used the measured diffusion rate parameters for Si adatoms on Si(100) to show that Si adatoms can easily move to step edges or islands during a desorption measurement. This argument neglects the possible effects of adsorbed H. However, more recent experiments have shown that adsorbed H reduces the adatom diffusion rate only at temperatures below 550 K, probably because the H atoms are mobile at high temperature [53]. Thus it appears that under desorption conditions, the deposited Si atoms form structures similar to the monohydride surface, rather than random defects.

Liehr et al. [47] and Greenlief and Liehr [49] dosed Si(100) wafers with silane, followed by rapid cooling to "freeze out" the surface species. At temperatures below about 840 K, cooling is rapid enough that the H coverage does not change substantially. H_2 desorption rates were then measured by TPD. These experiments found first-order kinetics, just as on the H-dosed surfaces, but the activation energy is 49 $\pm$ 3 kcal/mol and the prefactor is $8 \times 10^{13 \pm 1}\ s^{-1}$. A similar experiment by Kim et al. [48] determined an activation barrier of 46 kcal/mol. These values are closer to the activation energy determined by Sinniah et al. than to the consensus value determined from H-dosed surfaces. Rapid cooling of the surfaces after growth may prevent ordering of the newly adsorbed Si atoms, so that the lower activation energy on these surfaces may be due to a high density of defects.

Wu and Nix [51] observed similar TPD spectra for H_2 from disilane- and H-dosed surfaces, but they did not quantify the Arrhenius parameters. Okada et al. [50] compared H_2 desorption from disilane-dosed and H-dosed surfaces and found that the Arrhenius parameters are experimentally indistinguishable. In contrast to the silane-dosing experiments cited above, Okada et al. annealed the surface at 650 K for 1 min to remove any higher hydrides. The similar desorption kinetics support the proposal that annealing below the monohydride desorption temperature allows the disilane-dosed surfaces to form a (100)-2 × 1 structure. It is possible that silane-dosing and disilane-dosing lead to different activation barriers, even after annealing. For example, the adsorption mechanisms for silane and disilane may lead to different distributions of defects or higher hydrides on the surface. However, Gates and Kulkarni [54] have shown that disilane growth leads to higher surface hydrogen density than does silane growth. That is, if the Arrhenius parameters are different for silane and disilane dosing, and if defects or presence of higher hydrides account for the difference, then the disilane-dosed surface should have the lower activation energy. This contradicts observation.

7. *Other Covalent Surfaces*

Approximate first-order kinetics has been observed on the Ge(100)-2 × 1 surface, which also has a surface dimer reconstruction [44, 55]. D'Evelyn et al. [44] showed that the prepairing model reproduces deviations from first-order behavior on Ge, again with an E_{pair} of about 5 kcal/mol. This preliminary evidence hints at wider generality for the prepairing model. Several groups have found indications of first-order kinetics on the diamond (100)-2 × 1 surface [56–58] as well, though the kinetic parameters are not well determined.

There is some evidence that first-order desorption kinetics occur on many covalent surfaces. Desorption of H_2 from polycrystalline Te [59] and β-SiC [60] have also been analyzed in terms of first-order kinetics, but the TPD peaks are very broad and cannot decisively determine the kinetic order.

8. *Desorption from the Dihydride Phase*

Gupta et al. [20] found second-order kinetics for desorption from the dihydride phase on porous silicon. The corresponding activation energy from isothermal measurements is 43 kcal/mol. The TPD experiments of Flowers et al. [40] on Si(100)-2 × 1 also showed that desorption from the dihydride was second order, with an activation energy of 47 kcal/mol. This result is derived from a model that estimates the equilibrium density of dihydrides as a function of coverage and temperature. This model has been criticized [61], but it makes very weak assumptions [62]. The fit to the TPD data is model-dependent, but still establishes good evidence for second-order kinetics.

C. Energy Distribution of Desorbing H_2

The energy in desorbing H_2 molecules is a probe of adsorption and desorption dynamics. Once a molecule reaches the transition state for desorption, it generally has little opportunity to exchange energy with the surface, so that the final H_2 energy distribution reflects the energy of desorbing molecules at the transition state. In particular, a low-mass particle such as H_2 is unlikely to exchange energy with the surface after crossing the transition state. By detailed balance, a desorbing molecule must go through the same transition state as an adsorbing molecule, so that product energy distributions also indicate whether the adsorption probability depends on the energy of the incident molecule.

1. *Rotational and Vibrational Energy*

A series of papers by Kolasinski et al. [63–67] explored the vibrational and rotational energy distributions of H_2 desorbing from Si. The surfaces are dosed with either disilane (Si_2H_6) or H atoms. The surface is heated as in TPD, the desorbing molecules are ionized state-specifically, and the ions are detected. A resonance-enhanced multiphoton ionization process allows selective ionization of molecules in specific vibrational and rotational states (v, J). The basic experimental results for H_2 desorbing from the monohydride phase on Si(100) are as follows:

1. The rotational energy distribution of desorbing H_2 deviates from a Boltzmann distribution, but the average rotational energy (368 $\pm$ 67 K, measured in temperature units) is much less than expected for a Boltzmann distribution at the TPD peak temperature of 780 K.
2. The average rotational energy of H_2, HD, and D_2 are the same, within experimental uncertainty.
3. About 1% of desorbing H_2 is in the first excited vibrational state; this is about 20 times more than expected in a Boltzmann distribution at the surface temperature.
4. Higher fractions of HD and D_2 (2% and 8%, respectively) desorb in the first vibrational excited state, again corresponding to about 20 times the expected Boltzmann population in each case.
5. The average rotational energy in the ground and first vibrational states are the same, within experimental uncertainty.
6. The internal energy is the same whether the source of adsorbed H is disilane or atomic H.

The last fact is consistent with Boland's STM observation that surfaces dosed with disilane form islands in the (100)-2 $\times$ 1 structure and are covered with H [52]. More surprising are the following additional results:

7. The rotational energy distributions for molecules desorbed from the dihydride phase are indistinguishable from those for the monohydride.
8. A lower fraction (0.2%) of vibrationally excited molecules desorb from the dihydride phase. However, the desorption temperature is lower for this phase, so there is still a 20-fold enhancement of vibrationally excited molecules over that expected from a Boltzmann distribution;
9. The rotational and vibrational energy distributions for H_2 desorbing from Si(111)-7 $\times$ 7 and Si(100)-2 $\times$ 1 are indistinguishable.

These results indicate that the dynamics are remarkably insensitive to surface structure. This is surprising in light of the strong structure sensitivity of the kinetics. However, kinetics reflect the rate to approach the transition state, while the energy distribution is determined after the transition state is reached. Substantial insight into the nature of the transition state to desorption has been derived from these experiments. Vibrational excitation of desorbing H_2 must imply that the transition state occurs when the H–H distance is longer than the equilibrium H_2 bond length. The extreme frequency mismatch between the H–H vibration and all other modes makes any coupling unlikely. Rotational cooling is more subtle. Kolasinski et al. [64] rule out several possibilities. The molecule is unlikely to exchange energy with the surface by impact, due to the mass mismatch, and this is supported by the identical energies observed for all isotopes. Rotation-to-translation transfer is unlikely, since H_2 is nearly spherical and its interaction with the surface (after leaving the transition state) is nearly orientation-independent. The similar energy distribution of HD, which is more likely to change angular momentum, supports this expectation. This implies that the molecule cannot change its rotational energy after leaving the transition state, so Kolasinski et al. [64] conclude that rotational energy reflects properties of the transition state. In particular, they propose that the molecules experience low torques as they move across the transition state. However, lack of acceleration alone cannot explain rotational cooling, since at equilibrium a Boltzmann distribution of rotational energies would be expected at the transition state. Kolasinski et al. also propose that the recombining H atoms approach each other symmetrically along the nascent H–H bond, and that they carry little angular momentum into the transition state. The H* model of Sinniah et al. [33] is not necessarily inconsistent with this picture, but the mobile H atom must move in a way to yield a low angular momentum transition state.

Kolasinski et al. [64] concluded that the low rotational excitation is consistent with desorption from a prepaired state on a surface dimer. To obtain optimal orbital overlap, the H atoms and surface dimer atoms must lie in a plane. The H atoms can combine along a path where their axis stays parallel to the surface, leading to little angular momentum. The similar energy distributions for desorption from the dihydride and the (111) surface suggest that all three transition states have similar structure. Shane et al. [66, 67] suggest that this is a dihydride-like structure, which results, in the case of the (100) monohydride, from migration of one H atom to the opposite side of the dimer (Fig. 3). These arguments provide a detailed description of the transition state structure, based on the experimental work and qualitative theory alone. Later theoretical work has confirmed that transition states with this structure do indeed exist, though it is not certain that desorption from a doubly occupied dimer is the physically significant reaction path.

It is possible to propose plausible transition states that are similar on the (100) and (111) surfaces yet do not resemble an adatom dihydride. Suppose that the diffusing species on the (111) surface is not H but SiH. Adatoms

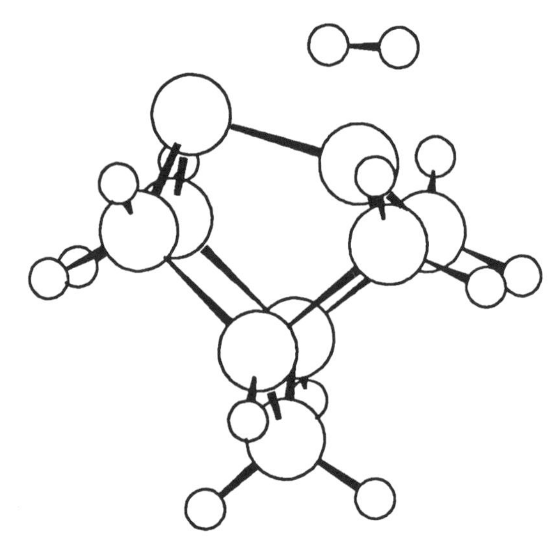

Figure 3. The transition state for desorption from a doubly occupied dimer. This geometry is taken from the work of Pai and Doren [103], but similar structures have been proposed on the basis of experimental results and other first principles calculations.

with H bound to their dangling bond should have weakened back bonds and be more mobile than bare adatoms. In this scenario, the monohydrides would migrate until two were adjacent. If they form a local structure similar to the (100) dimer, the desorption transition state would be like that of the (100) monohydride. Alternatively, the similarity among these surfaces may be due to a common impurity, such as oxygen or hydroxyl.

It may be too facile to conclude that transition states are similar on the basis of similar product energy distributions. Kinetic measurements by Gupta et al. [20] and Flowers et al. [40] have shown that desorption from the dihydride phase is second order, consistent with recombination of H atoms from each of two dihydrides. The transition state proposed by Shane et al. [67] involves recombination of the two H atoms from a single dihydride, leading to first-order kinetics. If desorption from the dihydride phase involves two dihydrides, it is difficult to imagine a transition state similar to that for desorption from the monohydride phase. Shane et al. felt that the evidence for second-order kinetics from the dihydride phase was not conclusive [67], though the more recent TPD measurements of Flowers et al. [40] add support for a second-order process. If the dihydride desorption mechanism is different from the monohydride mechanism, the similar energy distributions cannot be simply explained by a common transition state.

Sheng and Zhang [68] proposed a model that predicts, with suitably chosen parameters, rovibrational state distributions similar to those measured. The model assumes that the product molecule energy distribution is determined by Franck–Condon factors between an initial doubly occupied dimer state and a final free H_2. There is no direct connection between the product energy and the activation energy or transition state structure, so that this model contradicts the usual assumption that the product energy distribution is determined by the transition state properties. The model of Sheng and Zhang also does not describe how excitation to the continuum occurs, though the Franck–Condon model is most useful for a sudden transition from the ground state to the final state. Thermal excitation over high barriers does not occur by a sudden mechanism.

Despite these limitations, the model predicts cool rotational distributions and warm vibrational distributions [68]. However, Sheng and Zhang found that the predicted desorption rate and vibrational energy distribution are sensitive to small changes in the assumed surface structure, though they did not include surface degrees of freedom in their thermal averaging. The average rotational energy for the two vibrational states differ by about 50%, while measurements show they are the same. Sheng and Zhang did not report results from the (111) surface, but since the model predictions are determined by the structure of the adsorbed state, similar energy distributions on two surfaces of very different structures could only be coincidental. While the model only includes planar motion of the desorbing molecule and was not expected to be quantitatively accurate, its questionable foundations suggest that the points of agreement with experiment are fortuitous. Further tests of the sensitivity of the model predictions to parameter choices are needed to provide a firmer basis for the model.

2. *Translational Energy*

A complete description of the average energy of desorbing H_2 requires measurements of the translational energy. Park et al. [69] measured the angular distribution of D_2 desorbing from Si(100) and found that it was more peaked in the normal direction than expected at equilibrium. In other words, molecules have a larger-than-expected ratio of normal-to-lateral translational energy. This suggests a barrier to adsorption in the normal translational coordinate, which Park et al. estimate as about 3 kcal/mol.

Kolasinski et al. [70] have measured the translational energy of D_2 desorbing under the influence of laser heating. The time-of-flight for desorbing molecules to reach a detector is well fit by a Maxwellian distribution of a temperature equal (within a rather large uncertainty) to the surface temperature. As with the rovibrational energies, the mean translational energies for deuterium desorbing from the monohydride and dihydride phases on the (100) surface and from the (111)-7 $\times$ 7 surface are all indistinguishable.

Kolasinski et al. [70] found that there was no translational energy in excess of that expected at equilibrium, while Park et al. [69] found a higher ratio of normal translational energy than expected at equilibrium. These results appear to contradict each other, though this is not a strict logical necessity. Within their simplest interpretations, both experiments agree that any barrier to adsorption in the normal coordinate is at most a few kilocalories per mole.

3. *Surface Energy*

Kolasinski et al. [70] noted that the total rotational, vibrational, and translational energy of the desorbing molecules did not substantially exceed the value expected at equilibrium. On the other hand, the sticking probability for H_2 is low, suggesting an activation barrier to adsorption. By detailed balance, adsorption and desorption must occur through the same transition state. Thus, the desorption transition state must be higher in energy than the separated molecule and surface (Fig. 4). Since the molecule and surface are unlikely to exchange energy after the transition state is crossed, this creates an apparent paradox: Why isn't the excess energy at the transition state observed in the desorbing molecules? Kolasinski et al. [70] rhetorically questioned whether the principle of detailed balance is violated, but drew a more restrained conclusion: If the excess energy is not deposited in the desorbing molecules, it must be left in the surface. This could happen if the surface is distorted in the transition state, so that it can lower its energy after the molecule departs by relaxing to the equilibrium structure of the bare

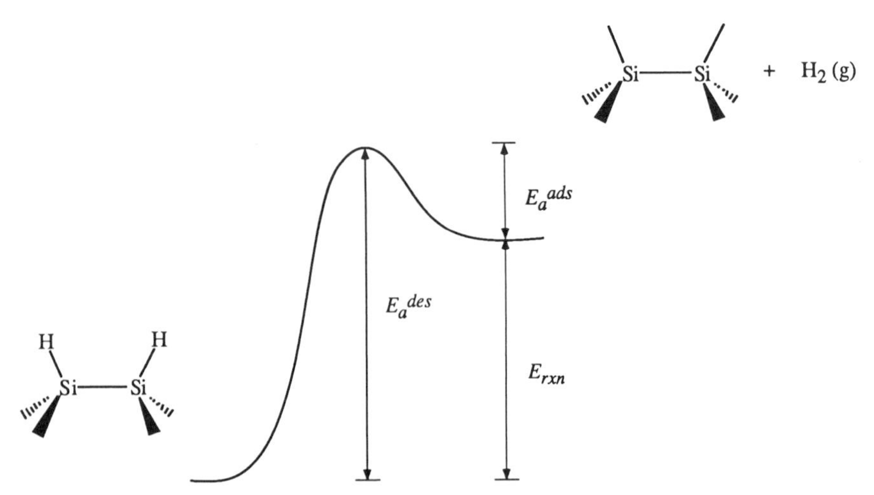

Figure 4. Relationship of the activation energies to adsorption (E_a^{ads}) and desorption (E_a^{des}) and the reaction energy, E_{rxn}.

surface. The same surface distortion in the transition state may account for the low sticking probability, since the bare surface is unlikely to be distorted into the optimal configuration for adsorption. Brenig et al. [71] have developed a model, described below, that shows that this mechanism can account for both low sticking and low product energies.

D. Sticking Probability

The sticking probability is determined by a combination of energetic, entropic, and dynamical effects. Molecules may be prevented from sticking if they do not have enough energy to get over a barrier to adsorption, if their impact parameter or orientation correspond to high-energy configurations, or if they reflect from the surface before losing enough energy to become trapped. If there is an entropic barrier, there may be low-energy pathways to adsorption, though they are available to a small fraction of trajectories. Dynamical effects depend on the forces an adsorbing molecule experiences as it reaches the surface. The molecular orientation and the distribution of energy among rotational and translational modes change while the molecule interacts with the surface, and these can produce subtle relations between the sticking probability and energy distribution in the incident molecules. In principle, experiments can distinguish between energy and entropy barriers [72]. For example, if the barrier is entropic, the average energy in desorbing molecules should depend on temperature. However, it may not be feasible to make the needed measurements for hydrogen desorption.

1. Measurements of Sticking Probability

Law [7] demonstrated many years ago that the sticking probability of H_2 is very low on Si surfaces, though he gave no quantitative value. Schulze and Henzler [26] found that H atoms have essentially unit sticking probability at low surface coverage on the Si (111)-7 $\times$ 7 surface, while the sticking probability of H_2 is at least six orders of magnitude smaller.

On Si(100) there are now several studies that quantify the sticking probability. Liehr et al. [47] exposed a surface to 10^7 Langmuir of H_2 and detected adsorption by a subsequent TPD experiment. They studied temperatures from 570 K to 970 K and found that there was "no hydrogen desorption related to adsorption from molecular hydrogen. . . ." At high temperatures, some adsorbed hydrogen will desorb before the TPD experiment is performed, but this experiment at least demonstrates that the sticking probability is low at temperatures up to 650 K. An exposure of 10^7 L corresponds to about 10^7 molecules striking each surface site. Even allowing that the TPD experiment is not sensitive to arbitrarily low coverage, the absence of any detected adsorbed H implies that the sticking probability is much less than 10^{-7}.

More recent experiments have detected somewhat higher sticking probabilities and have measured the temperature dependence of sticking. Kolasinski et al. [41] have used a normally incident molecular beam to dose the Si(100) surface with D_2. Heating the nozzle allows them to vary the average total energy of incident D_2 over a limited range (2–7 kcal/mol). Adsorption is detected by a TPD experiment (for surface temperatures where desorption is relatively slow) or by observation of HD produced from an initial mix of H_2 and D_2 (at temperatures where desorption is too rapid to establish significant surface coverage). Increasing the surface temperature at fixed nozzle temperature increases the sticking probability by a factor of three, up to 630 K, when desorption starts to compete with adsorption. Increasing the average incident energy (at surface temperatures below 630 K) increases the sticking probability by a factor of three, from 1.5×10^{-5} to 4×10^{-5}, at 630 K.

Detailed balance requires the energy distributions of desorbing molecules to be consistent with the energy dependence of the sticking probability. Since the translational energy distribution of desorbing molecules has no excess of high-energy molecules, one would expect that adsorption is not enhanced by increasing translational energy. If the measurements of translational energy in desorption are correct, even this weak dependence of sticking on beam temperature cannot persist over a wide range of incident energy. On the other hand, weak activation by normal translational energy is consistent with a normally peaked angular distribution of desorbing molecules [69]. The measurements of energy dependence of the sticking probability, translational energy distribution, and angular distribution may not be mutually consistent.

These results appear to provide evidence for activated adsorption in which either surface activation or molecular excitation can overcome an energy barrier. This supports the earlier suggestion [70] that the adsorption reaction coordinate involves surface motion, and Ref. 41 expands on this argument. However, the interpretation of these experiments is not straightforward. The range of sticking probabilities is never more than a factor of three, saturation sticking is never approached, the background contribution to surface coverage is often a significant fraction of the total, and the error bars are a large fraction (often more than 1/4) of the total range of variation. These factors make quantitative analysis difficult. Kolasinski et al. [41] have used an empirical model that assumes separate activation barriers for the molecular and surface degrees of freedom, each with an associated "width." They find barriers to adsorption of 23 kcal/mol (with a 14 kcal/mol width) in the molecular degrees of freedom and 7.2 kcal/mol (with a 2.5 kcal/mol width) in the surface degrees of freedom.

These values are model-dependent. In absolute terms, the variation of sticking with beam or surface temperature is very weak. No fundamental, molecular scale meaning is given to the separate activation energies for different coordinates or the width to each activation barrier. Fluctuating barriers and coupling between the reaction coordinate and surroundings are common to condensed phase reactions, yet rates are characterized by a single free energy of activation for the entire system. Indeed, a simple Arrhenius plot of the temperature-dependent sticking probability is linear, with a low activation barrier [73]. These experiments also find a much higher absolute sticking probability than determined by Liehr et al. [47], though this may partly be due to the higher normal translational energy and lower rotational energy in this beam experiment. An alternative explanation of this experiment is that adsorption occurs only at rare defects and is weakly activated. Kolasinski et al. [41] offer persuasive arguments that a defect-mediated mechanism cannot explain all the available experimental data. However, skeptics do not find these sticking measurements in themselves to be convincing evidence against a defect mechanism. Clearly, confirmation and extension of this work is needed.

Measurements by Bratu and Höfer [73] on both the Si(100)-2 $\times$ 1 and Si(111)-7 $\times$ 7 surfaces provide additional evidence for an activation barrier to sticking. These measurements use second-harmonic generation (SHG) to detect the surface coverage. This method is the most sensitive yet applied to measure sticking. This sensitivity permits measurements of sticking coefficients below 10^{-8}. Bratu and Höfer expose a surface, held at constant temperature, to H_2 or D_2 by backfilling their chamber with a low pressure (10^{-7} to 10^{-4} bar). The SHG signal is recorded during adsorption, and the sticking probability is derived from the initial rate of change in the SHG signal or from the measured equilibrium coverage. The sticking probability increases by three orders of magnitude between 550 and 1050 K [from 10^{-8} to 10^{-5} for H_2 on Si(100), and slightly lower values on Si(111)]. Simple Arrhenius plots yield activation barriers for adsorption of 17 kcal/mol for H_2/Si(100), 16 kcal/mol for D_2/Si(100), and 20 kcal/mol for H_2 on Si(111), each with an uncertainty of 2 kcal/mol.

The reasons for the differences between the results of Bratu and Höfer [73] and those of Kolasinski et al. [41] are presently unknown. Although Kolasinski et al. found a weak dependence of sticking on incident energy, the difference between beam dosing and exposure to an equilibrium gas can only account for a part of the difference between the sticking probabilities. The sensitivity of the SHG method, small error bars, and model-independent determination of kinetic parameters make the results of Bratu and Höfer appear more compelling at present. Clearly, additional tests of these results are needed.

2. *A Model of Adsorption and Desorption Dynamics*

Kolasinski et al. [41, 70] and Bratu and Höfer [73] have each suggested that the apparent conflict between an activation barrier to sticking and the low energy of desorbing molecules can be resolved if surface distortion accounts for most of the activation barrier to adsorption. Brenig et al. [71] have developed a model that qualitatively illustrates this behavior. There are two dimensions in the model, corresponding to a surface oscillator (such as the surface dimer stretch or tilt) and the distance of the molecule from the surface. A potential was created to describe the proposed model. It has a barrier to adsorption of 16 kcal/mol (to account for the low sticking probability), and the equilibrium position of the surface oscillator on the clean surface is displaced by 0.5 Å from its position on the doubly occupied dimer. This displacement is chosen so that the adsorption/desorption transition state is on an equipotential with a state in which the molecule is desorbed, but the surface is distorted. That is, surface distortion accounts for the entire energy difference between the desorbed state and the transition state. Numerical solution of the Schrödinger equation for this potential, including a model of thermal energy distributions, shows that this model reproduces the low sticking probability. The slope of the Arrhenius plot for sticking is close to that derived by Bratu and Höfer, though this has little significance since it is closely related to a parameter built into the model. The predicted translational excitation of the desorbing molecules is small, consistent with the observations of Kolasinski et al. [70]. Nevertheless, the sticking probability increases with incident translational energy, as observed.

Thus, a model in which surface vibrations absorb the excess energy at the transition state is qualitatively consistent with both the low sticking probability and the low energy in desorbing molecules. It is likely that a similar model with more degrees of freedom could be developed to model rotational, and perhaps vibrational, energy distributions as well. This begs the question of whether such models are realistic. The displacement in the oscillator coordinate of 0.5 Å is much larger than the ~0.1-Å displacements predicted by first principles theories (Section IV). The only experimental determination of the Si–Si dimer distance after hydrogen adsorption is an ion scattering experiment that is consistent with an increase of 0.5 Å in the dimer bond length [74]. This implies a strain in the dimer bond so large as to break the bond. This seems unrealistic, and the structure of the H-covered surface needs to be explored with other structural methods. More to the point than the geometry is the issue of whether the energy change in the surface coordinate caused by adsorption is realistic. Again, this is inconsistent with the results of first principles theory. If the proposed surface displacement is induced by adsorption of a single H atom, then the corresponding energy change will be approximately equal to the pairing energy. However, current theoretical and experimental estimates of E_{pair} are much smaller than the 16 kcal/mol value in the model of Brenig et al.

At present, this model suggests an intriguing possibility. It remains to be shown that a many-dimensional version will also make successful predictions. In its current form, it cannot address the rotational or vibrational state distributions. Additional microscopic details of the model must be specified before its relevance can be judged.

IV. FIRST PRINCIPLES THEORY

The mysterious experimental results on the kinetics and dynamics of H_2 adsorption and desorption on Si(100) have attracted attention from several theoretical groups. Papers have appeared, for the most part, in three ''waves'' in which similar papers from different groups appeared within a few months. These papers often use different methods and reach different conclusions. The calculations are so demanding that significant approximations are required. It is difficult to assess the absolute accuracy of any of these methods.

The first issue considered by theory was the thermodynamics of hydrogen adsorption from the prepaired state. Since surface bond energies are not directly measurable, theory could provide unique insight into the strength of Si–H bonds, the interaction between dimer atoms, and E_{pair}. The first calculations raised new questions about the validity of the prepairing mechanism. The second wave of papers revised the thermodynamic estimates and addressed the structure of the transition state and the activation energy of desorption. Work from three groups uniformly concluded that the activation energy for desorption from a doubly occupied dimer was too high to be consistent with experiment. Alternative, defect-mediated mechanisms were proposed. Finally, a third wave of papers using different theoretical methods has reached the opposite conclusion, finding that the reaction energy and activation energy for the prepairing mechanism may be consistent with experiment.

In the following discussion, the early calculations that disagree with the prepairing mechanism are treated first along with proposed alternatives. Calculations that support the prepairing mechanism are discussed last. No consensus has yet emerged about which theory is most accurate, and one goal of this work is to identify the points of agreement and issues remaining to be clarified. Many questions remain that are not satisfactorily answered by any of these theories.

A. Evidence Against the Prepairing Mechanism

1. Thermodynamics

The first results on reaction thermodynamics were calculations of Si–H bond energies, E_{pair}, and E_{rxn} by Nachtigall et al. [75] and by Wu and Carter [76] published in late 1991. Related results were reported by these groups and by Jing and Whitten over the following two years [77–81].

The Jordan and Carter groups used a cluster of nine silicon atoms as a model of the surface (Fig. 5). Similar clusters have been widely used as models for the surface because they constrain the surface dimer to a geometry similar to that of a realistic surface. Smaller clusters of Si require additional constraints to keep the dimer dangling bonds properly oriented. The model includes only one surface dimer, but includes all the near-neighbor atoms in the second layer. The two third-layer atoms and single fourth-layer atom simply constrain the geometry to the diamond lattice structure; it is unlikely that the atoms in these layers have a significant electronic effect on the surface sites. Since atoms not adjacent to the reactive site have little effect on the surface electronic structure, the Si_9 cluster is terminated by hydrogenic atoms. These saturate dangling bonds and prevent unrealistic reconstructions in the lower layers. Optimal geometries for such clusters show that the bond lengths and angles are similar to those on real surfaces and to those obtained from calculations with periodic boundary conditions. The terminating hydrogenic atoms may have different basis sets than the H atoms that adsorb on the surface to simplify the calculations or to model the differences between Si–H and Si–Si bonds. Some calculations also use different basis sets for the surface and subsurface layers, since the changes in bonding at the surface require a larger basis set for accurate representation. This approach presumes that the errors made in subsurface electronic structure will largely cancel out when energy differences between bonding states are calculated.

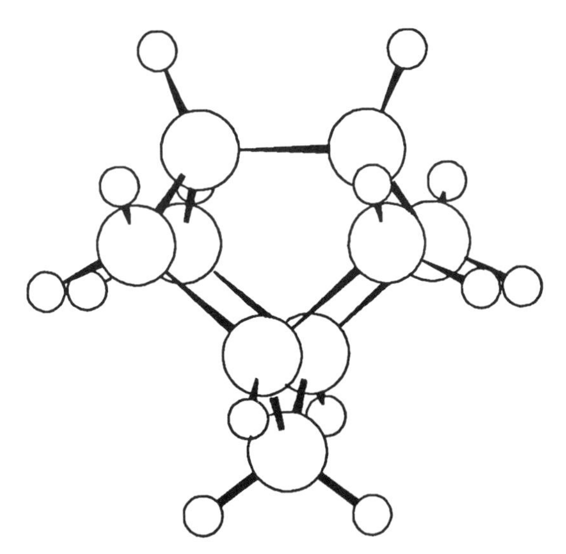

Figure 5. The hydrogen-terminated Si_9 cluster used to model the doubly occupied dimer in several first principles calculations.

Nachtigall et al. [75, 77] found that the energy to remove the first H atom from a doubly occupied dimer was 83.1 kcal/mol, including zero point energy (ZPE), while that for the second was 79.8 kcal/mol. These values are regarded by Nachtigall et al. as their most reliable results. They are corrected QCISD(T) values (see below) taken from the 1993 paper [77] and are higher by 2–4 kcal/mol than those in the 1991 paper [75].

$$\mathrm{HSi{-}SiH} \longrightarrow \mathrm{Si{-}SiH} + \mathrm{H} \quad (6)$$

$$\mathrm{Si{-}SiH} \longrightarrow \mathrm{Si{-}Si} + \mathrm{H} \quad (7)$$

$$\mathrm{HSi{-}SiH} \longrightarrow \mathrm{Si{-}Si} + \mathrm{H_2} \quad (8)$$

These bond energies are much lower than the 90 kcal/mol assumed by Sinniah et al. and seem to provide theoretical support for a concerted mechanism. These numbers imply a value for E_{pair} of 3.3 kcal/mol since pairing, Reaction (4), is obtained by subtracting Reaction (7) from Reaction (6). Nachtigall et al. predicted that the overall reaction energy for recombinative desorption was 62.1 kcal/mol [77]. They regard these values of E_{pair} and E_{rxn}, respectively, as a lower and an upper bound to the true values since the methods used are more likely to overestimate the energy of the bare surface model than the occupied surface models.

These predictions are consistent with the experimental value of 58 kcal/mol for the activation barrier to desorption, but it permits only a small barrier for the reverse adsorption reaction. Nachtigall et al. [75] pointed out that if there is a substantial activation barrier to adsorption, these calculations could not be consistent with desorption from a prepaired state. They suggested that a more complex, multistep mechanism might be responsible for desorption. For example, they drew an analogy with gas-phase elimination of H_2 from disilane, in which the 1,1-elimination mechanism has a lower activation barrier than the 1,2-elimination [75]. They suggested that desorption may occur by a 1,2-hydrogen shift (to form a dihydride) followed by 1,1-elimination (desorption of the dihydride). This speculation was supported by qualitative arguments about the difference in π-bond energies on the surface and in Si_2H_4. However, this neglects the very different geometries in the molecular and surface reactions.

Nachtigall et al. [75, 77] determined geometries using self-consistent field (SCF) methods: Hartree–Fock theory was used for the singly occupied (Si_9H_{13}) and doubly occupied (Si_9H_{14}) species, but two-configuration-SCF was used for the bare dimer (Si_9H_{12}) because it has π and π^* levels lying close in energy. Symmetry constraints were imposed on the cluster geometry, and only the positions of the surface H atoms, surface dimer atoms, and selected subsurface atoms were optimized in geometry searches. The most important subsurface degree of freedom considered was horizontal relaxation of the second-layer atoms in the direction parallel to the surface dimer, which lowered the energy of the bare dimer model by 1.3 kcal/mol. Deviations from C_{2v} symmetry in the bare dimer or relaxation of the terminating H atoms were not allowed, though other evidence indicates that these alter the energy by amounts on the order of a kilocalorie per mole. Nachtigall et al. noted that their optimized geometry for the bare surface showed less subsurface relaxation than that from LEED structure [9].

Nachtigall et al. [75, 77] used an effective core potential (ECP) on silicon to reduce the effort associated with calculating energies of core electrons. ECPs inherently neglect core polarization effects and may have other incidental deficiencies. Energies including electron correlation were calculated at the optimized geometries using a quadratic configuration interaction (QCISD(T)) method with a relatively small double-zeta plus polarization (DZP) basis set. To estimate errors in the energies caused by the ECP and small basis sets, Nachtigall et al. compared ECP and all-electron calculations using fourth-order Moller–Plesset theory and similarly estimated the effects of increasing the basis sets. They demonstrated (in Si_2H_x systems with constrained geometries) that a good approximation was achieved by estimating the all-electron and basis set corrections separately and adding them to the energy calculated with ECPs and a smaller basis. This allowed extensions of the calculations to higher levels than would be affordable if the corrections were not additive. The ECPs have a substantial effect on the energies, lowering the energies of Reactions (6)–(8) by 2.2–4.6 kcal/mol. Large basis sets for surface silicon and H atoms lower the Si–H bond energies by less than 1.1 kcal/mol; their biggest effect is to lower the energy of H_2.

In the 1991 paper [75], ZPE corrections were calculated at the Hartree–Fock level with a small basis in Si_2H_x model compounds, though in 1993 the same methods were applied to the full Si_9H_x cluster models [77]. The change in calculated ZPE accounts for 2–3 kcal/mol of the difference in bond energies reported in the two papers.

Despite the care taken to test and correct for the approximations in these calculations, there remains some uncertainty about the accuracy of the results. The energetic effects of using a cluster model have not been tested. The constraints imposed in the geometry search prevent some relaxation in the surface structure. For example, the buckled dimer structure for the bare surface could not be reached in these calculations. The errors from using Hartree–Fock or multiple configuration SCF (MCSCF) theory to optimize geometries of stable species are typically small. Despite the extensions of the basis set, it is unclear if the correlated calculations with the largest basis have converged: QCI calculations can be quite sensitive to the basis set. The errors from each of these approximations are likely to alter energies for each species by only a few kilocalories per mole. Depending on whether the errors cancel or reinforce one another, they may have a negligible or substantial effect on the energy differences between species.

Nachtigall et al. made a second important contribution in their 1993 paper [77], by showing that Density Functional Theory (DFT) calculations were in good agreement with the more "traditional" methods such as QCISD(T). Geometries optimized with a local exchange-correlation functional were close to those from the Hartree–Fock or MCSCF calculations. The nonlocal exchange-correlation functional of Becke and Perdew predicted relative energies of stable species in good agreement with the quadratic CI results. These are very useful results, because DFT calculations are much less costly than the QCI methods, though the accuracy of DFT methods was not as well documented. Calculations with DFT methods can remove some of the approximations that had limited earlier work, allowing larger clusters and larger basis sets, for example. DFT methods have been widely used in the most recent theoretical papers.

The work of Wu and Carter [76] is in many ways similar to that of Nachtigall et al. They used similar Si_9 cluster models, ECP and basis sets. However, Wu and Carter did not test or correct for errors due to ECPs or small basis sets. They optimized geometries and calculated frequencies using a self-consistent field method based on Generalized Valence Bond (GVB) Theory. The perfect pairing approximation (GVB-PP) was used to treat the surface dimer and Si–H bonds, with other orbitals treated at the Hartree–Fock level. Optimization was performed without constraints, and Wu and Carter noted the importance of subsurface relaxation. At the optimized geometry, more accurate energies were calculated with a correlation-consistent configuration interaction method known as GVB-CCCI. This method has been documented to give accurate bond energies for a number of small compounds. Direct tests of convergence with respect to basis set size or number of configurations have not been published for silicon surface models.

The GVB-CCCI method was too expensive to apply directly to a cluster of nine Si atoms, so Wu and Carter employed "geometry mapping." That is, after optimizing the geometry of the large cluster, they excised the two atoms in the surface dimer and the surface H atoms and saturated the newly created dangling bonds with pseudoatoms, denoted $\overline{\mathrm{H}}$ and sometimes called siligens. These pseudoatoms, first proposed by Redondo et al. [82], are one-electron atoms with a minimal basis chosen to mimic the electronegativity of Si. Wu and Carter used these pseudoatoms to terminate the large cluster model as well. They showed elsewhere [83] that the geometry mapped surface dimer had orbitals (at the GVB-PP level) similar, but not identical, to those from the larger cluster. One suspects that correlation effects are quite different in Si–$\overline{\mathrm{H}}$ and Si–Si bonds, but the calculations of Wu and Carter may be immune to this effect if errors in the correlation energy cancel when energy differences are calculated. Quantitative tests of the effects of geometry mapping on Si–H bond energies have not been published, but Wu and Carter [76, 83] have claimed that they introduce errors of only ~2 kcal/mol. Other authors [36, 84] have questioned the foundations of the pseudoatom method and their effectiveness at modeling bonds to the bulk in small clusters. The motivation behind siligens, of creating a saturating atom with the electronegativity of Si, is questionable, since Si–H bond energies in molecules do not correlate with the electronegativity of substituents [35, 36].

Wu and Carter's best estimates of the energy to remove the first H atom from a doubly occupied dimer was 87.9 kcal/mol, while that for the second was 86.1 kcal/mol (including ZPE) [76]. This implies that E_{pair} is only 1.8 kcal/mol. Wu and Carter interpret these bond energies as "essentially the same" and conclude that there is little energy advantage to pairing. Given Boland's observation of pairing by STM, they suggest that small energy differences are sufficient to cause pairing or that long-range interactions left out of their model may contribute.

Wu and Carter [76] also estimated that the overall reaction energy is 70.7 kcal/mol (including ZPE), which is substantially higher than the measured activation energy. However, later work by Wu et al. [78] included more configurations in the CI calculation and lowered the energy to 66 kcal/mol (without ZPE). The latter value is in good agreement with the comparable result of 64.4 kcal/mol from the QCISD(T) calculations of Nachtigall et al. The two values calculated by Carter and coworkers for the reaction energy without ZPE differ by 9 kcal/mol, illustrating the magnitude of error that can be caused by inadequate treatment of electron correlation. In either case, the reaction energy is too high to agree with measured activation energies, which led this group to conclude that desorption does not occur directly from the doubly occupied dimer and must take place by a multistep mechanism. Wu and Carter also reported a preliminary estimate of 120 kcal/mol for the activation energy of desorption from the prepaired state, though later work reduced this value considerably. They suggested that the reaction may be diffusion-limited, since they had estimated barriers for diffusion similar to the desorption energy [85]. Later work from this group discounted this latter possibility [86].

The 1991 calculations from the Carter and Jordan groups made qualitatively different predictions. Nachtigall et al. [75] predicted a pairing energy in close agreement with experiment, while Wu and Carter [76] predicted a much smaller value. The predictions of E_{rxn} differed by nearly 15 kcal/mol. Nachtigall et al. [77] have analyzed the sources of the differences. Differences in ZPE calculations make a substantial contribution. Wu and Carter's calculation did not recover all of the correlation energy, as demonstrated by the later work of Wu et al. [78]. Wu and Carter's work also differs from that of Nachtigall et al. in that they do not attempt to correct for errors due to small basis sets and ECPs, and they use geometry mapping in their calculation of correlation energy. The errors caused by these effects are unknown.

The values for E_{pair} and E_{rxn} reported by these two groups in 1993 were in much closer agreement. The close agreement may be accidental, since the calculations used different geometries and Nachtigall et al. used corrections for basis set size and ECPs. However, in 1992 Jing and Whitten [79] reported thermodynamic results from configuration interaction calculations using an $Si_{12}H_x$ cluster model. Their estimates of the energy to remove the first H atom from a doubly occupied dimer was 82.7 kcal/mol, while that for the second was 80.8 kcal/mol (including ZPE). Thus, E_{pair} is 1.9 kcal/mol and E_{rxn} is 60 kcal/mol. These values agree well with the best estimates from the Carter and Jordan groups, reducing the likelihood that agreement is purely accidental.

Despite potential quibbles about errors caused by basis sets, geometry optimization methods, and pseudoatoms, the theoretical methods used by the Jordan, Carter, and Whitten groups have proven reasonably accurate in other settings. Each group believed that their calculations were accurate enough to conclude that desorption was unlikely to occur in a single step from doubly occupied dimers. If there are flaws in these calculations, they are most likely due to the cluster model itself or the geometry optimization methods. Failures in the electronic structure calculations would have to be subtle, though they should not be ruled out.

Some effects that are left out of cluster calculations are illustrated in the work of Vittadini et al. [87]. They have suggested another prepairing mechanism, in which H atoms on the same side of adjacent singly occupied dimers (''interdimer'' airing) are more stable than independent singly occupied dimers. These are periodic slab calculations using a large unti cell and a local density functional (tests with the nonlocal Becke–Perdew functional do not change the results substantially). Vittadini et al. identify this stabilization as the result of weak overlap between dangling bonds on adjacent dimers. They find that the doubly occupied dimer (''intradimer'' pairing) is the most stable state, being 12 kcal/mol more stable than two isolated H atoms at low coverage. The interdimer pairs are 5 kcal/mol less stable than the doubly occupied dimer. Vittadini et al. suggest that their high estimate of the stability of doubly occupied dimers may be consistent with experimental determinations of E_{pair} if the measured pairing corresponds to a change from intradimer to interdimer pairing, rather than complete separation of the adsorbed H atoms. The effect that this might have on desorption kinetics has not been worked out. Vittadini et al. also use their calculated electronic energies to estimate pairing energies by the spectroscopic method of Boland [43], discussed above. They find good agreement with the values obtained from total energy differences. Thus, this work provides some theoretical support for both the method used by Boland and the high value of the pairing energy that he determined.

2. *Activation Energies and Alternative Mechanisms*

The next issue addressed by theory was the activation energy to desorption. Wu et al. [78] studied several mechanisms, beginning with the prepairing mechanism. In their earlier paper, Wu and Carter [76] reported an activation energy of over 120 kcal/mol, which cannot be consistent with experiment. However, this preliminary result was based on a highly constrained geometry search in which all of the substrate atoms were held fixed. An improved search, which imposes C_s symmetry (keeping the surface dimer and adsorbing H atoms in a plane) but is otherwise unconstrained, yielded a barrier of 94 kcal/mol for desorption from a doubly occupied dimer. This is substantially lower than the earlier estimate, but still far above the experimental value. The search for the transition state geometry was conducted within the Hartree–Fock approximation, so effects of electron correlation on geometry were ignored. There are many examples in the literature (e.g., Ref. 38) where this approximation has a large effect on the transition state geometry, though it may cause either an overestimate or underestimate of the activation energy.

Wu et al. [78] found no direct path on the potential surface from the adsorbed state to the transition state. The structure of the transition state (Fig. 3) is reminiscent of a dihydride. Indeed, the steepest descent path from this transition state led back to a dihydride local minimum. This dihydride is weakly stable, but is connected by a direct path to the doubly occupied dimer. This isomerization/desorption path, from the doubly occupied dimer to the dihydride and then to the transition state, is the 1,2-shift/1,1-elimination mechanism previously suggested by Nachtigall et al. [75] on the basis of an analogous gas phase reaction. Wu et al. [78] found that the dihydride is a short-lived intermediate in equilibrium with the doubly occupied state. Within the fast equilibrium approximation, this two-step process has the same effective activation energy, 94 kcal/mol, as the direct desorption pathway. Wu et al. provide a useful discussion of the differences in geometry and π-bonding that account for the differences between the gas-phase and surface desorption processes. They conclude that their calculated activation energies rule out both concerted desorption from the doubly occupied dimer and the isomerization/desorption mechanism.

As an alternative mechanism, Wu et al. [78] proposed that desorption occurs at defect sites, and the desorption rate is limited by surface diffusion to these sites. Using an isolated Si atom (with two dangling bonds) as a model defect, they estimated the activation barrier from a dihydride formed at the defect site as 50–53 kcal/mol (without ZPE). This is smaller than the measured desorption energy, and Wu et al. argued that desorption from the dihydride could not be rate-limiting. However, in other work, Wu and Carter [85] calculated activation barriers for diffusion of H atoms of 46 and 62 kcal/mol, for diffusion parallel and perpendicular to dimer rows respectively. Assuming that the barrier to migrate to a defect is at the high end of this range, the diffusion barrier is close to the measured activation energies for desorption. On this basis, Wu et al. [78] proposed that desorption occurred from the dihydride intermediate but that diffusion limited the rate of dihydride formation.

Despite using an isolated Si atom as a model defect site, Wu et al. [78] were not specific about the identity or detailed structure of the defect. Consequently, they could not directly calculate the activation energy to form the dihydride, or the barrier for conversion of the dihydride to more stable surface species. Lack of a detailed structure prevents a quantitative prediction of the kinetics of the proposed multistep mechanism, so this must be regarded as a qualitative proposal only. Nevertheless, this proposal has strong appeal. As described in Section III.C, the similar desorption dynamics on the (100) and (111) surfaces suggests that the transition states on the two surfaces are similar. This requirement would be satisfied if dihydrides form on isolated Si atom defects on both surfaces. However, the measured diffusion barrier of H on Si(111)-7 $\times$ 7 is much less than the desorption barrier on that surface. This means that the rate-limiting step cannot be diffusion, but might be desorption from the dihydride on (111). Different rate-limiting steps on the two surfaces would rationalize the difference in kinetic order.

Later work on diffusion rates led Wu et al. [86] to reject diffusion as the rate-limiting step on Si(100). They fit a potential surface to results of their calculations and used Monte Carlo transition state theory to calculate rate constants. Surface relaxation effects that were neglected in their first principles calculations lower the barrier for diffusion parallel to the dimer rows to 38 kcal/mol. The barrier for diffusion between dimer rows increases slightly, to 63 kcal/mol. The rate calculations predict that diffusion parallel to dimer rows is much faster than desorption at 800 K, while diffusion perpendicular to dimer rows is much slower, and they conclude that neither can limit desorption. This conclusion requires high confidence in the quality of the potential surface, since a change of only a few kcal/mol in the activation energy would increase the predicted diffusion rate across rows by enough to make it close to the measured value.

There have been two other calculations of the barrier to diffusion for hydrogen on Si(100), which find activation barriers quite different from those of Wu and coworkers [85, 86]. Vittadini et al. [88] used DFT in a periodic slab calculation with large unit cells (up to 4×4). They found activation barriers of 30 and 35 kcal/mol for hopping along rows and between rows, respectively. They concluded that diffusion is always much faster than desorption. Nachtigall and Jordan [89] used a different DFT functional in a cluster calculation and found barriers of 52 and 72 kcal/mol for diffusion parallel and perpendicular to dimer rows, respectively. Although their energies are higher than those of Wu et al., Nachtigall and Jordan reach the same conclusion: Diffusion along dimer rows is too fast to influence desorption, and diffusion between dimer rows is too slow. Nachtigall and Jordan [89] discuss the differences among these three calculations. No experimental activation barriers are available to test these predictions.

Having rejected a diffusion limited mechanism, Wu et al. [86] proposed that formation of the dihydride was rate-limiting. In order for this mechanism to obey first-order kinetics, the active defect site must bind one H atom, so that binding the second H atom is rate-limiting. The structure around the defect was not specified, so that the activation energy for this step could not be calculated. This modified proposal does not share the simplicity of the diffusion-limited model. If the isolated Si atoms have the same structure on the (111) and (100) surfaces (in order to explain the product energy distributions), one must explain why formation of the dihydride is rate-limiting only on the (100) surface. This proposal is general enough that it is difficult to convincingly prove or rule out. However, the calculated barrier to the dihydride desorption step is only a few kilocalories per mole below the measured desorption energy. Given the lack of detail about the defect site structure and the small cluster used as a model, the uncertainty of this calculation must be at least a few kilocalories per mole. Thus, the work of Wu et al. may be consistent with dihydride desorption as the rate-limiting step.

Jing and Whitten [80] studied a symmetric desorption pathway from the doubly occupied dimer, finding a barrier of 86 kcal/mol. This is lower than the barrier found by Wu et al. [78] for the asymmetric path, even though the geometrically symmetric path is forbidden by orbital symmetry (in the Woodward–Hoffmann sense). Jing and Whitten [80] reported that they had also found an asymmetric pathway with higher energy, but they provided no details. They concluded that direct desorption from the doubly occupied dimer could not account for the experimental facts. However, they did not demonstrate that their symmetric transition state was a first-order saddle point, so that lower-energy paths may be available. Their transition state search held the silicon surface geometry fixed, which raises the transition state energy substantially. Later work by Nachtigall et al. [90] showed that a similar geometry was actually a second-order saddle point and that allowing surface distortion would lower the energy. This implies that there is a first-order saddle point with even lower energy than that determined by Jing and Whitten. This also suggests that the higher activation energy (94 kcal/mol) found by Wu et al. [78] overestimates the true barrier.

In other work, Jing and Whitten [79] had proposed a multistep pathway in which the rate-limiting step was breaking the Si–Si dimer bond, with one H atom bound to each Si. This step was estimated to have a barrier of about 56 kcal/mol, without a direct first principles calculation. Once the dimer bond was broken, it was proposed that the H atoms would move closer together and recombine with a low barrier. No calculations are available that support this qualitative proposal.

Instead, Jing et al. [81] proposed another two-step mechanism in which desorption occurs from a dihydride. As with the proposal of Wu et al. [78], Jing et al. [81] suppose that the dihydride species is formed at an isolated, undimerized Si atom. Jing et al. simply assume that formation of the dihydride is fast and that desorption is rate-limiting. This is exactly the opposite of the assumption made by Wu et al. [78]. Jing and Whitten [91] calculated an activation barrier for desorption from the dihydride state of 53.1 kcal/mol (including ZPE) which is close to the value obtained by Wu et al. [78] using very different methods (and neglecting ZPE). However, Jing et al. [81] pointed out that if the dihydride formation step is not rate-limiting, the observed activation energy would not simply be the activation energy of this final step. Instead, the apparent activation energy would be the sum of the dihydride desorption activation energy with the energy difference between the monohydride and dihydride (assuming fast equilibrium between these species). They took the difference between the monohydride and dihydride energies as 7.4 kcal/mol using estimated bond energies, yielding an overall activation barrier of 60.5 kcal/mol. This result agrees well with experiment, though the monohydride–dihydride energy difference is not derived from first principles.

Nachtigall et al. [90] also predicted that the activation energy for direct desorption from the doubly occupied dimer is higher than the measured value, using both QCISD(T) and nonlocal DFT methods. However, their value of 74–75 kcal/mol for E_a^{des} is much lower than the values predicted by the Carter and Whitten groups. Again concluding that direct desorption could not occur, Nachtigall et al. proposed another defect-mediated mechanism, in which the rate-limiting step is desorption from a dihydride species on an isolated atom. In contrast to the earlier proposals, Nachtigall et al. propose that the dihydride formation does not involve movement of H atoms to a defect. Instead, the identity of the isolated atom changes by moving the position of an Si–Si dimer bond. When the isolated atom is adjacent to a doubly occupied dimer, isomerization to a dihydride and a bare dimer occurs, followed by desorption from the dihydride. This proposal is specific enough that Nachtigall et al. could calculate the activation energies for each step. They found that the rate-limiting step is desorption from the dihydride, with an overall activation energy of 59 kcal/mol, in excellent agreement with experiment. The cluster model neglects interactions between dimers in a row that will raise the barrier to defect migration, though this effect is not likely to be large enough to make migration rate-limiting.

3. *Critique of Defect-Mediated Mechanisms*

There are some common issues that arise with all the proposed defect-mediated models. Ideally, the defects involved should be shown to exist experimentally. While it is an often cited fact that the Si(100) surface can have many defects at room temperature [92–94], modern surface preparation methods [95] produce surfaces with defect densities on the order of 1%. Indeed, surfaces that have been exposed to hydrogen at 600 K have an even lower defect density [96]. Furthermore, the most common defects are not isolated atoms but are, instead, missing dimers, "Type C" defects, and steps [92–95]. Type C defects appear to have an atom missing from each of two adjacent dimers, though the structure is probably not the simple isolated atom postulated in the theories of defect-mediated desorption. Tunneling spectroscopy at Type C defects shows metallic behavior, which has been interpreted as evidence of subsurface geometry changes, or possibly strongly buckled dimers, subsurface vacancies, or impurities [94]. The presence of missing dimers and Type C defects has also been correlated with water adsorption, so that many apparent defects may be adsorbed impurities [97, 98]. Impurity defects have not been considered in any calculations to date, though they are ubiquitous. It is possible that the defect density and structure depends on temperature and H coverage, so that observations on clean surfaces at low temperature may not represent the structure relevant to a desorption measurement. STM observations of the surface structure and defect density at elevated temperature with adsorbed hydrogen would establish whether isolated atom defects exist under the conditions where desorption occurs. It is always possible that the defects involved are so rare that they have not been characterized, but this hypothesis borders on being untestable.

While the proposed defect-mediated mechanisms can successfully reproduce the desorption activation energy, none of the models has been used (as the prepairing model has) to reproduce desorption kinetics quantitatively. In particular, these models have not been shown to reproduce the observed deviations from first-order kinetics at low coverage. The model of Nachtigall et al. [90] permits the clearest analysis of this issue. Their mechanism requires the presence of doubly occupied dimers; and at coverages where there are a significant fraction of singly occupied dimers, this mechanism will deviate from first-order kinetics. However, if the pairing energy is as low as predicted by Nachtigall et al. [77], (2–3 kcal/mol), there would be a significant fraction of singly occupied dimers. Consequently, deviations from first-order behavior would be expected at higher coverage than with the parameters used to fit the data of Höfer et al. [39] or Flowers et al. [40]. The mechanisms proposed by Wu et al. [78] and Jing et al. [81] do not appear to permit any deviations from first-order behavior. A quantitative simulation of the desorption kinetics should be made with any model before it can be fully accepted.

Desorption prefactors provide another test of any mechanism. The measured values are about 10^{15} s^{-1}, at the high end of the typical range. Prefactors for defect-mediated models are likely to be much lower than the measured values. Assuming for illustration that the active defects have a density of 10^{-3} and that the monohydride–dihydride conversion is energy neutral, then the equilibrium constant for this isomerization step is $K_{eq} = 10^{-3}$. If these two species are in rapid equilibrium, and desorption from the dihydride is rate-limiting with rate constant $k_{SiH_2} = A_{SiH_2}e^{-E_{SiH_2}/k_BT}$, the observed rate constant will be

$$k_{obs} = K_{eq}k_{SiH_2} = 10^{-3} A_{SiH_2}e^{-E_{SiH_2}/k_BT} \tag{9}$$

Thus, the effective prefactor will be three orders of magnitude smaller than the prefactor for the desorption step, simply because the density of active defects is so small. If a defect desorption mechanism is to reproduce the observed prefactor, the rate-limiting step must have a prefactor correspondingly larger than the measured value. However, this will be difficult to achieve with a low defect density, since the observed prefactors of 10^{15} s^{-1} are at the upper end of the typical range for prefactors of elementary first-order reactions. This argument also implies that the prefactor will be proportional to defect density, while the activation barrier is constant. However, dependence of the prefactor on defect density probably does not allow a useful test of defect-mediated mechanisms, since defect density cannot be controlled over a wide range and prefactors are difficult to determine to high accuracy.

The defect-mediated mechanisms have desorption from the dihydride as a final step. Consequently, they all predict that the transition state for desorption is only a few (3–5) kilocalories per mole above the product state. This is consistent with the observation that product H_2 has little excess energy. On the other hand, the low excess energy of the transition state over the desorbed state means that the defect-mediated mechanisms have corresponding low barriers to adsorption. At the time these mechanisms were proposed, there were no measurements of the activation energy to adsorption. If the more recent evidence [41, 73] that adsorption has a high activation barrier is confirmed, these models must be ruled out or at least substantially modified.

A low sticking probability requires a barrier to adsorption, though not necessarily an energy barrier. An entropic barrier should be expected from a defect-mediated mechanism due to the low density of suitable defects. The loss of entropy in going from the free molecule to the transition state may also cause a reduction in sticking probability, though the high prefactor for desorption implies that the transition state is relatively unconstrained (compared to the adsorbed state). Finally, sticking may be prevented by a dynamical effect, such as recrossing the transition state due to inefficient loss of energy from the reaction coordinate. Note that energy transfer to the surface is not required to prevent transition state recrossing; conversion of energy from the reaction coordinate to other H modes may be sufficient. The low sticking probability puts constraints on the defect density and entropy changes that are consistent with a given activation barrier. For example, a simple one-dimensional barrier model predicts that at 700 K, less than 3% of incident molecules will have enough energy to cross a 5-kcal/mol activation barrier. If the density of active defect sites is 10^{-3}, then 3×10^{-5} molecules will stick. If the measured sticking probability at 700 K is less than 10^{-6}, then other entropy restrictions or dynamical effects must further reduce the sticking probability by over an order of magnitude. A lower activation barrier to adsorption, a larger density of defects, or a lower sticking probability each place even more stringent requirements on the entropy and dynamics. Given the geometry and frequencies calculated at the transition state and desorbed state, it is a simple matter to estimate the entropy changes from the harmonic oscillator–rigid rotor model. Such a calculation has not been reported, but it can provide a useful test of the consistency of any model. Direct calculations of sticking probabilities by molecular dynamics are much more difficult, but potentially feasible.

The above arguments in combination put fairly stringent limits on the desorption prefactors and sticking probability that are consistent with a given defect density. Consistency with the desorption prefactor is easier to understand with a high defect density, while consistency with the sticking probability is easier to explain with a low defect density. Calculation of the desorption prefactors and measurements of the sticking probability and the active defect density will provide stringent tests of the defect-mediated desorption mechanisms.

Any explanation of first-order kinetics should be general enough to account for the first-order kinetics (and deviations from first-order kinetics) on Ge (and perhaps C) surfaces. In particular, defect-mediated mechanisms that may appear feasible on Si surfaces would require modification since the defect density and bond energies on the Ge surface are very different. The prepairing model has already been shown to model the kinetics on Ge [44].

For philosophical reasons, defect-mediated mechanisms should be considered only as a last resort, with strong supporting evidence. Such mechanisms are inherently unsatisfying since there is a virtually endless set of possible proposals. For example, while the dihydride desorption mechanisms are reasonable, one could make other models that agree with the measured barrier to desorption. It is well known that surface structure is modified near defects [92–98], though details of the surface and subsurface structure are unknown. By postulating various combinations of surface strain, one could make models of a doubly occupied dimer near a defect with different adsorption energies. Adsorbed impurities may also alter the desorption barrier, either by providing a defect site or by interaction with an adjacent dimer site. Testable constraints on the imagination are essential, and defect-mediated mechanisms must meet a high standard of proof to be persuasive. The structure of the defect must be specified and shown to exist on the surface. The relative energies for binding at the defect and at majority sites, as well as the mechanism for populating the defect, must be given. A full theory will quantitatively reproduce experimental prefactors and desorption kinetics, with the defect density as the only adjustable (or measured) parameter. Without these requirements, a variety of plausible mechanisms could be examined until one was found with the desired values of calculated energies. A fortuitous combination of modeling and calculational errors could provide support for an unrealistic mechanism.

B. Evidence Supporting the Prepairing Mechanism

Calculations from five different groups have now appeared that conclude that the prepairing mechanism is consistent with experiment [99–103]. These groups have all used density functional theory to approximate the desorption energetics and activation energy from the doubly occupied dimer. This work has also tested some approximations used in earlier work, including the effects of the cluster model, approximations in the geometry optimization, and effects of limited basis sets. Four calculations [99–102] have used periodic boundary condition methods, while one [103] is a cluster calculation. Most [99–101, 103] come to similar conclusions about the structure and energy of the transition state.

The first of these papers to appear was by Kratzer et al. [99]. They used density functional calculations on a periodic slab model with a plane wave basis set and a norm-conserving pseudopotential for the silicon atoms. The slab model was six layers thick, with a 2×1 lateral unit cell. Hydrogen adsorbed symmetrically on both faces of the slab. Geometries were found within the local density approximation, and a higher-level Generalized Gradient Approximation was used to calculate energies from the density found with the LDA. The transition state found was similar to the asymmetric transition states found by Nachtigall et al. [90] and Wu et al. [78], but the activation energy of 57.8 kcal/mol for desorption (including ZPE) is in excellent agreement with experiment. The conclusion drawn from this work, that desorption can occur from the doubly occupied dimer by a concerted mechanism, contradicts the conclusion of the earlier cluster calculations. The difference may be due to the distinctions between cluster models and slab models, or to the different electronic structure methods. Of course, mundane issues like basis set convergence or inaccurate geometry optimizations may also affect one or more of these calculations.

Kratzer et al. [99] showed that motion along the reaction coordinate involves buckling of the surface dimer and that the configuration of the surface affects the barriers to adsorption and desorption. For example, an H_2 molecule incident from the gas phase encounters a barrier to adsorption of 9 kcal/mol on the minimum energy path. However, the difference in masses implies that the surface is unlikely to respond to the presence of the incident molecule on the time scale of the molecule's encounter with the surface. If the surface is frozen in the equilibrium geometry of the bare surface, the activation barrier to adsorption is 14 kcal/mol. Of course, thermal fluctuations will establish a variety of surface geometries and activation barriers. This picture also implies that in desorption, some of the energy in the transition state can be deposited in the surface, since the surface in the transition state is distorted with respect to the equilibrium geometry of the bare surface.

Pehlke and Scheffler [100] did a calculation very similar to that of Kratzer et al. [99], the main difference being the use of a 2×2 unit cell. This larger unit cell reduces the interactions between adjacent unit cells and allows more substrate relaxation, including anticorrelated buckling of dimers. The slab included five layers, and dangling bonds at the bottom of the slab were terminated with fixed H atoms. The predicted overall reaction energy for desorption from the doubly occupied dimer was 48 kcal/mol (without ZPE), compared to 53 kcal/mol with a 2 × 1 unit cell. Analogous comparisons of different size cluster models are not yet available, but the work of Pehlke and Scheffler suggests that surface relaxation and/or interactions between adjacent dimers can have a significant effect on the reaction energy. The high values of E_{rxn} calculated with cluster models may be artifacts of the small cluster size or differences between the optimized slab and cluster geometries.

Pehlke and Scheffler [100] found an activation energy to desorption of 55 kcal/mol (without ZPE). Including the ZPE correction of 3 kcal/mol (taken from a cluster calculation by Pai and Doren [103]) reduces this to 52 kcal/mol. The transition state is again asymmetric, and the reaction path is coupled to dimer buckling and stretching. The corresponding barrier to adsorption is 7 kcal/mol, but these authors argue that the sticking probability will be small due to the low probability that the surface atoms will sample the minimum energy configuration while a H_2 molecule is close to the surface. This is an entropic barrier to adsorption. No quantitative calculations of kinetic prefactors or product energy distributions have been made. However, Pehlke and Scheffler predict that only a small amount of energy (~3 kcal/mol) from the optimal transition state is deposited into surface vibrations. For desorbing molecules (which must stay close to the minimum energy path), this means that only a small amount of energy (about 4 kcal/mol) can be deposited in the desorbing molecule.

Vittadini and Selloni [101] used an even larger 4 × 4 unit cell, containing eight surface dimers. They used a smaller basis set and a slightly different nonlocal functional (Becke–Perdew) than the work of Kratzer et al. [99] or Pehlke and Scheffler [100], and they used only one **k** point. Vittadini and Selloni found two "prepairing" mechanisms for desorption with activation energies (55–62 kcal/mol, without ZPE) that are both close to the experimental value. They showed that the activation energy for the "intradimer" mechanism (the path studied in all the earlier work) can vary by 7 kcal/mol, depending on the phase of the buckling of adjacent dimers. This is another way to couple surface vibrations to adsorption and desorption that was previously not considered. Vittadini and Selloni also explored an "interdimer" path, in which H atoms recombine from the same side of two adjacent (in the same dimer row) singly occupied dimers. This mechanism was motivated by their earlier work [87] which showed that two H atoms on the same side of adjacent dimers were more stable than two H atoms on independent dimers. The barrier to adsorption on the intradimer path is higher than on the minimum energy interdimer path, but surface distortions can make them nearly equal.

A final slab calculation by Li et al. [102] used a slightly different basis set, but explored only symmetric dimer configurations and a local density functional. Errors in these two approximations appear to cancel, since they find an activation barrier to desorption of 60 kcal/mol, similar to the more thorough calculations.

Pai and Doren [103] have reexamined the cluster model using DFT. They found an activation energy for desorption of only 64 kcal/mol (including ZPE). This value is still higher than the measured value, but Pai and Doren concluded that the uncontrolled approximations in the cluster model, as well as the uncertainty in the DFT method, made the agreement close enough that the prepairing model should not be ruled out on the basis of such a calculation. This calculation is distinguished from prior cluster calculations in several ways. Geometry optimizations included no constraints or symmetry assumptions; hydrogen atoms were used to terminate dangling bonds; all equivalent atoms were given the same basis set; all electrons, including core electrons, were explicitly included (no pseudopotentials or effective core potentials); electron correlation effects were included in the geometry optimization, using DFT. Using local DFT to optimize the geometry and calculate energies gives a reaction energy of 49 kcal/mol and an activation barrier for desorption of 55 kcal/mol (including ZPE). However, including more accurate nonlocal corrections to the energy (with the B-LYP functional at the LSD geometry) raises the activation barrier to 65 kcal/mol. Using a nonlocal functional to optimize the geometries lowers the activation energy by 1 kcal/mol, to 64 kcal/mol.

C. Comparison of Theoretical Results

It is worth trying to understand the differences between the calculations of Pai and Doren [103] and the earlier DFT cluster calculations by Nachtigall et al. [90] that predicted activation barriers higher by 10 kcal/mol. The optimal cluster geometry of Pai and Doren showed an analogue of the anticorrelated buckling seen on real surfaces and in slab calculations with 2×2 or larger unit cells. This type of relaxation was prevented by constraints imposed by Nachtigall et al., and the constraints raise the activation energy by 2–3 kcal/mol. Differences in the functional and basis sets also account for a large part of the difference between the DFT calculations. It appears that the DFT calculations of Nachtigall et al. [90] were not converged with respect to the basis set. Pai and Doren recalculated the energy of their optimized geometries with the Becke3LYP functional and basis set used by Nachtigall et al., and also with a larger basis set. The larger basis set reduces the activation energy by 3 kcal/mol to 69 kcal/mol (including ZPE). Of course, even this result may not be converged. It is not clear whether the Becke3LYP functional is especially sensitive to the basis set, or if the bases used by Nachtigall et al. (which were optimized for Hartree–Fock calculations) are not well-suited to DFT calculations. The B-LYP estimate of E_a^{des} by Pai and Doren, using a basis designed for DFT, is only reduced by 0.5 kcal/mol when the basis is increased. Using the largest basis set, the Becke3LYP and B-LYP functionals predict desorption activation energies (69 and 64 kcal/mol, respectively, including ZPE) that differ by 5 kcal/mol. Other functionals give lower activation energies with the same geometry and basis. For example, the Becke–Perdew (BP) functional predicts that E_a^{des} is 60 kcal/mol (including ZPE).

Pai and Doren's cluster calculation can also be compared with the slab calculations. The BP functional is similar to those used in the slab calculations. The cluster calculation value for E_a^{des} of 60 kcal/mol (including ZPE) may be compared to 58 kcal/mol from the slab calculations with a 2×1 unit cell or 52 kcal/mol with a 2×2 unit cell. Apparently, the small unit cell and cluster calculations give similar values, but effects of surface relaxation and adjacent dimers that are included in the larger unit cell are neglected in the cluster calculation. As noted above, different functionals and basis sets can affect the results as much as the difference between slab and cluster boundary conditions.

The range of variation among functionals is somewhat surprising. Given that the local and BP functionals differ by only 5 kcal/mol, many workers would conclude that nonlocal effects are small, and their magnitude should be insensitive to the details of the functional. Yet, the local and Becke3LYP functionals differ by 15 kcal/mol, so that the different functionals clearly make very different estimates of the nonlocal exchange and correlation energy. Perhaps the diradical structure of the surface dimer is stabilized by subtle effects that are not well-modeled by these approximate functionals. If the variation among predictions in these functionals is caused by a subtlety in electronic structure, then other methods applied to this problem, such as configuration interaction methods, may have similar difficulties.

It is natural to ask which functional most accurately represents the electronic energy in this system. Nachtigall et al. [104] have made a systematic study of four H_2 elimination reactions from silanes and a constrained small molecule analogue of the surface reaction, using various density functionals. Some experimental data is available, and QCISD(T) calculations, with corrections to remove effects of ECPs and limited bases, are also used for comparison. QCISD(T) calculations are generally quite accurate if used with large bases. In the examples considered by Nachtigall et al., the BP, B-LYP, and Becke3LYP functionals all agree to within 8 kcal/mol for activation energies and reaction energies. Overall, the BP and Becke3LYP functionals have similar agreement with experiment, though BP predicts reaction energies more consistently while Becke3LYP predicts activation energies for H_2 elimination (though not for the reverse reaction) more consistently. Becke3LYP also agrees more closely with extrapolated QCISD(T). It is perhaps significant that the BP functional routinely underestimates activation barriers for H_2 elimination. Given the small number of examples and small variations among the theories, final conclusions about the quality of predictions for the surface reaction are premature.

Other effects that are not well-tested in these small molecule analogues may alter the accuracy of these functionals for surface reactions. For example, it was noted above that the Becke3LYP prediction of E_a^{des} required a very large basis set for convergence, while no such sensitivity to the basis was found in the small molecule reactions of Nachtigall et al. The differences may be related to the difference between Si–Si bonds (in the surface reaction) and Si–H bonds (in the calibration examples) or to special properties of the surface diradical structure. A conservative conclusion to draw would be that energy predictions of nonlocal DFT methods typically have errors of at least a few kilocalories per mole.

Quantitative differences in the predicted activation barriers are only part of the reason that different groups have reached different conclusions about the desorption mechanism. Expectations about the accuracy of the methods have also led to different interpretations. The Carter, Jordan, and Whitten groups have all believed that their calculations are accurate enough to rule out the prepairing mechanism. Indeed, they have argued that their methods are "chemically accurate," with errors of no more than 2 kcal/mol. However, their predicted activation energies [78, 80, 90] cover a range of 20 kcal/mol, indicating large errors in at least two of these calculations. The DFT calculations that support the prepairing mechanism also cover a significant range, but the authors who did this work have lower expectations for the accuracy of their methods. For example, the values of E_a^{des} found by Doren and Pai and by Pehlke and Scheffler were 64 kcal/mol and 52 kcal/mol (including an approximate correction for ZPE), respectively. Both considered their results to be in agreement with the experimental value of 58 kcal/mol. In contrast, Jing and Whitten [80] and Wu et al. [78] each calculated barriers to desorption from the dihydride of 53 kcal/mol, and they concluded that a single-step mechanism with this activation energy did not agree with experiment.

Clearly, activation barrier calculations alone will not settle the controversy about the desorption mechanism. Accurate comparisons to measured kinetics, desorbing molecule energy distributions, and sticking probabilities will make stringent tests of any theory. Kinetic modeling requires absolute rate constants, rather than just activation energies. These are easy to calculate with transition state theory in the harmonic approximation, but more accurate calculations require information about the potential at points other than the minima and transition state. Energy distributions and sticking probabilities can be determined from molecular dynamics simulations. This requires a global potential surface, defined at arbitrary configurations for all the involved atoms. Blank et al. [105] have shown that such a potential surface can be developed for the prepairing mechanism from first principles calculations, using a neural network model. However, Blank et al. used local DFT calculations to develop the method and have not yet incorporated nonlocal corrections into the model, nor have they run simulations to be certain that the model is accurate over a wide enough range of configurations. Nevertheless, the technology for dynamics simulations from first principles potentials is at hand.

D. Inferences About Kinetics and Dynamics

The only theoretical estimate of a prefactor for desorption is that of Pai and Doren [103], who used transition state theory in the harmonic approximation. Their predicted value for the prepairing mechanism was more than an order of magnitude below the measured value. Given the simplicity of the calculation and the uncertainty in experimental prefactors, it is premature to interpret the disagreement as evidence against the prepairing mechanism.

The calculations of Refs. 78, 90, 99–101, and 103 all determine similar transition state structures for the prepairing mechanism. These structures are qualitatively consistent with the observed energy of desorbing molecules. The H–H distance is longer than the equilibrium bond length in H_2, consistent with product vibrational excitation. The orientation of the hydrogen molecular axis is constrained at the transition state, consistent with the low product rotational energy. The surface is distorted, relative to the structure of the clean surface, so that some of the excess energy at the transition state can be deposited in surface vibrations. However, in contrast to the model of Brenig et al. [71], a large part of the excess energy resides in the H–H stretch, and surface distortion is a smaller contribution.

Arguments about dynamics that are based on the transition state structure are certainly not quantitative and may be misleading. The excess energy observed in H_2 vibrations illustrates some subtleties. Theory predicts that the H–H bond is stretched in the prepairing transition state, making vibrational excitation plausible. However, the vast majority of molecules leave the surface in the vibrational ground state. How is the excess energy in the molecular stretch coordinate transferred to another mode? Since other molecular modes do not have any excess energy after desorption, this stretching energy must be transferred to the surface. If this is the case, it must reflect a subtle coupling between the high-frequency hydrogen vibrational motion and surface modes. The difference in energy between the ground and first excited vibrational state of H_2 is about 12 kcal/mol. This means that molecules that depart the surface in the first vibrational state have much more energy in internal motion than do ground-state molecules, while ground-state molecules must transfer much more energy to other modes. Thus, molecules in the two vibrational states must follow very different trajectories, with qualitatively different coupling to other modes. This argument is independent of mechanism, so that the dynamics of vibrational excitation will be unusual in any case.

The possibility that some of the excess energy at the transition state may be deposited in surface vibrations may resolve the apparent conflict between the high activation barrier seen in adsorption measurements and the low barrier implied by desorption product energy distributions. This idea motivated the model of Brenig et al. [71] described in Section III.D. However, according to first principles calculations, the energy due to surface distortion at the optimal transition state is only about 5 kcal/mol, rather than the value assumed in the model of Brenig et al. of over 20 kcal/mol. Kratzer et al. [99] developed a model (using a quadratic approximation to the potential) derived from their first principles calculations. The H–H bond length and H_2–surface distance are presumed to be strongly coupled and they are treated quantum mechanically, within a WKB approximation. The remaining H_2 degrees of freedom are treated classically (in the sudden approximation). The dissociative sticking probability is determined as a function of incident energy and the dimer buckling angle. This sticking probability can be integrated with Boltzmann weighting to predict the temperature-dependent sticking probability. The calculated sticking probability for molecules with an incident energy of 7 kcal/mol is between 10^{-6} and 10^{-7} for temperatures between 300 and 1000 K. Increasing the incident energy to 9 kcal/mol raises the sticking probability by an order of magnitude. An Arrhenius plot of the sticking probability at different surface temperatures for molecules of fixed incident energy (as in the experiment of Kolasinski et al. [41]) yields an apparent activation energy of only 2 kcal/mol for incident energies below 7 kcal/mol. For higher incident energies, the apparent activation energy is even lower, suggesting that more energetic molecules are less dependent on surface configuration. This low apparent barrier to adsorption is consistent with the low internal energy of desorbing molecules.

There is a strong argument against the role of surface vibrations in determining the energy of desorbing molecules. The energy difference between the ground and first excited vibrational states of H_2 is 12 kcal/mol, but the phonons on the silicon surface all have energies less than 1 kcal/mol. Any mechanism that transfers excess energy from the H–H stretch coordinate to the surface would have to be a high order, many phonon process. Given the mass mismatch between hydrogen and silicon, and the distance of the H_2 molecule from the surface at the transition state, this seems extremely unlikely.

Although it has been given little attention, excess energy at the transition state may be transferred to surface electronic excitation, instead of surface vibrations. The surface density of states has a small band gap of about 7 kcal/mol, with the main $\pi - \pi^*$ transition at 18 kcal/mol [106]. Thus, a single surface electronic excitation can account for the excess energy in the transition state. In other words, the majority of molecules may desorb on a potential surface corresponding to an excited surface electronic state. This proposal has several implications:

1. By detailed balance, molecules incident on the surface with no excess energy should be more likely to adsorb if the surface is electronically excited. Indeed, the adsorption measurements of Bratu et al. [73] suggest that surface excitation alone is sufficient to activate adsorption. The activation energy they measure, 16 ± 2 kcal/mol, is about the energy of the $\pi - \pi^*$ transition on the bare surface.
2. Crossing of two surface electronic states could rationalize the observed vibrational energy distribution in desorbed H_2. Molecules observed in the ground vibrational state may desorb on a potential that leaves the surface electronically excited, while molecules observed in the first vibrational excited state may desorb on the potential that leaves the surface in its electronic ground state.
3. First principles calculations of activation energies have only examined the ground electronic state. A surface crossing could allow molecules to desorb at an energy below the calculated transition state of the ground state potential surface. This could explain the high activation barriers found in cluster calculations.
4. If there is a crossing of the ground and first excited surface electronic states, one expects that doping the surface with adsorbates that lower (or raise) the surface band gap would also lower (or raise) the desorption barrier. Experiments by Ning and Crowell have shown that electronic effects of Ge adsorption on Si(100) do lower the desorption barrier [107].

First principles calculations that locate a surface crossing would provide strong support for this proposal. DFT cluster calculations show that the electronic excitation energy is quite sensitive to small changes in the surface geometry [108]. Coupling to electronic states may also be an important contribution to H-atom sticking. These light atoms cannot transfer energy to surface vibrations yet they stick with nearly unit probability.

Several groups have made initial attempts to calculate dynamics with models of the prepairing mechanism. Convincing agreement with experiment remains elusive, though these calculations include many approximations.

No comparable calculations are available for other mechanisms. However, agreement between dynamical calculations and experiment will be essential before a complete consensus is reached for any mechanism.

E. Desorption from the Dihydride

Relatively little effort has been given to the mechanism of desorption from the dihydride. The dynamics measurements [66] suggest that the transition state is similar to that of the monohydride, so some theorists have taken this as evidence that the reaction is first order, removing both H atoms from a single dihydride. The kinetics measurements [20, 40] indicate that the reaction is second order, so other theories involve removal of one H atom from each of two adjacent dihydrides.

Ciraci and Batra [109] used local DFT to study the structure of the dihydride phase. They combined their results with empirical evidence to argue that the barrier to desorption should be less than 35 kcal/mol for a mechanism involving two dihydride units. Wu et al. [78] assumed that desorption involved only one dihydride and suggested that repulsion in the dihydride phase would lower the desorption energy that they had calculated for an isolated dihydride. This would make the activation barrier coverage-dependent, ranging from 38 kcal/mol for a full dihydride layer, to 55 kcal/mol as the monohydride phase is approached. Simulated TPD spectra should make a good test of this proposal. Jing and Whitten [91] found a first-order pathway with an activation barrier of 53 kcal/mol, calculated from first principles on a cluster model. The adsorption barrier is 5 kcal/mol. The transition state has an unusual, highly distorted geometry that would appear to be strongly affected by steric repulsion from nearby sites, though the cluster model does not include neighboring sites. Nachtigall et al. [90] have proposed that dihydrides may migrate by a variant of their defect diffusion mechanism. Dihydride migration would be rapid, allowing a second-order reaction. Nachtigall et al. calculate the barrier to the desorption step as 48 kcal/mol, in good agreement with experiment. Vittadini and Selloni [101] studied the fully saturated dihydride surface and found that desorption involving two dihydrides was exothermic. The activation barrier for a desorption path involving two dihydrides has a high barrier, but the barrier for elimination of H_2 from a single dihydride has a barrier of about 48 kcal/mol. The surface reconstructs following desorption to form a doubly occupied surface dimer. Vittadini and Selloni suggested that second-order kinetics is caused by an interaction between neighboring dihydrides, though they did not determine the coverage dependence of the activation energy.

Clearly, there is not even qualitative agreement among these theories, although the proposal of Nachtigall et al. agrees well with observations.

V. REMAINING ISSUES

The kinetic parameters for desorption from the monohydride phase on Si(100)-2 $\times$ 1 have been controversial, but consensus has now been reached on the reaction order and Arrhenius parameters. The rotational and vibrational energy distributions are also well-characterized. Almost everything else about this reaction requires clarification. In this section, some of the remaining issues for theory and experiment are outlined. The motivations for these issues have been addressed in more detail in the main body of this chapter.

A relatively simple test of theoretical predictions would be given by structural characterization of the monohydride surface. The only structural experiment, based on ion scattering [74], finds a dimer bond length that is 0.5 Å longer than predicted by any theory. Verification or correction of this measurement by electron or x-ray diffraction would be a useful guide to both structural theory and dynamical models.

A complete set of average energies for the desorbing molecules has been measured. However, it is not obvious how to reconcile a thermal distribution of translational energies in desorbing molecules with an angular distribution that is peaked in the normal direction and a sticking probability that increases with normal translational energy. The measurements of sticking probability disagree quantitatively with one another, though some of this may be due to the fact that the measurements were done with different incident molecule energy distributions. Further experiments to resolve the differences should be a high priority. The activation energy to adsorption is especially critical, since it makes a clear quantitative distinction between the proposed defect-mediated mechanisms (which have a low barrier to adsorption) and the prepairing mechanism.

The similarity of the average energies in molecules desorbing from the monohydride and dihydride phases on Si(100) and from Si(111) must be explained. The assumption that desorption occurs through a similar transition state in all three cases has led to the idea that the transition state must be structured like a dihydride. However, in the prepairing mechanism transition state on (100), the two H atoms and the surface dimer are confined to a plane, while no such constraint would be present for an adatom dihydride on (111). Differences between these transition states may appear in the rotational orientation of the desorbing molecules. It is still more difficult to understand how the transition state from the (100) dihydride could resemble the monohydride transition state if desorption from the dihydride is second order. Perhaps similarity of the three transition states is not the origin of the similar energy distributions. Further theoretical work on the (100) dihydride and (111) surfaces will constrain speculation about comparative dynamics.

Although most of the theoretical discussion of desorption mechanisms has concerned whether predicted activation energies agree with experiment, this issue can only be a guide to choosing candidate mechanisms for further testing. A consensus value for the barrier to adsorption will further help to distinguish good candidates. More refined tests are provided by predictions of the Arrhenius prefactors and the deviations from purely first-order kinetics. Among the current candidates, only the prepairing mechanism has been shown to reproduce the desorption kinetics. However, the energetics of prepairing make it difficult to explain the energy distribution of desorbing molecules, since there is a substantial excess of energy at the transition state. The defect-mediated mechanisms have little excess energy at the transition state, though it may be difficult to reconcile the low sticking probability with the high prefactor for desorption. Neither mechanism has been shown to be quantitatively consistent with the desorption product energy distributions or sticking probability. This will require development of accurate, many-dimensional, potential energy surfaces and classical or quantum dynamical simulations.

Since an excess of molecules desorb with vibrational excitation, it should be possible to enhance sticking by exciting vibrations in the incident molecules. An experiment to test a similar idea has been performed on Cu [110]. Although much attention has been given to the importance of the surface distortion, all transition state geometries from first principles calculations also have a stretched H–H bond. The excess energy in a vibrationally excited H_2 molecule is even larger than that predicted for surface distortion.

The controversy over defect-mediated mechanisms has raised hypotheses that might be tested by STM observations. For example, it is likely that the defect density depends on temperature and hydrogen coverage. If so, low-temperature observations of defects do not bear on the desorption process. Boland has shown that a surface saturated with H at 600 K has a very low defect density. At least within a small range of coverage near saturation, one expects that the defect concentration is constant. If desorption-mediating defects are created during desorption, the rate law should be coverage dependent. It would be interesting to begin with a surface saturated with H at 600 K and use STM to follow it in time at a slightly higher temperature to determine if there are structural changes as the coverage decreases by desorption, and if changes in the defect density are correlated with changes in desorption rate.

The difference between the desorption energy observed under growth conditions and in other measurements must be explained. If this is caused by differences in defect density, it should be possible to observe different structures by STM. This would confirm the idea that desorption from defects has a lower activation barrier than desorption from the ideal surface. However, it would also mean that the mechanism of desorption under growth conditions is not the same as on more ideal surfaces.

Our understanding of reactions of hydrogen on Si is far from complete. This system has yielded a number of surprising results. The confusing issues that remain will likely be resolved only after some additional surprises.

Finally, a related review by Kolasinski has recently been published [111]. This thorough review is written from an experimentalist's viewpoint, and is highly recommended.

Acknowledgments

Robert Konecny has provided invaluable assistance in preparing this chapter. I wish to thank him and my other coworkers, Kerwin Dobbs, Sharmila Pai, and Anita Robinson Brown. This work was supported by the National Science Foundation under grant CHE-9401312.

References

1. J. M. Jasinski and S. M. Gates, *Acc. Chem. Res.* **24,** 9 (1991).
2. S. M. Gates, C. M. Greenlief, S. K. Kulkarni, and H. H. Sawin, *J. Vac. Sci. Technol. A* **8,** 2965 (1990).
3. M. Liehr, C. M. Greenlief, S. R. Kasi, and M. Offenberg, *Appl. Phys. Lett.* **56,** 629 (1990).
4. S. M. Gates and S. K. Kulkarni, *Appl. Phys. Lett.* **60,** 53 (1992).
5. D. P. Adams, S. M. Yalisove, and D. J. Eaglesham, *Appl. Phys. Lett.* **63,** 3571 (1993).
6. M. Copel and R. M. Tromp, *Phys. Rev. Lett.* **72,** 1236 (1994).
7. J. T. Law, *J. Chem. Phys.* **30,** 1568 (1959).
8. J. T. Law, *J. Appl. Phys.* **32,** 600 (1961).
9. B. W. Holland, C. B. Duke, and A. Paton, *Surf. Sci.* **140,** L269 (1984).
10. N. Jedrecy, M. Sauvage-Simkin, R. Pinchaux, J. Massies, N. Greiser, and V. H. Etgens, *Surf. Sci.* **230,** 197 (1990).
11. Q. Liu and R. Hoffmann, *J. Am. Chem. Soc.* **117,** 4082 (1995).
12. J. A. Appelbaum and D. R. Hamann, in *Theory of Chemisorption*, J. R. Smith (ed.) (Springer-Verlag, 1980).
13. Y. J. Chabal, A. L. Harris, K. Raghavachari, and J. C. Tully, *Int. J. Mod. Phys. B* **7,** 1031 (1993).
14. J. J. Boland, *Adv. Phys.* **42,** 129 (1993).
15. R. A. Wolkow, *Phys. Rev. Lett.* **68,** 2636 (1992).
16. N. Roberts and R. J. Needs, *Surf. Sci.* **236,** 112 (1990); M. C. Payne, N. Roberts, R. J. Needs, M. Needels, and J. D. Joannopoulos, *ibid.* **211/212,** 1 (1989).
17. S. M. Gates, R. R. Kunz, and C. M. Greenlief, *Surf. Sci.* **207,** 364 (1989).
18. C. C. Cheng and J. T. Yates, Jr., *Phys. Rev. B* **43,** 4041 (1991).
19. Z. H. Lu, K. Griffiths, P. R. Norton, and T. K. Sham, *Phys. Rev. Lett.* **68,** 1343 (1992).
20. P. Gupta, V. L. Colvin, and S. M. George, *Phys. Rev. B* **37,** 8234 (1988).
21. K. Takayanagi, Y. Tanishiro, M. Takahashi, and S. Takahashi, *J. Vac. Sci. Technol. A* **3,** 1502 (1985).

22. A. Zangwill, *Physics at Surfaces* (Cambridge University Press, 1988).
23. C. Kleint, B. Hartmann, and H. Meyer, *Z. Phys. Chem. (Leipzig)* **250,** 315 (1972).
24. Yu. I. Belyakov, N. I. Ionov, and T. N. Kompaniets, *Sov. Phys. Sol. State* **14,** 2567 (1973).
25. K.-D. Brzóska and C. Kleint, *Thin Solid Films* **34,** 131 (1976).
26. G. Schulze and M. henzler, *Surf. Sci.* **124,** 336 (1983).
27. B. G. Koehler, C. H. Mak, D. A. Arthur, P. A. Coon, and S. M. George, *J. Chem. Phys.* **89,** 1709 (1988).
28. C. Kleint and K.-D. Brzóska, *Surf. Sci.* **231,** 177 (1990).
29. G. A. Reider, U. Höfer, and T. F. Heinz, *Phys. Rev. Lett.* **66,** 1994 (1991).
30. R. Gomer, *Rep. Prog. Phys.* **53,** 917 (1990).
31. M. L. Wise, B. G. Koehler, P. Gupta, P. A. Coon, and S. M. George, *Surf. Sci.* **258,** 166 (1991).
32. G. A. Reider, U. Höfer, and T. F. Heinz, *J. Chem. Phys.* **94,** 4080 (1991).
33. K. Sinniah, M. G. Sherman, L. B. Lewis, W. H. Weinberg, J. T. Yates, Jr., and K. C. Janda, *Phys. Rev. Lett.* **62,** 567 (1989); *J. Chem. Phys.* **92,** 5700 (1990).
34. K. Christmann, *Surf. Sci. Rep.* **9,** 1 (1988).
35. J. M. Kanabus-Kaminska, J. A. Hawari, D. Griller, and C. Chatgilialoglu, *J. Am. Chem. Soc.* **109,** 5267 (1987).
36. S. Pai and D. Doren, *J. Phys. Chem.* **98,** 4422 (1994).
37. J. Abrefah and D. R. Olander, *Surf. Sci.* **209,** 291 (1989).
38. K. D. Dobbs and D. J. Doren, *J. Am. Chem. Soc.* **115,** 3731 (1993).
39. U. Höfer, L. Li, and T. F. Heinz, *Phys. Rev. B* **45,** 9485 (1992).
40. M. C. Flowers, N. B. H. Jonathan, Y. Liu, and A. Morris, *J. Chem. Phys.* **99,** 7038 (1993).
41. K. W. Kolasinski, W. Nessler, K-H. Bornscheuer, and Eckart Hasselbrink, *J. Chem. Phys.* **101,** 7082 (1994).
42. R. Walsh, *Acc. Chem. Res.* **14,** 246 (1981); R. Walsh, in *The chemistry of Organic Silicon Compounds*, S. Patai and Z. Rappoport (eds.) (Wiley, 1989).
43. J. J. Boland, *Phys. Rev. Lett.* **67,** 1539 (1991); *ibid* **67,** 2591 (1991); *J. Vac. Sci. Technol. A* **10,** 2458 (1992).
44. M. P. D'Evelyn, S. M. Cohen, E. Rouchouze, and Y. L. Yang, *J. Chem. Phys.* **98,** 3560 (1993).
45. M. P. D'Evelyn, Y. L. Yang, and L. F. Sutcu, *J. Chem. Phys.* **96,** 852 (1992).
46. Y. L. Yang and M. P. D'Evelyn, *J. Vac. Sci. Technol. A* **11,** 2200 (1993).
47. M. Liehr, C. M. Greenlief, M. Offenberg, and S. R. Kasi, *J. Vac. Sci. Tech. A* **8,** 2960 (1990).
48. K.-J. Kim, M. Suemitsu, and N. Miyamoto, *Appl. Phys. Lett.* **63,** 3358 (1993).
49. C. M. Greenlief and M. Liehr, *Appl. Phys. Lett.* **64,** 601 (1994).
50. L. A. Okada, M. L. Wise, and S. M. George, *Appl. Surf. Sci.* **82/83,** 410 (1994).
51. Y. M. Wu and R. M. Nix, *Surf. Sci.* **306,** 59 (1994).
52. J. J. Boland, *Phys. Rev. B* **44,** 1383 (1991).
53. J. E. Vasek, Z. Zhang, C. T. Stalling, and M. G. Lagally, *Phys. Rev. B* **51,** 17207 (1995).

54. S. M. Gates and S. K. Kulkarni, *Appl. Phys. Lett.* **60,** 53 (1992).
55. L. B. Lewis, J. Segall, and K. C. Janda, *J. Chem. Phys.* **102,** 7222 (1995).
56. A. V. Hamza, G. D. Kubiak, and R. H. Stulen, *Surf. Sci.* **237,** 35 (1990).
57. R. E. Thomas, R. A. Rudder, and R. J. Markunas, *J. Vac. Sci. Technol. A* **10,** 2451 (1992).
58. Y. L. Yang, L. M. Struck, L. F. Sutcu, and M. P. D'Evelyn, *Thin Solid Films* **225,** 203 (1993).
59. D. A. Outka, *Surf. Sci. Lett.* **235,** L311 (1990).
60. M. D. Allendorf and D. A. Outka, *Surf. Sci.* **258,** 177 (1991).
61. P. Nachtigall and K. D. Jordan, *J. Chem. Phys.* **101,** 2648 (1994).
62. M. C. Flowers, N. B. H. Jonathan, Y. Liu, and A. Morris, *J. Chem. Phys.* **101,** 2650 (1994).
63. K. W. Kolasinski, S. F. Shane, and R. N. Zare, *J. Chem. Phys.* **95,** 5482 (1991).
64. K. W. Kolasinski, S. F. Shane, and R. N. Zare, *J. Chem. Phys.* **96,** 3995 (1992).
65. S. F. Shane, K. W. Kolasinski, and R. N. Zare, *J. Vac. Sci. Technol. A* **10,** 2287 (1992).
66. S. F. Shane, K. W. Kolasinski, and R. N. Zare, *J. Chem. Phys.* **97,** 1520 (1992).
67. S. F. Shane, K. W. Kolasinski, and R. N. Zare, *J. Chem. Phys.* **97,** 3704 (1992).
68. J. Sheng and J. Z. H. Zhang, *J. Chem. Phys.* **97,** 596 (1992).
69. Y.-S. Park, J.-Y. Kim, and J. Lee, *J. Chem. Phys.* **98,** 757 (1993).
70. K. W. Kolasinski, W. Nessler, A. de Meijere, and Eckart Hasselbrink, *Phys. Rev. Lett.* **72,** 1356 (1994).
71. W. Brenig, A. Gross, and R. Russ, *Z. Phys. B* **96,** 231 (1994).
72. D. J. Doren and J. C. Tully, *Langmuir* **4,** 256 (1988); *J. Chem. Phys.* **94,** 8428 (1991).
73. P. Bratu and U. Höfer, *Phys. Rev. Lett.* **74,** 1625 (1995); P. Bratu, K. L. Kompa, U. Höfer, *Chem. Phys. Lett.* (in press).
74. Y. Wang, M. Shi, and J. W. Rabalais, *Phys. Rev. B* **48,** 1678 (1993).
75. P. Nachtigall, K. D. Jordan, and K. C. Janda, *J. Chem. Phys.* **95,** 8652 (1991).
76. C. J. Wu and E. A. Carter, *Chem. Phys. Lett.* **185,** 172 (1991).
77. P. Nachtigall, K. D. Jordan and C. Sosa, *J. Phys. Chem.* **97,** 11666 (1993).
78. C. J. Wu, I. V. Ionova, and E. A. Carter, *Surf. Sci.* **295,** 64 (1993).
79. Z. Jing and J. L. Whitten, *Phys. Rev. B* **46,** 9544 (1992).
80. Z. Jing and J. L. Whitten, *J. Chem. Phys.* **98,** 7466 (1993).
81. Z. Jing, G. Lucovsky, and J. L. Whitten, *Surf. Sci. Lett.* **296,** L33 (1993).
82. A. Redondo, W. A. Goddard III, C. A. Swarts, and T. C. McGill, *J. Vac. Sci. Technol.* **19,** 498 (1981); A. Redondo, W. A. Goddard III, and T. C. McGill, *J. Vac. Sci. Technol.* **21,** 649 (1982); A. Redondo, W. A. Goddard III, and T. C. McGill, *Surf. Sci.* **132,** 49 (1983).
83. C. J. Wu and E. A. Carter, *Phys. Rev. B* **45,** 9065 (1992).
84. J. Sauer, *Chem. Rev.* **89,** 199 (1989).
85. C. J. Wu and E. A. Carter, *Phys. Rev. B* **46,** 4651 (1992).
86. C. J. Wu, I. V. Ionova, and E. A. Carter, *Phys. Rev. B* **49,** 13488 (1994).
87. A. Vittadini, A. Selloni, and M. Casarin, *Phys. Rev. B* **49,** 11191 (1994).

88. A. Vittadini, A. Selloni, and M. Casarin, *Surf. Sci. Lett.* **289,** L625 (1993).
89. P. Nachtigall and K. D. Jordan, *J. Chem. Phys.* **102,** 8073 (1994).
90. P. Nachtigall, K. D. Jordan, and C. Sosa, *J. Chem. Phys.* **101,** 8073 (1994).
91. Z. Jing and J. L. Whitten, *Phys. Rev. B* **48,** 17296 (1993).
92. R. J. Hamers, R. M. Tromp, and J. E. Demuth, *Phys. Rev. B* **34,** 5343 (1986).
93. B. S. Swartzentruber, Y.-W. Mo, M. B. Webb, and M. G. Lagally, *J. Vac. Sci. Technol. A* **7,** 2901 (1989).
94. R. J. Hamers and U. K. Köhler, *J. Vac. Sci. Technol. A* **7,** 2854 (1989).
95. H. Tochihara, T. Amakusa, and M. Iwatsuki, *Phys. Rev. B* **50,** 12262 (1994).
96. J. J. Boland, *Surf. Sci.* **261,** 17 (1992).
97. L. Andersohn and U. Köhler, *Surf. Sci.* **284,** 77 (1993).
98. M. Chander, Y. Z. Li, J. C. Patrin, and J. H. Weaver, *Phys. Rev. B* **48,** 2493 (1993).
99. P. Kratzer, B. Hammer, and J. K. Nørskov, *Chem. Phys. Lett.* **229,** 645 (1994).
100. E. Pehlke and M. Scheffler, *Phys. Rev. Lett.* **74,** 952 (1995).
101. A. Vittadini and A. Selloni, *Chem. Phys. Lett.* **235,** 334 (1995).
102. G. Li, Y.-C. Chang, R. Tsu, and J. E. Greene, *Surf. Sci.* **330,** 20 (1995).
103. S. Pai and D. Doren, *J. Chem. Phys.* **103,** 1232 (1995).
104. P. Nachtigall, K. D. Jordan, A. Smith, and H. Jónsson, *J. Chem. Phys.* **104,** 148 (1996).
105. T. B. Blank, S. D. Brown, A. W. Calhoun, and D. J. Doren, *J. Chem. Phys.* **103,** 4129 (1995).
106. A. W. Munz, C. Ziegler, and W. Göpel, *Phys. Rev. Lett.* **74,** 2244 (1995).
107. B. M. H. Ning and J. E. Crowell, *Surf. Sci.* **295,** 79 (1993).
108. R. Konecny and D. J. Doren, unpublished.
109. S. Ciraci and I. P. Batra, *Surf. Sci.* **178,** 80 (1986).
110. M. Gostein, H. Parhikhteh and G. O. Sitz, *Phys. Rev. Lett.* **75,** 342 (1995).
111. K. Kolasinski, *Int. J. Mod. Phys. B* **9,** 2753 (1995).

POTENTIAL ENERGY SURFACES OF TRANSITION-METAL-CATALYZED CHEMICAL REACTIONS

DJAMALADDIN G. MUSAEV AND KEIJI MOROKUMA

Cherry L. Emerson Center for Scientific Computation and Department of Chemistry, Emory University, Atlanta, Georgia

CONTENTS

I. INTRODUCTION

Chemical reactions catalyzed by transition-metal complexes are of a great interest from both fundamental and practical points of view and, therefore,

Advances in Chemical Physics, Volume XCV, Edited by I. Prigogine and Stuart A. Rice.
ISBN 0-471-15430-X

are one of the most active fields of chemistry. A full catalytic cycle usually contains several elementary steps of reactions such as oxidative addition, reductive elimination, insertion into metal–ligand bond, σ-bond metathesis, and nucleophilic addition to coordinated ligands. These elementary steps along with full catalytic cycles have been studied with various experimental and theoretical methods. Earlier experimental [1–5] and theoretical [6–9] studies have been subject of several interesting reviews. During the past several years the improvement of computational facilities, including better programs and faster computers as well as the development of better theoretical methods, has made it possible to apply high-level quantum chemical calculations for the study of mechanism of both elementary steps and full catalytic cycles of various fundamentally and the practically important reactions such as hydroboration, hydroformylation, bis-silylation, and olefin polymerization. Here we review the latest quantum chemical studies on oxidative addition, metathesis, and insertion elementary reactions as well as some full catalytic cycles.

II. OXIDATIVE ADDITION

The oxidative addition and its reverse, reductive elimination, of ligands are two of the most fundamental steps in reactions occurring at metal centers in organometallic and inorganic complexes and are extremely important in organic synthesis and catalysis [3, 4, 10, 11]. In typical oxidative addition reactions

$$L_nM + A{-}B \underset{\text{Reductive elimination}}{\overset{\text{Oxidative addition}}{\rightleftharpoons}} L_nM\begin{matrix} \diagup A \\ \diagdown B \end{matrix} \tag{II.1}$$

or

$$2L_nM + A{-}B \rightleftharpoons L_nM{-}A + L_nM{-}B \tag{II.2}$$

the single A–B bond is cleavaged and two metal–ligand bonds are formed with an increase of both the oxidative state and the coordination number of the metal by two and one, respectively. These processes proceed by a variety of mechanisms. In the overall process, whatever the mechanism may be, there is a net transfer of a pair of electrons from the metal into the σ^* orbital of the A–B bond, and of the A–B σ electrons to the metal, as illustrated in Fig. 1. In reductive elimination, the reverse process of oxidative addition, both the oxidative state and the coordination number of metal are decreased and a new A–B bond is formed.

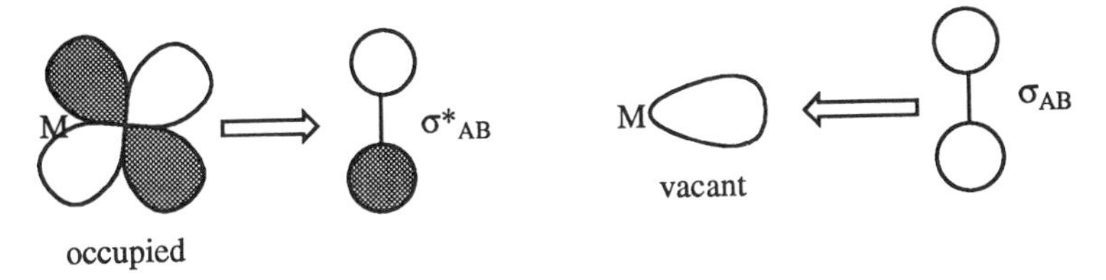

Figure 1. Schematic representation of charge transfer in oxidative addition reaction.

A. Oxidative Addition to d^8 Species

The oxidative addition of H_2 to d^8 square-planar transition-metal complexes such as Vaska's complex and $RhCl(CO)(PPh_3)_2$ has been extensively studied [11–16]. The generally accepted mechanism of the reaction involves the concerted addition of H_2 to form pseudo-octahedral products with a *cis* disposition of the hydride ligands. When the reactant has two isomeric forms, *cis*- and *trans*-$MX(CO)(PR_3)_2$, (X = Cl) with *cis* and *trans* phosphine ligands, respectively, the oxidative addition of H_2 to $MX(CO)(PR_3)_2$ may lead to four different isomeric products, as shown in Fig. 2. It has been found

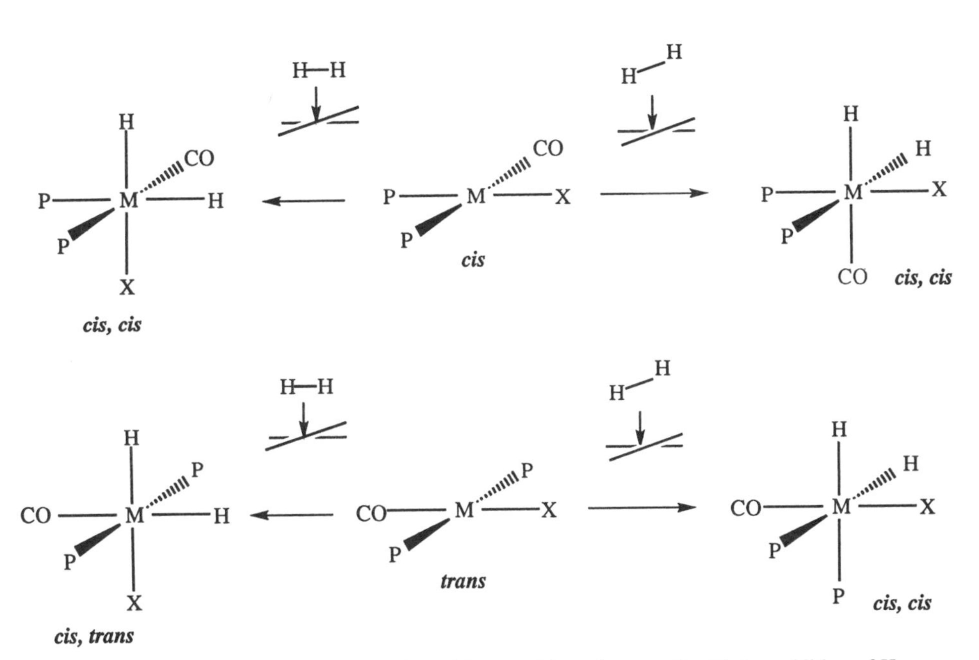

Figure 2. Schematic representation of the possible pathways of oxidative addition of H_2 to complex $M(PR_3)_2(CO)X$, where, for example, *cis*, *trans* refers to the *cis* disposition of the two hydrides and the *trans* disposition of the two phosphines.

experimentally for *cis*-$MCl(CO)(PR_3)_2$ that the isomer resulting from addition along the P–M–CO axis is formed with a >99% stereoselectivity at low temperature, while the other isomer from the addition along the P–M–X axis appears at higher temperatures and eventually dominates the final equilibrium for the M = Ir. The factors controlling the direction of H_2 addition to the d^8 square-planar transition-metal complex would include both steric and electronic contributions from the pairs of ligands along each axis. Sorting out which of these contributions is dominant is a complicated task. The task, however, is somewhat simpler for *cis*-$MX(CO)(PR_3)_2$, where one phosphine ligand is present in both planes of addition and the analysis is reduced to a comparison between CO and X ligands. The calculations carried out at the Hartree–Fock (HF) level with the effective core potential (ECP) for heavier atoms and double-zeta basis sets by Sargent et al. [12] for the M = Ir and X = Cl have shown that the difference in barrier heights between H_2 addition along the P–Ir–CO and the P–Ir–Cl axis is due to the difference in how the ligands in the plane of addition accommodate the electron–electron repulsive interactions due to the close encounter of these ligands as the complex evolves from a four-coordinate to a six-coordinate species. Addition in the P–Ir–CO plane is favored due to the electron-withdrawing nature of the CO ligand. The interelectronic repulsions between the concentration of charge around the metal atom and the electron density of the ligands in the plane of addition are reduced through the delocalization of some of this charge at the metal center by the low-energy occupied CO $2\pi^*$ orbitals. In the vicinity of the transition state, the CO ligand moves very little as the H_2 unit approaches and maintains an orientation aligned with the concentration of charge at the metal center. The electron-donor Cl^- ligand is unable to reduce these repulsions when H_2 adds in the P–Ir–Cl plane. In the vicinity of the transition state, the movement of the Cl^- ligand is large and indicates that the interaction between the ligand and the charge around the metal center are purely repulsive.

Sargent and Hall [13] have also studied oxidative addition of H_2 to d^8 square-planar Vaska-type complexes, *trans*-$IrX(CO)(PR_3)_2$, where X = Cl, H, Me, and Ph. It has been shown that weak electron donor ligands, such as Cl^-, favor addition of H_2 in the X–IR–CO plane, whereas for stronger electron-donor ligands, such as Ph^-, Me^-, and H^-, addition in the P–Ir–P plane is preferred. For Ph^- and Me^-, the more stable isomer of the final products corresponds to the species formed from addition in the X–Ir–CO plane. This isomer gains stability by orienting a weaker σ-donor and/or stronger π-donor ligand *trans* to a hydride ligand. The weak σ-donor character strengthens the strong Ir–H σ bond to which it is *trans*, while the strong π-donor character, which results in an antibonding interaction of π symmetry, is able to minimize the impact of antibonding interaction by being *trans* to the strongly *trans*-influencing hydride ligand.

Abu-Hasanayn, Goldman, and Krogh-Jespersen (AGK) [14] also have studied the oxidative addition reaction of H_2 to Vaska-type complexes, *trans*-$Ir(L)_2(CO)(X)$ for various L and X, as shown in Table I, with the HF geometry calculation and the MP4(SDTQ) energetics with double-zeta quality basis sets and ECP for heavier atoms. Although the product of the reaction depends on the L and X ligands and can be both *cis*, *trans*- and *cis*, *cis*-$(H)_2Ir(L)_2(CO)X$ isomers, as mentioned above with Fig. 2. AGK studied only the path leading to the *cis*, *trans* isomer. As seen in Table I, the barrier height of the reaction H_2 + *trans*-$Ir(L)_2(CO)(X)$ → *cis*, *trans*-$(H)_2Ir(L)_2(CO)X$ decreases and its exothermicity increases via the trend F < Cl < Br < I < CN < H. This trend is same with the π-donation capability of the halides, which decreases in strength going down in a group in the periodic table. The reaction energies given in the Table I, as well as simplified calculations carried out by AGK involving hypothetical *trans*, *trans*-$(H)_2Ir(PH_3)_2(CO)AH_2$ (A = B, N, and P), show clearly that the reaction energies are less sensitive to σ effects than to π effects. The π properties of the substituents play an important role in determining the thermo-

TABLE I
Energy of Reaction and Barrier Heights (in kcal/mol) for H_2 Addition to *trans*-$Ir(PH_3)_2(CO)(X)$[a]

	Barrier				Energy of Reaction			
X	$\Delta E\ddagger_{HF}$	$\Delta E\ddagger_{MP4}$	$\Delta H\ddagger_{MP4}$	$\Delta H\ddagger_{exp}$	ΔE_{HF}	ΔE_{MP4}	ΔH_{MP4}	ΔH_{exp}
F	21.9	20.4	21.4		−25.8	−13.6	−10.4	
Cl	13.9	13.3	14.1	11–12	−34.2	−22.0	−18.4	−14
Br	12.0	11.2	11.9	12	−36.2	−24.1	−20.1	−17
I	11.6	9.2	9.8	6	−39.3	−27.3	−23.3	−19
CN	14.0	7.6	7.4		−38.3	−28.3	−24.9	−18
H	7.8	6.0	6.1		−45.4	−38.8	−35.3	
CH_3			−36.1	−23.2				
SiH_3			−48.9	−35.1				
OH			−26.0	−12.8				
SH			−37.0	−23.7				
BH_4			−44.6	−32.2				
		L = NH_3						
Cl	16.7	18.7	19.9		−28.9	−12.8	−9.6	
		L = AsH_3						
Cl			−34.6	−22.4	−15			

[a] "HF" and "MP4" refer to Hartree–Fock and MP4SDTQ values, respectively; enthalpies (ΔH) were obtained by adding correction for zero-point energies and thermal excitations (298 K) to ΔE values [14].

dynamics of the addition reaction; the presence of occupied π orbitals on the substituents has a strongly disfavoring effect. Similarly, increased π donation from X seems to strongly contribute to the activation barrier to H_2 addition. The effect of the π donation from X on the magnitude of activation barrier has been explained by AGK in two related ways. First, four-electron/three-orbital interactions exist involving filled halide *p* orbitals, filled metal *d* orbitals, and empty CO(π*) orbitals. These are strongly directional and most favorable when the π-donating halide is *trans* to the π-accepting CO group. Since this effect is most important in the presence of strong π donors, increased π donation from X may increase the barrier to addition by resisting the necessary bending of the X–Ir–CO moiety of the reactant. Second, the lowest unoccupied molecular orbital (LUMO) in the four-coordinate square-planar complex in an Ir(p_z) orbital with some admixture of CO and halide π character. Increased π-halide donation to metal raises the energy of this virtual orbital and thereby enlarges the highest occupied molecular orbital (HOMO)–LUMO gap. Part of the electron redistribution and Ir–H_2 interaction in the transition state is expected to involve this orbital, which will be less accessible when X is a strong π donor. In general, an increased HOMO–LUMO gap in a molecule implies a reduced tendency to engage in chemical reaction.

For the above reasons, AGK [14] have proposed that the magnitude of π donation is the major determining the effect of X on the barrier to H_2 addition. This explanation is consistent with the decreased reactivity of complexes of the lighter halides as well as the high reactivity of the cyano and hydrido complexes. It also has been demonstrated the exothermicity of the addition reaction increases via the trend L = NH_3 < PH_3 < AsH_3 for X = Cl, as seen in Table I. When the PH_3 ligands in *trans*-Ir$(PH_3)_2$(CO)(X) are replaced by NH_3 (X = Cl), the activation barrier is computed to increase by about 5 kcal/mol. Thus, a comparison of the calculated exothermicity of the reaction and the magnitude of the kinetic barrier shows clearly that the NH_3 group may be considered a stronger π donor than PH_3. By analyzing the obtained structures of the transition states, reactants, and products in the halogen series (X = F, Cl, Br, I), as well as analyzing the exothermicity of the reaction, AGK have shown that with increasing reaction exothermicity, the transition states are located earlier along the reaction path and are more reactant-like in geometrical structure, which confirms the expectations based on the Hammond postulate [15].

The quantum chemical studies of the structure and stability of the complex RhCl(CO)$(PH_3)_2$, as well as its reactivity with H_2 and CH_4 molecules, were the subject of the Musaev and Morokuma's (MM) paper [16] performed at the MP2 level of theory with the double-zeta quality basis set and ECP for heavier atoms. It has been found that *trans*-$RhCl(CO)(PH_3)_2$, with two phosphine groups in *trans* position, is about 13 kcal/mol more stable than the *cis*-$RhCl(CO)(PH_3)_2$ isomer. The addition reaction of H_2 to the *trans*-$RhCl(CO)(PH_3)_2$ isomer has been shown to take place with a 16-kcal/mol activation barrier and is exothermic by 2 kcal/mol. The transition state corresponding to the addition of H_2 along the Cl–Rh–CO axis is a few kilocalories per mole more favorable than that along the P–Rh–P axis, as shown in Fig. 3. These results of MM obtained for the addition reaction of H_2 to *trans*-$RhCl(CO)(PH_3)_2$ are in good agreement with those obtained for *trans*-$IrCl(CO)(PH_3)_2$ by SH [13]. The addition reaction of H_2 to the less stable *cis*-$RhCl(CO)(PH_3)_2$ isomer is calculated to be exothermic by 11 kcal/mol and has an 18 kcal/mol activation barrier. For this reaction the more favorable path is the approach of H–H to the complex along the P–Rh–Cl axis, similar to the M = Ir case mentioned above. The transition state for H_2 addition along the P–Rh–CO bond is calculated to be 4 kcal/mol less favorable. This difference has also been explained in terms of the difference in electronic features between Cl^- and CO ligands.

The calculation of the structure and stability of various pseudo-octahedral products of H_2 addition reaction to $RhCl(CO)(PH_3)_2$ shows that the product isomers corresponding to species formed by the addition in the Cl–Ir–CO and P–Rh–P planes are energetically similar. MM have analyzed the *trans* influence in these isomers. As seen in Table II, all bond distances *trans* to H ligand are much longer than those *trans* to CO, PH_3, and Cl^- ligands. Among CO, PH_3, and Cl^-, the *trans* influence seems to decrease in the order CO $>$ PH_3 $\geq$ Cl^-.

MM have also studied the addition of a C–H bond of methane to $RhCl(CO)(PH_3)_2$ along the P–Rh–Cl and Cl–Rh–CO axes and found it to be a few kilocalories per mole less favorable than the addition of H_2. The C–H addition takes place with a barrier of 27 and 31 kcal/mol, higher than that for H–H addition, and is endothermic by 20 and 9 kcal/mol for *trans* and *cis* isomers, respectively, consistent with the Hammord postulate [15]. Thus, in spite of the similarity betwen the H–H bond strength of H_2 and C–H bond strength of CH_4, the H–H bond of hydrogen molecule is activated on the $RhCl(CO)(PH_3)_2$ complex more easily than the C–H bond of methane. The reason for the difference between H–H and C–H activation has been explained in terms of the directionality of the bond involving an alkyl group compared to the bond involving hydrogen atoms, as well as in terms of the stronger Rh–H bond compared to the Rh–CH_3 bond. The calculation of the structure and stability of the products of addition reaction of CH_4 to $RhCl(CO)(PH_3)_2$ shows that the isomers formed by addition in either P–Rh–P or OC–Rh–Cl plane are clustered in a 1-kcal/mol range, among

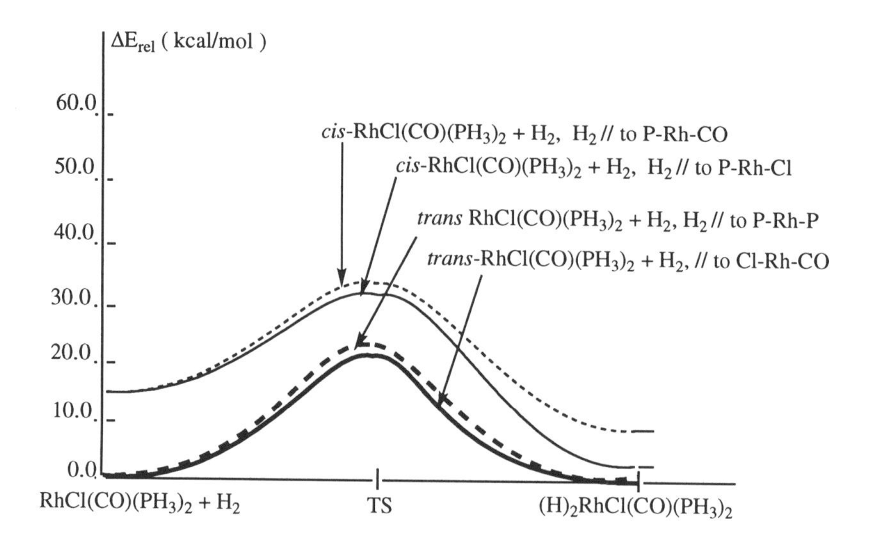

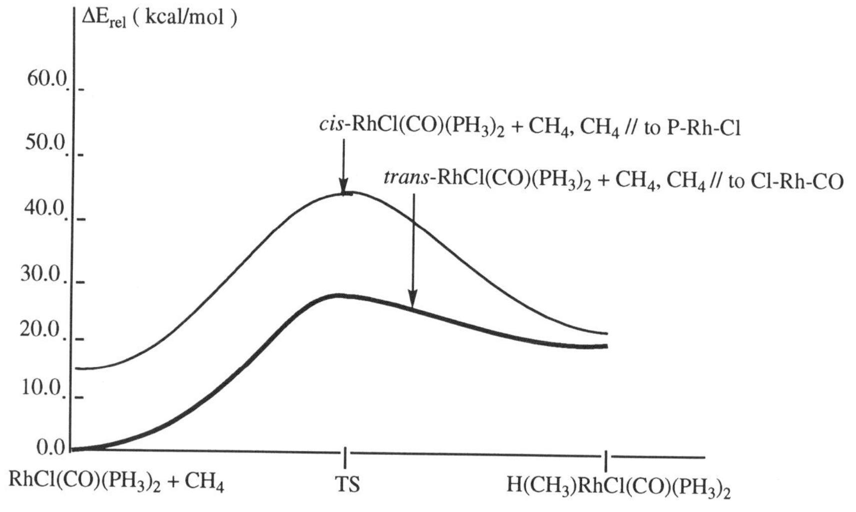

Figure 3. Potential energy profiles of reaction $RhCl(CO)(PH_3)_2$ + HR → $H(R)RhCl(CO)(PH_3)_2$ calculated at the MP2/II/MP2/I level, where R = H and CH_3 [16].

which the isomer with H atom *trans* to Cl and CH_3 group *trans* to CO is the most favorable. The calculated average Rh–ligand bond distances as functions of the *trans* ligands in Table II clearly show that all the bond distances *trans* to H and CH_3 ligands are much longer (by 0.05 Å for the Rh–Cl bond and 0.25 Å for the Rh–P bond) than those *trans* to CO, PH_3,

TABLE II
The Average Rh–X Bond Distances (in Angstroms) as Functions of *trans* Ligands in the Alternative Structures of $(H)_2RhCl(CO)(PH_3)_2$ and $H(CH_3)RhCl(CO)(PH_3)_2$ [16]

Bond	*trans* Ligand				
Rh–X	H	CH_3	CO	P	Cl
	For $(H)_2RhCl(CO)(PH_3)_2$ at the MP2/I Level				
Rh–Cl	2.521	—	2.483	2.420	—
Rh–P	2.549	—	2.426	—	2.341
Rh–CO	1.959	—	—	1.819	1.790
Rh–H	—	—	1.567	1.531	1.538
	For $H(CH_3)RhCl(CO)(PH_3)_2$ at the RHF/I Level				
Rh–Cl	2.513	2.510	2.446	2.420	—
Rh–P	2.700	2.626	2.441	2.396	2.385
Rh–CO	2.138	2.100	—	1.906	1.871
Rh–CH_3	2.160	—	2.022	2.033	2.064
Rh–H	—	1.623	1.516	1.506	1.528

and Cl^- ligands. Between H and CH_3, H seems to have a stronger *trans* influence than CH_3. CO and PH_3 seem to have similar *trans* influence, with CO slightly stronger than PH_3. The Cl^- ligand, though it qualitatively seems to have a *trans* influence similar to that of CO and PH_3, shows a peculiar bond dependence; for Rh–Cl, Rh–PH_3, and Rh–CO bonds, the *trans* influence of Cl^- is weaker than those of CO and PH_3 as in $(H)_2RhCl(CO)(PH_3)_2$, whereas for Rh–H and Rh–CH_3 bonds, the *trans* influence of Cl^- is definitely stronger than those of CO and PH_3.

Koga and Morokuma (KM) have studied [17] the addition reaction of the Si–H, Si–Si, and C–H bonds to the three-coordinate d^8 complex $ClRh(PH_3)_2$, using MP2 (mostly for geometries), MP4, and QCISD(T) (for energetics) methods with the double-zeta basis set and ECP for Rh. It has been found that the reaction of CH_4 with $ClRh(PH_3)_2$ passes through the η^2-CH_4 complex and a three-centered transition state. The overall reaction is exothermic by 23, 19, and 10 kcal/mol at the MP2, MP4, and QCISD(T) levels, respectively, as shown in Fig. 4. The activation barrier relative to the η^2-CH_4 complex is only 3–8 kcal/mol, and thus the C–H activation by this coordinatively unsaturated complex is quite easy. The potential energy profile for Si–H bond activation is quite different. The Si–H bond activation is downhill, and the η^2-SiH_4 complex is not an intermediate but is, instead, the transition state for intramolecular rearrangement between two hydrosilyl complexes, the products of Si–H bond activation. This difference of SiH_4 reaction from that of CH_4 originates from the strong Rh–Si bond and the weak Si–H bond, which results in the much larger exothermicity for Si–H

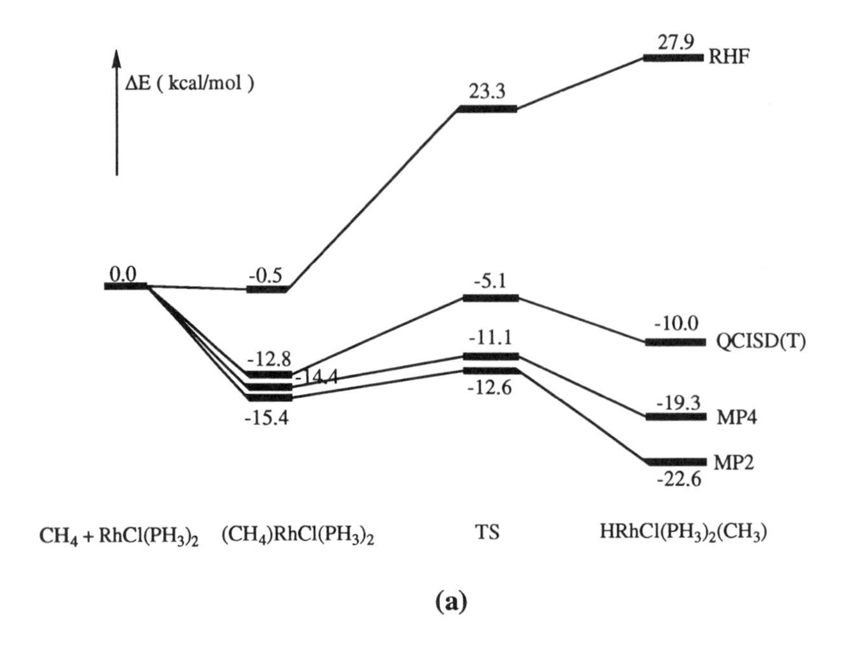

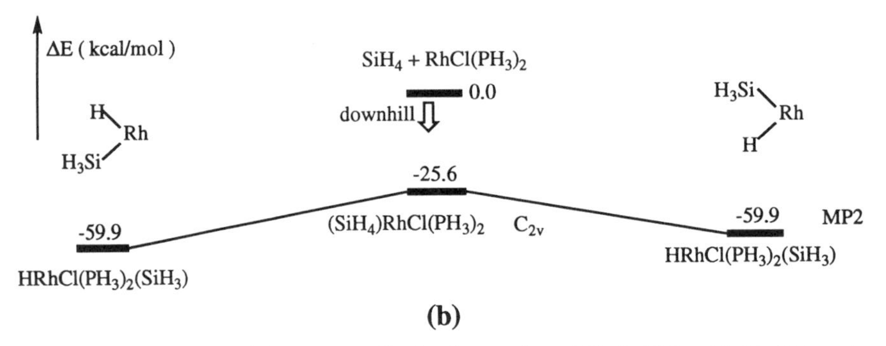

Figure 4. The potential energy profile for the reaction of $RhCl(PH_3)_2$ with (*a*) CH_4 and (*b*) SiH_4 [17].

bond activation than for C–H bond activation. The Rh–Si bond was calculated to be 20 kcal/mol stronger than the Rh–C bond, and the Si–H bond was found to be 16 kcal/mol weaker than the C–H bond. Analysis of the reason for this strong Rh–Si bond has shown that in $HRhCl(PH_3)_2(SiH_3)$, strong donation from SiH_3 to $RhCl(PH_3)_2$ takes place to stabilize the system. This donation is larger in the electropositive SiH_3 group than in the CH_3 group. The back-donation from Rh to Si *d* orbital is less important than what has been considered.

Since the Rh–Si bond is so strong, the reaction of Si_2H_6 with $RhCl(PH_3)_2$ giving $RhCl(PH_3)_2(SiH_3)_2$ is very exothermic by 82 kcal/mol at the MP2 level, and this disilyl complex is 16 kcal/mol more stable than $HRhCl(PH_3)_2(Si_2H_5)$. KM found the transition state for intramolecular rearrangement between these two products, where Si–H and Si–Si bond interact with the Rh atom in a similar fashion as the two Si–H bonds interact with the Rh atom in the transition state between two $HRhCl(PH_3)_2(SiH_3)$ molecules.

It is also shown by KM that the electron correlation effect on the energetics of the reactions is quite large; it makes reaction exothermic and lowers the activation barrier. In $HRhCl(PH_3)_2(CH_3)$, the Rh–C and the Rh–H bond are covalent with large *d* character, and thus the intrabond correlation and the correlation with the *d* electrons are substantial. In $HRhCl(PH_3)_2(SiH_3)$, the Rh–Si bond is more ionic with $Si^{\delta+}$ because of the strong Si $\rightarrow$ Rh donation, and thus the intrabond correlation is smaller. However, the donated electrons correlate with the *d* electrons, and thus the correlation effect is important to the stability of the silyl complex.

Thus, $RhCl(PH_3)_2$ with a vacant *d* orbital can interact with a molecule such as CH_4, SiH_4, and Si_2H_6 without any difficulty, leading to the low activation barrier or downhill reaction. As mentioned above, the low-lying vacant *d* orbital extends around the empty coordination site where CH_4, SiH_4, and Si_2H_6 attacks. This fact suggests that the vacant coordination site is electron-deficient and that, therefore, the repulsion between attacking molecules and $RhCl(PH_3)_2$ is small when reaction takes place at the vacant coordination site. This small repulsion and the electron donation from CH_4, SiH_4, and Si_2H_6 to the vacant *d* orbital lowers the activation barrier at the transition state and enhances the bonding interaction in the complexes.

B. Oxidative Addition of H–X Bonds on the CpML Species

The d^8 systems CpM(L) (M = Ir and Rh) are unique in that they are among a few late transition-metal complexes capable of activating H–H and C–H bonds. In early experimental studies, Janowicz and Bergman [18e,f] as well as Hoyano and Graham [18h] have succeeded in activation of C–H bonds of alkanes by the CpIr(L) complexes obtained by photolysis of 18-electron species $CpIr(CO)_2$ or $CpIr(PMe_3)(H_2)$. A similar reaction involving oxidative addition of CH_4 to $CpRh(PMe_3)$ was observed by Jones and Feher [18i]. Earlier experimental [18] and theoretical [19] studies have been summaries by different authors. However, recent experimental papers of Wasserman, Moore, and Bergman (WMB) [20, 21] have been an incitement for new theoretical studies in this field. WMB have carried out a gas-phase version of the H–H and C–H activation processes on the "naked" metal center in the coordinatively unsaturated 16-electron d^8 fragment CpRh(CO). They proposed the mechanism which involves formation of a weakly bound intermediate, CpRh(CO) · alkane. Because it is difficult to detect experimentally the assumed intermediate and the transition state, several groups have carried out theoretical studies of the potential energy profile of this reaction.

At first we would like to recall a few important conclusions from earlier studies. The electronic structure of the reactant CpML has been discussed by Hofmann and Padmanabhan [22] for various ligands L and M = Co, Rh, and Ir with the extended Hückel method. As shown in Fig. 5, the valence molecular orbitals are na', ma'', $(n + 1)a'$, $(m + 1)a''$, and $(n + 2)a'$ orbitals under C_s symmetry. One sees that the na' and ma'' orbitals are mainly d_{xx-yy} and d_{xz} orbitals stabilized by interaction with the π^* orbitals on L. At somewhat higher energy is the occupied metal-based d_{zz} orbital, $(n + 1)a'$, with a weak M–L antibonding σ interaction. The metal-based $(m + 1)a''$ (d_{yz}) and $(n + 2)a'$ (d_{xy}) orbitals are highest in energy, destabilized by interaction with occupied π orbitals on the Cp ring. In addition, the $(n + 2)a'$ orbital destabilized by interaction with the σ orbital on L. Thus, it

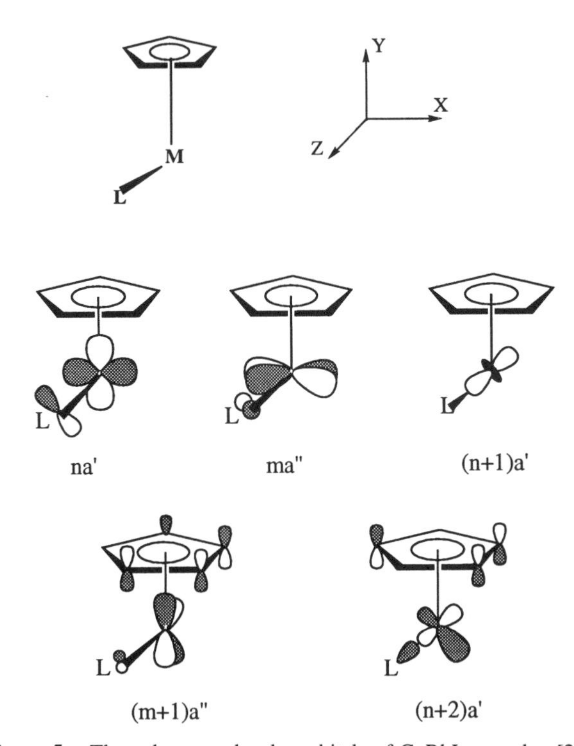

Figure 5. The valence molecular orbitals of CpRhL complex [22].

has been suggested that the singlet $^1A'$ state with the electronic configuration $(na')^2(ma'')^2((n+1)a')^2((m+1)a'')^2((n+2)a')^0$ and the triplet $^3A''$ state with $(na')^2(ma'')^2((n+1)a')^2((m+1)a'')^1((n+2)a')^1$ are candidates for the ground state of the CpML fragment.

The reaction for M = Rh and Ir, L = CO and PH_3 with H_2 and CH_4 was previously studied by Ziegler et al. [23] using a density functional theory with Hartree–Fock–Slater exchange (with Becke nonlocal correction for energy) and Stoll correlation functionals. In these calculations, geometries have been fixed both for the Cp ring with the local D_{5h} symmetry and for CO or PH_3. They find that the first stage of the reaction forms the molecular complex CpML · HR (R = CH_3) with the binding energy about 6 and 15 kcal/mol for the M = Rh and Ir, respectively. There is then a modest barrier of 8 and 2 kcal/mol for the Rh and Ir, respectively, before the final exothermic product CpML(H)CH_3. Though the H_2 activation takes place without activation barrier, a strange CpML · H_2 complex with r(H–H) = 1.11 Å was reported for the M = Rh, which might be due to the deficiency of the local density functional approximation used for geometry optimization.

The C–H oxidative addition of methane to the CpRh(CO) complex has been studied also by Song and Hall [24] at the restricted Hartree–Fock (RHF, for geometries) and MP2 (for energetics) levels of the theory with the double-zeta quality basis set and effective core potential for Rh. In these calculations the geometries of CO and of the Cp-ring with the local D_{5h}-symmetry were fixed. The shape of the potential energy surface obtained in this paper is similar to that obtained by Ziegler et al. [23]. At the first step the reactants give molecular complex CpRh(CO) · CH_4, with stabilization energy of 14.8 kcal/mol, and then activation of the C–H bond takes place with a 4.1-kcal/mol barrier, which leads to the product CpRh(CO)(H)(CH_3). The whole reaction is calculated to be exothermic by 31 kcal/mol.

Musaev and Morokuma [25] have applied the MP2 method with the double-zeta quality basis set and ECP to studies of the oxidative addition of H_2, H_2O, NH_3, CH_4, and SiH_4 molecules to the CpRh(CO) complex. In this paper they fully optimize the geometries of the reactants, transition states, an intermediate, and products. The paper had two main purposes: (i) to study the potential energy surface of the addition reaction of H_2, CH_4, NH_3, H_2O, and SiH_4 to CpRh(CO) and (ii) to test the trends and make predictions for oxidative addition reactions of H–R bonds. Their results are presented in Fig. 6. The ground state of CpRh(CO) has been found to be the singlet $^1A'$. The triplet $^3A''$ state lies only 3.9 kcal/mol higher. This is in good agreement with that (1–5 kcal/mol) obtained by Ziegler et al. [23]. The calculation shows that at gas-phase collisionless condition the oxidative addition reaction of H–SiH_3, H–H, and H–CH_3 bonds to CpRh(CO) ($^1A'$) takes place without barrier, while that of H–NH_2 and H–OH bonds requires

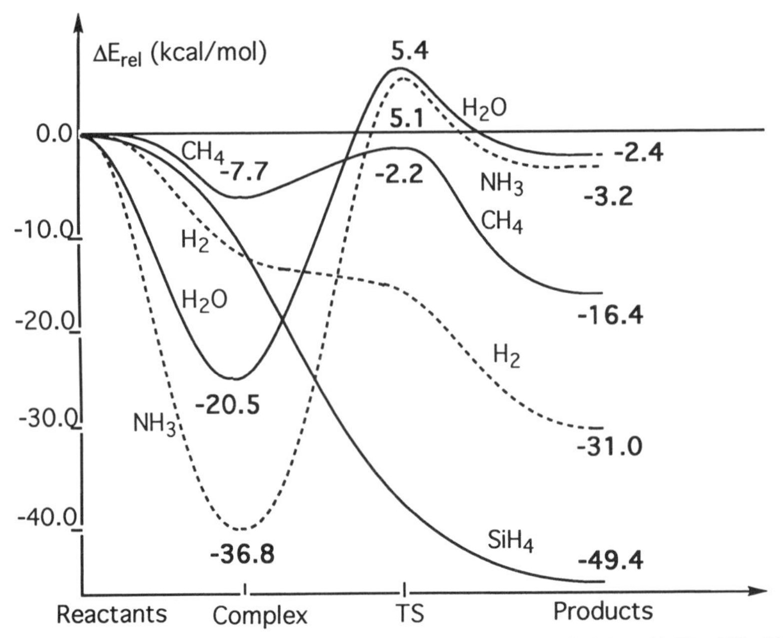

Figure 6. The potential energy profiles of the reaction of CpRh(CO) with H_2, CH_4, H_2O, NH_3, and SiH_4 [25].

an activation energy of about 5 kcal/mol. In solution or in gas phase when collisional energy equilibration is faster than the reaction itself, the reaction should be considered to start from the prereaction complex, CpRh(CO) · (HR). Here the addition of H_2 and SiH_4 should take place without barrier, while that of CH_4, NH_3, and H_2O requires an activation barrier of 5.5, 42, and 26 kcal/mol, respectively. Thus, under these conditions, it will be very difficult to activate H–R bonds of molecules that have a strong Lewis base character, such as NH_3 and H_2O. The activation barrier consists mostly of the energy required to dissociate the reactant coordinated with its lone pair. The only product experimentally detectable at not very rigorous conditions would be the molecular complex CpRh(CO) · (RH), the binding energy for which is calculated to be 7.7, 20.5, and 36.8 kcal/mol for the CH_4, H_2O, and NH_3, respectively. It has been shown that the ease of oxidative addition reaction—that is, existence of an activation barrier and its height—is correlated strongly to the H–R bond strength as well as to the Rh–R bond strength. Going from SiH_4 to H_2, CH_4, NH_3, and H_2O, the H–R bond energy increases, SiH_4 (90.3) $<$ H_2 (104.2) $<$ CH_4 (104.8) $<$ NH_3 (107.4) $<$ H_2O (119.0) kcal/mol, respectively [26], and the Rh–R bond energy decreases (except for the Rh–O case), 72.7 $>$ 65.0 $>$ 59.3 $>$ 46.9 $<$ 55.2

kcal/mol, for R = SiH_3, H, CH_3, NH_3 and OH, respectively, and the oxidative addition becomes more difficult. Based on this conclusion, a prediction has been made that the H–GeH_3 and H–SnH_3 bonds, which are weaker than the H–SiH_3 bond, may be activated without barrier by the CpRh(CO). Similarly, since the experimental value [26] of the H–SH bond strength (91.1 $\pm$ 1 kcal/mol) in SH_2 is weaker than that of H–OH (119 kcal/mol), an easier activation of H–SH than of H–OH by CpRh(CO) is predicted. The reactions studied are exothermic, by 49.4, 31.0, 16.4, 3.2, and 2.4 kcal/mol for SiH_4, H_2, CH_4, NH_3 and H_2O, respectively, and the exothermicity, related to the Rh–R bond energy minus the H–R bond energy, correlates well with the ease of reaction; the more exothermic the reaction is, the more easily it takes place.

Siegbahn and Svensson (SS) [27] have studied the addition reaction of CH_4 to the model systems LRh(CO) (L = Cl and H) at the MP2 (for geometries) and PCI-80 (for energetics) level with a larger basis set of double-zeta plus polarization quality. It has been shown that ClRh(CO) has a linear triplet ground state with an adiabatic excitation energy of 5 kcal/mol to a bent singlet. In contrast, HRh(CO) has a bent singlet ground state with an adiabatic excitation energy of 23 kcal/mol to the linear triplet. At the first stage of reaction the reactants give the molecular complex, LRh(CO) · CH_4, similar to that for the experimentally and theoretically studied complex CpRh(CO). The binding energies calculated relative to the singlet state of the reactants are 3.4 and 10.8 kcal/mol for L = Cl and H, respectively. Then activation of the C–H bond takes place with a 15-kcal/mol activation barrier calculated relative to the molecular complex for the HRh(CO). Unfortunately, SS were unable to find this transition state for ClRh(CO). The reaction is exothermic by 22.1 and 2.9 kcal/mol for L = Cl and H, respectively. Thus, this result shows that replacing the strongly electronegative ligand chlorine by covalently bound hydrogen destabilizes the prereaction complex LRh(CO) · CH_4 and decreases the exothermicity of entire reaction. On the basis of these results as well as the results for the reaction of RuH_2 and $RhH(NH_3)$ with CH_4, it has been concluded that for the formation of a strong prereaction complex the ground state singlet is important. In contrast, a low barrier for the oxidative addition reaction with methane requires a low-lying triplet state.

C. Oxidative Addition to d^{10} ML_2 and Related Systems

Oxidative addition of H–H, C–H, and other σ bonds to a d^{10} transition-metal complex is one of the most widely studied reactions in organotransition-metal chemistry both experimentally and theoretically [1–9]. In the oxidative addition of H_2 to d^{10} $Pt(PR_3)_2$, almost all hydrides assume a *trans* position except for the cases where the ligands are a chelating diphosphine

[1–3]. In the case of trimethylphosphine ligands, the presence of both *cis* and *trans* $Pt(PMe_3)_2H_2$ species has been found in equilibrium in solution. However, according to the molecular orbital correlation diagram, the *trans* addition is symmetry-forbidden, while the *cis* addition is symmetry-allowed [8, 9]. In the earlier investigations [1–9] in general, it has been shown that the oxidative addition of H_2 to the $Pt(PH_3)_2$ is exothermic, leads to the *cis*-$Pt(PH_3)_2H_2$ complex, and occurs with activation barrier ranging from 2 to 17 kcal/mol. The transition state resembles the reactants in that the H_2 bond was lengthened only slightly from the bond length in free H_2. For these reactions the energy difference between the s^1d^9 and d^{10} states is very important. Indeed, in the ML_2 fragment, the metal can be described qualitatively by a d^{10} configuration with the bonding to the ligands primarily involving donation of the ligand lone pairs to *s* and *p* metal orbitals. In order to form two more bonds, the promotion of the metal to the s^1d^9 state is required. During the process the relatively little charge is transferred to the hydrogen atoms and the M–H bond is a result of the coupling of the 1*s* H with the *sd* hybrid orbital of the M. With H_2 approaching the ML_2 fragment, the bending of the L ligands away from H_2 serves to decrease the separation between the d^{10} and s^1d^9 states to the point where the s^1d^9 state becomes accessible and formation of two M–H bonds can follow. Comparison of the addition of H_2 to Pt (which has the s^1d^9 ground state) and Pd (which has the d^{10} ground state) confirms the above statement; the reaction is exothermic for Pt by 34 kcal/mol but is endothermic for Pd by 4 kcal/mol [28].

The observation of C–H, Si–H, Si–I, and Si–Si bond activation by Pd(0) and Pt(0) complexes has prompted theoretical studies of Sakaki and coworkers [29] concerning the C–H, C–C, and Si–X (X = H, C, Si, and F) bond activation on the $Pt(PH_3)_2$ and $Pd(PH_3)_2$ complexes. In these calculations the MP2 and MP4(SDQ) methods with the double-zeta quality basis sets argumented by the polarization d_{Si} function and ECP for heavier atoms have been used. As shown in Table III, the oxidative addition of C–H and C–C bonds to $Pt(PH_3)_2$ is endothermic by 6.5 and 5.2 kcal/mol, respectively, while it is exothermic by 25.6, 28.6, 14.1, and 46.4 kcal/mol for Si–H, Si–F, Si–C, and Si–Si bonds, respectively. These differences in exothermicity of the reaction have been explained in terms of the Pt–X bond energies; metal–alkyl bonds are a few kilocalories per mole weaker than the metal–H bond, which in turn is 0–15 kcal/mol weaker than metal–silyl or metal–F bonds. The activation barriers increase via the trend SiH (0.7 kcal/mol) $<$ SiSi (17.0 kcal/mol) $<$ SiF (26.8 kcal/mol) $<$ SiC (28.1 kcal/mol) $<$ CH (30.4 kcal/mol) $<$ CC (68.5 kcal/mol). The easier activation of the Si–H bond compared to the CH bond, as well as the trend Si–Si $<$ Si–C $<$ C–C, has been explained again in term of the bond energies; the C–C

TABLE III
Energetics (in Kilocalories per Mole, Relative to the Reactants)[a] of the Oxidative Addition of Different Bonds to $M(PH_3)_2$ Calculated at the MP4(SDQ) Level [29]

Activated Bonds, Molecules	Precursor Complex	TS	Product
	M = *Pt*		
C–H, CH_4	−0.6(−0.4)	28.1(30.4)	6.5
Si–H, SiH_4	−2.3(−1.5)	−1.6(0.7)	−25.6
Si–F, $FSiH_3$	−6.0(−4.2)	19.8(26.8)	−28.5
C–C, C_2H_6	−1.1(−0.8)	66.0(68.5)	5.2
Si–C, SiH_3CH_3	−2.3(−1.4)	24.9(28.1)	−14.1
Si–Si, Si_2H_6	−3.7(−2.6)	13.7(17.0)	−46.4
	M = *Pd*		
Si–H, SiH_4	−2.4(−1.4)	−1.3(0.5)	−8.1
C–C, C_2H_6	−1.4(−0.8)	56.8(59.3)	30.5
Si–C, SiH_3CH_3	−2.5(−1.4)	16.3(20.1)	8.1
Si–Si, Si_2H_6	−3.0(−1.6)	8.2(11.9)	−18.1

[a]Numbers given in parentheses include basis set superposition error.

bond is the strongest, the C–H bond is about 15 kcal/mol stronger than Si–H, and the Si–Si bond is the weakest.

As seen from Table III, the potential energy surfaces of the Si–X (as well as C–C) addition reactions to $M(PH_3)_2$ are, in general, similar for M = Pt and Pd. However, the reactions for Pd are less exothermic (or more endothermic), and activation barriers are slightly lower than those of the corresponding reactions for Pt. This difference in the reaction energies has been explained in term of the stronger Pt–X bond relative to the Pd–X. The calculated Pd–X bond energies (48.9, 27.1, and 45.4 kcal/mol) are 10–15 kcal/mol smaller than the corresponding Pt–X bond energies (61.7, 39.7, and 61.5 kcal/mol, respectively) for X = H, CH_3, and SiH_3, which in turn have been attributed to the energy difference between low-lying s^1d^9 and d^{10} states of the atoms, as mentioned above. However, the difference in the activation barriers for Pd and Pt complexes cannot be explained only on the basis of bond energies of Si–X or C–C bonds and the energy difference between low-lying s^1d^9 and d^{10} states of the atoms. Therefore, Sakaki and coworkers [29] have used, beside the above mentioned factors, the flexibility of PMP bending required for lowering of the s^1d^9 state discussed above. $Pd(PH_3)_2$ is more flexible for PPdP bending and needs a smaller distortion energy (E_{dist}) to cause the bending than does $Pt(PH_3)_2$; when the PMP angle is taken to be 120°, E_{dist} is calculated to be 19.1 and 10.2 kcal/mol for the M = Pt and Pd, respectively.

We should note that theoretical studies of the widely studied process of oxidative addition of H–H and C–H bonds to bare transition-metal atoms are omitted from this chapter because they have been discussed in a recent review [7].

D. Special Topics: Dihydride–Dihydrogen Rearrangement in the Transition-Metal Polyhydride Complexes

The transition-metal polyhydride complexes have been a subject of considerable interest since the discovery by Kubas et al. [30] of $W(\eta^2\text{-}H_2)(CO)_3(PR_3)_2$ with a nonclassical structure containing molecular hydrogen. Intensive experimental [31] and theoretical [32] activities have resulted in the synthesis of many new transition-metal complexes containing dihydrogen and the observation of equilibrium between classical dihydride and nonclassical dihydrogen complexes. The experimental characterization of polyhydride complexes are difficult mainly due to the small size of the hydrogen atoms which makes the determination of its position quite complicated [33]. Still there exist a number of complexes of known stoichiometry but unknown coordination number, which can be referred to as "polyhydrides," regardless of their true nature as polyhydrides or molecular hydrogen complexes. Therefore the study of the factors influencing the relative stability of metal–dihydrogen forms compared to the metal–dihydride forms still remains an important issue.

Another interesting issue in the transition-metal polyhydride complexes is the frequent existence of an exchange reaction between the hydride (usually *trans*) and dihydrogen ligands which leads to their equivalence in nuclear magnetic resonance (NMR) spectra at room temperature. Based on latest experimental studies [31d], it has been postulated that the exchange reaction must start with oxidative addition of the molecular hydrogen ligand to give two hydrides, which then can exchange with other hydrides. The scarce data available about the reactivity of molecular hydrogen complexes consist essentially of its formation, substitution by other ligands, and oxidative addition to yield the dihydride complex [31, 32]. However, some process involving $\eta^2\text{-}H_2$ complexes have already been claimed to have a more elaborated chemistry. In particular, for $[Os(P\text{-}i\text{-}Pr_3)_2(CO)ClH(H_2)]$ [34] and $[Os(P(CH_3)_2Ph)_3H_3(H_2)]^+$ [35], the molecular hydrogen complex seems to act as a precursor of the reacting species after loss of hydrogen. Furthermore, the molecular hydrogen ligand has been postulated to play an active role in catalytic cycles. These and other issues make studies of polyhydride complexes interesting. High-level quantum chemical calculations, in addition to sophisticated experimental studies by NMR, neutron diffraction, and others, would be useful for solution of these problems.

During the past few years there has been a considerable theoretical activity on the transition-metal polyhydride complexes. Lin and Hall [36] have studied the structure and stability of a large number of model polyhydride complexes with the formula $ML_{7-n}H_n$ and $ML_{8-n}H_n$ (n = 2–7), where M ranges from group 6 to group 8 for second- and third-row transition metals and L can be PH_3 and CO. In these studies they used the HF (for geometries) and MP2 (for energetics) methods in conjunction with the double-zeta basis set for hydrogen and used the combination of the Stevens–Basch–Krauss ECP, double-zeta basis set for the $(n + 1)snp$ electrons and triple-zeta basis set for nd electrons for other atoms. According to these studies, $ReH_2Cl(PMePh_2)_4$ and $[OsH_5L_3]^+$ (L = PPh_3 and PMe_2Ph) should be reclassified as classical isomers. From these systematic studies, authors made several conclusions. (1) The influence of two H ligands *trans* to each other is significantly destabilizing. Thus, suggestions of the stability and structure of a polyhydride complex, particularly for that with an octahedral structure, must account for this influence. (2) For neutral complexes without strong π-accepting ligands, a diagonal line in the Periodic Table through Ru and Ir divides the classical (left side of the line) and nonclassical (right side of the line) forms, as shown in Fig. 7. Those complexes on the line may adopt either or both structures. (3) For monocationic and monochloride hydride complexes, the corresponding diagonal line shifts slightly toward early tran-

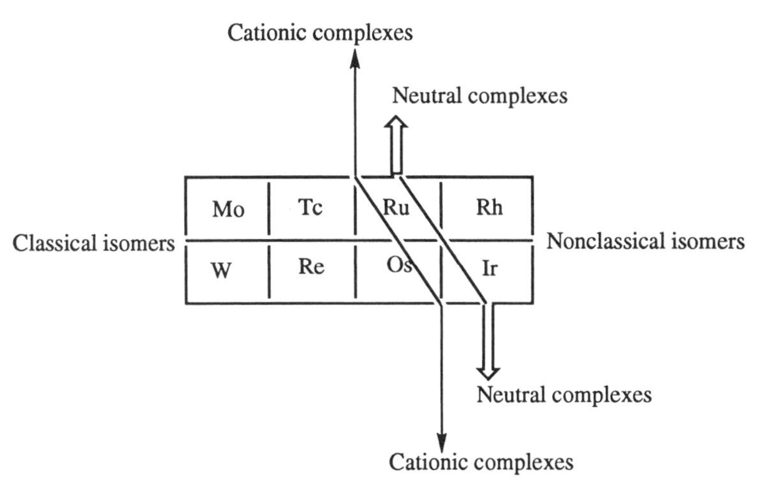

Figure 7. Schematic representation of the borders of classical and nonclassical isomers of polyhydride transition-metal complexes [36b].

sition metals and crosses between Tc/Ru and Os/Ir. (4) The stability of nonclassical complexes with strong π-accepting ligands depends on the number of strong π-accepting ligands and the diffuse nature of the transition metal *d* orbitals. The nonclassical isomers are more likely to exist in complexes with a larger number of strong π-accepting ligands and third-row transition metals.

Hay and coworkers [37] have studied the $ReH_7(PH_3)_2$ and $TcH_7(PH_3)_2$ complexes using different levels of theory. The electron correlation effect, as treated at the configuration interaction (CI) level, has been found to be important in obtaining relative ordering of stability of the various isomers. At the self-consistent field (SCF) level in the split-valance double-zeta (VDZ) basis set, the eight-coordinate $Re(H_2)H_5L_2$ and seven-coordinate $Re(H_2)_2H_3L_2$ structures lie 5 and 2 kcal/mol, respectively, below the classical nine-coordinate structure. At the CI level, in contrast, the classical structure is predicted to be the most stable, although only by 1 and 7 kcal/mol, relative to the eight- and seven-coordinate structures, respectively. For the Tc complexes, there are two major qualitative differences relative to the Re analogues. In the VDZ basis, the seven-coordinate structure is predicted to be the most stable form by 12 kcal/mol even at the CI level, just the reverse of the situation for Re, where the seven-coordinate structure was the less stable form. At the highest level of theory, CI with the Davidson correction in the valence double-zeta plus polarization (VDZP) basis set, this difference is reduced to 1.5 kcal/mol. The other major difference concerns the existence of a stable eight-coordinate structure for $Tc(H_2)H_5L_2$, since all attempts to obtain a local minimum corresponding to this form resulted in collapse to the seven-coordinate form. Thus, for Re the classical nine-coordinate structure is more favored relative to eight- and seven-coordinate nonclassical structures with the increasing order of energy:

$$ReH_7L_2\ (0.0\ \text{kcal/mol}) > Re(H_2)H_5L_2\ (1.0\ \text{kcal/mol}) > Re(H_2)_2H_3L_2\ (7.0\ \text{kcal/mol})$$

while for Tc the energy ordering

$$TcH_7L_2\ (1.5\ \text{kcal/mol}) < Tc(H_2)_2H_3L_2\ (0.0\ \text{kcal/mol})$$

is found. On the basis of these results, Hay and coworkers have concluded that dihydrogen-containing complexes should be more readily found for Tc than Re. These findings of Hay and coworkers are in good agreement with the above-mentioned prediction of Lin and Hall in Fig. 7 [36b].

Maseras et al. [38] also have studied the structure and stability of the model systems $[Os(PH_3)_3H_5]^+$ and $[Os(PH_3)_3H_4]$. In general, it has been found that the HF method is not enough and inclusion of the correlation effects is essential, consistent with previous findings [36, 37]. At the highest level of theory, QCISD(T), used in the paper for $[Os(PH_3)_3H_5]^+$, structure **1** in Fig. 8 with a dodecahedral eight-coordination is calculated to be the most stable isomer. All possible dihydrogen structures are energetically unfavorable. The seven- and six-coordinate structures, **2** and **3**, with one and two hydrogen molecules, respectively, are 3.8 and 11.4 kcal/mol higher. We should note once again that these results of Maseras et al. are also in good agreement with the previous predictions [36b].

Maseras et al. [38b] also have found that the most favorable dihydrogen structure, **4**, of $[Os(PH_3)_3H_4]$ lies about 12 kcal/mol higher than the most stable seven-coordinate structure, **5**, as shown in Fig. 8. These calculations confirm the pentagonal bipyramid as the preferred coordination for $[Os(PH_3)_3H_4]$, which is in agreement with the experimental data.

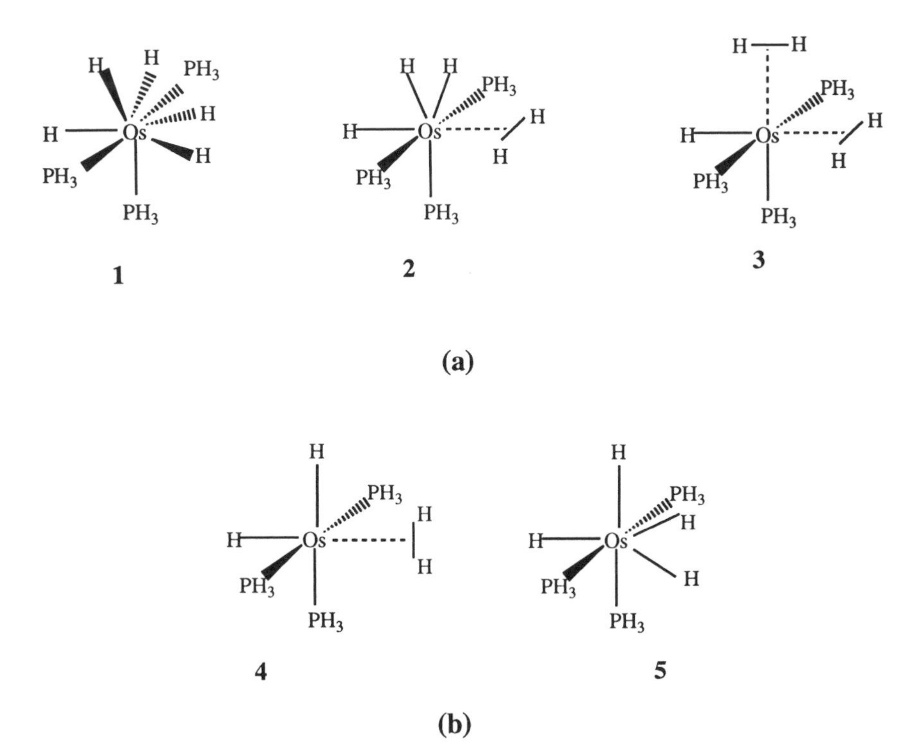

Figure 8. Calculated energetically more favorable structures of (*a*) $[Os(PH_3)_3H_5]^+$ and (*b*) $[Os(PH_3)_3H_4]$ [38].

Maseras et al. [39] have studied the model complex $[Ru(PH_3)_4H_3]^+$ at the MP2 level of theory and have found that the main isomer is an octahedral six-coordinate *trans* dihydrogen hydride complex, **6,** shown in Fig. 9. The six-coordinate *cis* dihydrogen hydride complex **7** lies only 6.4 kcal/mol higher and is a local minimum on the potential energy surface. The rotational barrier of the dihydrogen unit around the metal–mid(H_2) axis is calculated to be 0.2 and 2.3 kcal/mol for the *trans* and *cis* isomers, respectively. The higher rotational barrier for the *cis* isomer can be viewed as a proof of *cis* interaction between hydride and dihydrogen ligands. Another intermediate obtained on the potential energy surface is a trihydride structure, **8,** which lies only 5.5 kcal/mol higher than **6.** Structure **9** is calculated to be 2.1 kcal/mol higher than **8** and characterized to be the transition state between **7** and **8.** The calculation strongly suggests that the trihydrogen complex $[Ru(PH_3)_4(H_3)]^+$ is not likely to exist as an intermediate.

The presence of a trihydride complex, **8,** in the path of the exchange reaction for $[Ru(PH_3)_4H_3]^+$ species obtained at the MP2 level is a significant difference from previous RHF results on the analogous complex Fe, reported Maseras et al. [40]. The trihydride complex had been ruled out as very high

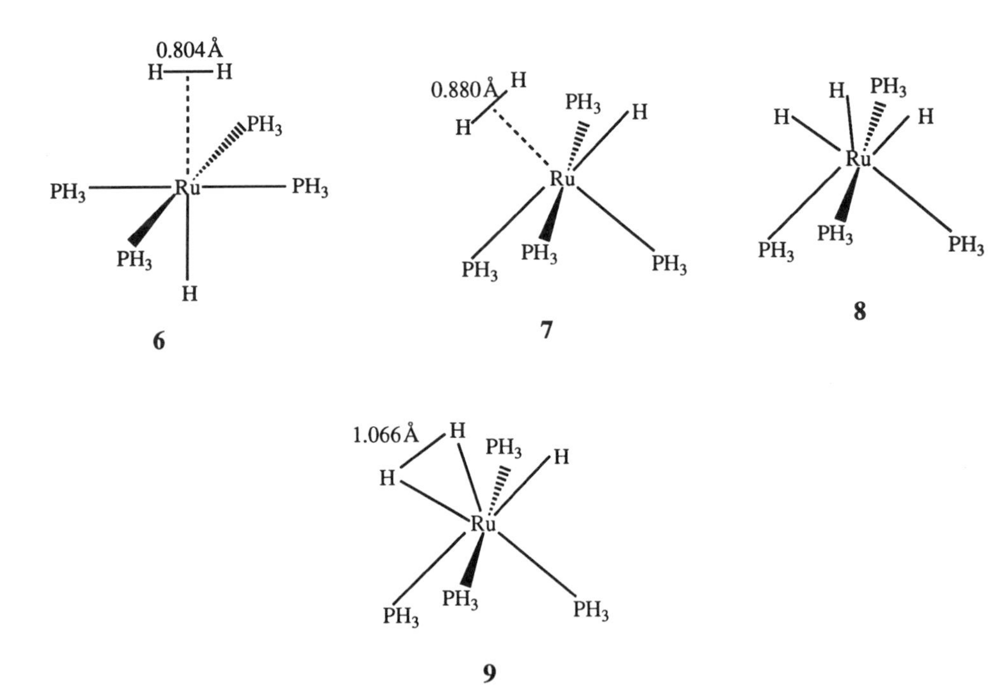

Figure 9. Calculated low-lying isomers (**6–8**) and transition state (**9**) for hydride dihydrogen exchange process for $[Ru(PH_3)_4H_3]^+$ species [39].

in RHF energy for Fe. However, since both the method and the metal are different in two papers, it is impossible to conclude whether this difference is real or not. In the same paper [40], the authors have studied various possible mechanisms, shown in Fig. 10, for the intramolecular exchange of a hydrogen atom between the hydride and molecular complex in the *cis*-$[Fe(PH_3)_4H(H_2)]^+$ complex and have concluded that the most likely is the so-called open direct transfer mechanism, which consists essentially of a single-step transfer of the hydrogen atom between the two ligands while the three hydrogen atoms are in a nearly linear arrangement. The calculated low barrier of 3 kcal/mol agrees with the experimental observation of an easy exchange between the molecular hydrogen and hydride ligands. Moreover, the barrier is so small that this mechanism can even be claimed to be operative for the *trans* octahedral structures, via small amount of the *cis* isomer in equilibrium with the *trans* isomer. The efficiency of this mechanism shown

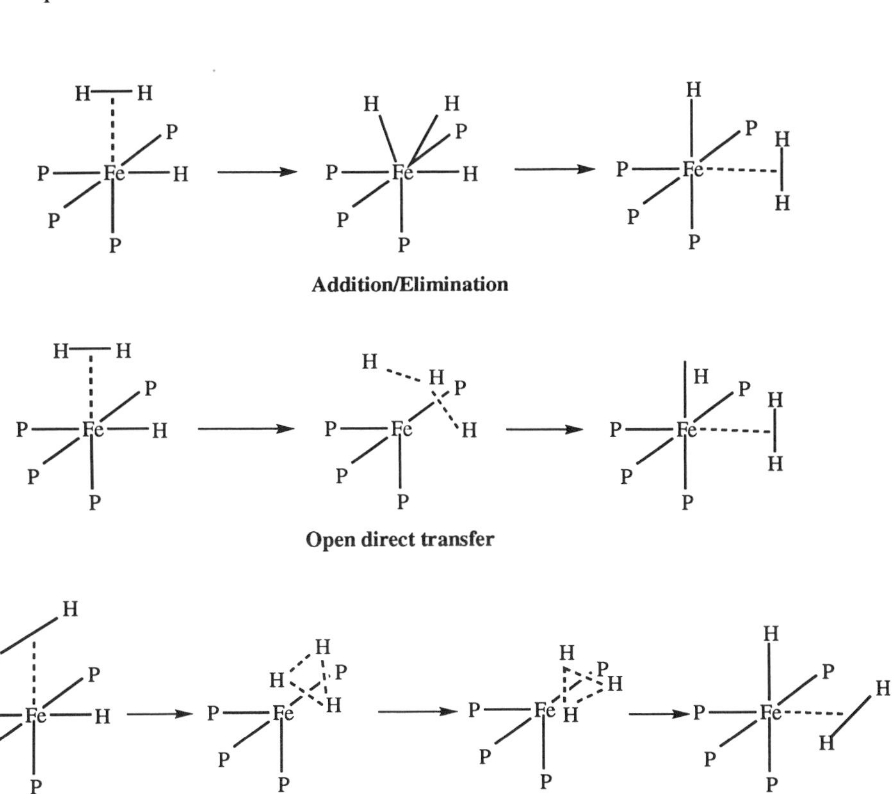

Figure 10. Schematic representation of different mechanisms of H and H_2 exchange [40].

in this paper rejoins a few pieces of evidence supporting an active role of molecular hydrogen complexes in breaking of the strong H–H bond.

E. Summary

In this chapter we reviewed theoretical studies on oxidative addition of single bonds into transition metal complexes. The addition of H–H and C–H bonds of H_2 and CH_4, respectively, to the Vaska-type complex, *trans*-$IrX(CO)(PR_3)_2$, takes place in the X–Ir–CO plane for a weak electron-donor ligand, such as Cl^-, whereas addition in the P–Ir–P plane is preferred for stronger electron donor ligands, such as Ph^-, Me^-, and H^-. The barrier height of the reaction H_2 + *trans*-$IrX(CO)(PR_3)$ → *cis,trans*-$IrX(CO)(PR_3)_2$ decreases and its exothermicity increases via the trend F–Cl–Br–I–CN–H, which is the same as the decreasing π-donation capability of the halides. The π properties of the substituents also play an important role in determining the thermodynamicity of the reaction; the presence of occupied π orbitals on the substituent has a strongly disfavoring effect.

In the gas phase and collisionless condition where the reaction is considered to take place from the reactants, the ease of oxidative addition (i.e., existence of an activation barrier and its height) is correlated strongly to the R–H bond strength as well as to the M–R bond strength. However, in solution or in gas phase when collisional energy equilibration is faster than the reaction itself, it should considered to start from the prereaction complex, $L_nM \cdot (HR)$, the stability of which strongly correlates with the Lewis base character of the addition molecule HR. Under these conditions it will be difficult to activate H–R bonds of molecules that have a strong Lewis base character. The activation barrier consists mostly of the energy required to dissociate the reactant coordinated with its lone pair.

Theoretical studies of the "polyhydride" complexes, $L_nM(H)_m$, have shown the existence of several channels of rearrangements and demonstrated the capability of high-level *ab initio* calculations properly to describe them. According to these studies, the stability of nonclassical complexes depends on the number of strong π-accepting ligands and the diffuse nature of the transition-metal *d* orbitals. Therefore, the nonclassical isomers are more likely to exist in complexes with a larger number of strong π-accepting ligands and third-row transition metals.

III. METATHESIS

Another fundamental reaction of organic synthesis in which bond breaking and formation takes place is the metathesis reaction [1–3]. The alkene metathesis reaction [41] exchanges alkildene groups between different alkenes, and it is catalyzed by a variety of high-oxidation-state early transition-metal complexes.

$$R^1CH{=}CHR^2 + R^3CH{=}CHR^4 \longleftrightarrow R^1CH{=}CHR^3 + R^2CH{=}CHR^4 \tag{III.1}$$

The reaction is of great interest because the strongest bond in alkene, the C=C bond, is broken during the reaction. The widely accepted so-called Chauvin mechanism [42] suggests that transition-metal carbene complex acts as catalyst by undergoing a $[2_\pi + 2_\pi]$ cycloaddition reaction with olefin, via a metallacyclobutane intermediate:

$$\begin{matrix} M{=}CHR \\ R^1HC{=}CHR^2 \end{matrix} \longrightarrow \begin{matrix} M - CHR \\ | \quad\quad | \\ R^1HC - CHR^2 \end{matrix} \longrightarrow \begin{matrix} M{=}CHR^1 \\ RHC{=}CHR^2 \end{matrix} \tag{III.2}$$

The subsequent decomposition of the intermediate metallacycle results in the formation of a new olefin. The σ-bond metathesis [43] in which a σ bond interacts with a transition-metal complex and is broken

$$L_nM{-}R + H{-}R^* \longrightarrow \begin{matrix} L_nM \cdots R \\ \vdots \quad\quad \vdots \\ R^* \cdots H \end{matrix} \longrightarrow L_nM{-}R^* + H{-}R \tag{III.3}$$

has also drawn the attention of experimentalists and theoreticians over the past 10 years. The general scheme of the σ-bond metathesis reaction includes hydrogen exchange (III.4a), methane exchange (III.4b), silane exchange (III.4c), hydrogenolysis (III.4d), and others.

$$L_nM{-}H + D_2 \longleftrightarrow L_nM{-}H + HD \tag{III.4a}$$

$$L_nM{-}CH_3 + {}^*CH_4 \longleftrightarrow L_nM{-}{}^*CH_3 + CH_4 \tag{III.4b}$$

$$L_nM{-}SiH_3 + {}^*SiH_4 \longleftrightarrow L_nM{-}{}^*SiH_3 + SiH_4 \tag{III.4c}$$

$$L_nM{-}R + H_2 \longleftrightarrow L_nM{-}H + RH \tag{III.4d}$$

$$L_nM{-}H + CH_4 \longleftrightarrow L_nM{-}CH_3 + H_2 \tag{III.4e}$$

In general, σ-bond metathesis reactions have been studied for complexes of electron-poor early transition metals and f-block elements. Early theoretical investigations on olefin metathesis and σ-bond metathesis have been reviewed [8, 9].

Folga and Ziegler [44] have studied the olefin metathesis by $L_2Mo(X)CH_2$ (L = Cl, OCH_3, OCF_3; X = O, NH) (III.5), using a Density Functional Theory (DFT) including the Hartree–Fock–Slater density functional, the Vosko local exchange-correlation functional, the Becke nonlocal-exchange correction, and the Perdew correlation correction.

$$L_2Mo(X)C^3H_2 + C^1H_2{=}C^2H_2 \longrightarrow \begin{array}{cc} H_2C^1 - & C^2H_2 \\ | & | \\ L_2Mo - & C^3H_2 \end{array} \qquad (III.5)$$

It has been shown that $L_2Mo(X)CH_2$ and C_2H_4 react along two alternative pathways, yielding square-pyramidal (SP) or trigonal-bipyramidal (TBP) metallacycles, respectively, as shown in Fig. 11. The reactions share the same path for the initial stage of olefin attack onto the carbene molecule until the formation of a transition state **TS** which is about 2.6 kcal/mol above the reactants for X = O and L = Cl. The reactions can then proceed along one of the two proposed routes; one (path **a**) leads to SP and the other (path **b**) leads TBP intermediates which are more stable than the reactants and products. The relative stability of SP and TBP structures depends on the nature of the coligand L and the spectator ligand X; SP is calculated to be more stable than TBP for X = O by 7.9, 5.2, and 4.2 kcal/mol for L = Cl, OCH_3 and OCF_3, respectively. However, for X = NH, SP is preferred to TBP only for L = OCF_3 (by 6.1 kcal/mol), but it lies 1.4 and 0.8 kcal/mol higher than TBP for L = Cl and OCH_3, respectively. In general, the obtained trends in the relative stability of SP and TBP structures agree with the experimental evidence that electron-withdrawing coligands (OCF_3) facilitate an early nucleophilic attack of an olefinic carbon atom on the metal center, leading to the formation of a TBP intermediate. The electron-donating coligands (OCH_3), on the other hand, delays the nucleophilic attack and result in the formation of an SP intermediate. On the basis of correlation diagrams, it has been shown that the involvement of Mo *d* orbitals makes the organometallic $[2_\pi + 2_\pi]$ reaction symmetry-allowed, in contrast to the analogous symmetry-forbidden organic $[2_\pi + 2_\pi]$ reactions between two olefin double bonds.

The metallacyclobutadiene formation from acetylene and molybdenum carbyne complex Cl_3MoCH was studied [45] along a postulated least-motion pathway as shown in Fig. 12. The Cl_3MoCH was brought together with the acetylene molecule such that the three carbon atoms and the metal center remained coplanar throughout the course of the reaction. It has been found

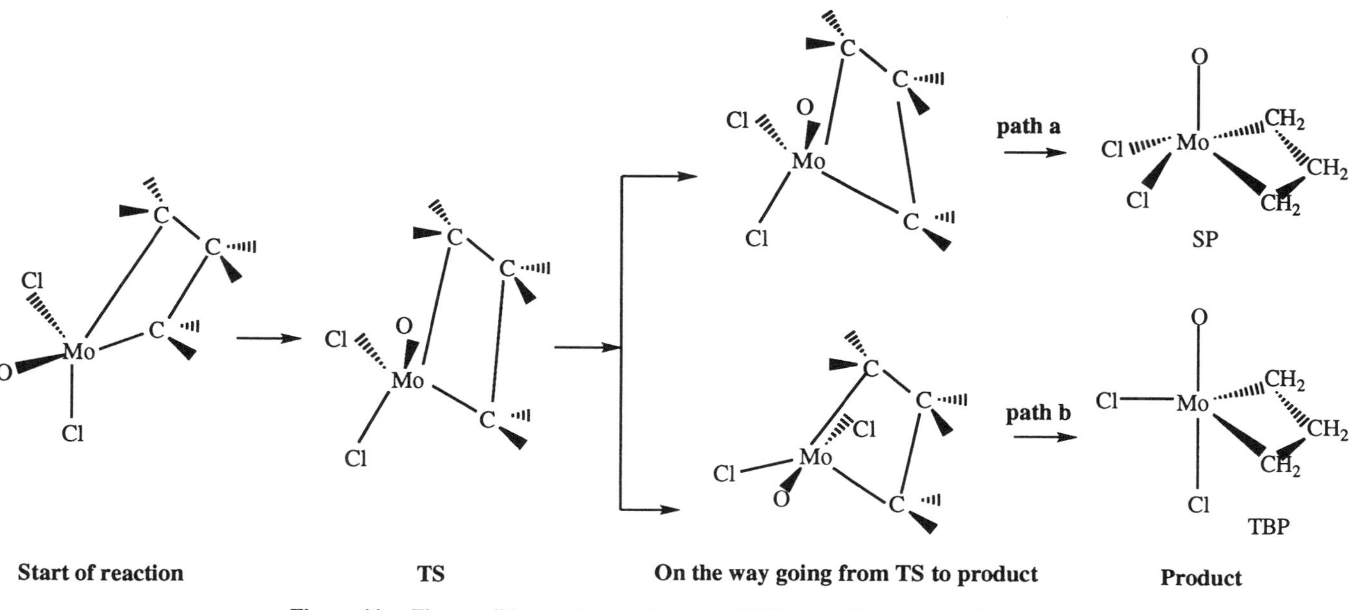

Figure 11. The possible reaction mechanism of $(Cl)_2OMoCH_2$ with ethylene [44].

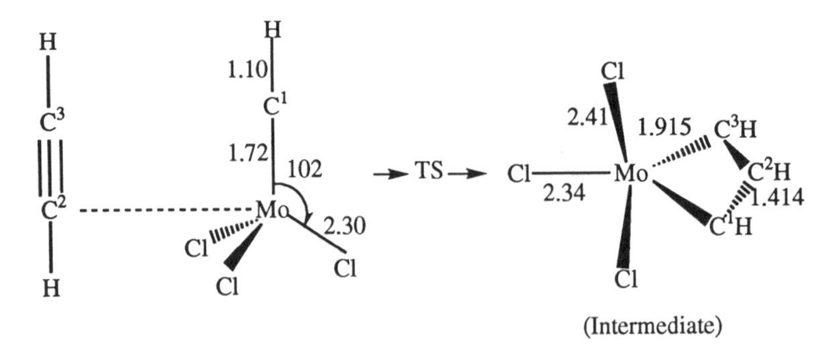

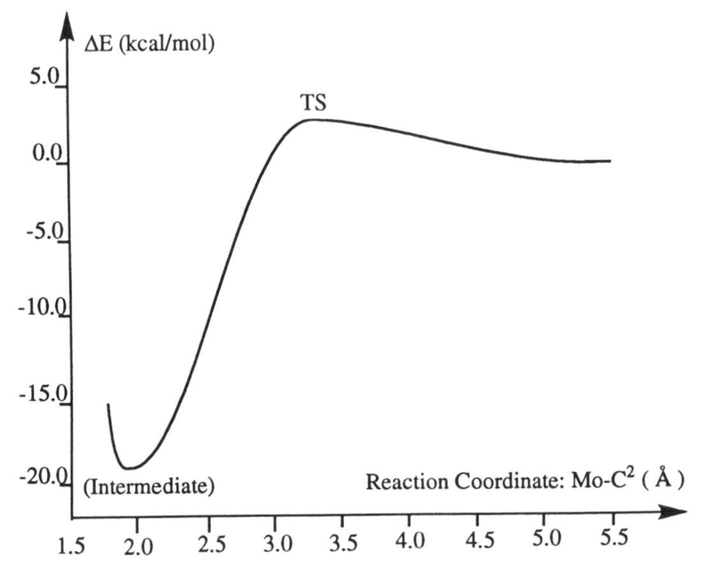

Figure 12. Calculated (*a*) reaction path with important geometric parameters (in angstroms and degrees) and (*b*) energy profile for the reaction of acetylene with Cl_3MoCH [45].

that the formation of metallacyclobutadiene $Cl_3MoC_3H_3$, initiated by nucleophilic attack of acetylene on the metal center, is symmetry-allowed, proceeds with a small barrier of 2.4 kcal/mol, and leads to the product which is 16.7 kcal/mol more stable than the acetylene and molybdenum carbyne reactants.

Recently, Ziegler and coworkers [46–48] have performed extensive studies on σ-bond metathesis reactions, such as hydrogen exchange, methane exchange, silane exchange, hydrogenolysis, methylation, and silylation by scandium and lutetium complexes using DFT. In general, it has been found

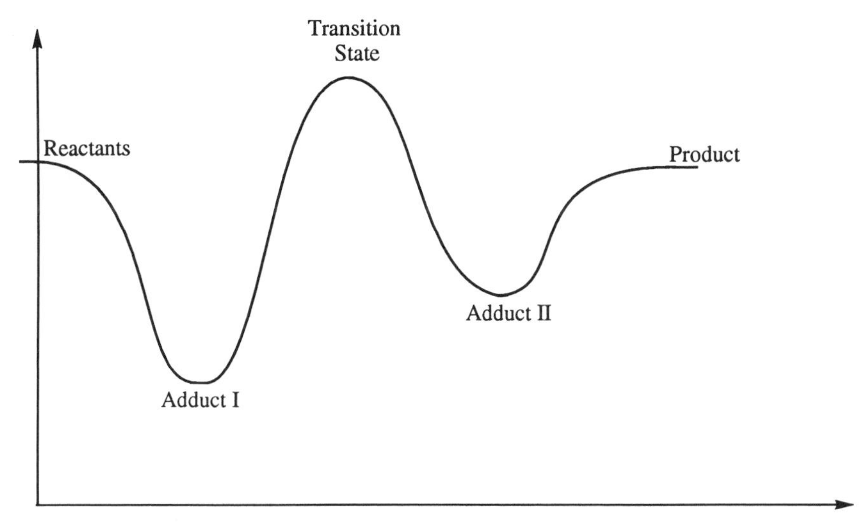

Figure 13. The schematic representation of the potential energy profile of the σ-metathesis reactions.

that all these reactions in the first step yield adduct complexes, schematically shown as Adduct I in Fig. 13, and then activation of the H–H, C–H, and Si–H σ bonds takes place with a four-center transition state, which leads to the second adduct complex, Adduct II, followed by formation of products. The calculated energetics for these reactions is shown in Table IV. As seen from Table IV, the hydrogen exchange reactions, T.1 and T.10, for Cl_2ScH and Cp_2LuH in first step lead to weak H_2 complexes with 3.8 and 1.2 kcal/mol stabilization energy relative to the reactants, respectively. The transition states for H–H bond activation are calculated to be about -1.7 and 10.6 kcal/mol, respectively, relative to the reactants. Thus, the hydrogen exchange reaction should go more easily for the scandium hydride complex than for the lutetium one. The transition state has a four-center structure with the H–H distance of 1.02 Å and 1.05 Å for M = Sc and Lu, respectively.

The methane exchange reaction, T.3 and T.12, also should occur more easily for scandium-methyl than for lutetium-methyl. As seen from Table IV, Adduct I, $Cl_2LuCH_3(CH_4)$, is calculated to be only kinetically stable and lies 4.6 kcal/mol higher than reactants. The transition state separating reactants and the CH_4 complex was not located. However, Adduct I for Sc, $Cp_2ScCH_3(CH_4)$, is thermodynamically stable by 6.0 kcal/mol relative to the reactants. The barrier for methane exchange is calculated to be 25.9 and 10.8 kcal/mol relative to the reactants, respectively, for the Lu and Sc compounds. In the transition state the distances of the two reacting CH bonds

TABLE IV
The Energies (Relative to Reactants, in Kilocalories per Mole) of σ-Bond Metathesis Reactions Calculated at the DFT Level by Ziegler et al.

N	Reactions	Adduct I	Transition State	Adduct II	Product	Reference
T.1	$Cp_2ScH + H_2 \rightarrow Cp_2ScH + H_2$	−3.8	−1.7	−3.8	0.0	46
T.2	$Cp_2ScCH_3 + H_2 \rightarrow Cp_2ScH + CH_4$	−4.1	1.9	−14.8	−10.0	46
T.3	$Cp_2ScCH_3 + CH_4 \rightarrow Cp_2ScCH_3 + CH_4$	−6.0	10.8	−6.0	0.0	46
T.4	$Cl_2ScCH_3 + H_2 \rightarrow Cl_2ScH + CH_4$	−6.7	1.0	−11.9	−1.9	47
T.5	$Cl_2ScSiH_3 + H_2 \rightarrow Cl_2ScH + SiH_4$	−1.9	1.0	−8.4	−0.7	47
T.6	$Cl_2ScCH_3 + CH_4 \rightarrow Cl_2ScCH_3 + CH_4$	−10.8	7.9	−10.8	0.0	47
T.7	$Cl_2ScCH_3 + CH_4 \rightarrow Cl_2ScH + C_2H_6$	−4.1	31.5	−10.3	6.0	47
T.8	$Cl_2ScSiH_3 + SiH_4 \rightarrow Cl_2ScSiH_3 + SiH_4$	−7.4	0.2	−7.4	0.0	47
T.9	$Cl_2ScSiH_3 + SiH_4 \rightarrow Cl_2ScH + Si_2H_6$	−4.8	−2.6	−5.5	2.4	47
T.10	$Cl_2LuH + H_2 \rightarrow Cl_2LuH + H_2$	−1.2	10.6	−1.2	0.0	48
T.11	$Cl_2LuCH_3 + H_2 \rightarrow Cl_2LuH + CH_4$	−0.4	19.6	−6.0	−4.1	48
T.12	$Cl_2LuCH_3 + CH_4 \rightarrow Cl_2LuCH_3 + CH_4$	4.6	25.9	4.6	0.0	48
T.13	$Cp_2ScH + C_2H_4 \rightarrow Cp_2ScC_2H_3 + H_2$	−2.9	−4.8	0.7	3.8	47
T.14	$Cp_2ScCH_3 + C_2H_4 \rightarrow Cp_2ScC_2H_3 + CH_4$	−2.9	9.3	11.5	−6.2	47
T.15	$Cp_2ScH + C_2H_2 \rightarrow Cp_2ScC_2H + H_2$	−8.8	−6.9	−25.8	−20.5	47
T.16	$Cp_2ScCH_3 + C_2H_2 \rightarrow Cp_2ScC_2H + CH_4$	−4.3	−0.9	−37.5	−30.6	47

are calculated to be 1.56 and 1.33 Å, respectively, for Lu and Sc, 0.44 and 0.19 Å longer than in the corresponding CH_4 complex.

Studies of the hydrogenolysis, T.11 and T.2, with Cl_2LuCH_3 and Cp_2ScCH_3 have shown that the incoming H_2 molecule forms a weak adduct, H_2 complex, in the early stage of reaction. The four-center transition state for the H–H bond-breaking and C–H bond-formation process for L = Cl is calculated to be 19.6 and 1.9 kcal/mol, respectively, relative to the reactants for Lu and Sc. After the transition state the reaction leads to the $L_2MH \cdot CH_4$ complex which is stable by 1.9 and 4.8 kcal/mol, respectively, relative to the dissociation limit $L_2MH + CH_4$. The entire reaction is exothermic by 4.1 and 10.0 kcal/mol for Lu and Sc, respectively, which is consistent with M–CH_3 and M–H bond strengths, the former being generally weaker than the latter. The reverse reaction of this hydrogenolysis, the methylation reaction, occurs with a large barrier, 25.6 and 16.7 kcal/mol, respectively, for Lu and Sc. The comparison between Sc and Lu complexes once again points out the ease of reaction for Sc compared to Lu. While the calculated energetics changes upon changing the coligand from Cl to Cp by several kilocalories per mole, as seen in Table IV, the qualitative conclusion about the easier reaction for Sc than for Lu does not change.

A comparison between hydrogenolysis reactions, T.4 and T.5, shows that they are similar between Cp_2ScCH_3 and Cp_2ScSiH_3 complexes. As seen in Table IV, the calculated activation barriers relative to the reactants are same (1 kcal/mol), and the exothermicities of the entire reactions are similar (2 and 1 kcal/mol). Therefore, one expects that in gas phase under collisionless condition the rates of these reactions will be very close. However, in solution the reaction with Cl_2ScSiH_3 will probably go faster than with Cl_2ScCH_3 because the calculated complexation energy for the former is 4.8 kcal/mol smaller.

Ziegler and Folga [47] also have studied the reactions

$$Cl_2ScCH_3 + CH_4 \rightarrow Cl_2ScCH_3 + CH_4 \quad (T.6)$$

$$\rightarrow Cl_2ScH + C_2H_6 \quad (T.7)$$

and

$$Cl_2ScSiH_3 + SiH_4 \rightarrow Cl_2ScSiH_3 + SiH_4 \quad (T.8)$$

$$\rightarrow Cl_2ScH + Si_2H_6 \quad (T.9)$$

As seen in Table IV, the reaction $Cl_2ScCH_3 + CH_4$ is most likely to go via methane exchange channel (T.6) because of a larger complexation energy and a smaller activation barrier than (T.7). The channel (T.7) leading to hydrogenolysis product is unlikely to take place because of a large activation barrier, 31.5 kcal/mol, and the endothermicity of 6.0 kcal/mol. However, the analogous reaction $Cl_2ScSiH_3 + SiH_4$ should take place very easily via both the silanation (T.8) and the hydrogenolysis (T.9) pathways. The activation barriers are calculated to be 7.6 and 2.2 kcal/mol, respectively, relative to the SiH_4 complex. The transition state is just 0.2 kcal/mol higher than the reactants for (T.8), while it lies 2.6 kcal/mol lower than the reactants for (T.9). In general, the above discussion and results given in Table IV show that for the σ-bond metathesis reactivity between the R–H and the M–R′ bond decreases in the order of R: $C_2H > C_2H_3 > H$ for $R' = H$, and $C_2H > H > CH_3 > C_2H_3$ for $R' = CH_3$ and $SiH_3 > H$ for $R' = SiH_3$.

Ziegler and Folga [47] also discussed the mechanism of the reaction of Cp_2Sc-R ($R = H$ and CH_3) with ethylene and actylene which may occur in two different ways: (i) by insertion into M–R bonds:

$$L_2M{-}R + H_2C{=}CR^1R^2 \rightarrow L_2M{-}CH_2{-}CRR^1R^2 \qquad \text{(III.6)}$$

and (ii) by activation of alkenylic and alkynylic C–H bonds by a σ-bond metathesis mechanism:

$$L_2M{-}R + H_2C{=}CR^1R^2 \rightarrow L_2M{-}C(H){-}CR^1R^2 + H{-}R \qquad \text{(III.7)}$$

Experimentally it has been established that for lutetium the simplest and least sterically hindered olefins, ethylene and propene, react by insertion [49], whereas more bulky olefins give rise to C–H activation. For scandium, only ethylene inserts into the Sc–R bond. The trend has been rationalized [49] by observing that the transition state for the insertion becomes sterically crowded with more bulky olefins, in particular for smaller metals such as scandium. The four-center transition state for C–H activation is sterically less demanding and thus accessible to bulkier olefins. It has further been observed that C–H activation in olefins invariably takes place at the stronger alkenylic C–H bond rather than at the weaker alkylic C–H bond. The DFT calculations by Ziegler and Folga [47] for the reaction of Cp_2Sc–R ($R = H$ and CH_3) with ethylene shows that the incoming ethylene forms an ethylene adduct with a stabilization energy of 3 kcal/mol both for $R = H$ and CH_3. Then activation of the alkenylic C—H bond takes place with a barrier of 7.6 and 12.2 kcal/mol relative to ethylene adduct for $R = H$ and CH_3, respectively. After the barrier the reaction gives the second adducts, H_2 adduct and CH_4 adduct, which are stable relative to the dissociation limit $Cp_2ScC_2H_3 + HR$ by 3.1 and 5.2 kcal/mol, respectively. The process is endothermic by 3.8 kcal/mol for $R = H$, but is exothermic by 6.2 kcal/mol for $R = CH_3$. Thus, vinylic C–H activation by Cp_2ScCH_3 is seen to be exothermic in contrast to the corresponding process mediated by Cp_2ScH.

However, vinylic C–H activation by Cp_2ScCH_3 has a larger barrier than by Cp_2ScH. By comparison of these results with the previously studied insertion of ethylene into the $Cp_2Sc{-}H$ and $Cp_2Sc{-}CH_3$ bond (which has been found to be 27 and 15.5 kcal/mol exothermic, respectively), the authors [47] have concluded that insertion is favored over σ-bond metathesis on thermochemical grounds for the reaction between Cp_2ScH or Cp_2ScCH_3 and ethylene.

Acetylene forms in the early stage of the reaction a tight adduct with Cp_2ScH and Cp_2ScCH_3, which is 8.8 and 4.3 kcal/mol lower, respectively, than the corresponding reactant. In these adducts a C–H bond on acetylene shows an agostic interaction with Sc, resulting in elongation of the C–H bond by 0.04 Å and 0.08 Å, respectively. The C–H activation barrier is calculated to be 1.9 and 3.4 kcal/mol relative to the adduct, for Cp_2ScH and Cp_2ScCH_3, respectively. The resultant complexes, $Cp_2Sc(H_2)C_2H$ and $Cp_2Sc(CH_4)C_2H$, are calculated to be 5.3 and 6.9 kcal/mol stable relative to the dissociation limits Cp_2ScC_2H + HR, respectively. The entire reactions of acetylene with Cp_2ScH and Cp_2ScCH_3 are exothermic by 20.5 and 30.6 kcal/mol, respectively. Comparison of these data with those for the alternative insertion process

$$Cp_2ScH + HCCH \rightarrow Cp_2ScC(H)CH_2 \qquad \text{(III.8)}$$

which is exothermic by 45.6 kcal/mol shows that insertion should be favored over C–H activations on thermodynamically grounds in reactions involving scandium. Thus, insertion of ethylene or acetylene into Sc–H and Sc–CH_3 bonds are preferred thermodynamically over the alternative alkenylic and alkynylic C–H bond activation. This is in line with experiment for ethylene [43b]. However, acetylenes have been observed to prefer alkynylic C–H activation over insertion for scandium [43b].

The comparison of the alkylic, alkenylic and alkynylic C–H bond activations reported by Ziegler et al. [46] shows that the barrier height and exotermicity of the reaction decreases via the trend alkynylic >> alkenylic > alkylic, and thus the strength of the Sc–R bonds should follow the same trend. This trend is roughly the same to that of the C–H bond strength in the corresponding H–R systems.

A. Summary

In this section we reviewed the latest theoretical studies on the metathesis reaction. In general, it has been shown that the metathesis reaction at first yields adduct complex; then activation of the H–R bond takes place with a four-center transition state, which leads to the second adduct. It occurs more easily for scandium complexes than for lutetium analogues. The σ-metathesis reactivity between R–H and M–R bonds decreases in the order of R: C_2H

> C_2H_3 > H for R′ = H; C_2H > H > CH_3 > C_2H_3 for R′ = CH_3; and SiH_3 > H for R′ = SiH_3. The comparison of the alkylic, alkenylic, and alkynylic C–H bond activation shows that the barrier heights and exothermicity of the reaction decreases via alkynylic >> alkenylic > alkylic.

IV. OLEFIN INSERTION

Insertion reactions of organic molecules into M–H and M–R bonds are the fundamental steps in industrially important catalytic processes such as polymerization, hydroformylation, hydrogenation, and isomerization and therefore are the focus of intensive theoretical and experimental investigations [1–10]. The current status of theory on insertion into a metal–hydride and metal–alkyl bonds has recently been reviewed by Koga and Morokuma [8], Yoshida, et al. [9], and Siegbahn and Blomberg [7]. The main orbital interactions involved in this reaction are well-characterized [7–9]. The π orbital of the incoming olefin interacts and donates some electron density to the LUMO of the M–R group, σ^*_{MR}. Meanwhile, the back-donation from the HOMO, with substantial hydride or alkyl contributions, σ_{MR}, to the antibonding π^* orbital of olefin takes place, as shown in Fig. 14.

Recently, several interesting quantum chemical studies have been performed in the field of olefin polymerization by transition metal complexes, where olefin insertion into the M–R bond is an important step. The main purpose of these studies is to elucidate different factors affecting the polymerization rate and selectivity. Extensive experimental studies have identified cationic metallocenes Cp_2MR^+, bridged cationic metallocenes $H_2SiCp_2MR^+$, half-sandwich metallocenes (or so-called constraint geometry catalysts) $H_2SiCp(NH)MR^+$ (M = group 4 metals), and complexes of f-block elements as active catalysts [50]. According to the widely accepted Cossee mechanism [51] the propagation step in the polymerization of olefins occurs via a prior coordination of the π bond to the vacant coordination site

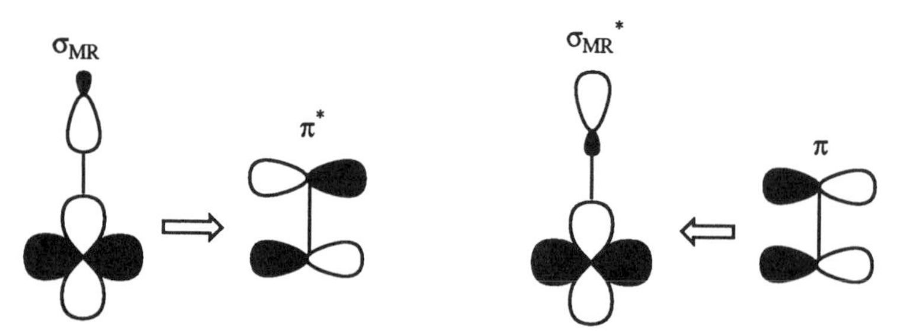

Figure 14. Schematic representation donation and back-donation interactions between metal–alkyl (or metal–hydride) bond and olefin.

of the active catalyst, followed by olefin insertion into the M–R bond. However, factors affecting the polymerization rate, including the role of both the bridged SiH_2 ligand and the half-sandwich structure as well as that of the nature of the transition metal atoms and its charge, need to be elucidated. Quantum chemical calculations have been very essential in providing insight into these important problems.

Extensive studies of the mechanism of ethylene polymerization reaction with homogeneous silylene-bridged metallocenes, $H_2SiCp_2MCH_3^+$, has been performed by Morokuma and colleagues. Yoshida, Koga, and Morokuma (YKM) [52] have studied, at the restricted Hartree–Fock (RHF) (mostly for optimize geometries), second-, third-, and fourth-order Møller–Plesset perturbation (MP2, MP3, and MP4), QCISD, and RQCISD levels of theory with split valence basis functions and effective core potential for transition metals, the mechanism of ethylene polymerization reaction with homogeneous silylene-bridged metallocenes, $H_2SiCp_2MCH_3^+$, where M = Ti, Zr, and Hf. At first, for M = Zr the optimized geometries at the RHF and MP2 methods are very similar, providing justification for RHF geometry optimization.

The optimized geometries of the reactants have C_s symmetry with the metal; Si and C (of CH_3) are nearly collinear, irrespective of the metal, as shown in Fig. 15. The energy required to bend this angle by 30° is less than 3 kcal/mol, and a strong influence of the counteranion or approaching ligand can easily change this angle. There is an α-agostic interaction between transition-metal center and one of H atoms of the CH_3 group which is stronger for Ti than for the others. However, even with Ti, the energy of α-agostic interaction is small and it is not likely to contribute significantly to the catalytic activity. This trend in the α-agostic interaction has been explained in term of the difference in the M–CH_3 bond distance which is calculated to be 0.2 Å shorter for Ti than for Zr or Hf. One needs to note that in the earlier paper of Kawamura-Kuribayashi, et al. [53] the authors could not find the α-agostic interaction on the $H_2SiCp_2ZrCH_3^+$, $Cl_2ZrCH_3^+$, and $Cl_2TiCH_3^+$ complexes which might be due to the use of smaller basis sets.

The coordination of the ethylene, as shown for complex **12** in Fig. 16, does not change the M–CH_3 bond distance but bends the Si–M–CH_3 angle by 53.6 (Ti), 46.8° (Zr), and 50.6° (Hf). The M–Cp centroid distances are slightly stretched by 0.031 Å (Ti), 0.017 Å (Zr), and 0.021 Å (Hf), to avoid steric contact between Cp and incoming ethylene. After coordination of ethylene, the reaction path follows the insertion of ethylene into the M–CH_3 bond with a four-center transition state, **13,** where the M–C–C–C torsional angle is small, 0.1° (Ti), 3.1° (Zr), and 0.1° (Hf), although a larger torsional angle would decrease the steric repulsion due to eclipsing

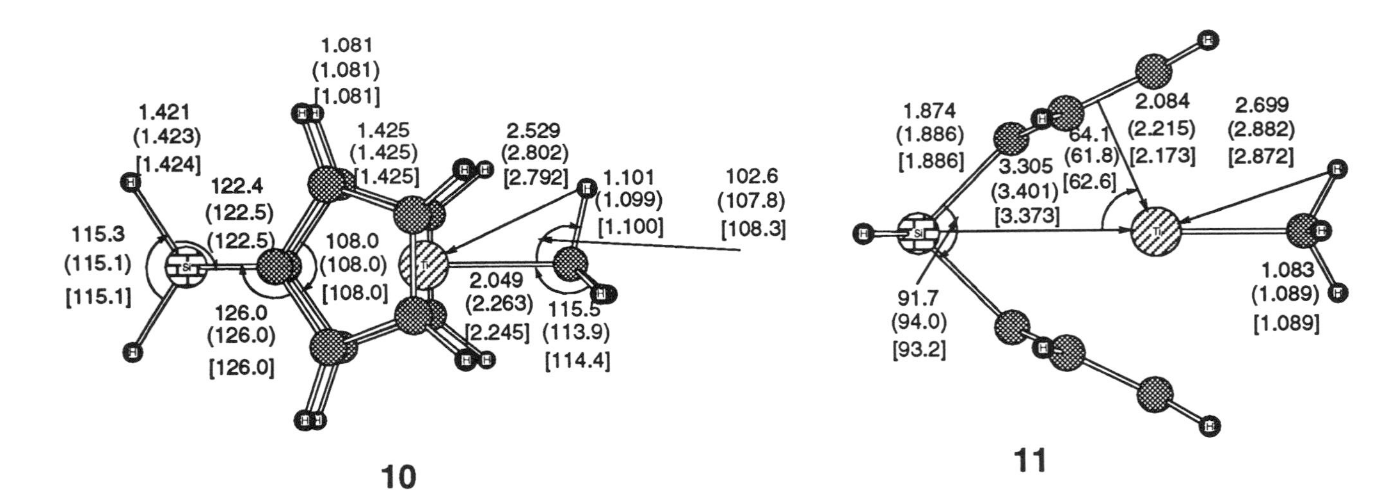

Figure 15. Calculated structures of the $SiH_2Cp_2MCH_3^+$ complex, where M = Ti, Zr (in parentheses), and Hf (in brackets) [52].

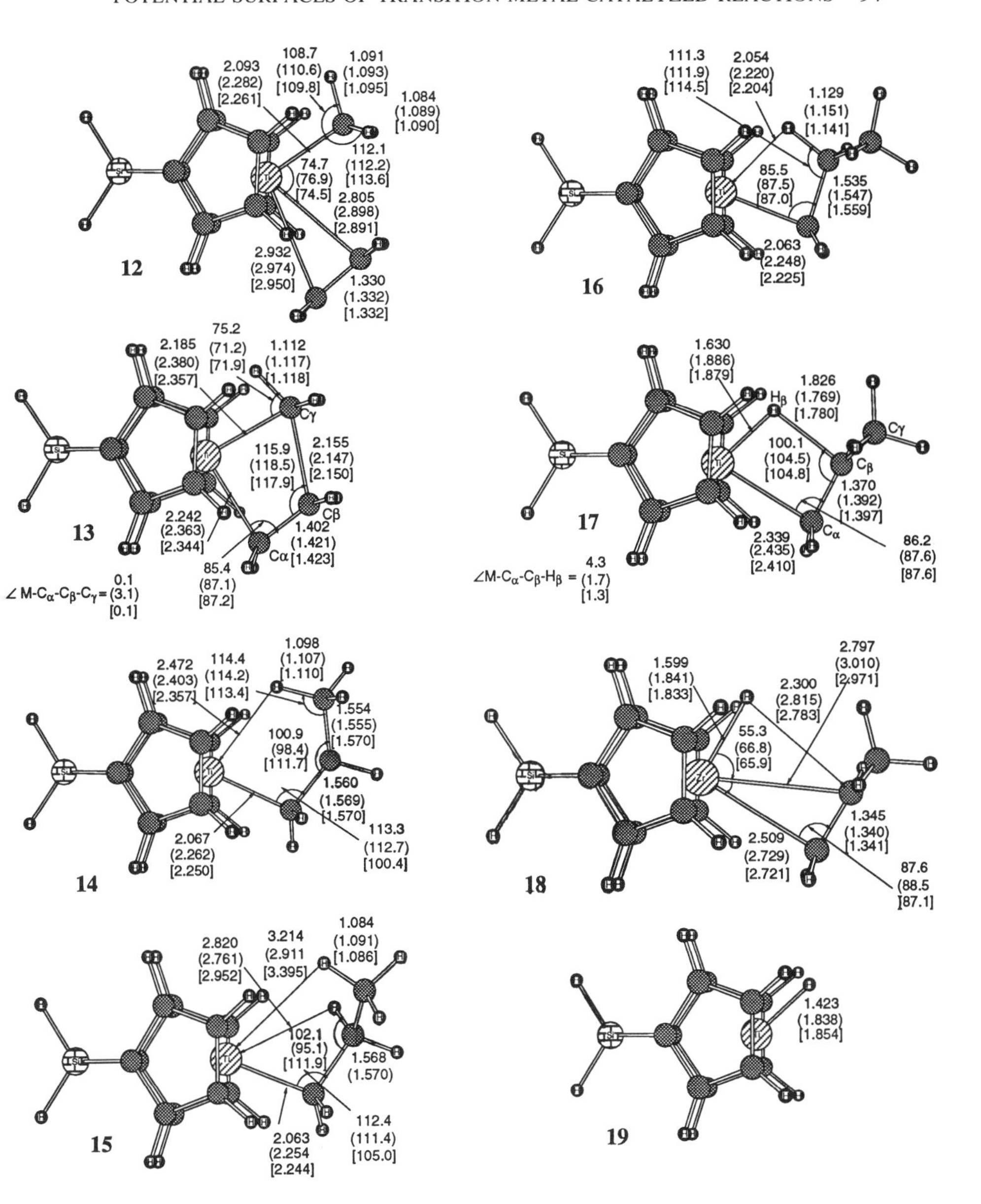

Figure 16. RHF optimized structures (in angstroms and degrees) of π complex (**12**), transition state (**13**), γ-agostic product (**14**), transition state (**15**), β-agostic product (**16**), transition state (**17**), π complex (**18**), and metal hydride (**19**) obtained by Yoshida, Koga, and Morokuma for complex of Ti, Zr (in parentheses), and Hf (in brackets) [52].

between the methyl and ethylene moieties. This transition state, where two bonds are being formed and two are being broken, is an early and tight transition state with asynchronous bond exchange. The reaction path from TS **13** leads to a local minimum on the potential energy surface, **14,** the insertion product with a γ-agostic interaction, which is weaker for Ti than for Zr and Hf. However, this γ-agostic intermediate rearranges into the β-agostic isomer, **16,** which is 6.1 (Ti), 2.0 (Zr), and 1.9 (Hf) kcal/mol more stable than the former. During this isomerization the agostic interaction migrates from the C_γ–H to the C_β–H bond, and these two agostic interactions might take place simultaneously in the transition state **15.** One should note that in the paper of Kawamura-Kuribayashi et al. [54] on the $Cl_2TiCH_3^+$ the γ-agostic product was found to be about 4 kcal/mol more stable than the β-agostic isomer at the RHF level.

As seen from Table V, where we summarized the calculated energetic parameters, the results for Ti are far from convergence in the perturbation series, which has been explained in term of triplet unstability of the RHF wave function of Ti alkyl complexes. According to the results obtained at the higher, RQCISD level [52], ethylene coordination is exothermic by 21.9 (Ti), 29.1 (Zr) and 27.6 (Hf) kcal/mol. Ethylene insertion from the π complex through the transition state, **13,** to γ-agostic, **14,** is exothermic by 7.4 (Ti), 4.3 (Zr), and 4.5 (Hf) kcal/mol with a low activation barrier of 7.1 (Ti), 9.4 (Zr), and 10.1 (Hf) kcal/mol from the π complex, showing that the elementary insertion reaction takes place easily.

The β-elimination, a side reaction leading to the chain transfer, as shown in Fig. 17, has a quite high endothermicity of about 52.8 (Ti), 45.6 (Zr), and 48.5 (Hf) kcal/mol and is an unfavorable path. The reverse reaction of olefin insertion into an M–H bond has a very low barrier. Based on the above-mentioned results, YKM [52] have concluded that ethylene polymerization with these cationic calalysts can occur rapidly, resulting in high-molecular-weight polymer, most favorably in the reaction with Hf.

Ziegler and coworkers [55–57] have studied the ethylene polymerization reaction with unbridged metallocene $Cp_2ZrCH_3^+$ and Cp_2ScCH_3, bridged metallocene $H_2SiCp_2ZrCH_3^+$, and constrained geometry catalysts $(SiH_2)(Cp)NH)MCH_3^+$ and $(SiH_2)(Cp)NH)MCH_3$ by using a DFT. In general, the qualitative picture obtained by Morokuma and coworkers [52–54] at the RHF or MPn levels and that by Ziegler and coworkers [55–57] at the DFT level are very similar. However, there are a several quantitative differences. At first, for both unbridged and bridged complexes of all metal atoms mentioned above except for Cp_2ScCH_3 and XMC angle (where X = none and Si for the unbridged and bridged complexes, respectively) is bent at the DFT level, while at the RHF level it is almost collinear. It should be noted that the bent and straight geometries were found to differ by only a

TABLE V
Relative Energies of Stationary Structures in Ethylene Insertion and β-Elimination with $[H_2SiCp_2MCH_3]^+$ (Kilocalories per Mole)[a] Calculated by Yoshida et al. [52]

Method	π Complex **12**	TS **13**	γ-Agostic **14**	TS **15**	β-Agostic **16**	TS **17**	π Complex **18**	M–H (+ Product) **19**
				$[H_2SiCp_2TiCH_3]^+$				
RHF	−13.0	1.3	−22.8	−19.9	−29.2	−10.3	−11.0	13.1
RMP2	−28.6	−27.3	−34.7	−27.9	−43.1	−24.5	−22.2	18.5
RMP3	−20.9	−10.0	−30.8	−25.6	−37.4	−17.7	−17.5	15.2
RMP4SDQ	−27.6	−24.5	−32.6	−26.3	−39.7	−21.7	−20.9	18.0
UHF		−0.4	−27.7	−25.9	−31.5	−10.4		
PUHF[b]		−19.2	−53.7	−52.1	−49.7	−15.7		
UMP2		2.5	17.6	26.3	−14.7	−17.4		
PUMP2[b]		−14.8	−7.2	1.2		−31.4	−22.2	
PUMP3[b]		−14.7	−29.3	−23.0	−41.0	−18.8		
RQCISD	−21.9	−14.8	−29.3	−23.8	−35.4	−17.0	−17.0	17.4
				$[H_2SiCp_2ZrCH_3]^+$				
RHF	−19.0	−2.4	−23.8	−21.1	−26.8	−10.5	−16.3	9.2
RMP2	−33.4	−27.4	−38.4	−33.5	−40.9	−27.2	−28.2	10.7
RMP3	−29.4	−19.5	−34.6	−30.2	−36.9	−22.1	−24.9	10.5
RMP4SDQ	−30.1	−20.9	−34.6	−30.1	−36.6	−22.5	−25.5	10.1
RQCISD	−29.1	−19.7	−33.4	−29.0	−35.4	−21.5	−24.5	10.2
				$[H_2SiCp_2HfCH_3]^+$				
RHF	−17.6	0.6	−22.0	−19.0	−25.3	−8.3	−14.4	11.5
RMP2	−31.7	−24.6	−36.6	−27.0	−39.2	−24.2	−26.0	15.5
RMP3	−28.5	−17.8	−33.7	−25.3	−35.8	−20.1	−23.5	14.2
RMP4SDQ	−28.4	−18.5	−33.1	−24.4	−35.1	−19.9	−23.2	14.5
RQCISD	−27.6	−17.5	−32.1	−23.5	−34.0	−19.0	−22.4	14.5

[a]Energy relative to the free reactant + ethylene at the RHF optimized geometries.
[b]Spin contaminants annihilated up to nonet.

Ethylene Insertion Reaction into Metal-Alkyl Bond

$[M]^+-R$ + CH_2CH_2 → $[M]^+-R(CH_2CH_2)$ → (TS) → $[M]^+-CH_2CH_2R$ → (TS) →
(Metal Alkyl) | π- complex | γ- Agostic

→ $[M]^+-R(CH_2CH_2)$
β- Agostic

β- Elimination Reaction

$[M]^+-R(CH_2CH_2)$ → (TS) → $[M]^+-H(CH_2CHR)$ → $[M]^+-H$ + CH_2CHR
β- Agostic | π- complex | (Metal Hydride)

Figure 17. Schematic representation of olefin insertion into the M–R bond and β-elimination reaction.

few kilocalories per mole at both RHF and DFT levels, and the potential energy surface is very flat. Second, at the DFT level the ethylene π complex, **12** in Fig. 16, has agostic interaction between the Zr atom and one of H atoms of the CH_3 group, while at the RHF and MP2 levels of theory the agostic interaction does not exist. YKM [52] have explained this discrepancy between their MP2 and Ziegler and coworker's DFT results [55] by the overestimating of the correlation energies at the transition state by the local density approximation (LDA) used in the latter. By the same reason the insertion barrier relative to ethylene complex obtained by YKM, 9.4 kcal/mol, is much larger than that, 1 kcal/mol, obtained by Ziegler and coworkers [55].

By comparing the results for unbridged $Cp_2ZrCH_3^+$ and bridged metallocenes $H_2SiCp_2ZrCH_3^+$, Ziegler and coworkers [55–57] have found that the monosilane bridge has two important effects. First, it opens up the coordination site of the complex by effectively "pulling back" the Cp rings. The Cp–Zr–Cp angle is 139° in the unbridged species and 127° in the monosilane bridged species. Second, by pulling back the Cp rings with the bridge, the electron deficiency of the Zr center is increased. This is because the Cp ring orbitals cannot interact optimally with the Zr *d* orbitals. However, the effect of the monosilane bridge on the energetics of the insertion is very small. The activation barrier is unchanged by the bridge, and reaction enthalpy is increased only by a few kilocalories per mole.

The profile for scandocene is very similar to those of its Zr^+ counterparts, as shown in Table VI, but is at higher energies relative to the reactants due to the lack of stabilization from the ion–dipole interaction present in the

TABLE VI

The Relative Energies (in Kilocalories per Mole) of Ethylene Insertion into the M–CH_3 Bond of $SiH_2Cp_2ZrCH_3^+$, $Cp_2ZrCH_3^+$, Cp_2ScCH_3, $(SiH_2)(Cp)(NH)M(IV)CH_3^+$ (Where M = Ti, Zr, and Hf), and $(SiH_2)(Cp)(NH)Ti(III)CH_3$ Calculated by Ziegler et al. [55–57]

	Metallocenes			$(SiH_2)(Cp)(NH)MCH_3^{+,0}$			
Structures	Zr[a]	Zr	Sc	Ti(IV)	Zr(IV)	Hf(IV)	Ti(III)
Reactant + ethylene	0.0	0.0	0.0	0.0	0.0	0.0	0.0
π Complex	−26.3	−22.9	−5.0	−20.8	−24.2	−25.7	−22.7
TS	−25.3	−22.2	−1.7	−17.8	−19.1	−10.0	−19.4
Insertion product (γ-agostic)	−34.0	−31.0	−14.0	−33.6	−32.4	−33.6	−26.6

[a]For the complex $SiH_2Cp_2ZrCH_3^+$.

cationic Zr systems. The insertion barrier is also calculated to be a few kilocalories per mole higher for Sc than for Zr. The higher barrier in the scandocene has been explained in terms of the weaker α-agostic interaction in the transition state of the Sc systems than in that of the Zr system since the neutral Sc is less electron-deficient than the Zr cation.

Although the constrained geometry catalyst (CGC) has a more open coordination site and potentially less steric hindrance to insertion, it has been found that the open nature of the catalyst actually increases the barrier. As seen in Table VI, the barrier heights for CGC are several kilocalories per mole larger than those for metallocenes. Although the large coordination site of the CGC increases the barrier, one may expect the more open site to better facilitate insertion of larger, more sterically demanding comonomers.

Extensive studies [57] of the reactivity of CGC such as $(SiH_2(Cp)NH)MCH_3^+$ (M = Ti, Zr, and Hf) and $(SiH_2(Cp)NH)TiCH_3$ at the DFT level apparently show, as shown in Table VI, that the difference between Zr- and Hf-based catalysts is minor with regard to the structure of the catalysts, the π complexes, the transition states, and the products as well as the relative energies. The exothermicity of ethylene insertion for M = Ti(IV), Zr(IV), and Hf(IV) is quite similar, about 16.6, 13.3, and 13.6 kcal/mol, respectively, and insertion barrier increases slightly in the order Ti (3.8 kcal/mol) < Zr (5.1 kcal/mol) < Hf (5.7 kcal/mol). Thus, the titanium-based catalyst was found to possess the smallest insertion barrier of the three. The higher activity of the titanium-based CGC has been explained in terms of a lower stability of the corresponding π complex due to steric repulsions between ethylene and the coligands at the metal center, notably the methyl group. The calculated binding energies of the corresponding π complexes are 20.8, 24.2, and 25.7 kcal/mol for the Ti-, Zr-, and Hf-based catalysts, respectively.

The Ti(IV)- and Ti(III)-based CGCs also show mostly similarities [57]. However, there are some differences in the structure of the reactants, transition states, and products as well as relative energies. In the reactant the charged species of Ti(IV) displays a large bending angle of Si–Ti–CH_3 = 63° versus 24° for neutral species of Ti(III). One should note that, as mentioned above, the bending mode is soft and less than 3 kcal/mol is required to move the methyl group from the in-plane to the out-of-plane position. The Ti(IV)-based catalyst has a slightly stronger agostic interaction between the methyl hydrogen and Ti center than does the Ti(III)-based catalyst. The Ti–C–$H^{agostic}$ angle is calculated to be 96° and 103° for the Ti(IV) and Ti(III), respectively, while the C–$H^{agostic}$ bond distance is essentially the same, 1.12 Å. There is also a considerable difference between the π complexes of charged Ti(IV) and neutral Ti(III). The bonding between ethylene and the d^0 metal center in Ti(IV) is basically established by an electrostatic stabilization of the ethylene π electrons due to the positively charged metal center. As a result, the ethylene double-bond distance in the π complex remains essentially the same as in the free molecule, 1.34 Å. However, for the neutral species the ethylene double bond is elongated by 0.04 Å to 1.38 Å. This elongation stems from a delocalization of the single d electron on Ti into the π^* orbital of ethylene. Thus, the π complex of the neutral species has been characterized as a real π complex in which both donation and back-donation are important. The calculated transition states are very similar with regard to the geometries and have been characterized as early transition states. The insertion barrier calculated from the corresponding π complex is 3.8 and 3.3 kcal/mol for Ti(IV) and Ti(III), respectively. Thus, although the active catalysts are generally assumed to be the positively charged methyl complexes of group IV, the neutral species of Ti(III) has been found to be as reactive as Ti(IV). The exothermicity of the entire reaction is about two times larger for Ti(IV) than for Ti(III), 16.6 and 7.2 kcal/mol, respectively.

Ahlrichs and coworkers [58, 59] also have studied the reaction of $Cp_2TiCH_3^+$ with ethylene at the MP2 level with a large basis set. Their MP2 geometry optimization has shown that the reaction is downhill, leading to $Cp_2TiC_3H_7^+$ without any barrier. However, these results seem to be an artifact in view of the latest development in this field. Indeed, YKM [52] have shown that due to the instability of the RHF wave function of $Cp_2TiCH_3^+$ the perturbation series is slow in convergence, and the MP2 method is apparently not enough to properly describe the reaction of $Cp_2TiCH_3^+$ with ethylene.

The insertion of ethylene into Pt–H and Pt–SiH_3 bonds of the complex $PtH(SiH_3)(PH_3)$ has been studied by Sakaki et al. [60] at the MP4SDQ level by calculating the geometries of the reactants, transition states, intermediates, and products at SCF and MP2 levels. The reactant $PtH(SiH_3)(PH_3)$ has three different isomers, **20a, 20b,** and **20c,** among which **20a** and **20b** are nearly degenerate and lie about 12 kcal/mol lower than **20c,** as shown in Fig. 18 and Table VII. However, it should be noted that it has not been tested whether these are real minima or not. Comparison of the calculated geometries apparently shows that the *trans* influence increases in the order $PH_3 \ll H < SiH_3$. The same trend has been found also by analyzing of

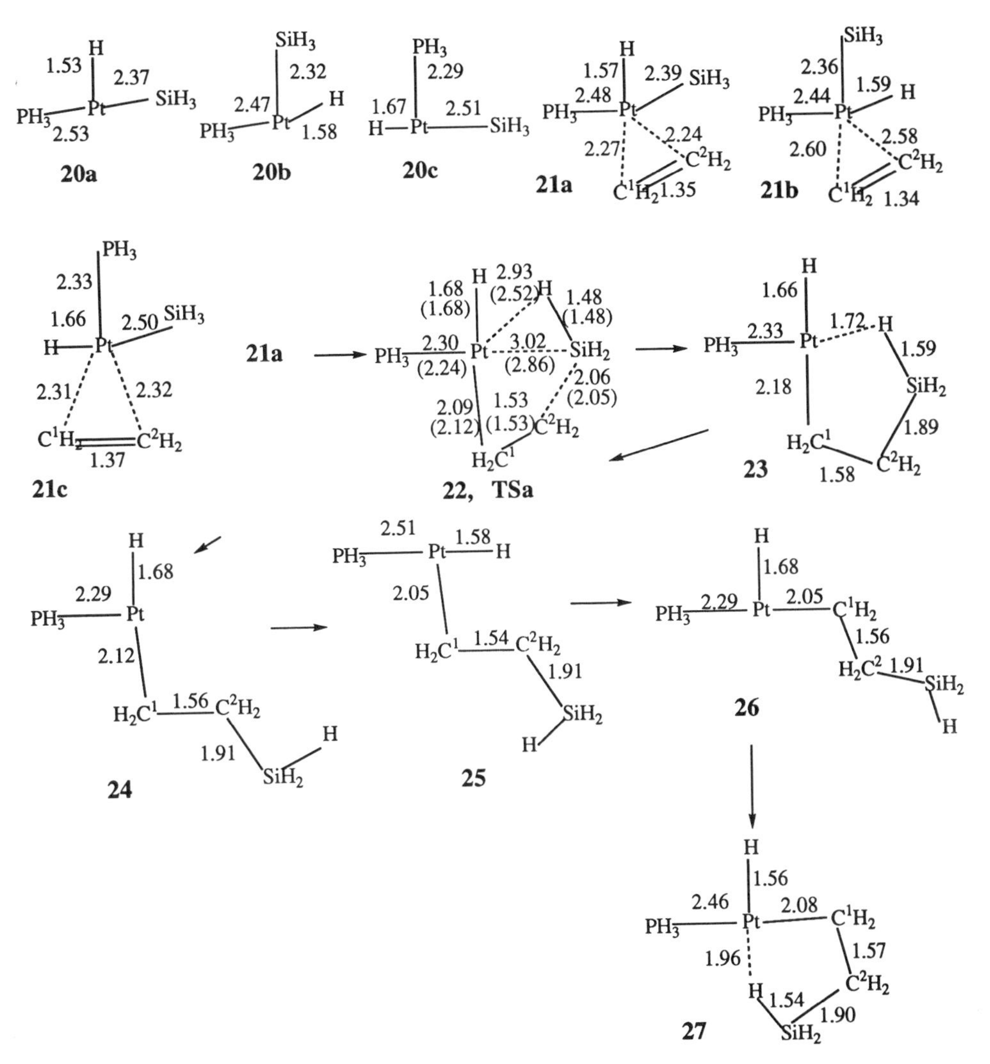

Figure 18. Optimized bond distances (in angstroms) of different isomers of $PtH(SiH_3)(PH_3)$ (**20a, 20b,** and **20c**), the transition state **TSa,** and products of C_2H_4 insertion into the Pt–SiH_3 bond of $PtH(SiH_3)(PH_3)(C_2H_4)$ (**21a**), taken from Sakaki et al. [60]. The MP2-optimized values are given in parentheses.

TABLE VII
The Relative Energies (in Kilocalories per Mole) of the Ethylene Insertion into Pt–SiH_3 or Pt–H bonds, Taken from Sakaki et al. [60]

Structures	HF	MP2	MP3	MP4DQ	MP4SDQ
		C_2H_4 Insertion into Pt–SiH_3 of ***20a***			
21a	0.0	0.0	0.0	0.0	0.0
22, TSa	45.9	54.9	50.2	53.4	53.9
23	21.7	31.3	27.0	29.8	30.5
24	32.8	51.4	44.7	48.4	49.3
25	12.5	23.0	18.0	20.6	21.6
26	8.6	20.4	15.0	17.9	18.9
27	6.4	13.8	9.7	12.1	12.8
		C_2H_4 Insertion into Pt–H of ***20b***			
21b	0.0	0.0	0.0	0.0	0.0
28, TSb	20.1	19.7	18.6	20.3	20.6
29	20.1	20.0	19.7	20.5	20.8
30	26.6	40.5	33.9	36.8	37.6
31	1.3	5.7	2.3	3.8	4.3
32	−8.7	−6.1	−7.9	−6.7	−6.5
33	−6.8	−4.8	−6.5	−5.3	−5.1
		C_2H_4 Insertion into Pt–SiH_3 of ***20c***			
21c	0.0	0.0	0.0	0.0	0.0
34, TSc	17.8	16.2	15.0	16.2	16.3(16.3)[a]
27	−6.3	2.4	−2.0	−0.2	0.6(2.9)
		C_2H_4 Insertion into Pt–H of ***20c***			
21c	0.0	0.0	0.0	0.0	0.0
35, TSd	5.1	4.0	4.6	4.6	4.4(4.2)
33	−21.6	−19.7	−19.4	−18.6	−18.4
32	−23.5	−18.0	−20.8	−20.0	−20.0(−17.2)

[a]In parentheses are values at the MP4SDTQ level.

the possible isomers of corresponding π complex, $PtH(SiH_3)(PH_3)(C_2H_4)$, for which ethylene is on the molecular plane. The isomer containing ethylene perpendicular to the plane is calculated to be a few kilocalories per mole less stable. The product-like transition state **22, TSa,** corresponding to ethylene insertion into the Pt–SiH_3 bond is calculated to be 53.9 kcal/mol relative to the ethlyene complex **21a** and contains the γ-agostic interaction. This transition state leads to an intermediate, **23,** which also has the γ-agostic interaction and lies about 30.5 kcal/mol higher than the prereaction ethylene complex. Then this intermediate **23** rearranges into the more stable product **27** which is only 12.8 kcal/mol higher than the ethylene complex. Ethylene insertion into the Pt–H bond, as shown in Fig. 19 and Table VII,

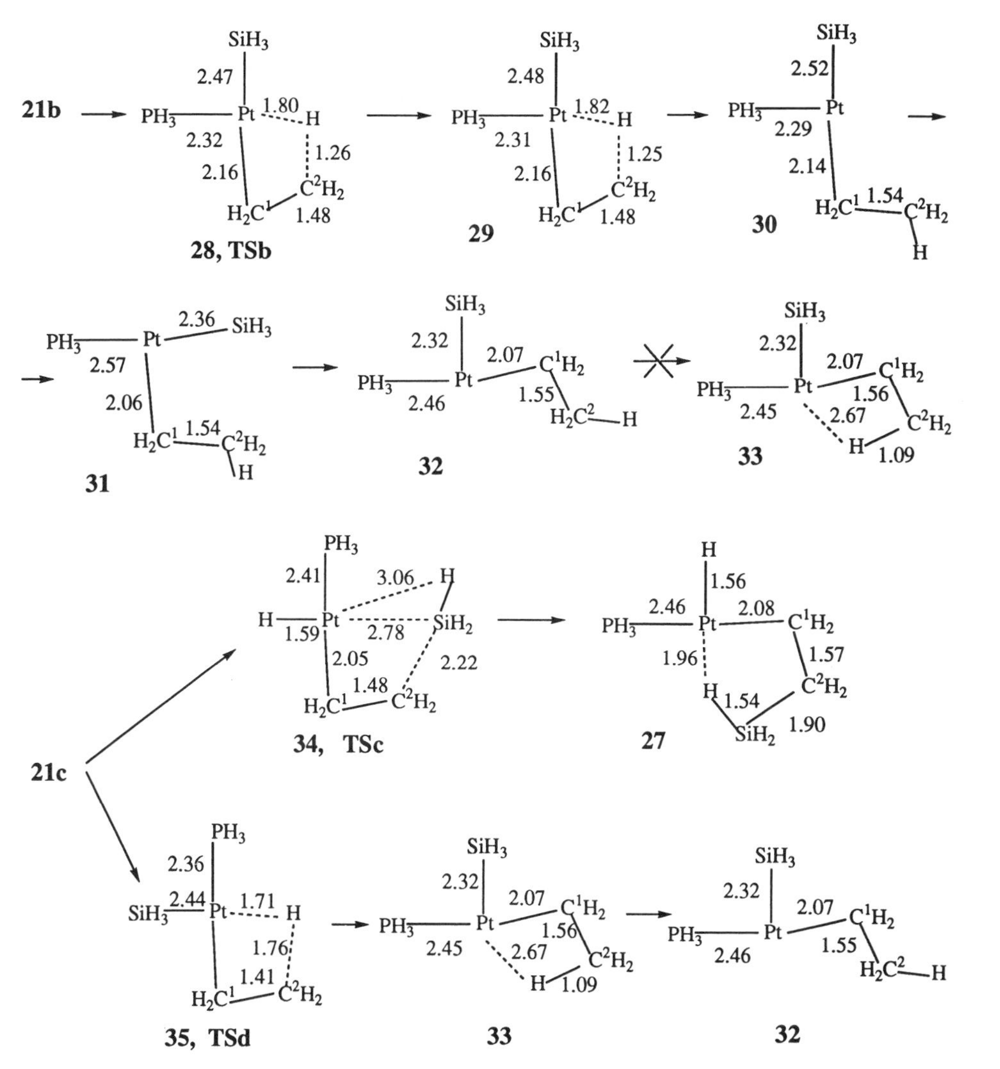

Figure 19. Optimized bond distances (in angstroms) of the transition states and products of the C_2H_4 insertion into the Pt–H and Pt–SiH_3 bonds of $PtH(SiH_3)(PH_3)(C_2H_4)$ (**21b** and **21c,** respectively), taken from Sakaki et al. [60].

occurs with a relatively smaller barrier of 20.6 kcal/mol, and leads to the product **29,** which lies about 21 kcal/mol higher than the π complex and rearranges into the more stable product, **32,** which is lower than the ethylene complex by 6.5 kcal/mol. The relative stability of **23** and **27** isomers have been explained in terms of stronger *trans* influence of hydride relative to PH_3.

By using RHF (to optimize geometries) and MP2 (for energetics) methods in conjunction with split-valence basis sets, Hyla-Kryspin et al. [61] have shown that insertion of acetylene into the Zr–H bond of Cl_2ZrH^+ should occur much easier than that into the Zr–CH_3 bond of $Cl_2ZrCH_3^+$. The insertion barriers are calculated to be 0.2 and 5.0 kcal/mol, respectively, relative to the corresponding the π complexes. This difference in the insertion barriers for Zr–H and Zr–CH_3 bonds is consistent with the previous finding for oxidative addition in the previous section and also can be explained in terms of directionality of the Zr–CH_3 bond. The acetylene binding energies are calculated to be 51.6 and 41.5 kcal/mol for the Cl_2ZrH^+ and $Cl_2ZrCH_3^+$, respectively. The exothermicity of the entire reaction is calculated to be 86.9 and 53.3 kcal/mol, respectively. The exothermicity of the entire reaction is calculated to be 86.9 and 53.3 kcal/mol, respectively. Thus, the small insertion barrier and large exothermicity of the present reactions suggest that the insertion reaction of acetylene into the Zr–H and Zr–CH_3 bonds of Cl_2ZrH^+ and $Cl_2ZrCH_3^+$, respectively, should be irreversible.

Siegbahn [62] has studied the energetics of olefin insertion into the transition-metal–H bond for the entire sequence of second-row transition-metal hydrides using the Hartree–Fock (for geometry optimization) and MCPF (for energetics) methods with large basis sets including f_M, d_C, and p_H polarization functions. Results shown in Table VIII indicate that the repulsion between nonbonding metal electrons and electrons on the olefin plays a dominant role in the height of insertion barrier. First, there is a sharp increase in the barrier height between Nb and Mo both with and without additional hydride ligands present. The explanation for this increase is that Nb can make use of its empty 4*d* orbital in an *sd* hybridization, which effectively reduces the repulsion toward the olefin. This type of hybridization cannot be used for Mo since all nonbonding *s* and *d* orbitals have the same occupation and the energy is invariant to rotations among these orbitals. The second point where the consequences of metal–ligand repulsion effects can be clearly seen in Table VIII is in a comparison for $x = 1$ between Rh on one hand and Ru and Pd on the other; Rh has a dramatically lower barrier, which can be explained by the s^0 state. For Rh the s^0 state is low-lying and has the ability to form the covalent *d* bond required at the transition state. For Ru a substantial promotion energy is required to reach the s^0 state, and for Pd the s^0 state has a closed *d* shell and can therefore not form any *d* bond. The reason why the s^0 state is dominant at the transition state must be that repulsive effects play a significant role for the size of the reaction barrier. Another general trend in Table VIII that has been explained by dominant M–ligand repulsion effects is the tendency for the atoms to the left in the row to have lower barriers than those to the right; there are fewer repulsive 4*d* electrons for the atoms to the left.

TABLE VIII

Energies[a] for the Olefin Coordination $MH_x + C_2H_4 \rightarrow MH_x(C_2H_4)$ and for the Olefin Insertion $MH_x + C_2H_4 \rightarrow MH_{x-1}(C_2H_5)$ Reactions (in Kilocalories per Mole) Calculated by Siegbahn [62]

	$x = 1$			$x = 2$			$x = 3$		
M	π Complex	TS	Product	π Complex	TS	Product	π Complex	TS	Product
Y	−15.1	−5.2	−32.2	−25.1	−6.9	−30.5	—	−7.9	−29.4
Zr	−43.3	−11.7	−36.4	−53.3	−3.1	−32.4	−18.6	−15.9	−33.9
Nb	−30.9	−11.7	−29.4	−43.3	−10.7	−33.3	−40.3	−17.7	−32.2
Mo	−10.6	16.1	−29.1	−18.3	4.3	−31.1	−15.6	−6.0	−30.1
Tc	−33.1	19.5	−30.9	−13.4	2.4	−28.3	−10.1	−10.1	−27.0
Ru	−21.4	6.7	−26.1	−18.4	−2.4	−26.4	−5.6	−12.3	−27.0
Rh	−27.5	−12.8	−24.9	−18.0	−0.9	−24.8	—	−3.6	−25.7
Pd	−17.8	6.9	−27.0	−36.8	−1.4	−25.7	—	—	—

[a]The energies are calculated relative to ground state of the corresponding reactants.

It has been shown that the main effect of adding hydride ligand (going from $x = 1$ to $x = 2$ and 3) is a general reduction of the size of the insertion barriers, which has been explained in terms of the additional H ligands helping to remove some of the repulsive nonbonding electrons from the metal. Another trend noted as hydride ligands are added is a reduction of the variation of barrier heights across the periodic in Table VIII. Since the main origin of the differences between the metals is the varying degree of repulsion between nonbonding metal electrons and olefin electrons, the removal of some of these metal electrons by the ligand also removes part of the difference between the metals.

These studies of Siegbahn also shows that some well-established principles do not work. One relationship that might have been expected is a correlation between the M–H bond strength and the ease of olefin insertion [62]. However, this kind of relationship is contradicted by a large number of the presently studied reactions. For example, TcH has the weakest bond of all the second-row hydrides and yet it has the highest barrier for olefin insertion. Second is the relationship between the exothermicity of insertion reaction and the barrier height. It is well established that an increase of the exothermicity of reaction leads to a decrease of the barrier height [15]. However, as seen in Table VIII, there is a very small variation of exothermicities between the different systems, while the barrier heights and the binding energies of the π-bonded olefins vary much more.

The insertion of acetylene into the Pd–CH_3 bond of the complex $PdCl(NH_3)(CH_3)$ has been studied by de Vaal and Dedieu [63] by using the valence double-zeta basis sets. The geometries of the prereaction complex $[PdCl(NH_3)(CH_3)(C_2H_2)]$, the transition state, and product have been optimized at the SCF level, and their energetics has been improved at the CASSCF and CI level. It has been shown that acetylene is quite weakly bound (5.8 kcal/mol) in the square-planar Pd(II) complexes because of weak π back-donation from Pd to the π^* orbital of C_2H_2. The insertion barrier calculated relative to the acetylene complex is 20.5, 22.6, and 17.1 kcal/mol at the SCF, CASSCF, and CI levels of theory, respectively, and the transition state corresponding to this barrier displays monohapto coordination of acetylene. The entire insertion reaction is calculated to be exothermic by 26.0, 19.3, and 22.4 kcal/mol at the SCF, CASSCF, and CI levels, respectively, relative to the acetylene complex.

Insertion of ethylene and acetylene into the Zr–R bond of the $Cp_2Zr(R)Cl$ complex (where R = H and CH_3), the so-called hydrozirconation reaction, has been studied by Endo, Koga, and Morokuma (EKM) [64] at the RHF and MP2 levels of theory. They have investigated two different paths, path 1 and 2, of the reaction shown in Fig. 20. In path 1, olefin attacks between the hydride and the chloride ligand, whereas in path 2, it attacks from the opposite side of the chloride. In both paths the ethylene or acetylene C–C bond is assumed to be coplanar with the HZrCl plane. It has been found that path 1 is the more favorable reaction path at all the used levels of theory. This is mainly determined by the difference in deformation energy of the Zr complex between two paths; the distortion of the Zr complex for path 1 is easier than that for path 2 because of a smaller repulsion between the hydride and the chloride ligand and also because of a stronger Zr–H bond. At the more reliable MP2 level, path 1 has no activation barrier while path 2 has a 14-kcal/mol barrier. The direct product has an agostic interaction, which is only realized when the geometry is optimized at the correlated MP2 level. The entire reaction $Cp_2Zr(H)Cl + C_2H_4 \rightarrow Cp_2Zr(C_2H_5)Cl$ is calculated to be exothermic by 30 kcal/mol. The preference of path 1 is unchanged upon replacement of Cp by Cl, path 1 being downhill. In reactions of $Cl_2Zr(H)Cl$ with C_2H_4, however, the smaller size of the Cl ligands makes the approach through path 2 not very unfavorable and the ethylene complex is a stable intermediate, by 11.5 kcal/mol relative to reactants, from which this is a moderate activation energy of 3.6 kcal/mol. The studies of the path 1 of the reaction of ethylene with $Cp_2Zr(CH_3)Cl$ shows that the insertion into the Zr–C bond is much more difficult and has a 27-kcal/mol activation barrier. This has been explained in term of directionality of the M–C bond compared with that of M–H.

As shown by EKM [64], the hydrozirconation of acetylene also has two paths, as also shown in Fig. 20, among which the path 1 is kinetically more favorable by 14 kcal/mol. While the insertion barrier of acetylene into the Zr–H bond is small, 3.8 kcal/mol, it is positive, which is different for the reaction with ethylene. These results of EKM suggest that hydrozirconation of alkyne is intrinsically more difficult.

Nakamura et al. [65] have applied the RHF and MP2 methods in conjunction with split-valence basis sets to study the insertion of acetylene into the $M–CH_3$ bond of the complexes LiMe, CuMe, and Me_2Cu^-. As shown in Fig. 21, the reactions of LiMe and CuMe with parent acetylene have been found to proceed via (a) and π complex, **37,** which is, by 11 kcal/mol, stable relative to reactants at the HF level for both complexes, and (b) the transition state, **38,** which is calculated to be 20 and 46 kcal/mol relative to the π complex at the HF level, respectively, to the product, **39,** which is 38 and 35 kcal/mol lower than that of the reactants, respectively. Thus, both reactions are much more exothermic than the addition to ethylene [66]. However, the activation energies of insertion of acetylene into the M–Me bond obtained in this paper is unrealistically high. This has been explained by the deficiency of the monomeric, ligand-free model used in the study. The qualitatively similar results have been found also for $CuMe_2^-$, except that the reaction proceeds through the π complex, which does not exist in

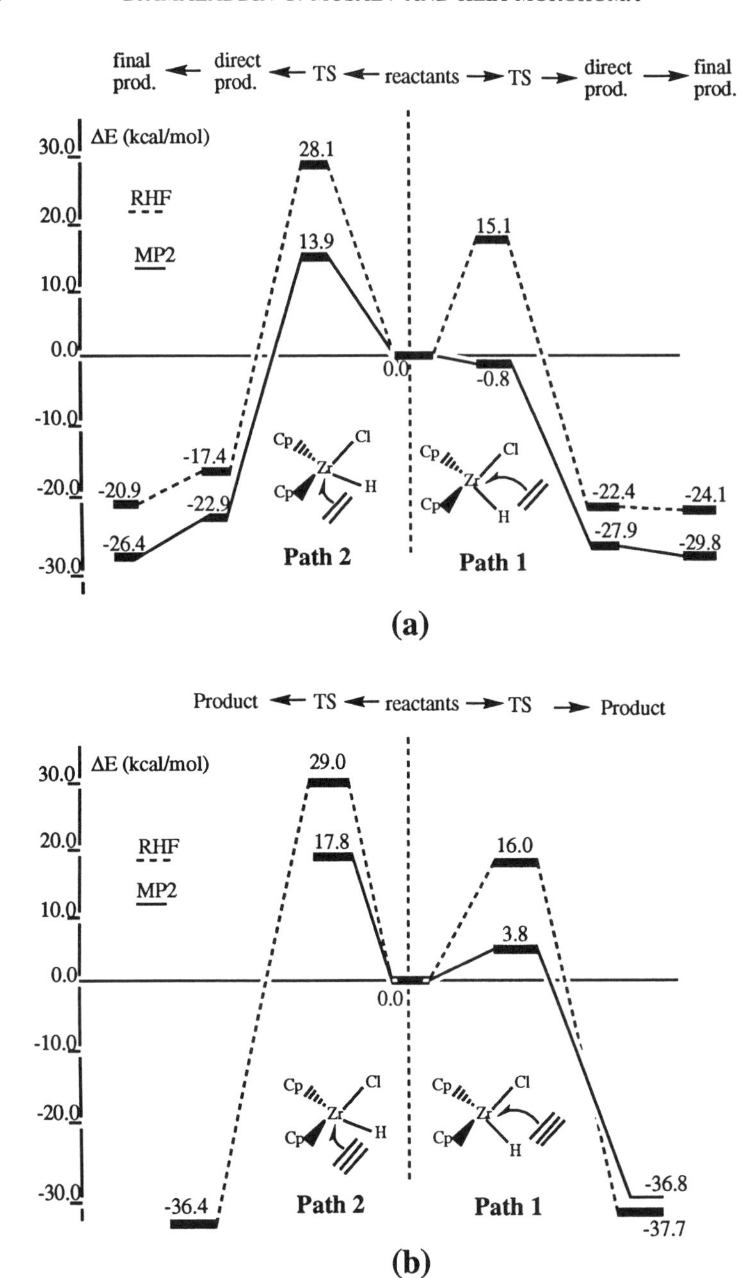

Figure 20. Relative potential energy profiles from the reactants for hydrozirconation of (*a*) ethylene and (*b*) acetylene [64].

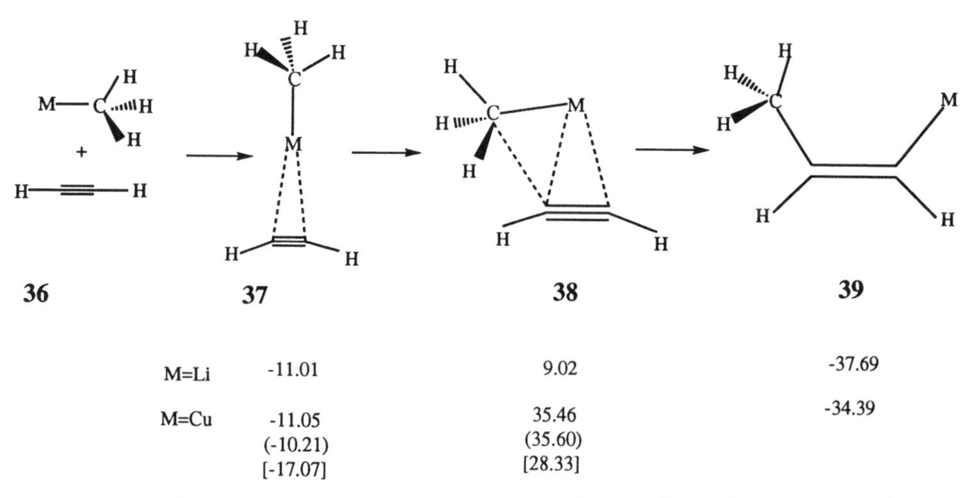

Figure 21. Addition of MeM to acetylene. Energies (in kilocalories per mole), relative to reactants, are at the HF/3-21G, HF/6-31G*//HF/3-21G (in parentheses), and MP2/6-31G*//HF/3-21G (in brackets) levels [65].

the reaction with parent acetylene. The substitution of one H atom by OH, SH, and SiH_3 groups in acetylene only slightly affects the structures, energetics, and mechanism of the reaction with both MeCu and MeLi.

A. Summary

In this section we reviewed recent theoretical studies of olefin and acetylene insertion into metal–ligand bonds of transition-metal complexes. Investigations of the reaction of ethylene with bridged and unbridged metallocenes have shown that the monosilane bridge opens up the coordination site of the complex by effectively "pulling back" the Cp ring and increases the electron deficiency of the metal center. Although the constrained geometry catalyst has a more open coordination site than the corresponding metallocenes, and potentially less steric hindrance to insertion, it actually increases the barrier. Qualitative analysis of the insertion of alkenes into M–R bonds shows that the interaction between occupied π orbitals of alkenes and vacant d orbitals of metal makes the activation barrier low. Therefore, the early transition-metal complexes with vacant d orbitals, as well as the complexes with electron withdrawing ligands, require a lower activation barrier.

V. FULL CATALYTIC CYCLES

The full catalytic cycle can be considered to be a series or multiple series in different order of many "elementary" reactions, each of which can consist of several alternative or parallel paths involving different isomeric species. A careful molecular orbital study can unveil detailed structural and energetic changes associated with these individual steps and can provide insight into the mechanism of the reactions. However, since there are nearly infinite possibilities of such series, it is impossible to theoretically explore all of them and determine *a priori* the mechanism of the catalytic cycle. However, one can select a few reasonable paths of reactions or "proposed mechanisms" and perform molecular orbital studies for each step of reactions or "proposed mechanisms" and perform molecular orbital studies for each step of reactions involved and examine whether these paths are reasonable or not from a theoretical point of view. Recently, the results of *ab initio* studies have been published for three homogeneous catalytic cycles: hydroboration, hydroformylation, and silastannation with transition-metal complexes.

A. Olefin Hydroboration Catalytic Cycle by $Rh(PR_3)_2Cl$

Recently Musaev, Mebel, and Morokuma (MMM) [67, 68] and Dorigo and Schleyer (DS) [69] have presented the results of *ab initio* study of the mechanism of Rh(I)-catalyzed olefin hydroboration catalytic cycle. These studies have been prompted by a recent surge of experiments in this field [70–73]. A number of transition-metal complexes have been found to catalyze the olefin hydroboration with catecholborane, HBcat (where cat = 1,2-$O_2C_6H_4$) and TMDB (4,4,6-trimethyl-1,3,2-dioxaborinane). This process has shown a variety of promising features, including regio-, diastereo-, and chemoselectivity, and has been attracting a substantial interest in connection to organic synthesis [70–73]. In general, it has been found that (1) the reductive elimination step is the slowest step in the overall transformation, (2) Rh complexes are most suitable catalysts, among those the Wilkinson catalyst appears to be the most efficient, (3) boron hydrides bearing oxygen ligands are the most successful reagents, and (4) the rate of catalyzed hydroboration reaction is very sensitive to the olefin substitution pattern, with terminal alkenes more reactive than highly substituted olefins.

However, the mechanism of this practically important catalytic reaction remains unclear. The mechanism proposed in early papers [70, 71] for the Wilkinson catalyst involves oxidative addition of a B–H bond to the metal center, followed by olefin coordination to the metal center accompanied with dissociation of one of the two PPh_3's, further followed by migratory insertion of olefin into the M–H bond and subsequent reductive elimination of the B–C bond. However, still unresolved are several important questions: (1) whether the reaction occurs with phosphine dissociation or not, (2) which of M–B and M–H bond insertions of olefin is energetically more favorable, and (3) how competitive is the "σ-bond metathesis" pathway involving coordination of the HBcat and olefin to the complex followed by simulta-

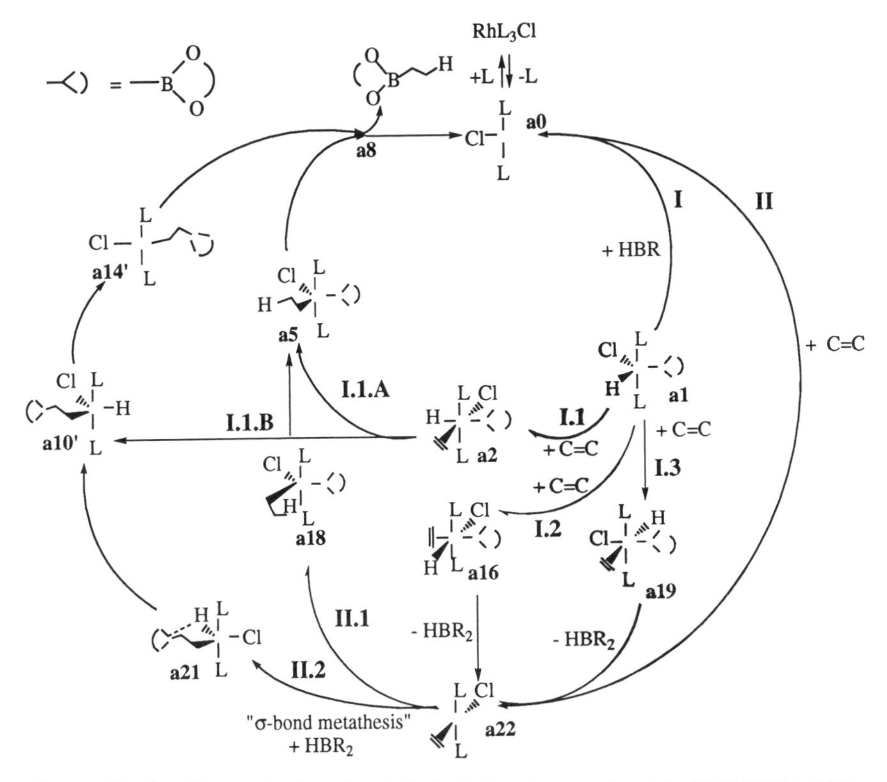

Figure 22. Possible mechanisms for olefin hydroborations mediated by $RhCl(PPH_3)_2$ [67].

neous cleavage of the M–C and B–H bonds with formation of the M–H and B–C bonds. Important proposed mechanisms of the rhodium(I)-catalyzed olefin hydroboration are shown in Fig. 22.

Thus, detailed experimental and theoretical studies are highly desirable on the mechanism of the transition-metal-catalyzed olefin hydroboration reactions, as well as on the role of the transition-metal center, substrates, and electronic and steric factors in the mechanism. MMM [67] have presented the first detailed *ab initio* molecular orbital (MO) study of possible reaction pathways illustrated in Fig. 22 for the reaction of C_2H_4 with the boranes $HB(OH)_2$ and $HBO_2(CH_2)_3$ catalyzed by the model Wilkinson catalyst $RhCl(PH_3)_2$. The reaction of BH_3 with C_2H_4 catalyzed by the $Rh(PH_3)_2Cl$ have been studied by MMM [68] and DS [69].

The calculated overall potential energy profile (PEP) of the reaction with $HB(OH)_2$ is given in Fig. 23 [67]. The mechanism (**I.1.B**) has been found to be the most favorable pathway of the catalytic hydroboration of C_2H_4 by

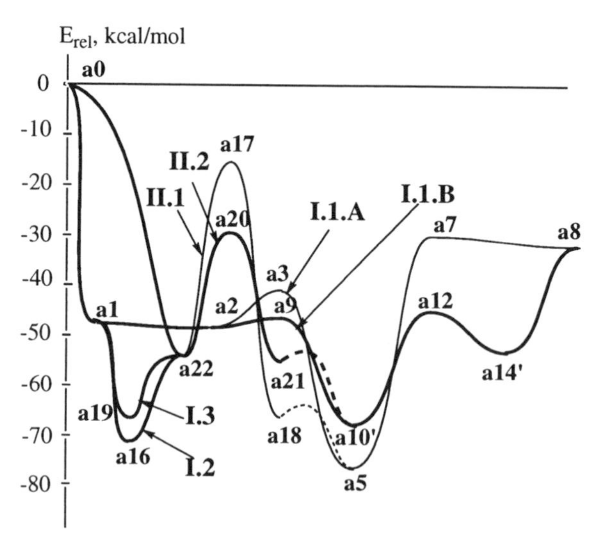

Figure 23. The overall profiles of the reaction $HB(OH)_2$ + C_2H_4 + $ClRh(PH_3)_2$ → $ClRh(PH_3)_2$ + $C_2H_5B(OH)_2$. Labels such as **I.1.A** refer to different mechanisms, and labels such as **a1** refer to transition states and intermediates. The bold curves show the most favorable reaction mechanisms. The dashed curves mean that the transition states were not calculated, but the barriers are expected to be low [67].

$HB(OH)_2$ with the model Wilkinson catalyst, $RhCl(PH_3)_2$. Since the PEP for the (**I.1.B**) mechanism for $HBO_2(CH_2)_3$ is nearly quantitatively the same as that for $HB(OH)_2$, the same conclusion should be applicable for reactions of real boranes experimentally studied. It involves oxidative addition of the B–H bond of borane to the catalyst (**a0** → **a1**), followed by coordination of olefin to the complex between B and H ligands (→ **a2**). The reaction further proceeds by insertion of C=C into the Rh–B bond (→ **a10′**), followed by the coupling of H and $C_2H_4BR_2$—that is, dehydrogenative reductive elimination of $C_2H_5BR_2$ to give the product complex, **a14′**, and eventual dissociation of $C_2H_5BR_2$. The activation energy for the last two steps of the mechanism, reductive elimination and dissociation, is calculated to be about 20 kcal/mol. Since in solution the endothermicity for the dissociation would be reduced by solvating of the regenerated catalyst, the coupling of H and $C_2H_4BR_2$ should be the rate-determining step. The conclusion agrees with experimental observation that the reductive elimination step is the slowest in overall transformation [71]. The coordination of olefin to the complex **a1** between Cl and H ligands or between Cl and B ligands gives a stable complex, **a16** or **a19,** and cannot proceed further without reductively eliminating borane (→ **a22**) without dissociating olefinback to **a1.**

The other competitive mechanism, (**I.2**), begins with coordination of olefin to the catalyst, (**a0** → **a22**). The next step is "σ-bond metathesis"—that is, coordination of borane to the complex, **a22,** accompanied by simultaneous cleavage of Rh–C and B–H bonds with formation of B–C and Rh–H bonds. After an internal rotation (**a21** → **a10′**), which does not require high activation energy, dehydrogenative reductive elimination of $C_2H_5BR_2$ takes place. The final steps for the mechanism (**II.2**) coincide with those for the mechanism (**I.1.B**), and the rate-determining barrier for (**II.2**) corresponding to the metathesis process, 23.9 kcal/mol, is not much higher than the barrier for (**I.1.B**), 22.4 kcal/mol. However, for the (**II.2**) pathway, the system has to overcome both of these barriers.

In the complex **a2,** the olefin has a choice of inserting into the Rh–B bond or the Rh–H bond. The latter process (**I.1.A, a2** → **a5** → **a8**) requires higher barriers than the former (**I.1.B**) discussed above. The insertion into the Rh–B bond (mechanism (**II.2**)) is preferred to the Rh–H bond (mechanism (**II.1**)) also in the σ-bond metathesis pathway.

MMM [68] have also studied the reaction mechanism for $RhCl(PH_3)_2$ + C_2H_4 + BH_3 → $RhCl(PH_3)_2$ + $C_2H_5BH_2$ catalytic reaction without dissociation of phosphine ligands and have found that (i) the mechanism involving olefin insertion into the Rh–B bond is a few kilocalories per mole more favorable than that for insertion into Rh–H bond and (ii) in the preferable pathway, BH_3 reacts with the catalyst before olefin does. Thus, for this reaction occurring *without* dissociation of a PH_3 group the initial formation of a C–B bond is more favorable than initial formation of a C–H bond. DS [69] have studied the mechanism of this reaction *with* dissociation of one of PH_3 ligands upon coordination of olefin and have shown that the insertion of olefin into the Rh–H bond is a few kilocalories per mole more favorable than that into the Rh–B bond.

B. Silastannation of Acetylene with a Palladium Catalyst

Recently, Hada et al. [74] have reported a theoretical study on the reaction mechanism and regioselectivity of silastannation of acethylene with a palladium catalyst. Experimentally, terminal acetylenes react with silylstannanes to give highly regio- and stereoselective 1-silyl-2-stannylalkenes with tetrakis(triphenylphosphine)palladium as a catalyst [75]. The products are always *cis* adducts; tin adds to the internal position as follows:

$$R{-}C{\equiv}C{-}H + Me_3Si{-}SnMe_3 \xrightarrow{Pd(PPh_3)_4} (R)(Me_3Sn)C{=}C(H)(SiMe_3) \quad (V.1)$$

where R is a phenyl, alkyl, or silyl group. Alkoxyacetylene also reacts with various silylstannanes to give regio- and stereoselective 1-silyl-2-stannylalkenes with the palladium acetate-tert-alkyl isocyanidffe catalyst [76]:

$$\mathrm{EtO{-}C{\equiv}C{-}H} + \mathrm{R_3Si{-}SnR_3} \xrightarrow[\text{tert-alkyl-NC}]{\mathrm{Pd(OAc)_4}} \mathrm{(EtO)(R_3Si)C{=}CH(SnR_3)} \tag{V.2}$$

However, the regioselectivity in this reaction is opposite to that in Eq. (V.1). Hada et al. [74] have adopted the mechanism of the reaction given in Fig. 24. In this scheme, silylstannane, $H_3Si–SnH_3$, oxidatively adds to the $Pd(PR_3)_2$ complex, then acetylene coordinates with dissociation of one of PR_3 ligands and inserts into the Pd–Sn bond, and finally the SiH_3 group on Pd and the C=C group are coupled and reductively eliminated from the Pd.

Calculation has been performed mainly at the HF level with the double-zeta plus polarization quality basis sets and ECP for Pd, P, Si, and Sn. The energetics for R = Me has been recalculated at the MPn (n = 2–4) level at the HF geometries. The overall reactions for various RCCH (R = CN, H, CH_3, and OCH_3) and H_3SiSnH_3 with the model catalyst $Pd(PH_3)_2$ have been studied. It has been shown that $H_3Si–SnH_3$ easily, with a few kilocalories per mole barrier, adds oxidatively to the catalyst $Pd(PH_3)_2$ and the resultant complex lies only 5.9 kcal/mol lower than the reactants. The next step, acetylene coordination to the catalyst, causes with the dissociation of

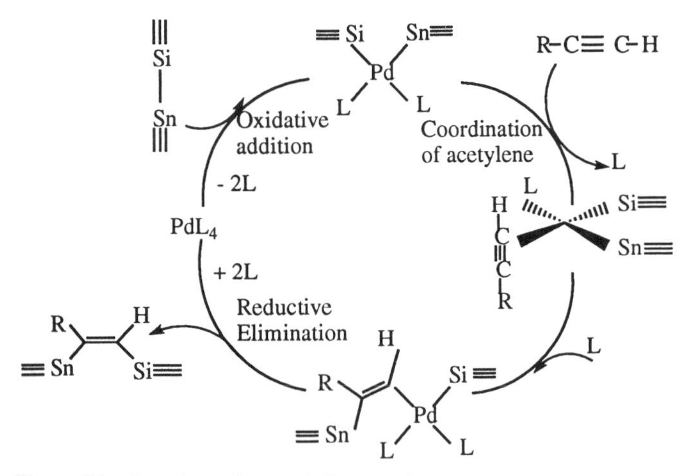

Figure 24. Reaction scheme of silastannation adopted by Hada et al. [74].

one of the PH_3 groups and is only 2.6 kcal/mol exothermic. The rate-determining step is the insertion of acetylene into Pd–Sn or Pd–Si bonds, which is calculated to have a barrier of 16–24 or 23–37 kcal/mol, respectively, that increases in the order of R: CN < OCH_3 < H < CH_3. The entire reaction for R = CH_3 is calculated to be exothermic by 21 kcal/mol.

Three factors are shown to be important in determining the reactivity and regioselectivity of this reaction:

(a) *Electronic Effects.* (i) The orientation of acetylene is determined by the electron-donating interaction from the HOMO (π) of acetylene to the LUMO of the Pd complex which is localized on Sn. Monosubstituted acetylene prefers to orient with the terminal carbon to Sn. (ii) The back-donating interaction from the HOMO of Pd complex, which is mainly the *d* orbital of Pd, to the LUMO (π^*) of acetylene is energetically important but is insensitive to the orientation of acetylene. (iii) This interaction controls the reactivity of substituted acetylenes but does not cause the regioselectivity. The insertion of acetylene into Pd–Sn is preferable to that into Pd–Si because of the larger electrophilicity of Sn.

(b) *Steric Repulsion.* The steric repulsion between the PPh_3 ligand of $Pd(PPh_3)_4$ complex and CH_3 in CH_3CCH is large enough to determine the orientation of the acetylene. Therefore, in the reaction using $Pd(PPh_3)_4$ as a catalyst, the steric effect controls the regioselectivity, so that Sn always adds to the internal carbon of the substituted acetylenes.

(c) *Thermodynamic Control.* This factor becomes important when the reaction intermediates after the rate-determining TS barrier are less stable than the initial compounds. Therefore, in the case of methylacetylene, the regioselectivity is determined by the relative stability of the intermediates after the TS.

C. Hydroformylation Catalytic Cycle by $RhH(CO)_2(PR_3)_2$

Another catalytic cycle studied by Matsubara, Morokuma, and coworkers [77] is the hydroformylation of olefin by an Rh(I) complex. Hydroformylation of olefin by the rhodium complex [78–80] is one of the most well known homogeneous catalytic reactions. Despite extensive studies made for this industrially worthwhile reaction [81, 82], the mechanism is still a point of issue. The active catalyst is considered to be $RhH(CO)(PPh_3)_2$, **47,** as presented in Fig. 25. The most probable reaction cycle undergoes CO addition and phosphine dissociation to generate an active intermediate **41.** The intramolecular ethylene insertion, CO insertion, H_2 oxidative addition, and aldehyde reductive elimination are followed as shown with the surrounding dashed line. Authors have optimized the structures of nearly all the relevant transition states as well as the intermediates to determine the full potential-

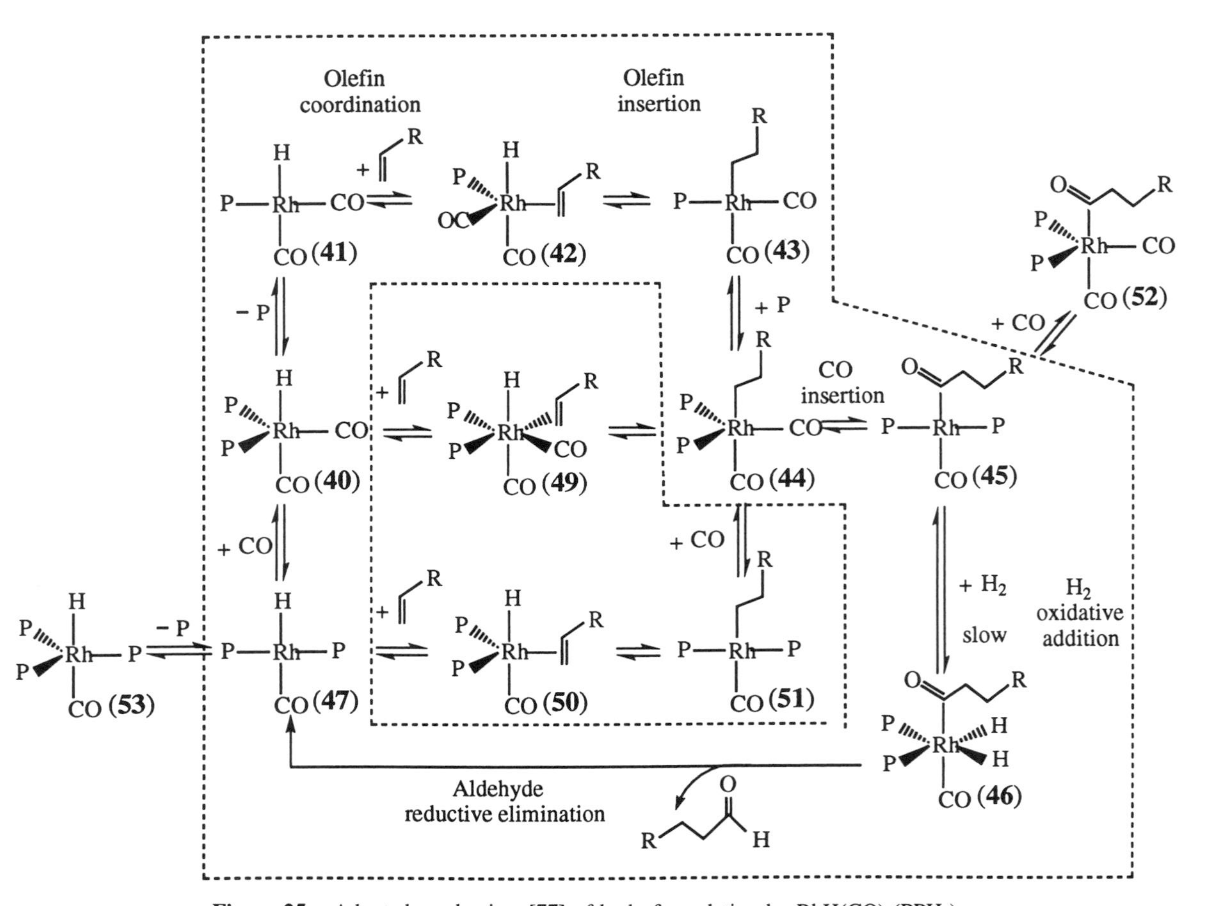

Figure 25. Adopted mechanism [77] of hydroformylation by $RhH(CO)_2(PPH_3)_2$.

energy surface and have shown that the solvent plays a critical role in determination of the potential-energy profile. They have chosen RhH-(CO)$(PH_3)_3$ as a model catalyst and have adopted ethylene as the reacting olefin.

At first the structure of the intermediates and the transition states involved in the catalytic cycle have been determined at the RHF level. Although the aldehyde-coordinate intermediate **48** is not included in the catalytic cycle proposed by Wilkinson and coworkers [78–80], it has been found to exist between **46** and **47.** The energetics has then been determined at the MP2 level at the RHF determined structures. The MP2 potential-energy profile for the entire catalytic cycle, presented in Fig. 26, is strange-looking. In addition to ethylene and CO insertion steps, the first phosphine dissociation and the last aldehyde dissociation step have a large barrier. The hydrogen oxidative addition, supposedly the rate-determining step, has a very small barrier. Experimentally the reaction takes place experimentally in solution, mostly with alkenes as solvent. Such solvent molecules can coordinate to coordinatively unsaturated intermediates to stabilize them. For a nonpolar solvent like alkene, the direct interaction of one solvent molecule (or a few) is much more important than the long-range interaction via a solvent reaction field. For instance, if a solvent ethylene molecule coordinates to the four-coordinate intermediate **41,** the solvated species will be only the ethylene complex **42.** Therefore, they have performed optimization of the 16e four-coordinate intermediates **41, 43, 45,** and **47** and the five-coordinate transition states **TS(42–43), TS(44–45), TS(45–46),** and **TS(46–47)** with one ethylene molecule as solvent. All 16e four-coordinate intermediates were by more than 20 kcal/mol, stabilized by the η^2-coordination of a solvent ethylene, with their square-planar structures transformed to the trigonal-bipyramidal structures. On the other hand, the five-coordinate transition states were stabilized only a few kilocalories per mole by the weak interaction of an ethylene molecule, with the Rh–ethylene bond distance of about 5 Å, without large change of the structure. Furthermore, the free-energy profile in ethylene solution has been estimated, by adding and subtracting 10 kcal/mol for entropy contribution for solvation, upon each coordination and decoordination, respectively.

The free-energy profile with solvent interaction taken into account is shown in Fig. 27. The feature of the entire potential surface is dramatically changed. The large barriers in the reaction from **40** to **42** and from **48** to **40** have disappeared because of the stabilization of the four-coordinate intermediates **41** and **47** by the solvation. The endothermic ethylene and CO insertion reactions became exothermic and the exothermic H_2 oxidative addition became endothermic, because the four-coordinate intermediate **43** and

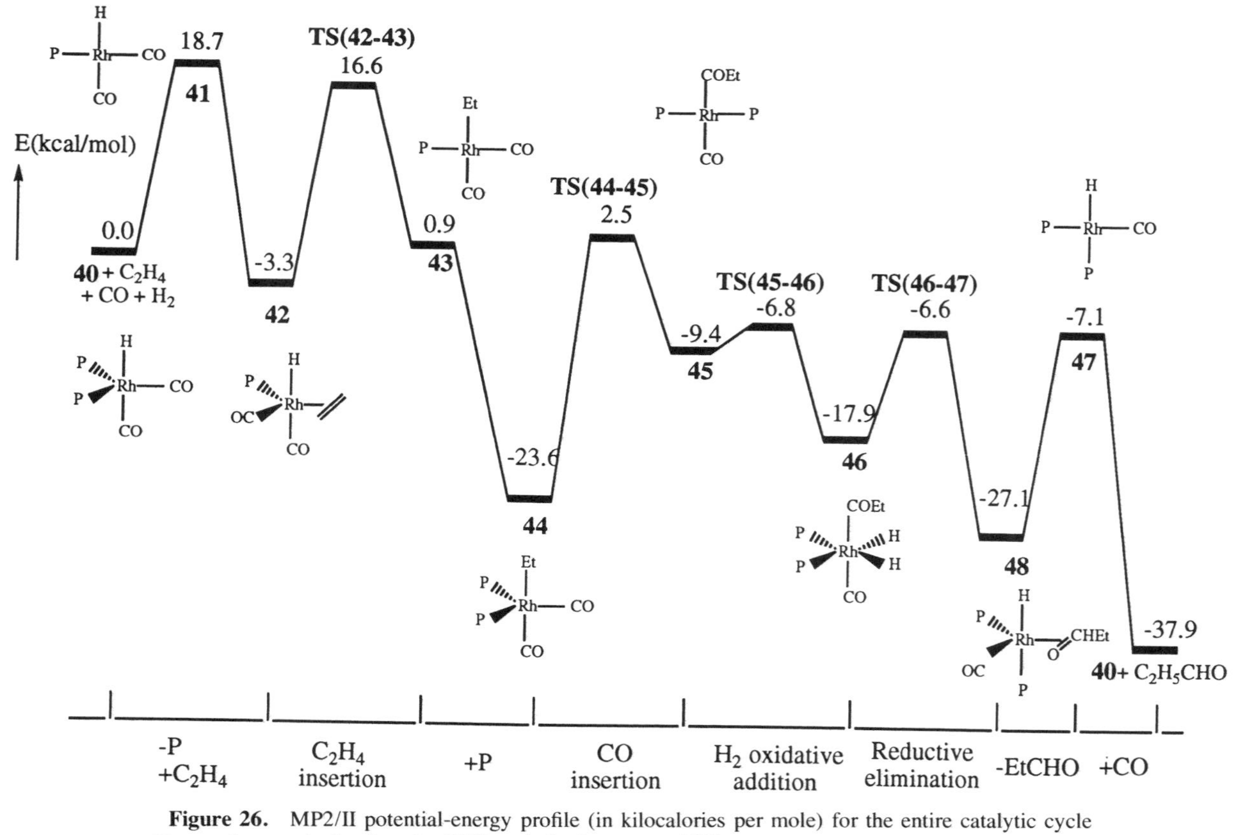

Figure 26. MP2/II potential-energy profile (in kilocalories per mole) for the entire catalytic cycle without solvent calculated at the HF/I optimized structure [77].

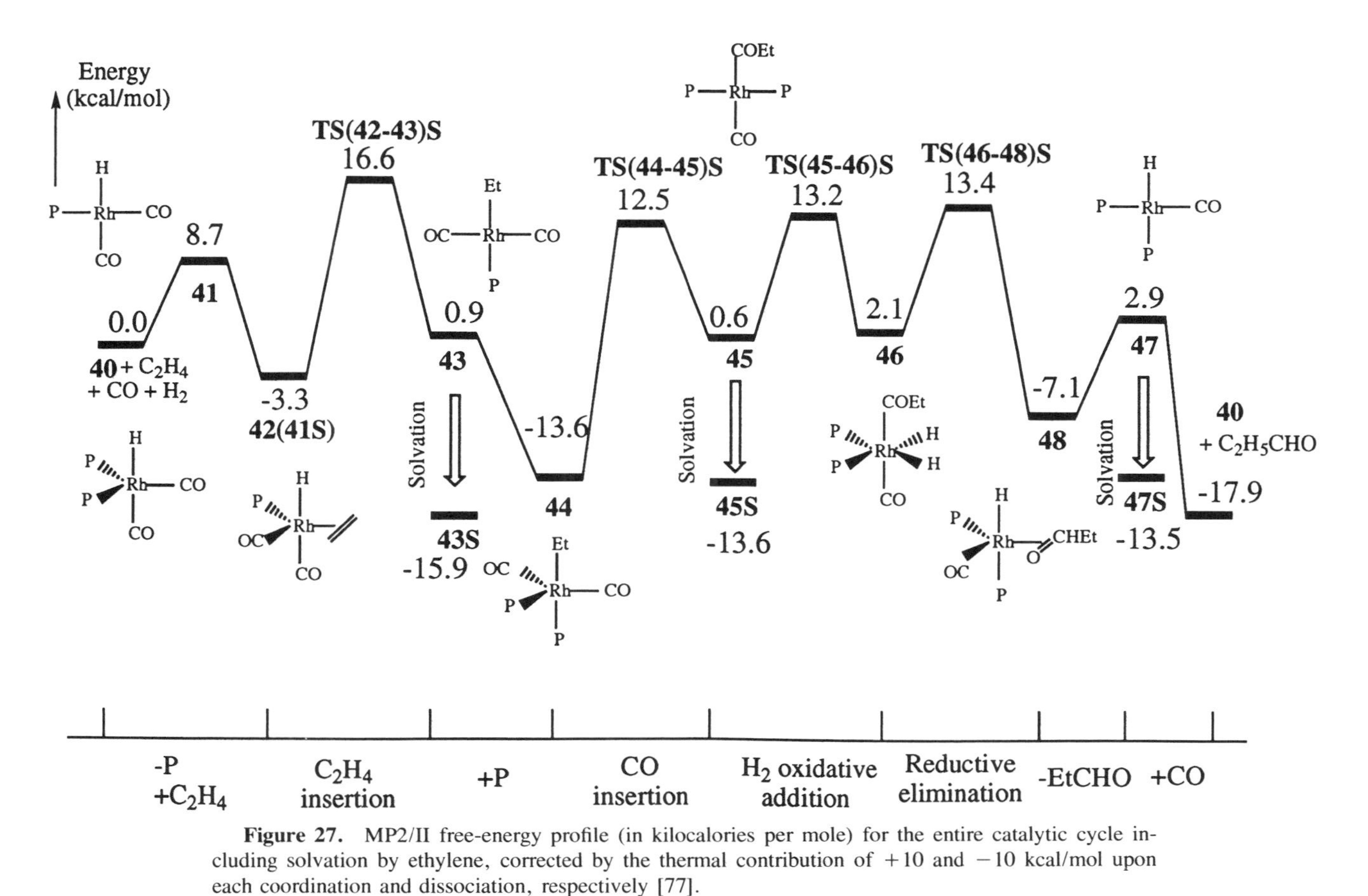

Figure 27. MP2/II free-energy profile (in kilocalories per mole) for the entire catalytic cycle including solvation by ethylene, corrected by the thermal contribution of +10 and −10 kcal/mol upon each coordination and dissociation, respectively [77].

45 were also stabilized by solvation. The largest barrier is at the H_2 oxidative addition step, in agreement with the experimental proposal that the H_2 oxidative addition is rate-determining, and consists of the desolvation energy of the four-coordinate intermediate **45** and the activation energy for H_2 addition to the desolvated intermediate. There are relatively large barriers at the ethylene insertion and the carbonyl insertion step.

D. Summary

In this section we reviewed the results of the latest *ab initio* studies of the full catalytic cycles: (a) Rh(I)-catalyzed olefin hydroboration and hydroformylation and (b) Pd(0)-catalyzed acetylene silastannation. For the Rh(I)-catalyzed olefin hydroboration occurring without dissociation of a PPh_3 ligand, the initial formation of a C–B bond is more favorable than that of a C–H bond. This favorable mechanism invovles the oxidative addition of the B–H bond of borane to the catalyst, followed by the coordination of olefin to the complex between B and H ligands, the insertion of C=C into the Ph–B bond, and the reductive elimination of $C_2H_5BR_2$ to give the product complex. Although the last two steps, reductive elimination and dissociation, have similar calculated activation energy, in solution the endothermicity for the dissociation step is expected to be reduced by solvation of the regenerated catalyst and thus the coupling of H and $C_2H_4BR_2$ should be the rate-determining step.

For Rh(I)-catalyzed olefin hydroformylation reaction the CO insertion step takes place via a trigonal-pyramidal transition state, while the olefin insertion proceeds via a square-pyramidal transition state. In the trigonal-bipyramidal transition states for H_2 oxidative addition and aldehyde reductive elimination, as well as in trigonal-bipyramidal equilibrium structures of intermediates, the axial preference of hydride has been demonstrated. The effect of solvation by alkene solvent to coordinatively unsaturated four-coordinate intermediates drastically changes the shape of the entire energy profile. The estimated free-energy profile of the entire catalytic cycle indicates that the H_2 oxidative addition step is rate-determining and that about a half of the barrier of this step comes from the desolvation of a resting stage intermediate.

The reaction of silastannation of acetylene catalyzed by a Pd complex involves the oxidative addition of the $H_3Si–SnH_3$ to the catalyst $Pd(PR'_3)_2$, followed by coordination of acetylene to the catalyst accompanied with dissociation of one of the PR'_3 bond and reductive elimination of the product. The rate-determining step of the reaction is the insertion of substituted acetylene RCCH into the Pd–Sn or Pd–Si bond, the activation energy of which increases in the order of R: $CN < OCH_3 < H < CH_3$.

VI. PERSPECTIVES OF QUANTUM CHEMICAL STUDIES OF ORGANOMETALLIC REACTIONS AND HOMOGENEOUS CATALYSIS

In this review we summarized the results of the latest *ab initio* studies of the elementary reaction such as oxidative addition, metathesis, and olefin insertion into metal–ligand bonds, as well as the multistep full catalytic cycles such as metal-catalyzed hydroboration, hydroformylation, and silastannation. In general, it has been demonstrated that quantum chemical calculations can provide very useful information concerning the reaction mechanism that is difficult to obtain from, and often complementary to, experiments. Such information includes the structures and energies of unstable intermediates and transition states, as well as prediction of effects of changing ligands and metals on the reaction rate and mechanism.

However, the desired level of calculation (i.e., geometry optimization at low correlated levels such as MP2 and the energetics at higher correlated levels such as coupled cluster) for real compounds studied experimentally is quite expensive and often impossible due to computational limitation. Therefore, almost all quantum chemical calculations have been performed for "model" systems with simplified ligands (e.g., with PH_3 instead of PPh_3) and without consideration of solvent effects. Therefore, comparison of the calculated results with experiments had to be indirect and qualitative.

One possible solution to this problem is a combination of the MO and the empirical molecular mechanics (MM), treating the active and important part of the molecule with the MO method and the remainder, such as bulky substituents or other chemical environments, with the MM method. Such treatments have been made for some organometallic systems [83–86]. However, in these treatments, only the geometry of the MM part is optimized under the assumption that the MO part is frozen at the optimized geometry of the small model system. This frozen assumption can result in a substantial overestimation of the MM energy. Recently, Maseras and Morokuma have proposed a new integrated MO + MM scheme, called IMOMM, in which both the MO part and the MM part of geometry are simultaneously optimized [87]. This method can combine any MO approximation with any molecular mechanics force field. The application of this method at the IMOMM(HF : MM3) and IMOMM(MP2 : MM3) levels to the oxidation addition reaction of H_2 to $Pt(PR_3)_2$, where R = H, Me, *t*-Bu, and Ph [88], has shown a promise that more realistic models of elementary reactions and catalytic cycles may be studied in the near future with this method.

Another important improvement of the methodology seems to come from the Density Functional Theory (DFT). Recent extension of the correlation-correction functions and nonlocal gradient correction have dramatically improved the reliability of the density functional approach [89–91]. Recent studies of transition-metal complexes [92–94] at modified DFT levels show a good agreement with experiment and more sophisticated methods in geometries and, in some cases, in bonding energies. DFT methods are much cheaper than *ab initio* correlated methods, and we should be able to use them as an alternative for low-level correlated calculations to study more realistic models of organometallic reactions and processes. IMOMM(DFT : MM) approaches may also be very promising.

With the above-mentioned new developments, theoretical calculations should be able to use more realistic models of elementary reactions and homogeneous catalysts and utilize many important aspects of transition-metal-catalyzed reactions, including the role of different ligands, the role of different transition-metal atoms, chemo-, and stereoselectivity, and other factors controlling the reactions. Predictions of catalytic reactivities based on theoretical calculations should be forthcoming. Collaboration between experimentalists and theoreticians is strongly encouraged.

References

1. E. L. Muetterties, *Chem. Soc. Rev.* **12,** 283 (1983).
2. A. E. Shilov, *Activation of Saturated Hydrocarbons by Transition Metal Complexes* (Reidel, Dordrecht, Holland, 1984).
3. R. H. Crabtree, *The Organometallic Chemistry of the Transition Metals* (John Wiley & Sons, New York, 1988).
4. A. Yamamoto, *Organotransition Metal Chemistry* (John Wiley & Sons, New York, 1986).
5. J. P. Collman, L. S. Hegedus, J. R. Norton, and R. G. Finke, *Principles and Applications of Organotransition Metal Chemistry* (University Science Books, Mill Valley, CA, 1987).
6. *Transition Metal Hydrides*, A. Dedieu (ed.) (VCH Publishers, New York, 1992).
7. P. E. M. Siegbahn and M. R. A. Blomberg, in *Theoretical Aspects of Homogeneous Catalysts, Applications of Ab Initio Molecular Orbital Theory*, P. W. N. M. van Leeuwen, J. H. van Lenthe, and K. Morokuma (eds.) (Kluwer Academic Publishers, Hingham, MA, 1995).
8. N. Koga and K. Morokuma, *Chem. Rev.* **91,** 823 (1991).
9. S. Yoshida, S. Sakaki, H. Kobayashi, *Electronic Processes in Catalyst* (VCH, New York, 1994).
10. (a) P. B. Chock and J. Halpern, *J. Am. Chem. Soc.* **88,** 3511 (1966). (b) J. Halpern, *Acc. Chem. Res.* **3,** 386 (1970). (c) L. Vaska, *Acc. Chem. Res.* **1,** 335 (1968). (d) R. Ugo, A. Pasini, A. Fusi, and S. Cenini, *J. Am. Chem. Soc.* **94,** 7364 (1972). (e). J. P. Collman, *Acc. Chem. Res.* **1,** 136 (1968). (f) M. J. Burk, M. P. McGrath, R. Wheeler, and R. H. Crabtree, *J. Am. Chem. Soc.* **11,** 5034 (1988). (g) Y. Jean and A. Lledos, *Nouv. J. Chim.* **10,** 635 (1986).
11. (a) B. J. Fisher, R. Eisenberg, *Inorg. Chem.* **23,** 3216 (1984). (b) C. E. Johnson, B. J. Fisher, and R. Eisenberg, *J. Am. Chem. Soc.* **105,** 7772 (1983). (c) C. E. Johnson and R. Eisenberg, *J. Am. Chem. Soc.* **107,** 6531 (1985). (d) C. E. Johnson and R. Eisenberg,

J. Am. Chem. Soc. **107,** 3148 (1985). (e) S. B. Duckett, R. Eisenberg, *J. Am. Chem. Soc.* **115,** 5292 (1993).

12. A. L. Sargent, M. B. Hall, and M. F. Guest, *J. Am. Chem. Soc.* **114,** 517 (1992).
13. A. L. Sargent and M. B. Hall, *Inorg. Chem. Soc.* **31,** 317 (1992).
14. (a) F. Abu-Hasanayn, K. Krogh-Jespersen, and A. S. Goldman, *Inorg. Chem.* **32,** 495 (1993). (b) F. Abu-Hasanayn, A. S. Goldman, and K. Krogh-Jespersen, *Inorg. Chem.* **33,** 5122 (1994). (c) F. Abu-Hasanayn, A. S. Goldman, and K. Krogh-Jespersen, *J. Phys. Chem.* **97,** 5890 (1993).
15. G. S. Hammord, *J. Am. Chem. Soc.* **77,** 334 (1955).
16. D. G. Musaev and K. Morokuma, *J. Org. Chem.* **504,** 93 (1995).
17. N. Koga, K. Morokuma, *J. Am. Chem. Soc.* **115,** 6883 (1993).
18. (a) R. G. Bergman, *Science,* **223,** 902 (1984). (b) R. H. Crabtree, *Chem. Rev.* **85,** 245 (1985). (c) J. Halpern, *Inorg. Chim. Acta* **41,** 100 (1985). (d) W. D. Jones and F. J. Feher, *Acc. Chem. Res.* **22,** 91 (1989). (e) A. H. Janowicz and R. G. Bergman, *J. Am. Chem. Soc.* **104,** 352 (1982). (f) A. H. Janowicz and R. G. Bergman, *J. Am. Chem. Soc.* **105,** 3429 (1983). (g) R. H. Crabtree, M. F. Mellea, J. M. Mihelcie, and J. M. Quick, *J. Am. Chem. Soc.* **104,** 107 (1982). (h) J. K. Hoyano and W. A. G. Graham, *J. Am. Chem. Soc.* **104,** 3723 (1982). (i) W. D. Jones and F. J. Feher, *J. Am. Chem. Soc.* **104,** 4240 (1982). (j) D. M. Roundhill, *Chem. Rev.* **92,** 1 (1992). (k) R. H. Crabtree, in *Selective Hydrocarbon Activation: Principles and Progress*, J. A. Davies et al. (eds.) (VCH Publishers, New York, 1990), p. 1.
19. (a) M. R. A. Blomberg, P. E. M. Siegbahn, and M. Svensson, *New J. Chem.* **15,** 727 (1991). (b) M. R. A. Blomberg, P. E. M. Siegbahn, and M. Svensson, *J. Phys. Chem.* **95,** 4313 (1991). (c) P. E. M. Siegbahn, M. R. A. Blomberg, and M. Svensson, *J. Phys. Chem.* **97,** 2564 (1993). (d) P. J. Hay, in *Transition Metal Hydrides*, A. Dedieu (ed.) (VCH Publishers, (1992) p. 127. (e) B. C. Guo, K. P. Kerns, and A. W. Castleman, Jr., *J. Phys. Chem.* **96,** 4879 (1992).
20. E. P. Wasserman, C. B. Moore, and R. G. Bergman, *Science,* **255,** 315 (1992).
21. R. H. Schultz, A. A. Bengali, M. J. Tauber, B. H. Weiller, E. P. Wasserman, K. R. Kyle, C. B. Moore, and R. G. Bergman, *J. Am. Chem. Soc.* **116,** 7369 (1994).
22. P. Hofmann and M. Padmanabhan, *Organometallics* **2,** 1273 (1983).
23. T. Ziegler, V. Tschinke, L. Fan, and A. D. Becke, *J. Am. Chem. Soc.* **111,** 9177 (1989).
24. J. Song and M. B. Hall, *Organometallics* **12,** 3118 (1993).
25. D. G. Musaev and K. Morokuma, *J. Am. Chem. Soc.* **117,** 799 (1995).
26. *CRC Handbook of Chemistry and Physics*, 72nd edition (CRC Press, Boca Raton, FL, 1991–1992).
27. P. E. M. Siegbahn and M. Svensson, *J. Am. Chem. Soc.* **116,** 10124 (1994).
28. J. P. Hay, *New J. Chem.* **15,** 735 (1991).
29. (a) S. Sakaki and M. Ieki, *J. Am. Chem. Soc.* **115,** 2373 (1993). (b) S. Sakaki, M. Ogawa, Y. Musashi, and T. Arai, *Inorg. Chem.* **33,** 1660 (1994). (c) S. Sakaki and M. Ieki, *J. Am. Chem. Soc.* **113,** 5063 (1991).
30. G. J. Kubas, R. R. Ryan, B. I. Swanson, P. J. Vergamini, and H. J. Wasserman, H. J. *J. Am. Chem. Soc.* **106,** 451 (1984).
31. See, for instance, the following reviews and the references therein: (a) G. J. Kubas, *Acc. Chem. Res.* **21,** 120 (1988). (b) R. H. Crabtree, D. G. Hamilton, *Adv. Organomet. Chem.* **28,** 299 (1988). (c) R. H. Crabtree, *Acc. Chem. Res.* **23,** 95 (1990). (d) P. G.

Jessop and R. H. Morris, *Coord. Chem. Rev.* **121,** 155 (1992). (e) R. H. Crabtree, *Angew. Chem. Int. Ed. Engl.* **32,** 767 (1993). (f) D. M. Heinekey, and W. J. Oldham, Jr., *Chem. Rev.* **93,** 913 (1993). (g) W. T. Klooster, T. F. Koetzle, G. Jia, T. P. Fong, R. H. Morris, and A. Albinati, *J. Am. Chem. Soc.* **116,** 7677 (1994).

32. See, for instance, the following papers and references therein: (a) D. G. Musaev and O. P. Charkin, *Russ. J. Inorg. Chem.* **35,** 389 (1990). (b) D. G. Musaev, V. D. Makhaev, and O. P. Charkin, *Koord. Chem. (Russ.)* **16,** 749 (1990). (c) D. G. Musaev and O. P. Charkin, *J. Struct. Chem. (Russ.)* **30,** 11 (1989). (d) D. G. Musaev, O. P. Charkin, *J. Struct. Chem. (Russ.)* **31,** 8 (1990). (e) D. G. Musaev and O. P. Charkin, *Russ. J. Inorg. Chem.* **34,** 1372 (1989). (f) Y. Jean, O. Eisenstein, F. Volatron, B. Maouche, and F. Sefra, *J. Am. Chem. Soc.* **108,** 6587 (1986). (g) P. J. Hay, *J. Am. Chem. Soc.* **109,** 705 (1987). (h) J. K. Burdett and M. R. Pourian, *Inorg. Chem.* **27,** 4445 (1988). (j) J. Eckert, G. J. Kubas, J. H. Hall, P. J. Hay, and C. M. Boyle, *J. Am. Chem. Soc.* **112,** 2324 (1990). (l) F. Maseras, M. Duran, A. Lledos, and J. Bertran, *J. Am. Chem. Soc.* **113,** 2879 (1991). (k) F. Maseras, M. Duran, A. Lledos, and J. Bertran, *J. Am. Chem. Soc.* **114,** 2922 (1992).

33. G. C. Hlatky and R. H. Crabtree, *Coord. Chem. Rev.* **65,** 1 (1985).

34. M. A. Esteruelas, E. Sola, L. A. Oro, U. Meyer, and H. Werner, *Angew. Chem. Int. Ed. Engl.* **27,** 1563 (1988).

35. T. J. Johnson, K. G. Huffman, S. A. Caulton, S. A. Jackson, and O. Eisenstein, *Organometallics* **8,** 2073 (1989).

36. (a) Z. Lin and M. B. Hall, *J. Am. Chem. Soc.* **114,** 2928 (1992). (b) Z. Lin and M. B. Hall, *J. Am. Chem. Soc.* **114,** 6102 (1992). (c) Z. Lin and M. B. Hall, *J. Am. Chem. Soc.* **114,** 6574 (1992). (d) Z. Lin and M. B. Hall, *Inorg. Chem.* **31,** 4262 (1992). (e) Z. Lin and M. B. Hall, *Inorg. Chem.* **30,** 2569 (1991).

37. G. R. Haynes, R. L. Martin, and P. J. Hay, *J. Am. Chem. Soc.* **114,** 28 (1992).

38. (a) F. Maseras, N. Koga, and K. Morokuma, *J. Am. Chem. Soc.* **115,** 8313 (1993). (b) F. Maseras, X. K. Li, N. Koga, and K. Morokuma, *J. Am. Chem. Soc.* **115,** 10974 (1993).

39. F. Maseras, N. Koga, and K. Morokuma, *Organometallics* **13,** 4003 (1994).

40. F. Maseras, M. Duran, A. Lledos, and J. Bertran, *J. Am. Chem. Soc.* **114,** 2922 (1992).

41. (a) T. J. Katz, *Adv. Organomet. Chem.* **16,** 283 (1977). (b) N. Calderon, J. P. Lawrence, and E. A. Ofstead, *Adv. Organomet. Chem.* **17,** 449 (1979).

42. J. L. Herisson and Y. Chauvin, *Makromol. Chem.* **141,** 161 (1970).

43. (a) B. J. Burger, M. E. Thompson, W. D. Cotter, and J. E. Bercaw, *J. Am. Chem. Soc.* **112,** 1566 (1990). (b) M. E. Thompson, S. M. Buxter, A. R. Bulls, B. J. Burger, M. C. Nolan, B. D. Santarsiero, W. P. Schaefer, and J. E. Bercaw, *J. Am. Chem. Soc.* **109,** 203 (1987). (c) P. L. Watson, *J. Chem. Soc. Chem. Commun.* 276 (1983). (d) P. L. Watson and G. W. Parshall, *Acc. Chem. Res.* **18,** 51 (1985). (e) P. L. Watson, *J. Am. Chem. Soc.* **112,** 1566 (1983).

44. E. Folga and T. Ziegler, *Organometallics* **12,** 325 (1993).

45. T. Woo, E. Folga, and T. Ziegler, *Organometallics* **12,** 1289 (1993).

46. T. Ziegler, E. Folga, and A. Berces, *J. Am. Chem. Soc.* **115,** 636 (1993).

47. T. Ziegler and E. Folga, *J. Organomet. Chem.* **478,** 57 (1994).

48. E. Folga and T. Ziegler, *Can. J. Chem.* **70,** 333 (1992).

49. P. L. Watson, in *Selective Hydrocarbon Activation* J. A. Davis, P. L. Watson, J. F. Liebman, and A. Greenberg (eds.) (VCH Publishers, New York, 1990).

50. (a) W. Kaminsky, K. Kulper, H. H. Brintzinger, and F. R. W. P. Wild, *Angew. Chem. Int. Ed. Engl.* **31,** 1347 (1992). (b) R. F. Jordan, R. E. LaPoint, N. C. Baenziger, and G. D. Hinch, *Organometallics* **9,** 1539 (1990). (c) S. L. Borkowsky, R. F. Jordan, and G. D. Hinch, *Organometallics* **10,** 1268 (1991). (d) G. C. Hlatky, R. R. Ecsman, and H. W. Turner, *Organometallics* **11,** 1413 (1992). (e) J. A. Ewen, M. J. Elder, R. L. Jones, L. Haspeslagh, J. L. Atwood, S. G. Bott, and K. Robsinon, *Makromol. Chem. Macromol. Symp.* **48/48,** 253 (1991). (f) A. Zambelli, P. Long, and A. Grassi, *Macromolecules* **22,** 2186 (1989). (g) J. J. Eshuis, Y. Y. Tan, A. Meetsms, and J. H. Teuben, *Organometallics* **11,** 362 (1992). (h) X. Yang, C. L. Stern, and T. Marks, *J. Am. Chem. Soc.* **113,** 3623 (1991). (i) A. D. Horton and A. G. Orpen, *Organometallics* **10,** 3910 (1991). (j) J. C. W. Chien, W. M. Tsai, and M. D. Rausch, *J. Am. Chem. Soc.* **113,** 8570 (1991). (k) J. J. Eisch, K. R. Caldwell, S. Werner, and C. Kruger, *Organometallics* **10,** 3417 (1991). (l) W. Kaminsky and R. Steiger, *Polyhedron* **7,** 2375 (1988).
51. (a) P. Cossee, *J. Catal.* **3,** 80 (1964). (b) E. J. Arlman and P. Cossee, *J. Catal.* **3,** 99 (1964).
52. T. Yoshida, N. Koga, and K. Morokuma, *Organometallics* **14,** 746 (1995).
53. H. Kawamura-Kuribayashi, N. Koga, and K. Morokuma, *J. Am. Chem. Soc.* **114,** 2359 (1992).
54. H. Kawamura-Kuribayashi, N. Koga, and K. Morokuma, *J. Am. Chem. Soc.* **114,** 8687 (1992).
55. T. K. Woo, L. Fan, and T. Ziegler, *Organometallics* **13,** 2252 (1994).
56. T. K. Woo, L. Fan, and T. Ziegler, *Organometallics* **13,** 432 (1994).
57. L. Fan, D. Harrison, T. K. Woo, and T. Ziegler, *Organometallics* **14,** 2018 (1995).
58. H. Weiss, F. Haase, and R. Ahlrichs, *Chem. Phys. Lett.* **194,** 492 (1992).
59. H. Weiss, M. Ehrig, and R. Ahlrichs, *J. Am. Chem. Soc.* **116,** 4919 (1994).
60. S. Sakaki, M. Ogawa, Y. Musashi, and T. Arai, *J. Am. Chem. Soc.* **116,** 7258 (1994).
61. I. Hyla-Kryspin, S. Niu, and R. Gleiter, *Organometallics* **14,** 964 (1995).
62. P. E. M. Siegbahn, *J. Am. Chem. Soc.* **115,** 5803 (1993).
63. P. de Vaal and A. Dedieu, *J. Organomet. Chem.* **478,** 121 (1994).
64. J. Endo, N. Koga, and K. Morokuma, *Organometallics* **12,** 2777 (1993).
65. E. Nakamura, Y. Miyachi, N. Koga, and K. Morokuma, *J. Am. Chem. Soc.* **114,** 6686 (1992).
66. K. N. Houk, N. G. Rondan, P. v. R. Schleyer, E. Kaufmann, and T. Clark, *J. Am. Chem. Soc.* **107,** 2821 (1985).
67. D. G. Musaev, A. M. Mebel, and K. Morokuma, *J. Am. Chem. Soc.* **116,** 10693 (1994).
68. D. G. Musaev, A. M. Mebel, and K. Morokuma, unpublished results.
69. A. E. Dorigo and P. v. R. Schleyer, *Angew. Chem. Int. Ed. Engl.* **34,** 115 (1995).
70. D. manning and H. Noth, *Angew. Chem. Int. Ed. Engl.* **24,** 878 (1985).
71. K. Burgess and M. J. Ohlmeyer, *Chem. Rev.* **91,** 1179 (1991) and references therein.
72. S. A. Westcott, T. B. Marder, and R. T. Baker, *Organometallics* **12,** 975 (1993).
73. P. R. Rablin, J. F. Hartwig, and S. P. Nolan, *J. Am. Chem. Soc.* **116,** 4121 (1994).
74. M. Hada, Y. Tanaka, M. Ito, M. Murakami, H. Amii, Y. Ito, and H. Nakatsuji, *J. Am. Chem. Soc.* **116,** 8754 (1994).
75. B. L. Chenard and C. M. Van Zyl, *J. Org. Chem.* **51,** 3561 (1986).

76. M. Murakami, H. Amii, N. Takizawa, and Y. Ito, *Organometallics* **12,** 4223 (1993).
77. T. Matsubara, N. Koga, Y. Ding, D. G. Musaev, and K. Morokuma, submitted for publication
78. D. Evans, J. A. Osborn, and G. Wilkinson, *J. Chem. Soc. A* 3133 (1968).
79. D. Evans, G. Yagupsky, and G. Wilkinson, *J. Chem. Soc. A* 2660 (1968).
80. C. K. Brown and G. Wilkinson, *J. Chem. Soc. A* 2753 (1970).
81. C. A. Tolman and J. W. Faller, *Homogeneous Catalysis with meatal Phosphine Complexes*, (Plenum Press, New York, 1983).
82. R. T. Pruett, *Adv. Organomet. Chem.* **17,** 1 (1979).
83. J. Eckert, G. J. Kubas, J. H. Hall, P. J. Hay, and C. M. Boyle, *J. Am. Chem. Soc.* **112,** 2324 (1990).
84. H. Kawamura-Kuribayashi, N. Koga, and K. Morokuma, *J. Am. Chem. Soc.* **114,** 8687 (1992).
85. F. Maseras, N. Koga, and K. Morokuma, *Organometallics* **13,** 4008 (1994).
86. T. Yoshida, N. Koga, and K. Morokuma, *Organometallics*, in press.
87. F. Maseras and K. Morokuma, *J. Comp. Chem.* **16,** 1170 (1995).
88. T. Matsubara, F. Maseras, N. Koga, and K. Morokuma, *J. Phys. Chem.*, in press.
89. R. G. Parr and W. Yang, *Density Functional Theory of Atoms and Molecules* (Oxford University, New York, 1989).
90. B. G. Johnson, P. M. W. Gill, and J. A. Pople, *J. Chem. Phys.* **97,** 7846 (1992).
91. B. G. Johnson, P. M. W. Gill, and J. A. Pople, *J. Chem. Phys.* **98,** 5612 (1993).
92. C. Heinemann, R. H. Hertwig, R. Wesendrup, W. Koch, and H. Schwarz, *J. Am. Chem. Soc.* **117,** 495 (1995) and references therein.
93. A. Ricca and C. W. Bauschlicher, Jr., *J. Phys. Chem.* **98,** 12899 (1994).
94. D. G. Musaev and K. Morokuma, *J. Phys. Chem.*, in press.

HIGH-RESOLUTION HELIUM ATOM SCATTERING AS A PROBE OF SURFACE VIBRATIONS

SANFORD A. SAFRON

Department of Chemistry, Florida State University, Tallahassee, Florida

CONTENTS

Advances in Chemical Physics, Volume XCV, Edited by I. Prigogine and Stuart A. Rice.
ISBN 0-471-15430-X

I. INTRODUCTION

A. Overview

The motion of atoms bound together is governed by the interaction potential between them and is typically periodic or quasiperiodic in nature. Various spectroscopic probes have been enlisted to investigate this vibrational motion and thereby to enable the understanding of the forces well enough to model them. The probe of choice needs to satisfy the criteria required for selectivity, sensitivity, resolution, ease of interpretation of data, and cost of equipment, among others. For molecular systems of atoms the probe most often chosen, and by far the best known, is the *photon* employed through infrared or Raman spectroscopy. For crystalline systems, where the atoms are correlated with each other through translational symmetry, such methods are at a definite disadvantage when compared with inelastic neutron scattering for bulk vibrations and with inelastic helium atom scattering for surface vibrations as will be explained below.

In 1912 Born and von Kármán [1, 2] proposed a model for the lattice dynamics of crystals which has become the "standard" description of vibrations in crystals. In it the atoms are depicted as bound together by harmonic springs, and their motion is treated *collectively* through traveling displacement waves, or lattice vibrations, rather than by *individual* displacements from their equilibrium lattice sites [3]. Each wave is characterized by its frequency, wavelength (or wavevector), amplitude and polarization.

In the quantum mechanical treatment of this model, the equations of motion in the harmonic approximation become analogous to those for electromagnetic waves in space [2–4]. Thus, each wave is associated with a quantum of vibrational energy $\hbar\omega$ and a "crystal" momentum $\hbar q$. By analogy to the *photon* for the electromagnetic quantum, the lattice vibrational quantum is called a *phonon*. The amplitude of the wave reflects the phonon population in the vibrational mode (i.e., the mode with frequency ω and wavevector q) as given by the statistical mechanical relation for harmonic oscillators [3, 4]. The polarization of the phonon is the direction of the displacement of the wave.

With the development of nearly monochromatic, thermal energy neutron sources in the 1950s, the detection of single vibrational quanta (i.e., single phonons) was realized [5–7]. The neutrons in this energy regime have de Broglie wavelengths comparable to the dimensions of the lattice and have energies and momenta comparable to those of the bulk phonons. From measurements of the energy and momentum transfer to the crystal under conditions where phonon creation and/or annihilation events are probable, the relation between phonon frequency and wavevector can be determined for the bulk phonons over the entire range of crystal momentum (i.e., the phonon dispersion relations can be determined over the Brillouin zone (BZ)) [8]. These results can then be compared with model calculations so that the fits to the lattice dynamics yield force constants or potential parameters which describe the atomic interactions [9].

Prompted by theoretical treatments of atom–surface scattering adapted from neutron scattering [10–12], experiments were begun in the 1970s to measure the *surface* phonon dispersion [13–15]. First reported in 1980, the experimental approach developed by the Toennies' group in Göttingen, Germany, set the standard for virtually all subsequent work in this field [16, 17]. This group's experimental design succeeded by employing a high-pressure nozzle expansion of helium to produce a nearly monoenergetic beam ($\Delta v/v \approx 1\%$), many differential pumping stages to reduce the background helium pressure at the detector, a time-of-flight (TOF) technique for energy analysis of the scattered atoms, and a mass spectrometer with very high resolution for masses in the vicinity of helium. Since this work, several groups in Europe, in the United States, and recently in Japan have followed the Göttingen lead to take up high-resolution helium atom–surface scattering (HAS) experiments.

The properties of the helium atoms are unique and of central importance in the atom–surface scattering approach to surface structure and dynamics. First of all, the atoms have very large cross sections at low relative energy [18], which makes possible the production of highly monoenergetic ($\Delta E/E \approx 2\%$), low-energy beams (10–75 meV) [19]. Next, for these energies the helium atoms have momenta (or wavevectors 4–12 Å^{-1}) comparable to those of the phonons and thus can provide information on the phonon dispersion over the entire surface Brillouin zone (SBZ) [20–24]. Furthermore, because they do not charge, react with, or penetrate the surface, they can be used with all surface types, including surfaces with adsorbates [20]. Finally, as the atoms are virtually insensitive to the presence of electric and/

or magnetic fields, the HAS technique can be employed to study the effects of these fields on surfaces and on processes occurring at surfaces.

Two additional points are also worth noting. First, theoretical calculations have shown that in HAS the atoms are scattered by the electrons in the *surface* layer of the crystal, typically 3–4 Å out from the plane of the atomic cores [25]. For other common surface probes the interaction with the atoms of the crystal is markedly different. Electrons, ions, and x-rays have energies at least 10^3 higher for the same wavelengths, or wavevectors, and are scattered by the core electrons of the atoms [26, 27]; neutrons are scattered only by the atomic nuclei themselves [7]. These probes, therefore, can penetrate into the bulk (to a greater or lesser extent), and hence are not strictly surface probes. A direct consequence of this is that helium atom scattering is much more sensitive to the coupling of the lattice vibrations with the least tightly bound electrons, and thus much more sensitive to the most accessible vibronic interactions affecting processes at the surface, than are other surface scattering techniques.

The second point is that surface scattering experiments can in some cases be more sensitive to the details of the interatomic forces within a crystal than bulk scattering experiments are. This is because certain features of the interactions may "cancel out" in the higher symmetry of the crystalline bulk which do not cancel out in the lower symmetry at the surface [28].

B. Scope

It is intended for this chapter to be a brief introduction to surface lattice dynamics and to some of the kinds of information that one is able to obtain about the electronic and vibrational properties of surfaces through high-resolution helium atom scattering. Such information is critical to the understanding of many aspects of surface chemistry and to the design of novel materials with specifically desired properties.

Section II is devoted to setting up elementary concepts associated with collective vibrational motion of atoms for reference later in the discussion of experimental results. Section III deals with atom–surface scattering theory so that the reader unfamiliar with this kind of spectroscopy can relate to the quantities which are measured directly. A brief description of the helium scattering instrument then follows in Section IV.

In Sections V, VI, and VII some experimental results are presented. Since our laboratory has focused primarily on ionic insulator materials, this work is presented in much greater detail. For other materials, only a small sampling of results is presented to show the range of applicability of the HAS technique and the new insights into the physics of materials which it has provided in a relatively short span of time.

II. LATTICE VIBRATIONS

A thorough treatment of lattice vibrations is beyond the scope of this chapter. However, many characteristic features which arise in the discussion later in the chapter can be illustrated by referring to the analogous behavior of one-dimensional model crystals [8, 29]. These models are also a good starting point for those not very familiar with the concept of lattice vibrations and are briefly sketched out here [8, 29].

A. One-Dimensional Models

The simplest model is the one-dimensional "monatomic" crystal consisting of N atoms, each with mass M and separated by lattice spacing a along the x-axis at sites $x_0^\circ, x_1^\circ, \ldots, x_l^\circ, \ldots, x_{N-1}^\circ$, as shown in Fig. 1. We consider only nearest-neighbor interactions and have every atom bound to its next-door neighbors by harmonic springs with force constant f. The Born–von Kármán treatment applies periodic boundary conditions so that the atom at x_0° is bound to the atoms at both x_1° and at x_{N-1}°. The instantaneous position of the lth atom is given by $x_l = x_l^\circ + u_l$, where u_l is the displacement from the equilibrium site x_l°, and Newton's Second Law for the motion of the atom can be written as

$$M\ddot{x}_l = M\ddot{u}_l = f(u_{l+1} - u_l) + f(u_{l-1} - u_l) \tag{1}$$

where the "double dots" stand for second derivative with respect to time. The displacements of the atoms are described as traveling waves, which can be written as

$$u_l(q) = (NM)^{-1/2}B(q) \exp [i(qx_l^\circ - \omega t)] \tag{2}$$

where $(NM)^{-1/2}$ is a convenient normalization, q is the wavevector, and $B(q)$ is the wave amplitude, and for positive q the wave propagates in the positive

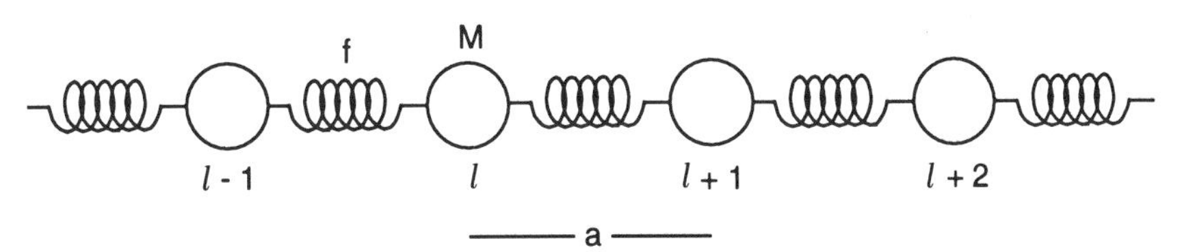

Figure 1. Schematic representation of a one-dimensional crystal showing the region of the lth unit cell, with atoms of mass M, force constants between neighboring atoms f, and lattice spacing a.

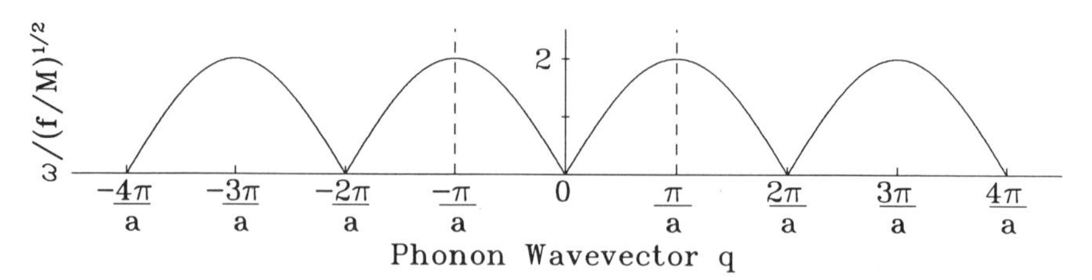

Figure 2. Dispersion relation $\omega(q)$ for the one-dimensional crystal in Fig. 1. The relation is given by Eq. (3) in the text. The ordinate scale is in units of $(f/M)^{1/2}$. The dashed vertical lines show the first Brillouin zone.

x sense. When Eq. (2) is substituted into Eq. (1), one obtains the following relation, the dispersion relation, between ω and q,

$$\omega(q) = 2(f/M)^{1/2}|\sin(qa/2)| \tag{3}$$

with the absolute values ensuring only positive frequencies. As shown in Fig. 2, this function is symmetric about $q = 0$, and it has maxima at odd multiples of $\pm(\pi/a)$ and zeroes at even multiples of $\pm(\pi/a)$ including $q = 0$. For this one-dimensional crystal the reciprocal lattice consists of the points $g_n = 2\pi n/a$, where $n = 0, \pm 1, \pm 2, \ldots$, so that whenever q approaches a reciprocal lattice vector, the wave frequency goes to zero.

We note that the wavelength of the displacement wave in Eq. (2) is $\lambda = 2\pi/|q|$. Thus, a wave with a value of $|q|$ larger than π/a has a wavelength smaller than $2a$ and therefore implies a displacement of the atoms which is unphysical because the oscillation length of such a wave is smaller than the lattice spacing. The region of reciprocal space with physically meaningful wavelengths, $-\pi/a < q \leq \pi/a$, is called the *first* Brillouin zone, or just *the* Brillouin zone (BZ). As can be seen in Fig. 2, the dispersion relation is periodic in the reciprocal lattice vector g; namely, $\omega(q') = \omega(q + g_n) = \omega(q)$. Although it will be seen later that it is sometimes useful to use an "extended zone" representation for the relation between ω and q, one must keep in mind that for $|q| > \pi/a$ the values of ω are to be "folded" back into the first Brillouin zone by adding or subtracting a reciprocal lattice vector to q. In the wavevector range from $-\pi/a < q \leq \pi/a$ there are N possible values for q corresponding to the allowed number of independent vibrational modes for N atoms constrained to move in one dimension, which are given by $2\pi s/Na$, where $s = 0, \pm 1, \pm 2, \ldots$, from $-(N/2) + 1$ up to $(N/2)$ for even N or from $-(N - 1)/2$ to $(N - 1)/2$ for odd N.

So far the motion that has been allowed is a displacement of the atoms along the axis of the crystal. This is called a *longitudinal vibration* since the displacement occurs (or is polarized) in the direction of wave propagation. However, if we also permit motion perpendicular to the axis, one can easily imagine a second set of springs harmonically coupling the atoms for these displacements. The equations of motion for the resulting transverse waves are of just the same form as Eq. (1). With this modification of the crystalline dynamics, one finds that there are now two frequencies for each q, ω_L and ω_T, given by Eq. (3), which differ only in the size of the force constants; normally $\omega_L > \omega_T$. Furthermore, since there are two equivalent directions perpendicular to the crystal axis, there are two degenerate transverse branches to the dispersion relations for this crystal, or three branches in all, one longitudinal and two transverse. Because these N atoms are now allowed motion in three dimensions, there are $3N$ states $\omega(q, j)$, where j (= 1, 2, or 3) is the label for the branches of the dispersion relation.

The three-dimensional displacive motion of any given atom in this crystal is the superposition of all the waves, or of all the normal modes [8, 29],

$$\mathbf{u}_l(t) = (NM)^{-1/2}\Sigma_{q,j}B(q, j)\mathbf{e}(q, j) \exp [i(qx_l^\circ - \omega(q, j)t)] \tag{4}$$

where $\mathbf{e}(q, j)$ is a unit vector which gives the polarization of the wave.

We move on now to consider a "diatomic crystal" where each unit cell has two different atoms G and H with masses M_G and M_H. Again, the one-dimensional crystal has N atoms of each type with lattice spacing a, and within each cell the atoms are separated by distance b. Furthermore, we take force constants f_1 and f_2 for the binding of the atoms within each cell and between nearest-neighbor atoms outside the cells, respectively, as illustrated in Fig. 3. By analogy to Eq. (2) we expect the form of the wave to be

$$u_{lG} = (NM_G)^{-1/2}B(q)e_G(q) \exp [i(qx_{lG}^\circ - \omega t)] \tag{5a}$$

$$u_{lH} = (NM_H)^{-1/2}B(q)e_H(q) \exp [i(qx_{lH}^\circ - \omega t)] \tag{5b}$$

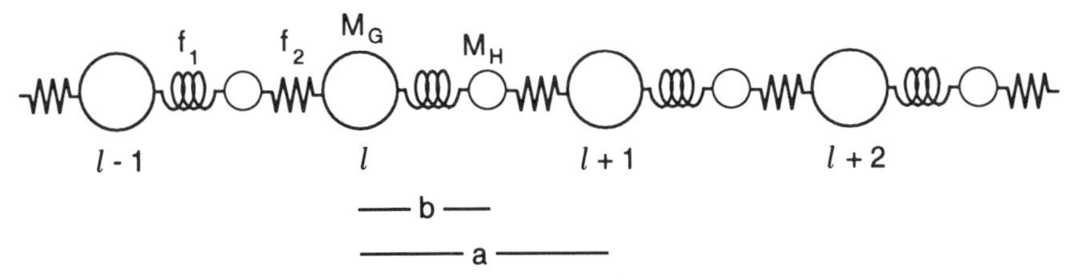

Figure 3. Schematic representation of a one-dimensional "diatomic" crystal with atoms of mass M_G and M_H in the region near the lth unit cell. The force constant between atoms within the cell is f_1, and that between the cells is f_2. The lattice spacing for the crystal is a and the separation between atoms within each cell is b.

The quantities e_G and e_H represent the portion of the amplitude associated with the displacement of each type of atom, and it is convenient to require that $e_G^2 + e_H^2 = 1$. Newton's Law [Eq. (1)] for nearest-neighbor interactions becomes the coupled pair of equations

$$M_G \ddot{u}_{lG} = f_1(u_{lH} - u_{lG}) + f_2(u_{l-1,H} - u_{lG}) \tag{6a}$$

$$M_H \ddot{u}_{lH} = f_2(u_{l+1,G} - u_{lH}) + f_1(u_{lG} - u_{lH}) \tag{6b}$$

Substituting Eq. (5) into Eq. (6) leads to a set of simultaneous equations which can be written in matrix form

$$\omega^2(q)\mathbf{e}(q) = \mathbf{D}(q)\mathbf{e}(q) \tag{7}$$

where $\mathbf{e}$ is the vector with components e_G and e_H, and $\mathbf{D}$ is a two-by-two square Hermitian matrix, called the *dynamical matrix*. Equation (7) is in the form of an eigenvalue equation where the $\omega^2(q)$ are the eigenvalues and the $\mathbf{e}(q)$ are the eigenvectors. With some algebraic effort one can show that the solutions for this one-dimensional diatomic crystal are

$$\omega(q) = [(f_1 + f_2)/(2\mu)]^{1/2}\{1 \pm [1 - 16f_1f_2M_GM_H \sin^2(qa/2)/((f_1 + f_2)^2 \cdot (M_G + M_H)^2)]^{1/2}\}^{1/2} \tag{8}$$

where μ is the reduced mass $M_GM_H/(M_G + M_H)$.

For each q there are two values of ω, as shown in Fig. 4 for a range of values of masses and force constants, and the two sets of solutions lead to different kinds of vibrational dispersion. In each of the examples shown in the figure, the curves found by using the *negative* sign in the square root term go to zero as q goes to zero. This behavior appears similar to the dispersion for the monatomic case (above) which goes over to the usual propagation of sound in the long wavelength limit and is called the *acoustic* dispersion branch. For the solutions with the *positive* sign, the ω go to a finite value as q approaches zero. These vibrations are optically active in the infrared or Raman spectroscopy of some materials and are called the *optical* branch [3].

To develop one's sense of how the behavior of the system depends on the model parameters, it is instructive to examine the solutions for some trends. First, we can see what happens as f_1 becomes much larger than f_2. This is the situation when the *intra*cellular force, or binding of the diatomic GH, is much greater than the intercellular force. Again, with a bit of algebraic manipulation one finds the leading terms in the ratio f_2/f_1 to be

$$\omega = [(f_1)/\mu]^{1/2}[1 - (2(f_2/f_1)\mu \sin^2(qa/2))/((M_G + M_H))] \tag{9a}$$

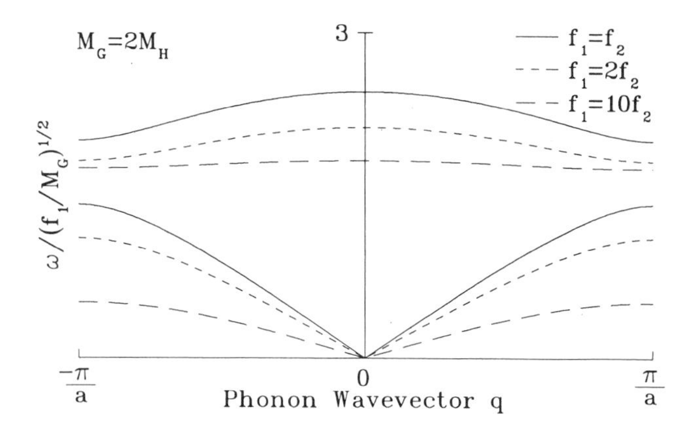

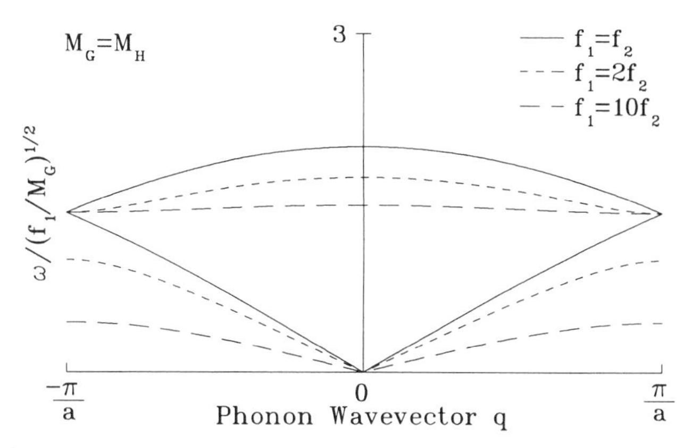

Figure 4. Dispersion relation $\omega(q)$ for the one-dimensional "diatomic" crystal in Fig. 3. The relation is given by Eq. (8) in the text with (*top panel*) $M_G = 2M_H$ and (*lower panel*) $M_G = M_H$ for $f_1 = f_2$ (solid lines), $f_1 = 2f_2$ (short-dashed lines), and $f_1 = 10f_2$ (long-dashed lines). In each panel the lower curves which go to zero for $q = 0$ are the acoustic branches of the dispersion relation [negative sign in Eq. (8)] and the upper curves which have finite values for $q = 0$ are the optical branches [positive sign in Eq. (8)]. The ordinate scale for all curves is in units of $(f_1/M_G)^{1/2}$. Note that the gap between the optical and acoustic bands increases as f_1 increases relative to f_2 and as the difference in masses of G and H increases.

for the optical modes and

$$\omega = 2[f_2/(M_G + M_H)]^{1/2}\,|\sin(qa/2)|[1 + (f_2/f_1)]^{1/2} \tag{9b}$$

for the acoustic modes. As this ratio decreases, the optical branch becomes flatter and flatter, or dispersionless like an Einstein mode, and the two atoms in each unit cell just vibrate against each other like independent diatomic molecules with a frequency approaching that of the diatomic GH molecule, $(f_1/\mu)^{1/2}$. (See Fig. 4.) At the same time, the acoustic mode has dispersion just like the monatomic crystal case above, but with each atom replaced by the diatom of mass $M_G + M_H$, the total mass. This behavior is general in the sense that the optical vibrational modes tend to be more sensitive to intracellular interactions while the acoustic vibrational modes tend to reflect the intercellular interactions.

This contrasting behavior of the two branches also can be seen in Fig. 5 where the displacements of the atoms G and H at points in the BZ are compared; for examples, we have taken $q = 0$ (also called the *Brillouin zone center*), $q = \pi/4a$, $q = \pi/2a$, $q = 3\pi/4a$ and $q = \pi/a$ (the zone boundary). From Eq. (5) the ratio of displacements for G and H in the lth cell with the same value of q is given by

$$u_H/u_G = (M_G/M_H)^{1/2}(e_H/e_G)\exp(iqb) \tag{10}$$

and the ratio (e_H/e_G) is found by substituting $\omega^2(q)$ into Eq. (7). For an alkali halide type of crystal where $f_1 = f_2$, every atom bonds equally to each neighboring atom. (Also see Fig. 4b.) In general for such a crystal, at the zone center the acoustic mode frequency, $\omega_A(0)$, equals zero and the optical mode has $\omega_O(0) = (2f_1/\mu)^{1/2}$, while at the zone boundary $\omega_A(\pi/a) = (2f_1/M_G)^{1/2}$ and $\omega_O(\pi/a) = (2f_1/M_H)^{1/2}$, if $M_G > M_H$.

At the zone center the corresponding displacement ratio for the optical branch from Eq. (10) is $u_H/u_G = -M_G/M_H$, which is just equivalent to vibrational motion about the center of mass of a GH molecule. (See Fig. 5.) On the other hand, for the very-long-wavelength acoustic phonons at the zone center, $u_H/u_G = 1$ so that both types of atoms in the crystal are displaced equally from their equilibrium sites. That is, the wavelength is so long that the crystal as a whole appears displaced during this motion. At the zone boundary the calculation from Eq. (10) shows that in the optical mode only the lighter of the two atoms (H) moves while the heavier atom (G) remains stationary, but that for the acoustic mode the reverse situation obtains.

Another interesting result from this simple picture of lattice dynamics for an alkali-halide-type crystal is that when the atoms have the same mass, $M_G = M_H = M$, and there are no differentiating forces between the atoms (such as different next-nearest-neighbor interactions), then G and H are identical within the context of this dynamical model. The crystal is then indistinguishable from a monatomic crystal that has a unit cell length $a' = a/2$ and corresponding Brillouin zone boundary at $q = \pi/a' = 2\pi/a$. (See Fig. 6.) The consequence for the lattice dynamics is that the optical and acoustic mode frequencies become the same at $q = \pm\pi/a$, $\omega = (2f_1/M)^{1/2}$, and there is no acoustic-optical band gap just as in Fig. 4 (lower panel). From a different perspective, one could think of the optical branch for this "isobaric" diatomic crystal as just the extended portion of the acoustic branch of the pseudomonatomic crystal from π/a to $2\pi/a$ folded back to the zone center, as indicated in Fig. 6. This "folding," or "extended symmetry," does seem to occur, in part, for the dispersion curves of alkali halide crystals with the rocksalt structure when the alkali and the halide ions are nearly of the same mass, for example, for NaF, KCl, and RbBr [30, 31]. In related behavior, alkali halides seem to display "mirror symmetry" [32]—that is, very similar phonon dispersion—for pairs of alkali halides in which the alkali mass of one compound matches the halide mass of the other compound and vice versa. Particularly striking are the similarities in the bulk phonon dispersions of KBr and RbCl which are shown in Fig. 7 [32].

Without going into the details here, it should be clear how to extend this treatment for increasing complexity in the crystal. For example, one can deal with longitudinal and transverse vibrations in the diatomic crystal by using vector polarizations for each atom type as in the monatomic case of Eq. (4). The dynamical matrix for the diatomic crystal would then be six-by-six, but no new principles need be invoked. Similarly, one can increase the basis, or number of atoms per unit cell, to three or more by following the simple procedure described above for two atoms. For n atoms per unit cell there are always $3n$ dispersion branches; three of these are acoustic branches and the rest of the $3n - 3$ branches are optical in character [8, 29]. Finally, one can extend the model of the dynamics to include next-nearest-neighbor interaction forces and next, next-nearest-neighbor forces, and so on, by adding the appropriate terms to the equations of motion for each atom type [29]. This just increases the complexity of the elements of the dynamical matrix **D**, but does not otherwise change the form of the eigenvalue equation, Eq. (7), which must be solved for the frequencies and polarization vectors. Force laws other than the harmonic springs between neighboring atoms can similarly be employed to define the dynamical matrix elements, and some of these will be described later in the chapter when specific experimental measurements are discussed.

B. Extension to Surface Dynamics

Behavior remarkably similar to that revealed by the one-dimensional model crystals is generally observed for lattice vibrations in three dimensions. Here the dynamical matrix is constructed fundamentally in the same way, based on the model used for the interatomic forces, or derivatives of the crystal's potential energy function, and the equivalent of Eq. (7) is solved for the eigenvalues and eigenvectors [2–4, 29]. Naturally, the phonon wavevector in three dimensions is a vector with three components, $\mathbf{q} = (q_x, q_y, q_z)$, and both the frequency of the wave, $\omega(\mathbf{q})$, and its polarization, $e(\mathbf{q})$, are functions

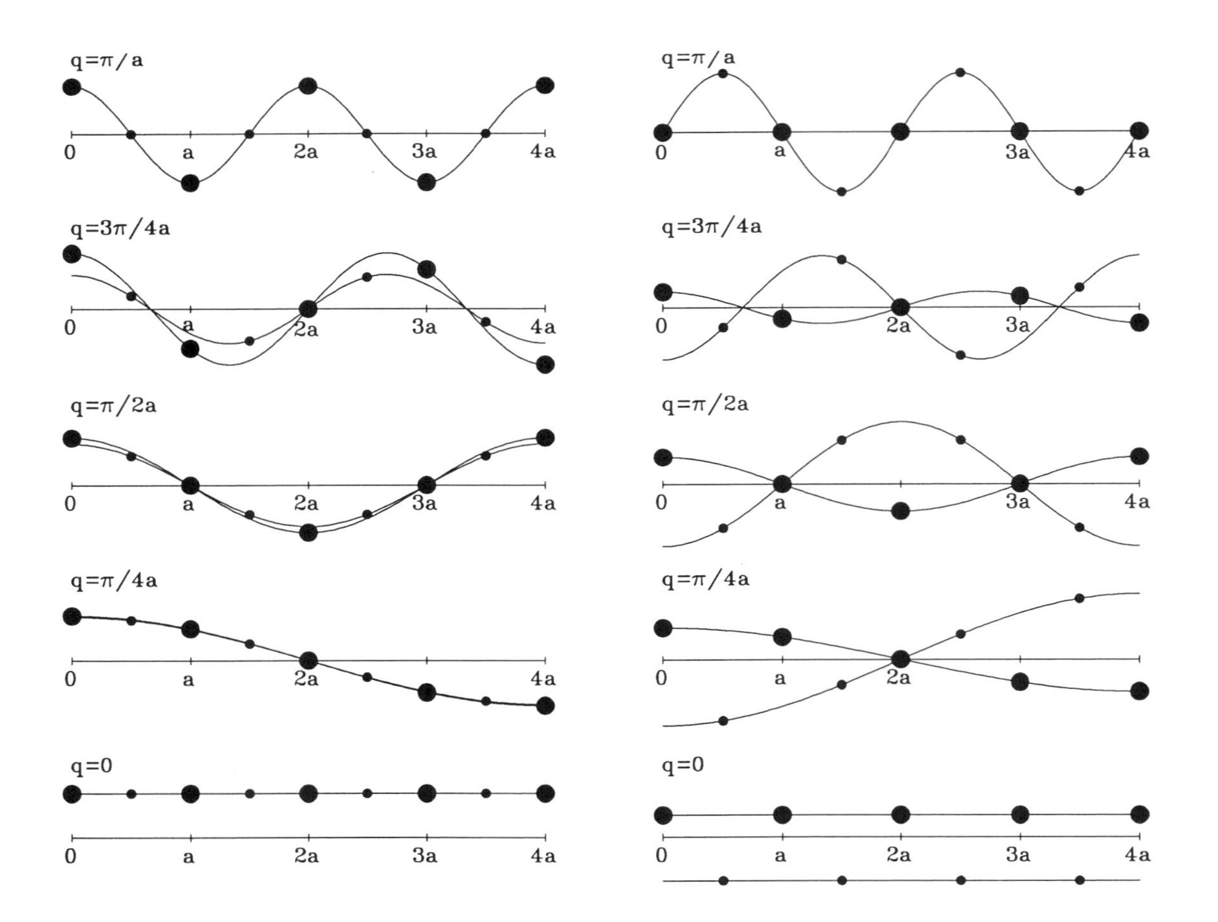
q=π/a
0
a
2a
3a
4a
q=3π/4a
q=π/2a
q=π/4a
q=0

Figure 5. Relative displacements of the atoms in the diatomic crystal with $M_G = 2M_H$ and $f_1 = f_2$ sampled across the Brillouin zone for wavevectors $q = 0, \pi/4a, \pi/2a, 3\pi/4a$, and π/a. The acoustic branch displacements are shown on the left side, and the optical branch displacements are shown on the right. In each case the heavier atom, G, is depicted by the larger filled circles and the lighter atom, H, is depicted by the smaller filled circles. The abscissa scale labels the equilibrium sites of the G atoms in units of the lattice constant a, and the separation of the atoms in each cell is taken to be $b = a/2$. The ordinate scale is relative for different values of q. The lines connecting the atoms represent the sinusoidal waves for the atomic displacements at each wavevector q and are just to guide the eye. Note that, for clarity of display, the displacements are shown here as *transverse* waves whereas the spring model in Fig. 3 is drawn for *longitudinal* waves.

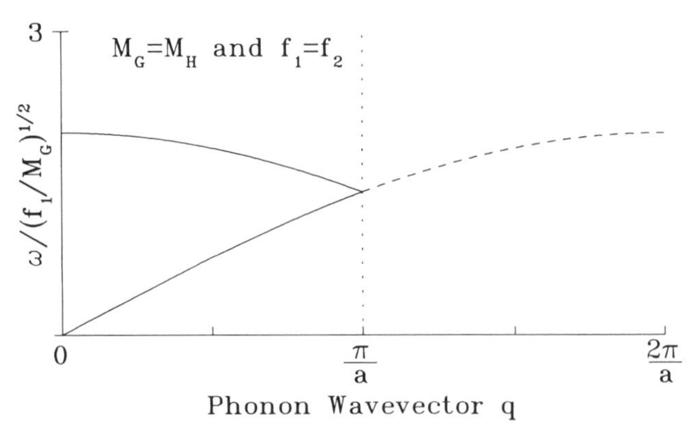

Figure 6. Dispersion relation for an "isobaric" crystal, where $M_G = M_H$. When $f_1 = f_2$ for this force constant model, the diatomic crystal is equivalent to a monatomic crystal with lattice spacing $a/2$ or Brillouin zone boundary at $2\pi/a$. Hence the optical branch (solid line) appears as the portion of the acoustic branch from π/a to $2\pi/a$ (dashed line) folded back to the zone center from the real Brillouin zone boundary (dotted vertical line).

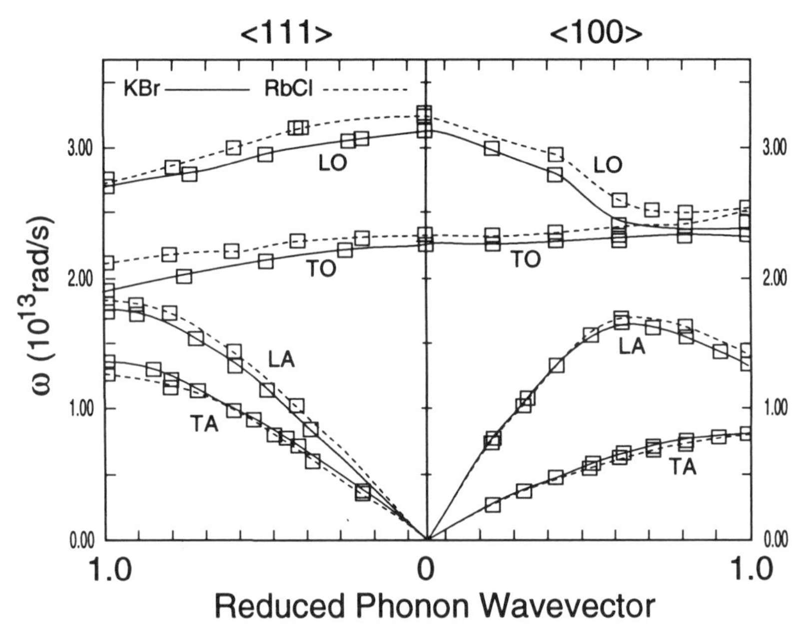

Figure 7. Bulk phonon dispersion curves for KBr and RbCl in their ⟨100⟩ and ⟨111⟩ high-symmetry directions. Both crystals have fcc lattices and rocksalt structures. Note that the transverse branches, labeled TA (transverse acoustic) and TO (transverse optical), are doubly degenerate in these directions. (Adapted from Fig. 3 of Ref. 32.)

of **q** except that in the high-symmetry directions of the crystal, the polarization of the phonon generally remains fixed across the BZ [2–4, 29]. In many cases the symmetry of the crystal requires that the two perpendicular transverse branches be degenerate (for both acoustic and optical branches, if there is more than one atom per unit cell). This situation occurs for the alkali halides with rocksalt structures in the ⟨100⟩ and ⟨111⟩ directions, but not in the ⟨110⟩ direction [9]. (See Fig. 7 for an example.) Finally, as in Equation (4), the displacement of any atom is given by the superposition of all possible lattice waves of the crystal.

The situation at surfaces is more complicated, and richer in information. The altered chemical environment at the surface modifies the dynamics to give rise to new vibrational modes which have amplitudes that decay rapidly into the bulk and so are "localized" at the surface [33]. Hence, the displacements of the atoms at the surface are due both to surface phonons and to bulk phonons projected onto the surface. Since the crystalline symmetry at the surface is reduced from three dimensions to the two dimensions in the plane parallel to the surface, the wavevector characterizing the states becomes the two-dimensional vector $\mathbf{Q} = (q_x, q_y)$. (We follow the conventional notation using uppercase letters for surface projections of three-dimensional vectors and take the positive sense for the z-direction as outward normal to the surface.) Thus, for a given **Q** there is a whole band of bulk vibrational frequencies which appear at the surface, corresponding to all the bulk phonons with different values of q_z (which effectively form a continuum) along with the isolated frequencies from the surface localized modes.

Theoretical calculations of surface phonon dispersion have been carried out in two ways. One method is to use a Green's function technique which treats the surface as a perturbation of the bulk periodicity in the z-direction [34, 35]. The other is a slab dynamics calculation in which the crystal is represented by a slab of typically 15–30 layers thick, and periodic boundary conditions are employed to treat interactions outside the unit cell as the equations of motion for each atom are solved [28, 33, 35, 37]. In the latter both the bulk and the surface modes are found and the surface localized modes are identified by the decay of the vibrational amplitudes into the bulk; in the former the surface modes can be obtained directly. When the frequency of a surface mode lies within a bulk band of the same symmetry, then hybridization can take place. In this event the mode can no longer be regarded as strictly surface localized and is referred to as a *surface resonance* [24]. Figure 8, adapted from Benedek and Toennies [24], shows how the bulk and surface modes develop as more and more layers are taken in a slab dynamics calculation.

Because of symmetry constraints at the surface, phonons propagating in a high-symmetry direction must be polarized either in the sagittal plane

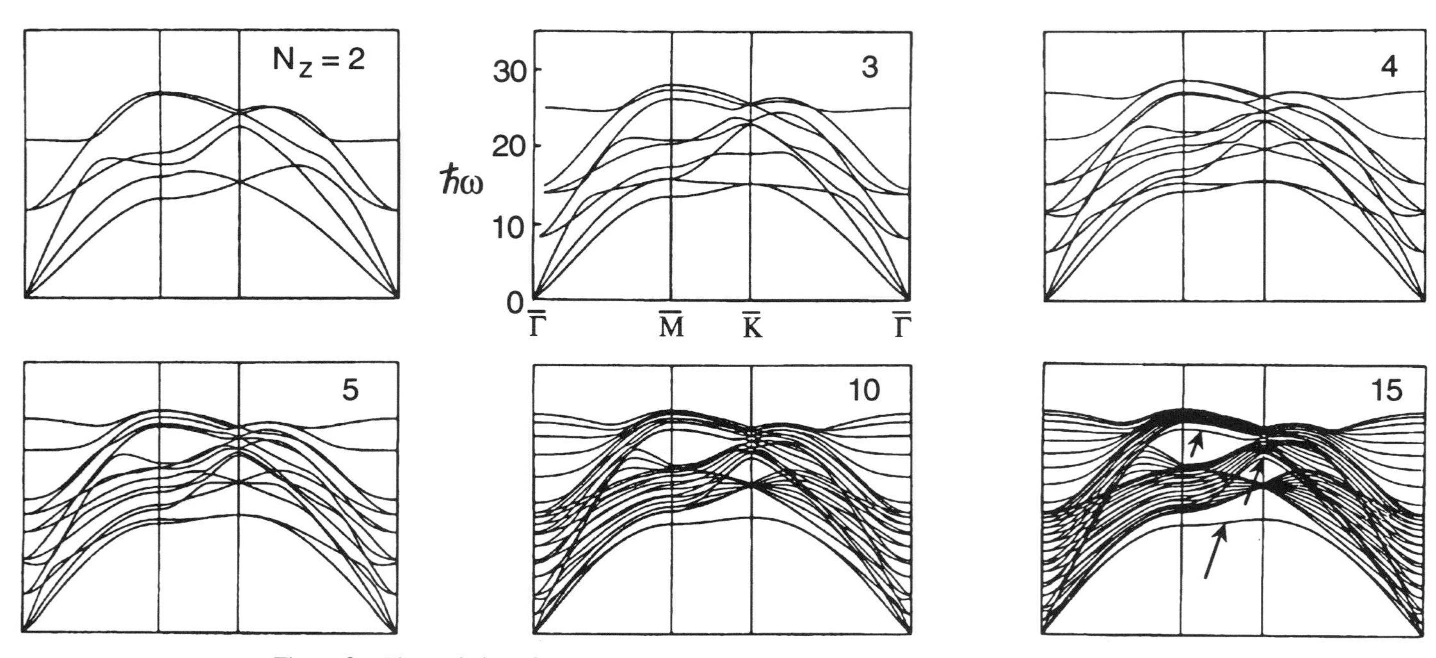

Figure 8. The evolution of surface phonon dispersion curves for a monatomic fcc (111) surface in slab dynamics calculations as a function of the number of layers N_Z in the slab. The surface localized modes, marked by arrows in the last panel ($N_Z = 15$), lie below the bulk "bands" across the entire surface Brillouin zone and appear between the bands in the small gap near $\bar{K}$ in the $\overline{\Gamma K}$ region and in the larger gap in the $\overline{MK}$ region. (Reproduced from Fig. 1 of Ref. 24, with permission of Elsevier Science Publishers.)

(defined by the direction of propagation and the surface normal) or in the plane of the surface which is perpendicular to the sagittal plane [21]. Moreover, in the reduced symmetry at the surface, the sagittal plane modes are generally regarded as "elliptically" polarized, having both transverse and longitudinal components which vary with **Q** whereas the modes in the plane perpendicular to the sagittal plane are referred to as "shear horizontal" modes [21]. Furthermore, there can be surface phonons with acoustic or optical characteristics with either polarization, and for many substances a surface localized mode having optical character occurs in the gaps between the bulk acoustic and optical bands.

In most materials, however, the modification of the forces at the surface is such that the surface localized modes have frequencies which lie below the frequencies of an associated bulk band with the same symmetry; they have the appearance of having been "peeled" down from this bulk band [24]. In the usual case, the lowest energy of all these "peeled"-down modes derives from the bulk transverse acoustic band and is normally sagittally polarized. This dispersion branch is called the *Rayleigh wave* (RW) because it was predicted by Lord Rayleigh from continuum wave theory over a century ago [38]. Helium atom scattering experiments on virtually every material so far investigated have detected the RW on clean crystalline surfaces.

III. INELASTIC ATOM–SURFACE SCATTERING

The theory of atom–surface scattering, applied to the inelastic scattering of helium atoms from crystalline surfaces, is sketched out only briefly enough here to describe the conceptual framework and to identify the quantities and issues which experimentalists deal with in carrying out the measurements [39].

Before discussing atom–surface scattering theory, a comment is offered on why atom scattering is better suited than electromagnetic waves for studying interatomic forces at crystalline surfaces. The first point is that to obtain the most complete account of the microscopic behavior, one needs to carry out measurements of the dispersion across the entire surface Brillouin zone. Thus, the probe should have a wavevector k and energy E large enough to create phonons across the SBZ. At the zone boundary the phonon wavevector is π/a or ~ 1 Å^{-1} and the RW phonon energies are typically 10–20 meV. The relation between k (Å^{-1}) and E (meV) for electromagnetic waves is $E \approx 2 \times 10^6 k$, while for helium atoms $E \approx 0.52k^2$. Hence, for k's in the range of several Å^{-1} the electromagnetic wave would have to lie in the x-ray region of the spectrum which is not very surface-sensitive. At the other limit, infrared or visible light photons have very small k's and can create optical phonons only very close to the zone center. For comparison, electrons have $E \approx 3.8 \times 10^3 k^2$ and electron energy loss spectroscopy (EELS) [26] has been used to measure surface phonon dispersion, but primarily for the high-energy phonons in metals. For helium atoms the match between the phonon momentum and energy and the atom momentum and energy is very good over a wide range of these quantities [22, 23].

In molecular spectroscopy, experiment provides two kinds of information, the spectroscopic values for the frequency and/or wavelength of the inelastic event and the intensity of the spectroscopic bands. The former generally are analyzed or interpreted through dynamical models for the system, whereas the latter provides an intimate glimpse into the fundamental workings of the system and its coupling with the probe. Selection rules, for example, represent extreme restrictions on the coupling due to the symmetries of the system in relation to the probe. A similar situation exists in atom–surface scattering. Until recently, however, comparatively little has been made of intensity data; most of the reported results consist of spectroscopic data in the form of surface dispersion relations. Future investigations are likely to make better use of both kinds of data for the reasons discussed briefly at the end of this section.

A. Atom–Single-Phonon Scattering

Because of the two-dimensional translational symmetry of the surface, the interaction of the helium atom with a perfect crystalline target, $V(\mathbf{r})$, can be written as a Fourier series [40].

$$V(\mathbf{r}) = V(\mathbf{R}, \mathbf{z}) = \Sigma_G v_G(z) \exp(i\mathbf{G} \cdot \mathbf{R}) \tag{11}$$

where $\mathbf{r} = (\mathbf{R}, \mathbf{z})$ is the distance of the helium atom to an origin on the surface, with $\mathbf{R}$ and $\mathbf{z}$ the vectors representing the surface parallel and perpendicular coordinates, respectively, $v_G(z)$ is the two-dimensional Fourier transform of $V(\mathbf{r})$ with respect to $\mathbf{R}$, and $\mathbf{G}$ is the surface reciprocal lattice vector. The summation goes over all possible positive and negative surface reciprocal lattice vectors. The periodic, static interaction potential of a perfect surface with the probe, as expressed in Eq. (11), gives rise to coherent elastic scattering of the atoms in which the parallel component of the incoming helium wavevector $\mathbf{k}_i = (\mathbf{K}_i, \mathbf{k}_{zi})$ can change only by a reciprocal lattice vector; that is [40],

$$\mathbf{K} \equiv \mathbf{K}_f - \mathbf{K}_i = \mathbf{G} \tag{12}$$

where the outgoing helium wavevector $\mathbf{k}_f = (\mathbf{K}_f, \mathbf{k}_{zf})$ and the magnitudes of the final and initial wavevectors are equal; that is, $k_f = k_i$.

Implicit in the interaction potential are the positions of each atom of the crystal, $\mathbf{r}_l$; namely, $V = V(\mathbf{R}, \mathbf{z}, \mathbf{r}_0, \ldots, \mathbf{r}_l, \ldots, \mathbf{r}_{N-1})$. When a lattice wave propagates along the crystal surface with wavevector $\mathbf{Q}$ and frequency ω, the positions of the atoms are displaced as in Eq. (2) and the interaction potential of Eq. (11) is modulated spatially and temporally. For small displacements this potential can be expressed in the form [41]

$$V(\mathbf{R}, \mathbf{z}, t) = V_0(\mathbf{R}, \mathbf{z}) + \mathbf{V}'(\mathbf{R}, \mathbf{z}) \cdot \mathbf{u}(t) + \cdots \tag{13}$$

where V_0 is the atom–static-surface potential, $\mathbf{u}(t)$ is the displacement of the wave [summed over all the wavevectors and branch indices as in Eq. (4)], and V' is the gradient of the potential with respect to the displacement. The perturbation term allows for coherent atom scattering with the simultaneous creation or annihilation of single phonons with wavevector $\mathbf{Q}$. The "selection rule" for these scattering events takes on the same appearance as Eq. (12), namely [40],

$$\mathbf{K} \equiv \mathbf{K}_f - \mathbf{K}_i = \mathbf{G} + \mathbf{Q} \tag{14a}$$

while at the same time energy conservation demands that

$$\hbar\omega = E_f - E_i = (k_f^2 - k_i^2)(\hbar^2/2m) \tag{14b}$$

Here we use the convention that $\omega > 0$ for phonon annihilation and < 0 for phonon creation. For the differential reflection coefficient, or the fraction of incident atoms which are scattered into final solid angle $d\Omega_f$ with energy E_f to $E_f + dE_f$, the result from perturbation theory—as, for example, from the distorted wave Born approximation [39, 41]—is that

$$d^3R/d\Omega_f dE_f = C\, \Sigma_{Qj} |B(\mathbf{Q}, j)|_{fi}^2\, |\{\mathbf{e}(\mathbf{Q}, j) \cdot \mathbf{V}'\}_{fi}|^2 \cdot \delta(\mathbf{K} - \mathbf{Q} - \mathbf{G})\delta(E_f - E_i - \hbar\omega(\mathbf{Q}, j)) \tag{15}$$

where C is a coefficient which depends on initial and final wavevectors and the target atomic mass M, $|B_{fi}|^2$ is the square of the wave amplitude matrix element which is essentially proportional to the average number of vibrational quanta $\langle n \rangle$ in the mode at temperature T (actually proportional to $\langle n \rangle + 1$ for creation, $\langle n \rangle$ for annihilation, or the Bose factor) [41], and $\mathbf{e}(\mathbf{Q}, j)$ is the displacement polarization. In addition, a Debye–Waller-like factor, $\exp(-2W)$, must be appended to the right-hand side of Eq. (15) by analogy with the coherent scattering of x-rays and neutrons [41]. (Also see below.)

The form of the result in Eq. (15) is very reasonable. The energy exchange with a particular vibrational mode is enhanced by strong coupling between initial and final states through the potential gradient and becomes more likely as the displacement amplitude of the mode increases. However, one should note from Eq. (15) that when the polarization of the displacement wave is perpendicular to the potential gradient arising from the displacement, then the coupling between states is zero. Furthermore, while the amplitude or Bose factor increases with increasing surface temperature, the correspondingly larger displacements also enhance the probability of multiphonon scattering which can eventually overwhelm the contribution due to single-phonon events, and the Debye–Waller factor decreases exponentially with T. This point is discussed further below. For detailed calculations the interaction potential is normally approximated as a sum of pair potentials between the atoms in the surface and the helium probe [42]. Good results for scattering intensities have also been obtained more recently by using a parameterized Born–Mayer potential for the atom–surface short-range repulsive interaction, together with a cutoff factor for parallel momentum transfer, as in Eq. (21) below [41].

Helium atom scattering experiments are generally not very sensitive to the bulk phonons even though, in principle, they contribute to the vibrational amplitude at the surface [43]. This is because the bulk states become ''spread out'' in energy bands at the surface so that the state densities are very much lower than those of the surface states. Put in a different way, the time-of-flight energy analysis experiments are carried out at fixed incident and final angles (see the discussion in the next section) so that only a few surface modes, but a wide range of bulk modes, can be observed in each arrival time spectrum. The surface modes in these spectra then appear as sharp peaks, while the bulk modes become spread out as a broad background.

In a more rigorous approach, Manson starts the scattering treatment with the generalized golden rule for the transition rate [44],

$$w(\mathbf{k}_f, \mathbf{k}_i) = (2\pi/\hbar) \langle\langle \Sigma_{nf} |T_{fi}|^2 \delta(E_f^s - E_i^s) \rangle\rangle \tag{16}$$

where T_{fi} is the matrix element of the transition operator which couples the initial state i of the combined atom–crystal system having energy E_i^s to the final state f with energy E_f^s through the atom–crystal interaction potential, nf represents the final states of the crystal, and the double brackets signify the ensemble average over the initial states. Since only the probe's states are measured, one has to sum over all the final crystal states and average over the initial crystal states.

Then, using the standard transformation to the interaction representation to obtain the time-dependent transition operator $\mathbf{T}(t)$ and expressing the

δ-function over energy as the Fourier transform of the time-dependent correlation function, he arrives at [44]

$$w(\mathbf{k}_f, \mathbf{k}_i) = (1/\hbar^2)\int_{-\infty}^{\infty} dt \exp(-i(E_f - E_i)t/\hbar) \langle\langle ni|T_{kf,ki}(0)T_{kf,ki}(t)|ni\rangle\rangle \tag{17}$$

where the $T_{kf,ki}$ are now matrix elements over the probe states, and E_f and E_i are the final and initial probe atom translational energies. In this formalism the differential reflection coefficient can be written as

$$d^3R/d\Omega_f dE_f = [L^4/(2\pi\hbar)^3][(m^2|k_f|)/k_{iz}]\ w(\mathbf{k}_f, \mathbf{k}_i) \tag{18}$$

where m is the probe atom mass and L is a normalization length.

To proceed further, three major approximations to the theory are made [44]: First, that the transition operator can be written as a pairwise summation of elements $\tau^{l,\kappa}$ where the index l denotes surface cells and κ counts units of the basis within each cell; second, that the element $\tau^{l,\kappa}$ is independent of the vibrational displacement; and, third, that the vibrations can all be treated within the harmonic approximation. These assumptions yield a form for $w(\mathbf{k}_f, \mathbf{k}_i)$ which is equivalent to the use of the Born approximation with a pairwise potential between the probe and the atoms of the surface, as above. However, implicit in these three approximations, and therefore also contained within the Born approximation, is the physical constraint that the lattice vibrations do not distort the cell, which is probably true only for long-wavelength and low-energy phonons.

The result of these approximations is that the time dependence of the integral in Eq. (17) enters only through the exponential of the displacement correlation function $\exp(\langle\langle ni|\mathbf{k}\cdot\mathbf{u}_{l',\kappa'}(0)\mathbf{k}\cdot\mathbf{u}_{l,\kappa}(t)|ni\rangle\rangle)$, where k is defined through $\mathbf{k} \equiv \mathbf{k}_f - \mathbf{k}_i = \mathbf{K} + \Delta\mathbf{k}_z$. The integral can then be evaluated by expanding the exponential in a power series. The leading term is independent of the displacement $\mathbf{u}$ and leads to expressions for the elastic diffractive scattering. The first-order term which goes roughly as $\langle\langle(\mathbf{k}\cdot\mathbf{u})^2\rangle\rangle$ gives rise to the single-phonon scattering intensities. Although the resulting expressions in Eq. (18) are not generally so transparent, one can see that when the displacement is perpendicular to $\mathbf{k}$, the expression becomes zero. Hence, there is no contribution to the single-phonon scattering intensity for phonons polarized perpendicular to the scattering plane of the helium atoms, defined by $\mathbf{k}_i$ and $\mathbf{k}_f$.

In HAS experiments (see the next section) the crystal target is usually aligned so that the scattering plane contains a high-symmetry direction of the crystal. Since phonons propagating in a high-symmetry direction retain their symmetry character with respect to the sagittal plane for all Q's in the SBZ, each mode can be labeled either as a sagittal plane mode polarized in the scattering plane or as a shear horizontal mode polarized perpendicular to the sagittal plane. The latter cannot be observed, according to the above argument, in this experimental configuration. Moreover, the scattering intensity depends roughly on the square of $\mathbf{k} \cdot \mathbf{u} = K u_{\parallel} + \Delta k_z u_{\perp}$. In the typical HAS geometry the magnitudes of the parallel and perpendicular components of the wavevector most often are such that $K^2 << \Delta k_z^2$, so that the larger scattering intensities generally are associated with the perpendicularly polarized vibrational modes.

Further analysis of Eq. (17) demonstrates, as noted for Eq. (15), that the intensities one observes for the probe–surface energy transfer through single-phonon creation and annihilation events are modified by a surface-temperature-dependent factor analogous to the Debye–Waller factor that appears in neutron and x-ray scattering [44]. In HAS the Debye–Waller exponent $2W$ is often approximated as $3k^2 k_B T/M\omega_D^2$ [44], where k_B is the Boltzmann constant and ω_D is the Debye frequency which characterizes the simplified Debye phonon distribution function [8]. To a large extent this decrease in intensity can be attributed to the higher-order terms in the power series which give rise to multiphonon scattering. In this connection, Weare [45] has suggested a criterion for single-phonon processes to dominate the multiphonon process which is equivalent to

$$2W \approx 3k^2 k_B T/M\omega_D^2 << 0.24 \tag{19}$$

Clearly, the scattering intensities should be helped by carrying out the experiment at low surface temperatures and with relatively small wavevectors. In practice, however, one finds that the Weare criterion is overly severe and one is often able to discern distinct peaks in time-of-flight spectra due to single-phonon scattering events on top of a broad multiphonon background even for Debye–Waller exponents $2W \approx 1$.

B. Atom–Multiphonon Scattering

The atom–multiphonon component of the inelastic scattering can be obtained from the theory of Manson [44] by subtracting off the zero- and first-order terms from the power series expansion of the exponential displacement correlation function. With the assumption that the major contribution to the multiphonon scattering is due to low-energy, long-wavelength phonons, he is able to arrive at an expression with a very simple form for the transition rate,

$$w(\mathbf{k}_f, \mathbf{k}_i) = (N/\hbar^2)|\tau_{fi}|^2 \exp(-2W(\mathbf{k}))S(\mathbf{K}, \omega)I(\mathbf{K}, \omega) \tag{20}$$

where $|\tau_{fi}|^2$ is the form factor which depends on the helium–surface interaction potential, $S(\mathbf{K}, \omega)$ is the structure factor, $I(\mathbf{K}, \omega)$ is the energy exchange factor, and $2W$ is again the Debye–Waller exponent for which an approximate expression is given above. Semiclassical expressions for the $S(\mathbf{K}, \omega)$ and $I(\mathbf{K}, \omega)$ factors have been derived by Manson [44], and the form factor can be approximated with reasonable accuracy by [41, 44, 46, 47]

$$|\tau_{fi}|^2 = |v_{fi}|^2 \exp(-K^2/2Q_C^2) \tag{21}$$

where v_{fi} is the matrix element with respect to the distorted wave Born approximation states of the repulsive part of the pair potential $v_0 \exp(-\beta z)$ [46], and the second factor accounts for the decreasing probability of transferring parallel momentum as $K = K_f - K_i = Q + G$ increases [46]. This cutoff factor arises primarily because the helium atoms scatter from the electrons several angstroms from the surface and therefore interact with more than one surface atom simultaneously [25]. The comparisons of this multiphonon scattering model with experiment have been rather good, and they are described in a later section.

There are two important additional consequences of this treatment. First, for very "soft" materials such as organic films, which have very low vibrational frequencies, the Weare criterion in Eq. (19) is not satisfied at all and the inelastic scattering is dominated by multiphonon annihilation and creation events. Expression (20) can still be fit to the data, in its classical limiting form, which provides some usable information about the film's vibrational behavior. Second, even for materials for which the Weare criterion is approximately fulfilled, the background to the single-phonon peaks in time-of-flight arrival spectra is mainly due to multiphonon scattering. Thus, fitting the parameters of the theory to this background allows one to find the form factor for the interaction potential and thereby to determine the polarization of the surface vibrational states from the single-phonon scattering intensity through Eq. (15). This, in turn, would enable one to evaluate the density of surface states which could be directly compared with models of the surface forces. Taken together with the spectroscopic data, this would provide for a much more stringent test of the models of the lattice dynamics than just the surface phonon dispersion alone. In addition, fitting the multiphonon background would indirectly improve the determination of the spectroscopic quantities because one could subtract that component of the scattering from the time-of-flight spectra with some confidence and thereby enhance the signal-to-noise for weak single-phonon peaks.

IV. TIME-OF-FLIGHT SCATTERING INSTRUMENT

High-resolution HAS instruments employed to measure the surface vibrational modes generally follow the basic design pioneered by the Göttingen group [17, 48]. More thorough analyses and reviews of the design criteria are given elsewhere [21–23]. The FSU instrument [49], shown schematically in Fig. 9, is composed of several vacuum chambers connected together. The helium beam is produced by continuous expansion of helium gas from a nozzle in the source chamber, which then passes through a skimmer into the chopper chamber. For inelastic scattering experiments to determine the energy transfer, the beam is chopped into ~7-μs pulses which are employed in the time-of-flight (TOF) technique. The chopper can be displaced out of the beam path for angular distribution measurements of the total scattered intensity which is predominantly elastic diffractive scattering. In either arrangement the atoms then pass through two stages of differential pumping to the scattering chamber where they collide with the cyrstal.

For this instrument the angle between the incoming beam and the detector axis is fixed at 90°. To reach the detector the helium atoms must scatter from the surface such that the incident and scattering angles measured with respect to the surface normal sum to 90°; $\theta_i + \theta_f = 90°$. These atoms pass

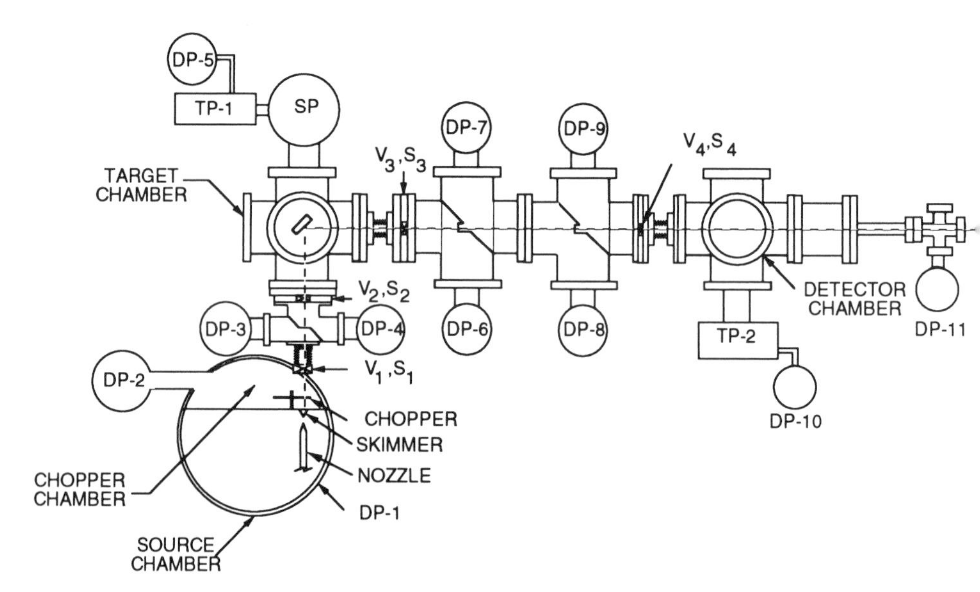

Figure 9. Schematic of the HAS instrument at the Florida State University. The symbol DP refers to diffusion pumps, TP to turbomolecular pumps, SP to titanium sublimation pumps, V to valves, and S to defining slits.

through four stages of differential pumping to the detector chamber where their flux is measured by a quadrupole mass spectrometer (QMS) operated in a pulse-counting mode. The geometry of the scattering region is shown in Fig. 10. The instrument employs a ''sump'' chamber which follows the detector. Helium atoms not ionized in the detector pass through the QMS into the sump chamber where they are pumped away, reducing the background by minimizing the number of atoms reflected from the detector chamber walls back into the ionizer region. The data acquisition system consists of a computer-controlled CAMAC interface which interacts directly with the instrument.

In brief, the purpose of having the instrument composed of so many differentially pumped sections is to maintain the 19-orders-of-magnitude difference in pressure between the ~40 atm in the nozzle and the 10^{-15}-torr partial pressure of helium in the detector required for sufficiently low background to carry out the experiment.

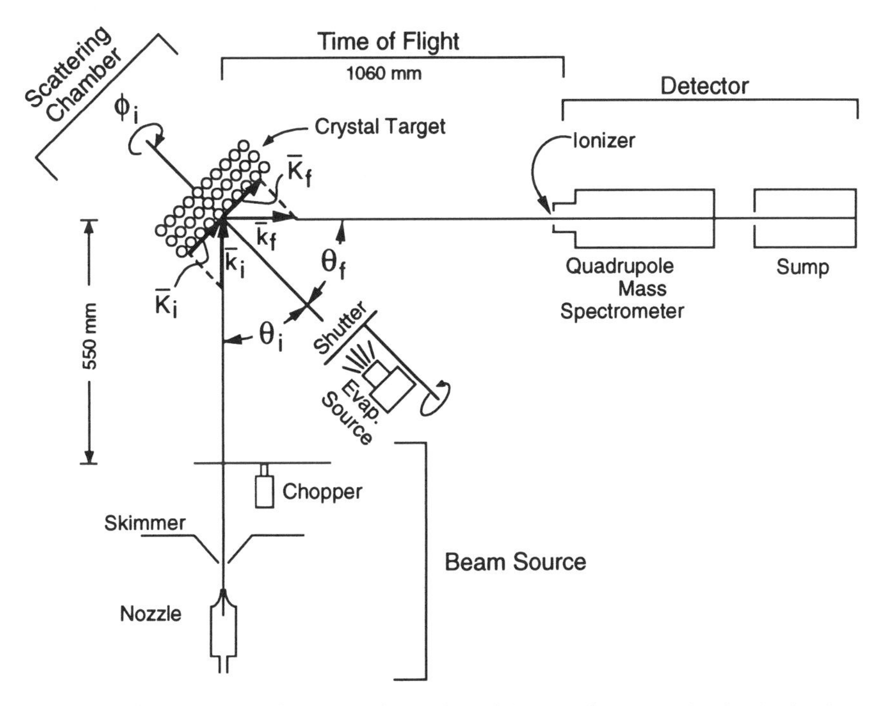

Figure 10. Schematic of the scattering region of the HAS instrument in Fig. 9, showing the dimensions from the chopper to the target and from the target to the ionizer region of the quadrupole mass spectrometer. Also shown is the deposition source for growing alkali halides, as well as the sump chamber, as described in the text.

The ionizer region of the QMS is estimated to be about 10 mm long, and thus for helium atoms with a wavevector of ~ 7 Å^{-1} (or speed ~ 1100 m/s) it contributes ~ 9 μs to the width of the flight times. If one assumes that the time distributions due to the intrinsic helium beam speed distribution ($\Delta v/v \approx 1\%$), the chopper function, and the ionizer region function are approximately Gaussian in shape, then the observed TOF width should be the square root of the sum of the squares of the contributing widths. For our pathlength of 1060 mm under these conditions the full width at half-maximum (FWHM) for the specularly reflected beam ought to be ~ 15 μs. In most of the experiments carried out with the FSU instrument, the collimation of the He beam is set so that its calculated cross-sectional area is roughly 5 mm^2 at the target and the angular resolution (FWHM) of the detector is $\sim 0.15°$, very close to the value expected from the instrument geometry.

The incident helium beam energies in this instrument currently can be varied from about 15 to 60 meV by suitably cooling the nozzle source. These values correspond to incident wavevectors of ~ 5 to 11 Å^{-1}, or de Broglie wavelengths ($2\pi/k_i$) of about 0.6 to 1.3 Å, and provide a good match with the energies and momenta of the surface phonons for determining the surface phonon dispersion.

The crystal target is mounted onto a manipulator which allows the surface to be aligned in the desired orientation (usually a high-symmetry direction) by permitting translation in the x, y, and z directions, azimuthal rotation, and several degrees of tilting. The heating and refrigerator cooling attachments to the manipulator also permit variation of the temperature of the crystal holder from approximately 40 K to 1000 K. The manipulator itself is attached onto a differentially pumped, rotatable platform so that the incident angle of the helium beam onto the crystal, θ_i, may be changed without disturbing the alignment controls. A stepper motor under computer control is used to drive the platform.

Preparation of the target surface is of paramount importance and generally follows the long list of "tried and true" recipes which vary from material to material. For metallic surfaces, after an initial preparation stage the normal procedure in vacuum is to sputter with ions and then anneal; sometimes there are also "cleaning" treatments with O_2, followed by sputtering or/and annealing [50]. Typically, Auger electron spectroscopy (AES) and/or x-ray photoelectron spectroscopy (XPS) are used to gauge the chemical composition of the surface and low-energy electron diffraction (LEED) is employed to check the surface quality and preliminary alignment [27].

For alkali halides our group has had equally good success either by cleaving the target in air and then heating gently under vacuum in the scattering chamber to drive off adsorbed water and gases or by cleaving the target *in situ* [51, 52]. In both cases these surfaces usually need to be cleaned by flashing to about 400°C on a daily basis to maintain their high quality. At this temperature sublimation takes place, effectively producing a new surface. For other ionic insulators such as the oxides NiO and MgO, the results are far, far better after cleaving in vacuum [51, 53]. However, for CoO the surface quality deteriorates so rapidly even at very low pressures ($\sim 10^{-10}$ torr) that TOF measurements longer than a few days are problematic [51]. Generally, the oxide surfaces have resisted restoration through heating, although it is possible that this is only because high enough temperatures are too difficult to achieve (>1300 K). There is some evidence from this laboratory that new surfaces of oxides can be grown *in situ* by molecular beam epitaxy. High-quality surfaces of layered materials like $TaSe_2$ have been produced by "peeling" with an attached level arrangement *in situ*, followed by very gentle heating to avoid decomposition [54].

For growth studies, films on substrates can be produced by a variety of sublimation, vaporization, and dosing procedures depending very much on the materials. Under conditions where the overlayer grows two-dimensionally (i.e., layer-by-layer), one is able to follow the growth in real time by monitoring the intensity oscillations of the specular helium beam during the deposition procedure [55]. These oscillations arise because the specular scattering is a coherent process and as material is being deposited onto the surface, defect sites accumulate which diminish the coherence of the surface. Hence, the specular intensity drops until about a half-monolayer is grown; at this point, further deposition begins to improve the surface quality and the specular beam intensity increases. This oscillatory behavior is also seen in reflectance high-energy electron diffraction (RHEED) during growth; however, helium scattering is a much gentler probe of the surface, particularly for insulators. By regulating the parameters of the film growth, such as target temperature and deposition rate, one can optimize the layer-by-layer growth of the new surface material and thereby explore the lattice dynamics of these thin films for each layer and over a range of stability regimes. In some overlayer studies where the surface geometry is modified, as in hydrogen adsorption on metals, it may be more convenient to follow the growth of the new structure by monitoring the growth of a new diffraction beam. The chemical inertness of the helium probe and its insensitivity to the presence of electric and magnetic fields makes such studies feasible for a wide variety of materials and growth conditions.

The dispersion relations are obtained through analysis of the peaks in the TOF spectra. To start, each arrival time spectrum is first converted into an energy transfer distribution. When the path length from the target to the detector is d and the transit time for this distance is t, then the wavevector of the arriving helium atom is $k = m(d/t)/\hbar$ and its translational energy is

$E = \frac{1}{2}m(d/t)^2$. The initial values k_i and E_i are found from the TOF of the elastic scattering, as for example, at the specular angle (45°). For the inelastic scattering the energy transferred is $\Delta E = E_f - E_i$. Since the detector is a QMS which transmits ions produced by electron bombardment, the number of counts recorded in time interval (or window) Δt at time t in the TOF spectrum, $T(t)$, is proportional to the *number density* (atoms per unit volume) of helium atoms in the ionizer region. This signal is converted to the helium *flux density* (atoms per unit area per second) for comparison with theory simply by multiplying by the speed of these atoms, $\Phi(t) = T(t)(d/t)$. Through the standard Jacobian transform between distribution functions, one obtains the energy transfer distribution

$$P(\Delta E) = \Phi(t)|dt/d\Delta E| = T(t)(t^2/md) \tag{22}$$

An example of this conversion is shown in Fig. 11 for a TOF spectrum from a two-monolayer KBr film grown onto RbCl(001) [52].

Analysis of the peak positions in Fig. 11 yields values of E_f and k_f. Since $\omega = (E_f - E_i)/\hbar$, this measurement determines the point $\omega(K)$ for the KBr film. Note that $K = K_f - K_i = k_f \cos\theta_i - k_i \sin\theta_i$ since $\sin\theta_f = \cos\theta_i$ for the 90° instrument. To construct an entire dispersion curve from such data requires TOF spectra for a range of incident angles θ_i. One can see how this works in Fig. 12, where a set of scan curves, the combination of Eqs. (14a) and (14b) [17].

$$\hbar\omega/E_i = [(K_i + K)/(k_i \cos\theta_i)]^2 - 1 \tag{23}$$

has been drawn for several incident angles in an extended zone plot. Only those phonons which satisfy Eq. (23) can be observable in the TOF spectrum obtained at a given incident angle for a given incident wavevector. Conversely, the single-phonon peaks found in the spectrum must correspond to intersection points of the scan curve with the dispersion curves. Thus, by carrying out measurements over a wide range of angles and folding back all the analyzed points $\omega(K)$ into the first Brillouin zone, one can map out the dispersion across the SBZ to obtain $\omega = \omega(\mathbf{Q})$.

To recover the scattering intensities due to single phonon creation or annihilation events, which are proportional to the differential reflection coefficients through an instrument factor requires subtracting off the background in the TOF spectra. As discussed in the previous Section, it is now possible in principle to do this for the multiphonon scattering. Other contributions to the background, such as the diffuse inelastic scattering, depend on the surface quality and can be minimized by careful surface preparation.

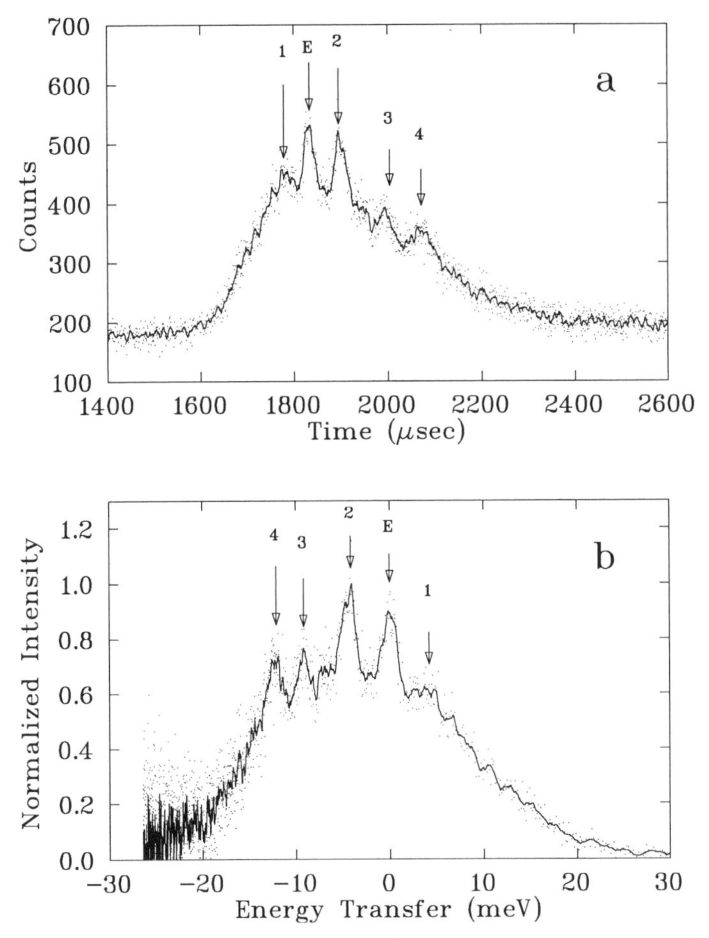

Figure 11. (*a*) TOF arrival spectrum for helium scattering from a two-monolayer film of KBr deposited onto RbCl(001). The total counts of arriving helium atoms in each 1-μs window of the multichannel scaler are shown as the small dots and the jagged solid line is the Savitzky–Golay average of these points. The peak labeled E corresponds to diffuse elastic scattering from surface defects; the peaks labeled by numbers correspond to phonon creation (to the right of peak E) and to phonon annihilation (to the left of peak E). (*b*) The spectrum in (*a*) has been transformed to an energy transfer distribution as described in the text, and the labels on the peaks match those in (*a*). Note that positive energy transfer corresponds to phonon annihilation and negative energy transfer to creation. The large background in both panels is primarily due to multiphonon scattering. (Reproduced from Ref. 52.)

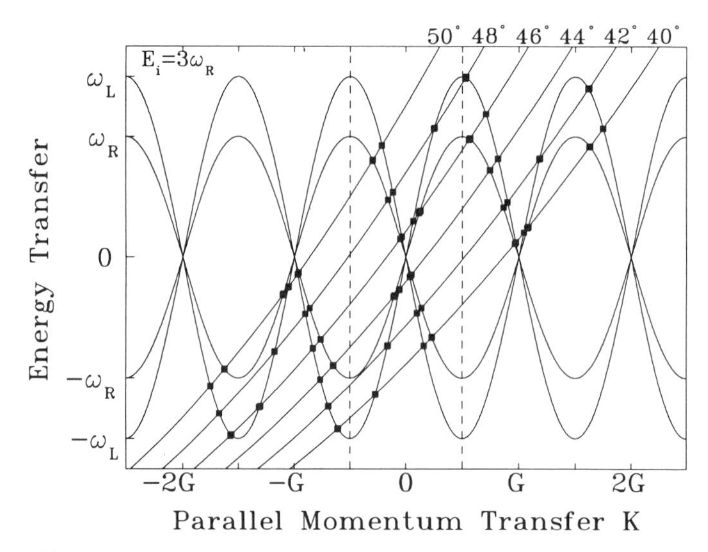

Figure 12. Intersection of scan curves with surface phonon dispersion curves drawn in an extended zone plot. For this figure a Rayleigh wave and a longitudinal resonance are represented by sine curves with maximum frequencies $\omega_R = 12$ meV and $\omega_L = 18$ meV, respectively. The lattice constant for the surface is taken to be 1 Å so that $G = \pi$ Å^{-1}. The scan curves for incident energy $E_i = 36$ meV, Eq. (23), are drawn for even incident angles from 40° to 50°. The intersection points (solid squares) along each scan curve are those single-phonon creation (negative energy transfer) and annihilation (positive energy transfer) events which are kinematically possible to observe in a time-of-flight measurement at that angle with a 90° instrument, such as the one shown in Figs. 9 and 10.

V. CLEAN CRYSTALLINE SURFACES: IONIC INSULATORS

A. Alkali Halides

The first experimental determination of surface phonon dispersion by high-resolution helium atom scattering was of the cleaved LiF(001) surface [17]. It was selected for several reasons, among them that it is relatively easy to prepare and clean, and that Estermann and Stern [56] had shown that the diffraction pattern obtained from this surface by H_2 and He scattering satisfied the de Broglie relation for matter waves. More pertinent perhaps is that the (001) surfaces of LiF and the other rocksalt-structured alkali halides had been investigated theoretically by slab dynamics calculations [36] and by a Green's function method [34] since the early 1970s so that the measured dispersion curves could be directly compared with theoretical predictions.

The alkali halides are the prototypical ionic insulator materials. The ions all have closed shell configurations with essentially localized electronic distributions, which leads to large electronic band gaps and insulator behavior [8]. The forces of interaction between ions governing the lattice dynamics are predominantly (a) the long-range Coulombic attraction and repulsion of the ions and (b) the short-range overlap repulsions of the electronic "clouds." The chemistry of these materials is rather similar, although there are differences in some properties such as aqueous solubility, which can in part be ascribed to the variation in size and in polarizability in the series from F^- to I^- and Li^+ to Cs^+.

The face-centered cubic (fcc) unit cell common to the alkali halides except for CsCl, CsBr, and CsI is shown in Fig. 13; these crystals all cleave fairly easily along the (100) planes. One should note that the unit surface cell for the (001) plane has lattice vectors rotated by 45° from those of the bulk. The generic direct and reciprocal lattices for the (001) surfaces of these materials are shown in Fig. 14.

The dynamical model employed in the theoretical calculations is the *Shell Model*, of which there are several variants [6, 9]. It is designed to approximate the physical situation of the ions in the crystalline environment more realistically than does the Born–von Kármán treatment with harmonic force constants between neighboring atoms as discussed in Section II, but its handling of the forces is not so very different. The Shell Model was developed for these materials to account for the bulk phonon dispersion that was measured by neutron scattering experiments as well as for their dielectric

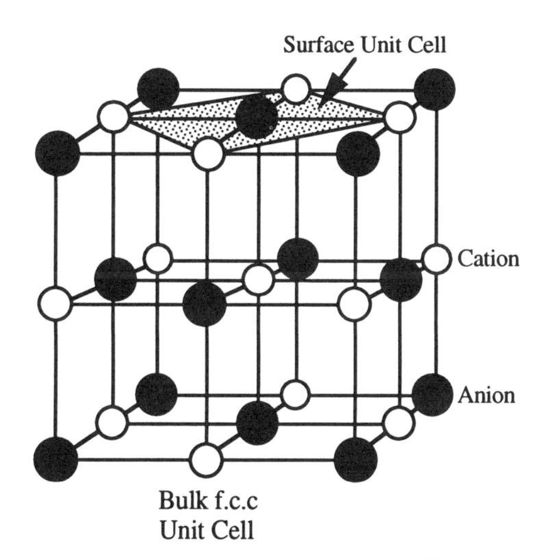

Figure 13. Schematic drawing of the fcc unit cell for rocksalt structure. The surface unit cell for a (001) surface is shown as the shaded area and is rotated by 45° with respect to the bulk cell.

a) Direct Lattice

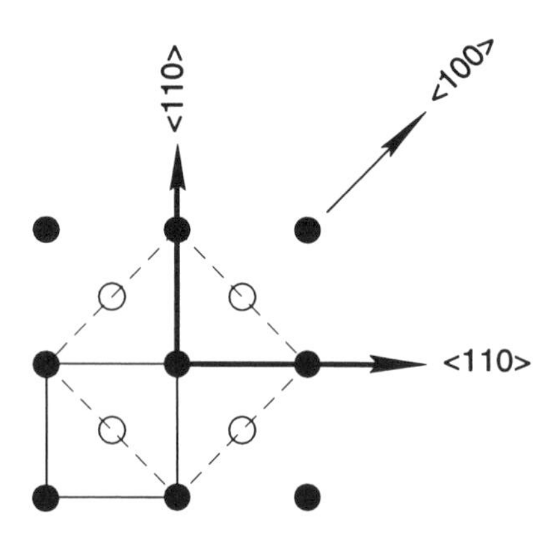

b) Reciprocal Lattice

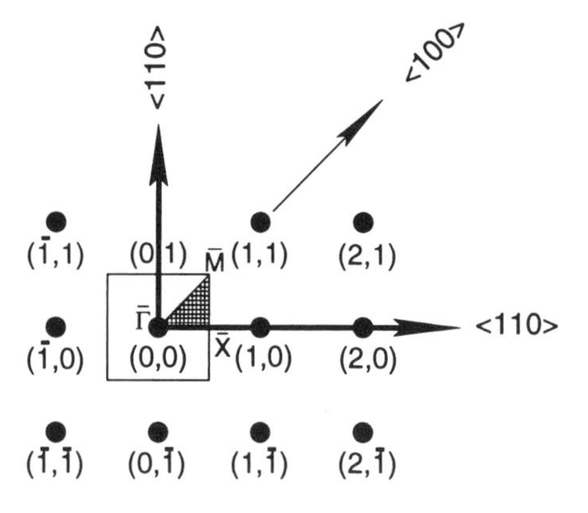

Figure 14. (*a*) Schematic of the direct lattice for the (001) surface of the rocksalt structured crystal of Fig. 13 with the bulk unit cell indicated by dashed lines and a surface unit cell by solid lines. The high-symmetry directions for this surface, ⟨100⟩ and ⟨110⟩, are indicated with arrows. (*b*) The reciprocal lattice corresponding to the direct lattice in panel *a* is shown with the first surface Brillouin zone (SBZ) indicated by solid lines. The shaded triangular region, $\bar{\Gamma}\bar{X}\bar{M}$, is the irreducible portion of the SBZ, from which the entire SBZ can be constructed by symmetry.

properties [6]. In its simplest form, shown in Fig. 15, each ion is depicted by a core having the total ionic mass but only part of its charge and a massless shell representing the valence electrons with the rest of the charge. The point of the model is to allow for the distortion of the electronic distribution (especially of the outer electrons) around the core due to the vibrations of neighboring ions—that is, to treat the polarizability of the ions in the crystal. (An earlier model for the alkali halides, the Kellermann or Rigid Ion Model, did not consider the ionic polarizabilities [57].) Each core is coupled to its shell by a harmonic spring, and in the simplest version shown in Fig. 15 the ions are coupled together through springs between their shells.

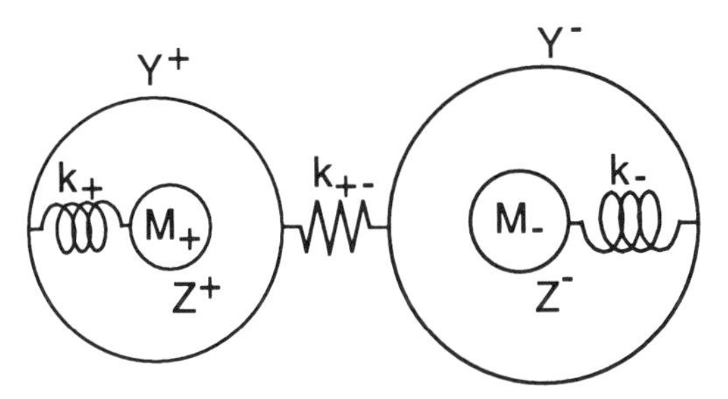

Figure 15. Schematic of a simplified version of the Shell Model. The positive ions with mass M_+ and charge Z^+ are bonded to nearest-neighbor negative ions which have mass M_- and charge Z^- through their shells representing polarizable valence electrons. The portions of the total ionic charges at the massless shells are Y^+ and Y^-, respectively. The force constants between the positive ion core and its shell, between the negative ion core and its shell, and between shells are k_+, k_-, and k_{+-}, respectively. Analogous force constants, not shown, are used between shells of the next nearest neighbors (i.e., between like-charged ions).

Typically, only six spring constants are used to represent the central force overlap repulsion. They are the three pairs of radial and tangential force constants between neighboring positive ions, between neighboring negative ions, and between nearest-neighbor positive and negative ions [33]. Next nearest neighbors, and so on, can be similarly coupled if desired. The calculation for the long-range Coulombic interactions between the ions needs to be carried out carefully because it is only slowly convergent. For most theoretical calculations of the surface dynamics, model parameters have been taken essentially unchanged from fits to the neutron scattering experiments for the bulk phonons. It is worth pointing out that the determination of the surface phonon dispersion is, in principle, a more stringent test of the applicability of the dynamical model than fitting the bulk dispersion relations in that the higher symmetry of the bulk lattice may allow some cancellation of errors which cannot be overlooked in the lower symmetry of the surface [28].

Alkali halide surfaces are all highly corrugated because of the localized charge distributions, and diffractive helium atom scattering typically reveals several Bragg diffraction peaks with intensities comparable to the specular peak. This is sharply contrasting with the diffraction patterns obtained from metallic surfaces which are rather flat to helium scattering, consisting of large specular peaks and very small first-order Bragg peaks.

1. LiF

Bracco et al. [58] extended the range of measurements first reported by Toennies and coworkers [17] to cover the entire SBZ of LiF, as shown in Fig. 16. In the original work the Rayleigh wave (RW), labeled S_1 in the

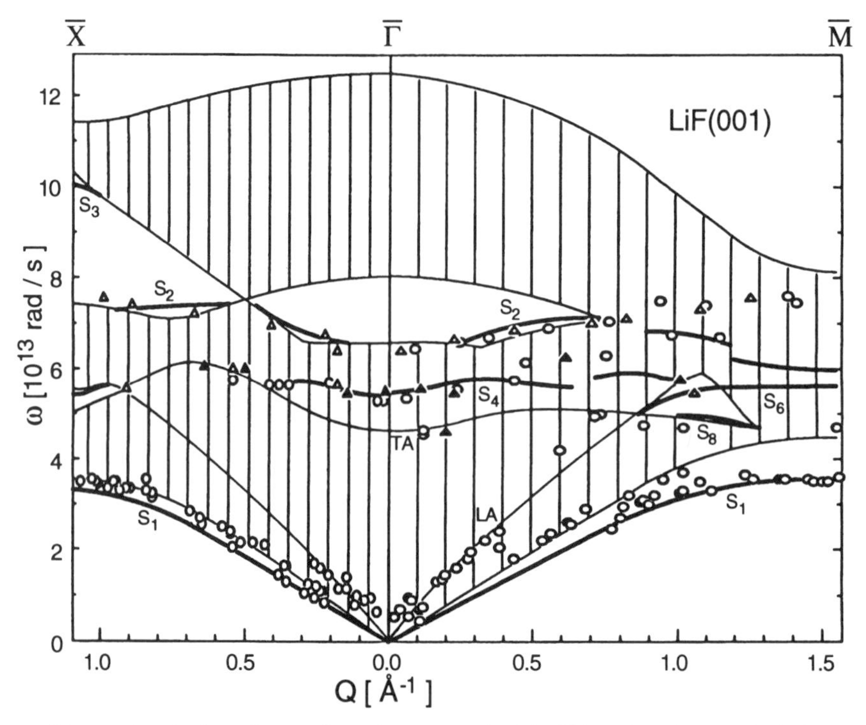

Figure 16. Surface phonon dispersion curves for LiF(001). The calculated bulk bands are indicated by the vertical-striped regions. The surface localized modes are shown by heavy solid lines, whereas the resonances lying within bulk bands are given by thinner solid lines. The mode label S_1 refers to the Rayleigh wave, S_6 to the longitudinal resonance, S_8 to the crossing resonance, and S_2, S_3, and S_4 to optical modes. (Reproduced from Fig. 2 of Ref. 58, with permission of Elsevier Science Publishers.)

figure, was measured out to the zone boundary in the ⟨100⟩ high-symmetry direction ($\overline{\Gamma}\overline{M}$ region of the SBZ). The agreement with the Shell Model calculations was found to be rather good except near the zone boundary (or $\overline{M}$ point), where the frequencies were found to lie below the calculated values by about 11%. The theoretical groups proposed that this discrepancy could be removed by either increasing the polarizability of the F^- [59] ion at the surface or by the inclusion of noncentral forces in the interaction [60].

The more extensive results of Bracco et al. [58] include data in both high-symmetry directions, ⟨100⟩ and ⟨110⟩ ($\overline{\Gamma}\overline{M}$ and $\overline{X}\overline{\Gamma}$ regions of the SBZ). In addition to the RW, they observed points that reasonably well match the calculated longitudinal acoustic resonance, S_6, and the optical surface modes S_4 (or Lucas mode) and S_2. The S_6 mode originates from the lower edge of the bulk longitudinal acoustic (LA) band, but partially hybridizes with the transverse acoustic bulk (TA) band modes, because both have sagittal plane polarizations in the reduced symmetry of the surface. The Lucas mode is one of a pair with S_5 which originates from the bulk transverse optical modes; S_5 is polarized shear horizontal, normal to the sagittal plane, and cannot be observed in this helium scattering alignment [24, 35]. Going to small Q, or toward the zone center, the calculations show that the S_4 mode becomes polarized longitudinally and that the S_2 mode, which stems from the bulk longitudinal optical branch, is polarized perpendicular to the surface.

The data obtained by Bracco et al. for the S_2 mode appear in the TOF spectra as *phonon-assisted bound state resonances* rather than as single-phonon creation/annihilation events. This phenomenon was first shown to occur by Evans et al. [61], but has since been observed in a number of systems, some of which are mentioned later in this section. A bound state resonance refers to a temporary trapping of the atom in a bound state of the atom–surface potential due to a diffractive scattering event involving particular values of the surface reciprocal lattice vector. It often shows up as a resonance feature in the angular distribution measurements away from the diffraction peaks. When not complicated by other structure in the scattering intensity, such features can sometimes be analyzed to determine the bound state energies of the atom–surface potential [62, 63]. A similar resonance effect can occur with phonon assistance; that is, the helium atom scatters into a bound state as the result of the diffraction event and a simultaneous phonon creation or annihilation event. When the momentum transfer K of the process and the phonon frequency $\omega(Q)$ satisfy the scan curve equation [Eq. (23)], then such resonances can appear as peaks in TOF spectra. For LiF the bound state energies had been measured previously [61, 63] so that the S_2 phonon frequencies and wavevectors could be determined.

The results of this work compared with the newer calculations [59] suggest that the better explanation for the discrepancy in the RW energy at the zone boundary is the enhanced polarizability of the F^- ions at the surface, although there are still some discrepancies for the S_2 mode in the ⟨100⟩ direction. What also makes these results interesting is just the fact of observing the optical modes at all. As discussed in Section II, the displacement in the optical vibrational modes is predominantly that of the lighter ion—in this case Li^+, which has only about one-third the mass of F^-. For this material, the expectation was that since F^- is very much larger than Li^+ (The ratio of ionic radii, fluoride to lithium, is nearly 2! [64]), the perturbation of the helium–surface interaction potential due to the vibrating Li^+ would be very weak and, hence, very few single-phonon scattering events involving the optical phonons could be observed. As the data show, this qualitative analysis and prediction turned out to be wrong. Evidently, the vibrating Li^+ modulates the charge distribution of the polarizable F^- ions sufficiently greatly that the probability for optical phonon creation scattering events is much larger than one might have thought.

2. *KBr and RbCl*

KBr and RbCl form a mirror pair in that the masses of Rb and Br (85 and 80 amu, approximately), and K and Cl (39 and 35 amu) are close [32]. In the first approximation, then, one would expect that their dispersion relations should be similar as is the case for the bulk (see Fig. 7); this is, in fact, largely the situation. However, there remain important differences in their surface vibrational properties.

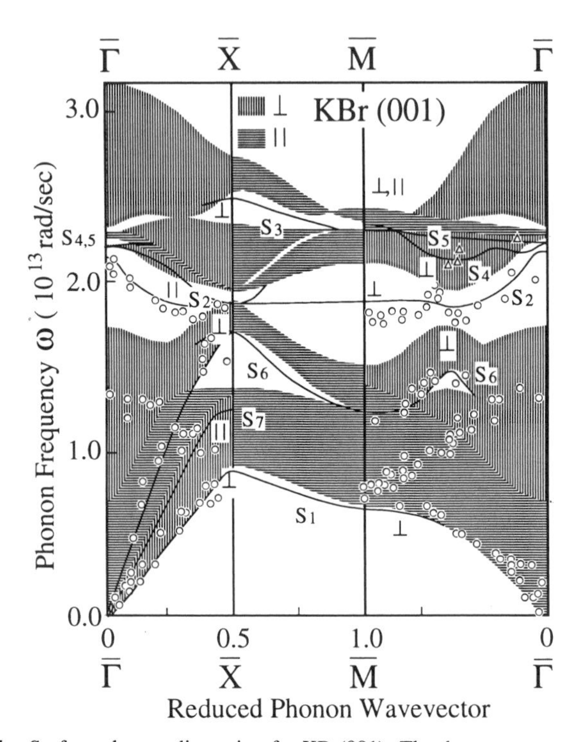

Figure 17. Surface phonon dispersion for KBr(001). The data are compared to a Green's function calculation used to determine the bulk bands (shown by the shaded regions with polarizations perpendicular or parallel to the surface as indicated in the figure) and the surface localized modes (shown as solid lines). The predominant polarizations of the modes are indicated by perpendicular and parallel symbols, and the labels of the modes follow the notation in Fig. 16. Note that modes S_7 and S_5 are polarized shear horizontal and cannot be observed in this scattering arrangement. The data plotted as triangles are obtained from weaker peaks in the TOF spectra than the points represented by open circles. (Reproduced from Fig. 8 of Ref. 49, with permission.)

KBr (001) is unique among the surfaces studied by this group in that more branches of the surface dispersion relations, or vibrational modes of more varied character, were found than in any other ionic insulator. The measured dispersion points from TOF spectra are shown in Fig. 17 along with the Green's function calculation by Benedek et al. [49]. Here one finds the RW, the lowest-energy sagittal plane mode, in both the ⟨100⟩ and ⟨110⟩ high-symmetry directions, $\overline{\Gamma}\overline{M}$ and $\overline{X}\overline{\Gamma}$, respectively. In addition, one sees across each of these regions of the SBZ the longitudinal resonance S_6 and the S_2 optical mode which lies in the gap between the bulk acoustic and optical bands. Furthermore, there is a crossing resonance in each region of the SBZ which goes from the top of the Rayleigh wave at the zone boundaries upwards to higher energy at the zone center. Interestingly, this resonance is not indicated in the calculated dispersion relations because it is not a true surface mode; rather it shows up in the calculation as a high density of states projected onto the surface [65]. Finally, there are a few points which seem to be associated with the S_4 optical mode which lies at even higher energies for this crystal although these appear with much lower intensity in the TOF spectra.

For RbCl(001), shown in Fig. 18, in the ⟨100⟩ direction one finds the RW and the optical S_2 gap modes which nearly span the $\overline{\Gamma}\overline{M}$ region of the SBZ, but only a few points appear to be associated with the longitudinal resonance S_6 and the crossing resonance [52, 66]. In the ⟨110⟩ direction, only the RW is measurable across the $\overline{\Gamma}\overline{X}$ region of the SBZ, with only a

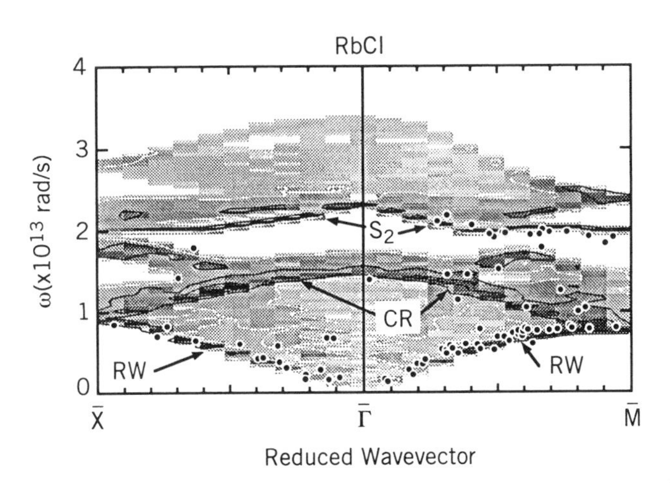

Figure 18. Surface phonon dispersion for RbCl(001). The shaded regions correspond to the surface projected density of states, with the darker shades representing higher state densities. The Rayleigh wave, crossing resonance, and optical mode are indicated by RW, CR, and S_2, respectively. (Reproduced from Ref. 119.)

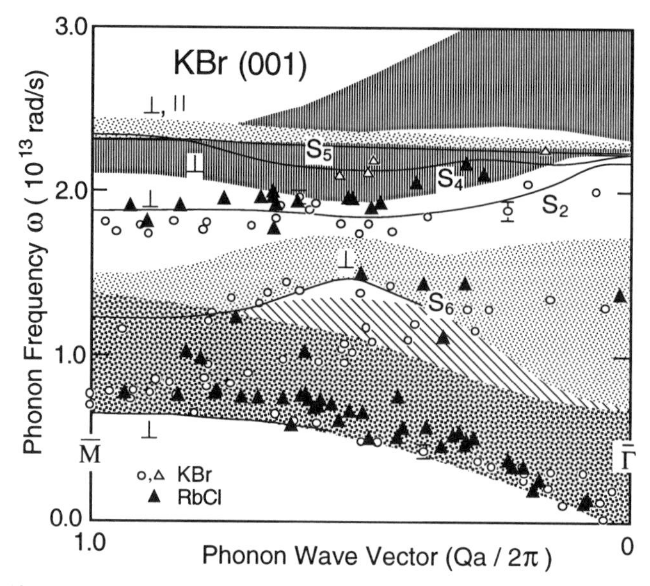

Figure 19. Comparison of the surface phonon dispersion of KBr(001) and RbCl(001) in the $\overline{\Gamma}\overline{M}$ region of the SBZ. The open triangles and open circles are for KBr and the closed triangles for RbCl. The shaded regions are adapted from the calculation in Fig. 17 for KBr. (Reproduced from Fig. 6 of Ref. 66, with permission.)

few points possibly associated with the longitudinal resonance. The data for the two surfaces in the ⟨100⟩ direction, $\overline{\Gamma}\overline{M}$, are overlaid in Fig. 19 for comparison [66].

In the bulk the phonon dispersion for RbCl and KBr are remarkably similar. For the surface vibrations the similarity is there, but it is not particularly remarkable. For the RW and the optical gap modes, Fig. 19 shows that they largely overlap, except perhaps that the RbCl vibrations are at slightly higher energy toward the zone center ($\overline{\Gamma}$) in S_2. The major difference between the two substances lies with the other acoustic vibrations, the longitudinal and crossing modes. Since longitudinally polarized displacements do not couple as well in atom scattering as the displacements normal to the surface (see Section III.A), and since the motion in the acoustic modes is predominantly that of the heavier ion, the implication is that the static helium–surface interaction potential is more perturbed by the Br^- displacements in KBr than by the Rb^+ displacements in RbCl, perhaps because of the larger polarizability of the Br^- than of the Cl^-.

The crossing resonances are primarily polarized normal to the surface, originating with *bulk* zone boundary transverse and longitudinal modes which have hybridized acoustic and optical character and appear as "folded" extensions of the Rayleigh wave [65]. The pattern of motion of the acoustic vibrational RW (see Fig. 5) is of the two ions displacing "in-phase" for small Q values, but becoming increasingly just the displacement of the heavier ion from the zone center out to the zone boundaries (from $\bar{\Gamma}$ to $\bar{M}$ and from $\bar{\Gamma}$ to $\bar{X}$). In the crossing resonance from the zone center to the zone boundaries, the lighter ion again decreases in displacement relative to the heavier one, but the pattern of motion is "anti-phase," or optical in character, in contrast to the RW. There appear to be two plausible explanations for the differences between KBr and RbCl: Either the extent of hybridization is less for these vibrational modes in RbCl than in KBr so that they are not as well-defined or this vibrational motion affects the helium–surface interaction less in RbCl. From the calculations by Schröder and Bonart [67] the projected densities of states normal to the surface seem to be comparable so that the second explanation appears more likely. However, at present there is no simple experimental test to decide the question.

3. *RbI*

The results for the scattering from RbI(001) are summarized in Fig. 20, where the peaks in the TOF spectra have been converted to points of the surface phonon dispersion. In the lower panel the data are compared with a slab dynamics calculation for which the surface has been allowed to relax; in the upper panel the surface has not been allowed to relax, with the spacing between layers being kept fixed to the bulk value [68]. According to the calculation by de Wette and coworkers [28, 37, 69], RbI has the largest relaxation by far of all the alkali halides, roughly 6% of the lattice constant, with the Rb^+ and I^- ions relaxing by differing amounts, Rb^+ more than I^-, which causes the surface to "rumple" by about 11%. They find that the relaxation pattern reverses in the second layer, I^- relaxing more than Rb^+, and the total extent of relaxation is less; by the fourth layer the relaxation becomes negligible. One can see by comparing the calculated curves in the figure that the dispersion has been modified substantially by the surface relaxation, especially the optical modes. The most notable prediction of this calculation is that the relaxation "stiffens" the S_2 mode so that it should lie above the bulk bands. (Some authors label this mode S_3 rather than S_2 [24, 35].)

The data in Fig. 20 seem to bear this prediction out; in the $\bar{\Gamma}\bar{M}$ region of the SBZ, several points were found which match the calculated S_2 mode rather well. Other vibrational modes which are observed in RbI are the RW and a few points that appear to be associated with the crossing and longitudinal resonances. One should point out that because of the large mass of the I atom, the RW frequencies are among the lowest, if not the lowest, so far measured for surface vibrations (<3.5 meV). Also seen in this region

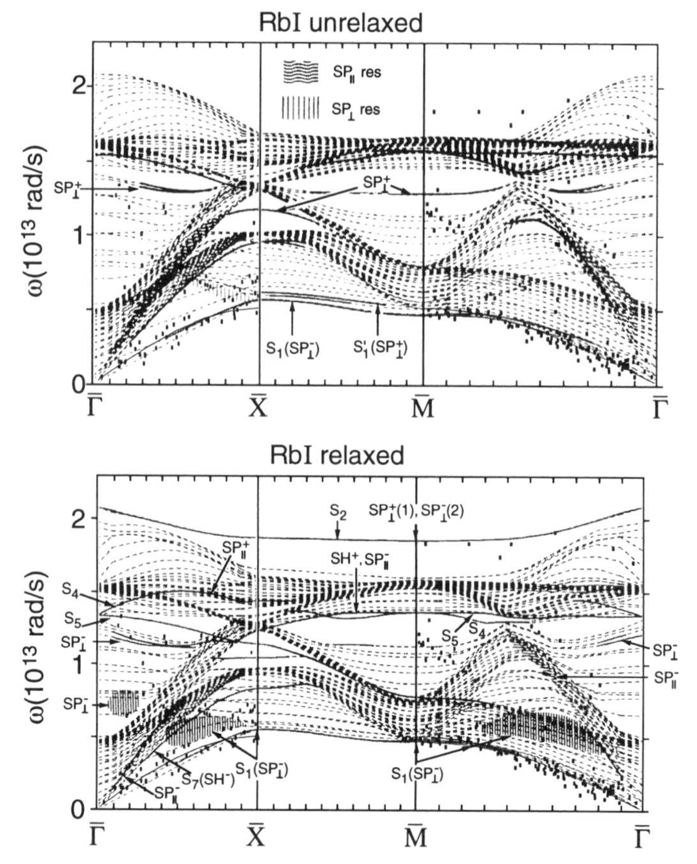

Figure 20. Surface phonon dispersion for RbI(001). The upper panel shows a comparison of the HAS data with a slab dynamics calculation for the unrelaxed surface, while the lower panel is a comparison of the same data with a similar calculation for a relaxed surface. The sagittal plane and shear horizontal modes are labeled by SP and SH, respectively, and the superscripts indicate which ion (Rb^+ or I^-) is predominantly involved in the motion of the mode. The other labels follow the notation of Figs. 16 and 17. (Reproduced from Fig. 3 of Ref. 68, with permission.)

of the SBZ is a series of points which decrease in energy from the top of the acoustic band at $\overline{M}$ to about one-third of the way toward the zone center. The origin of these points is not understood at this time.

In the $\overline{\Gamma}\overline{X}$ region of the SBZ, no points corresponding to the S_2 mode were found. However, the relaxation appears to introduce a sagittal plane mode with perpendicular polarization just below the shear horizontal mode S_5, for which a number of data points are rather close. Again, there are some points that appear to be associated with crossing and longitudinal resonances, but these vibrations do not appear as well-defined in the helium scattering for this high-symmetry direction.

4. *NaF, KCl, and RbBr*

These three salts are "isobaric" because the masses of the anions and cations are nearly the same. Although crossing resonances were somewhat unexpectedly observed in the studies of KBr and to a lesser extent in RbI, for these compounds an S_8 mode, the "folded" extension of the Rayleigh wave as in Fig. 6, was anticipated [70]. For the isobaric case, this vibration appears to be a true surface mode at least for part of the way across the SBZ and not just a resonance with a high density of states at the surface [70].

The experimental results for NaF and RbBr are shown in Figs. 21 and 22, respectively. In the data reported by Benedek et al. [70] for NaF, how-

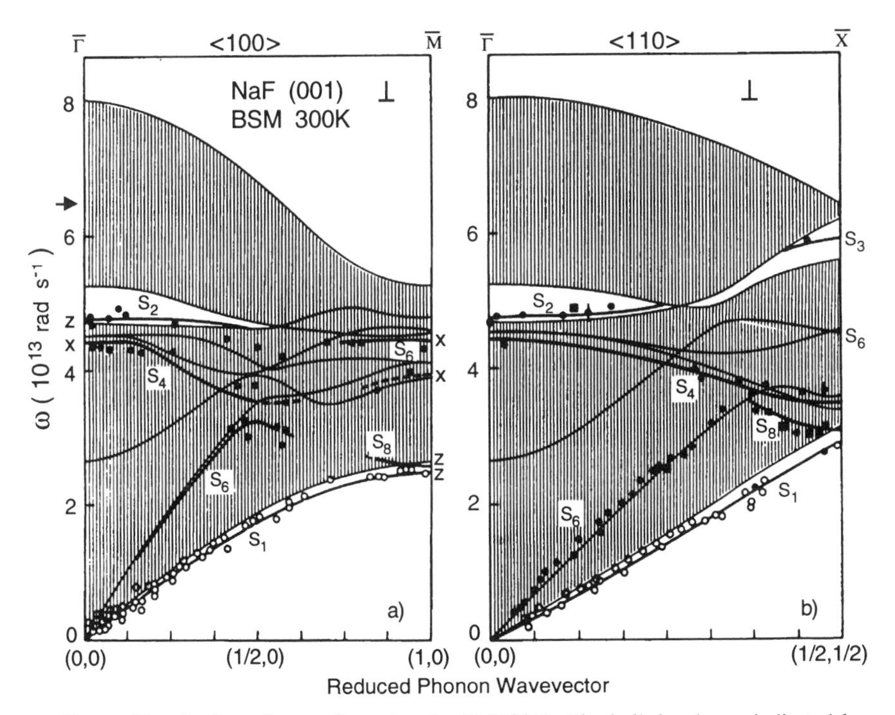

Figure 21. Surface phonon dispersion for NaF(001). The bulk bands are indicated by shaded regions and the surface localized modes by heavy solid lines, as determined by a Green's function calculation. The mode labels follow the notation of Fig. 16. The z and x designations indicate the predominant mode polarization, perpendicular and parallel, respectively. (Reproduced from Fig. 4 of Ref. 70, with permission).

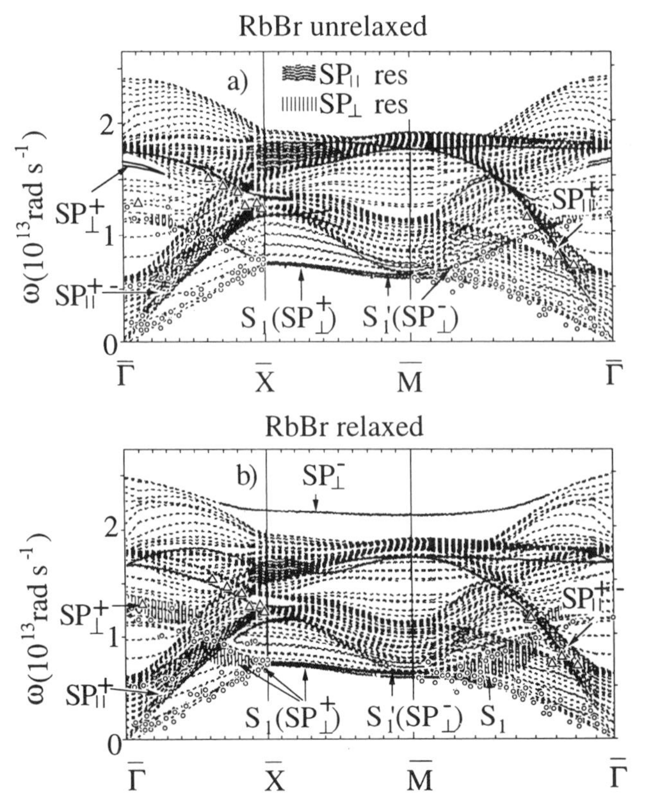

Figure 22. Surface phonon dispersion for RbBr(001). The upper panel shows a comparison of the HAS data with a slab dynamics calculation for the unrelaxed surface, while the lower panel shows the comparison for the same data with a similar calculation for a relaxed surface. The superscripts indicate which ion (Rb^+ or Br^-) is predominantly involved in the motion of the mode. The open triangles correspond to weaker peaks in the TOF spectra than the points represented by open circles. (Reproduced from Fig. 7 of Ref. 71, with permission.)

ever, the most prominent vibrational branch beside the RW is the longitudinal resonance S_6 in both high-symmetry directions and, especially in the ⟨100⟩ direction, the S_2 optical mode. The crossing mode for this compound, according to the calculations, quickly mixes with the longitudinal resonance and the S_4 optical mode and so does not exist across the entire SBZ. The experimental results from the Toennies' group on KCl (not shown) is limited to the $\overline{\Gamma}\overline{M}$ region [17]. They basically show only the RW and part of the crossing mode, although recent calculations by Schröder and Bonart [67] suggest that the crossing mode should have a high density of states across both high-symmetry regions of the SBZ. Calculations by de Wette and coworkers [28, 36] show that the (001) surfaces of NaF and KCl do not exhibit sufficient surface relaxation that the surface phonon dispersion should be measurably affected, and, as the figure for NaF demonstrates, the fits to the data are quite good.

For RbBr, however, the calculations predict an effect similar to that found for RbI; that is, the relaxation, predicted to be significant, should stiffen the S_2 mode energies, shifting them above the bulk bands partway across the SBZ [69]. Despite considerable effort, Chern et al. [71] (Fig. 22) could not find these modes and, in fact, concluded that, overall, the fit of the data to the calculated dispersion curves was better for the calculation which did not allow for surface relaxation. A more recent calculation with a modified set of Shell Model parameters indicates that the relaxation of this surface may not be as large as in the original prediction [67]. For this salt the only optical modes observed lie in the region near $\overline{X}$ and extend back about halfway to the zone center, and these data were not obtained from very-well-defined peaks in the TOF spectra. It is possible that they derive from a "folded" longitudinal mode [71]. The crossing modes for RbBr, however, are very prominent and extend across the entire SBZ similar to the RW.

For these three materials, all have nearly equal anion and cation masses, similar closed shell electronic configurations, and nearly the same ratios of anion to cation ionic radii, ~ 1.35 [64]. Yet aside from the RW vibrational modes, the characteristic vibrational patterns differ significantly. This is especially true for the crossing modes which appear to exist across the Brillouin zone only for RbBr and for the optical modes which are seen in NaF but not in KCl and RbBr. The differences certainly lie with the quite different balance of forces, Coulombic versus short-range repulsions, attributable to the range of sizes and polarizabilities of the ions, particularly of the anions. This absence of similarity also illustrates the importance of close collaborations between theoretical and experimental groups in the analysis and interpretation of HAS data.

5. *NaCl, NaI, and CsF*

As with LiF, the anion-to-cation radius ratios are quite large for NaCl and NaI, 1.85 and 2.23, respectively [64]. However, unlike LiF no optical modes were observed for NaI, which has a large gap between the acoustic and optical branches, as shown in Fig. 23 [72]. As usual there is a well-defined Rayleigh wave, but there are only a few points along the calculated lines marking the longitudinal resonance and a few more for the crossing resonance, labeled S_8 in $\overline{\Gamma}\overline{M}$. Examination of the surface projected density of states from the Green's function calculation [34] shows that the highest-lying four points in $\overline{\Gamma}\overline{X}$ are probably associated with the crossing resonance. For NaCl in Fig. 24 [73] the RW and the longitudinal resonance can be seen in both high-symmetry directions, and a few points lie along the cal-

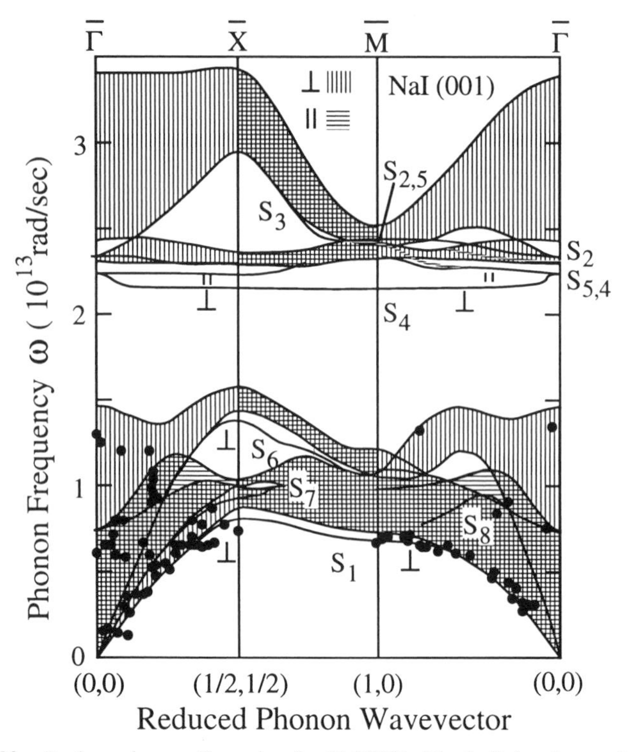

Figure 23. Surface phonon dispersion for NaI(001). The bulk bands are shown by the shaded regions with polarizations as indicated in the figure. The surface mode labels follow the notation of Fig. 16. Note that the modes labeled S_7 and S_5 are polarized shear horizontal and cannot be observed in this scattering arrangement. (Reproduced from Fig. 3 of Ref. 72, with permission.)

culated crossing resonance positions. More recent measurements also appear to have found points associated with optical modes [74].

For NaI, the large group of points in the center of the bulk band in $\bar{\Gamma}\bar{X}$ near the $\bar{\Gamma}$ point are probably due to phonon-assisted bound state resonances which were also found for NaCl and for LiF [58, 61, 63]. In the case of NaCl, the bound state energies had been determined by other scattering experiments [75, 76] so that the peaks in the TOF spectra due to bulk phonon resonances could be reliably removed from the phonon dispersion diagram in Fig. 24. For NaI the values of the bound states still need to be established.

CsF is the only alkali halide with rocksalt structure in which the cation makes the greater contribution to the electronic polarizability. Although there have not been any Shell Model calculations on this material, its mirror compound NaI should provide at least some guide as to the energetics of

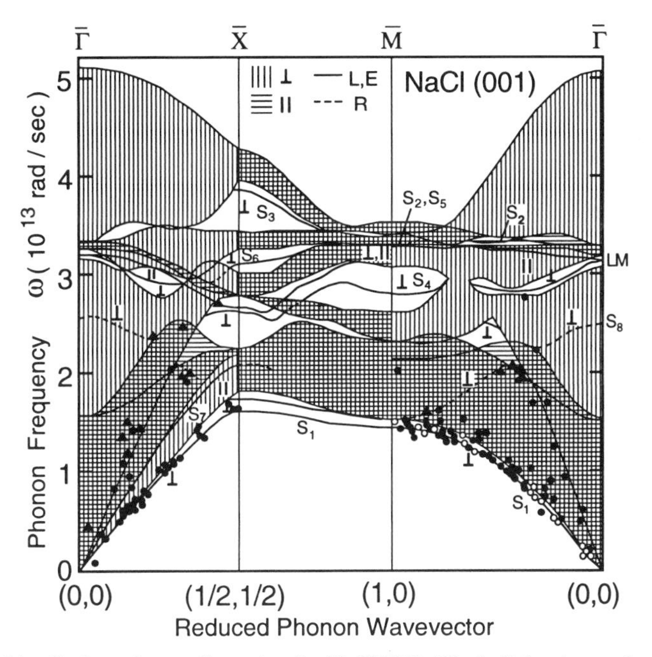

Figure 24. Surface phonon dispersion for NaCl(001). The bulk bands are shown by the shaded regions with polarizations as indicated in the figure. Note that the modes labeled S_7 and S_5 are polarized shear horizontal and cannot be observed in this scattering arrangement. The dashed curve labeled S_8 is the crossing resonance. The longitudinal resonance (the acoustic mode at higher energies than the Rayleigh wave, S_1) is not marked in this Green's function calculation, but several points fall very close to the calculated line. The remaining surface mode labels follow the notation of Fig. 16. (Reproduced from Fig. 3 of Ref. 73, with permission.)

the phonon modes. Figure 25 shows the available dispersion data [77]. As always the RW phonons seem to show up strongly in the TOF spectra. However, there are some distinct points which lie higher and are most probably longitudinally polarized vibrations. Finally, there are a handful of weak points which seem to lie at about the energy predicted for the S_4 optical mode in NaI, and it is tempting to assign these to that mode. Why these modes should be observable for CsF and not for NaI appears to require additional model calculations to sort out.

At this point it may be useful to summarize the major features of the lattice dynamics which have been discovered and explored with the alkali halides. For all of these materials the most prominent peaks in the TOF spectra are associated with the creation and sometimes annihilation of Rayleigh wave phonons. Phonons associated with the longitudinal resonance are usually also observed. Optical phonons have clearly been found for LiF,

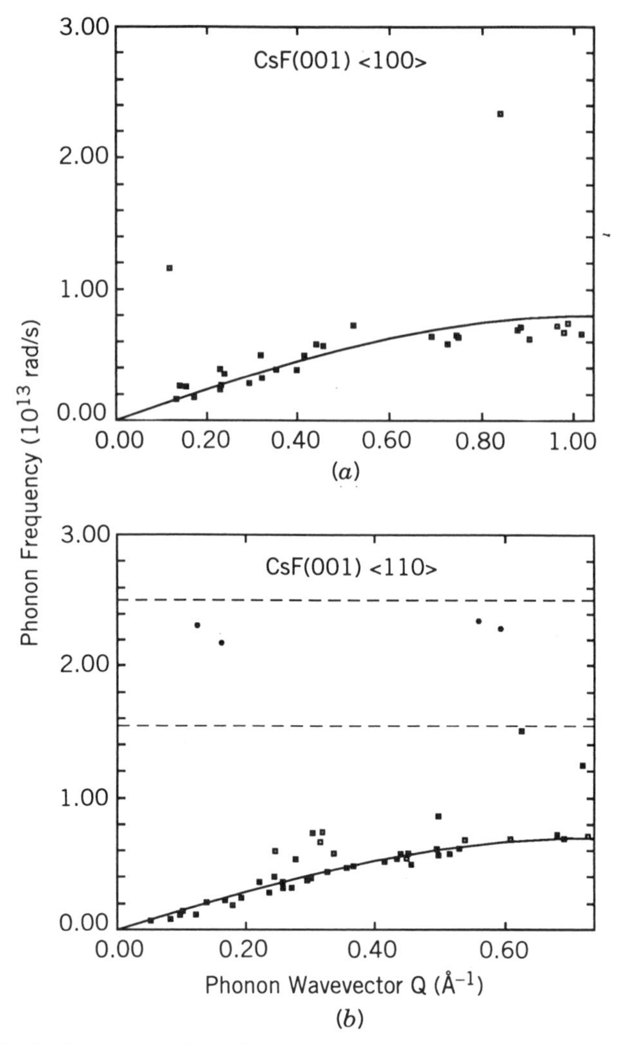

Figure 25. Surface phonon dispersion for CsF(001). The solid curve is a sine function which has been drawn to fit the data corresponding to the RW. The dashed horizontal lines in the ⟨110⟩ direction (panel *b*) are estimates of the lower and upper limits of the expected bulk band gap. The four points near the upper limit lie in the energy region expected for the gap mode, based on energies for the corresponding gap mode, S_4, in the "mirror" compound NaI, as in Fig. 23. (Reproduced from Figure 2 of Ref. 77, with permission of Elsevier Science Publishers.)

NaF, KBr, RbCl, and recently NaCl. For RbI, the peaks in the TOF spectra corresponding to the optical phonons lying above the bulk bands are weak but consistent; for CsF the measured points, also weak, occur at about the expected energies and are therefore suggestive.

Crossing resonances have been found in nearly all the compounds, but most prominently and across the entire SBZ for KBr and RbBr. The data clearly indicate that nearly equal anion and cation masses are *not* required for this kind of vibrational motion to exist at the surface. Rather, the existence of these modes is more likely connected with the hybridization of acoustic and optical bulk modes which can occur in isobaric and sometimes also in nonisobaric crystals.

The effects of relaxation on the calculated surface phonon dispersion in RbI have apparently been verified, particularly by the observation of a surface optical mode which lies above the bulk phonon optical bands. Except for the mysterious acoustic band mode in RbI, the Shell model calculations have generally been quite accurate in predicting surface vibrational mode energies in both high-symmetry directions of the alkali halide (001) surfaces.

The next phase for the theorists in connection with this work lies in predictions of helium atom scattering intensities associated with surface phonon creation and annihilation for each variety of vibrational motion. In trying to understand why certain vibrational modes in these similar materials appear so much more prominently in some salts than others, one is always led back to the guiding principle that the vibrational motion has to perturb the surface electronic structure so that the static atom–surface potential is modulated by the vibration. Although the polarizabilities of the ions may contribute far less to the overall binding energies of alkali halide crystals than the Coulombic forces do, they seem to play a critical role in the vibrational dynamics of these materials.

B. Metal Oxides

1. NiO, CoO, and MgO

Three metal oxides with fcc lattices and rocksalt structures have been examined by HAS: NiO [78, 79], CoO [51, 80], and MgO [81–83]. Because the ions in these materials are divalent, the atomic interaction forces are very much stronger than those for the alkali halides. This results in substantial differences in a number of physical properties, including the very high melting and boiling points of the oxides and the hardness of their crystals. Nonetheless, they still cleave along (100) planes like the alkali halides to form high-quality surfaces.

NiO and CoO are of interest, largely because the Ni^{2+} and Co^{2+} ions have unpaired electrons, and the oxides behave as Mott insulators, having antiferromagnetic-to-paramagnetic phase transitions at their respective Néel temperatures of 523 K and 291 K [84]. Figure 26, taken from the work by Toennies et al. [79], shows the measured surface phonon dispersion compared with a Shell Model calculation. As with the alkali halides, the agreement here between the data and the model calculations which just use the parameters obtained by a fit to the bulk phonon measurements seems to be quite satisfactory. HAS is able to map out the RW (S_1) fully, and enough of the S_4 optical modes to confirm the calculation. The S_2 optical modes for this very hard substance appear to lie at the upper limits of what may be possible with HAS, at about 50 meV [85], and have only been seen in EELS experiments [86]. The only unexpected result seems to be a vibrational branch lying below the top of the bulk acoustic bands in the ⟨110⟩ direction. However, the calculations by Toennies et al. seem to be able to account for this mode without modifications to the Shell Model [79].

There was a careful attempt by the Göttingen group to find a shift in the Rayleigh phonon energies as a function of temperature over the range 100–600 K, which spans the Néel temperature, but no significant changes could be observed that might correlate with the phase transition [79].

The results for CoO(001) are not shown here, but exhibit very similar

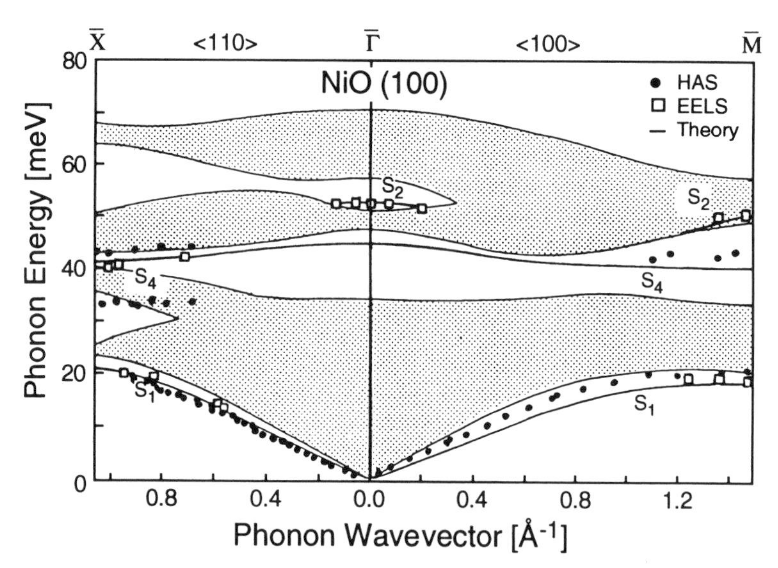

Figure 26. Surface phonon dispersion for NiO(001). The HAS data (solid points) and EELS data (open squares) are compared with a slab dynamics calculation. The bulk bands are shown as the shaded regions, and the surface localized modes are indicated by solid lines and labeled as in Fig. 16. (This figure has been reproduced from Fig. 5 of Ref. 79, with permission.)

behavior to that found for NiO, including the "extra" branch near the top of the bulk acoustic bands in the ⟨110⟩ direction [80].

It is perhaps also worth mentioning that there have been two helium atom diffractive scattering experiments on NiO which have attempted to observe the spin ordering in the antiferromagnetic phase. The (2×1) magnetic lattice–that is, with a unit cell twice the chemical cell—has been seen in LEED [87, 88]. In one experiment ^{3}He was employed as the probe with the hopes that there would be sufficient nuclear-spin–electronic-spin coupling to observe the magnetic lattice, but it did not succeed [89]. In the other experiment a beam of metastable triplet (^3S) helium atoms was produced to probe the surface. What was found is an unexpected oscillatory "survival" pattern in the quenching of the metastable helium atoms through the interaction with Ni^{2+} spins [90].

MgO was actually the first of these oxide materials to be studied [81] because Shell Model calculations had been carried out [36] and because it is a commonly used substrate for supported catalysts, among other uses. There is also considerable interest in the structure of this surface because of the suspected instability of a doubly charged anion outside the crystalline bulk [82, 83]. (This is also a question for NiO and CoO.) The results of Jung et al. [83], shown in Fig. 27 as an extended zone plot, do not indicate any significant surface rumpling due to the oxygen anion. However, the possibility that there is some charge transfer or other process occurring which stabilizes the surface ions cannot be ruled out. From the energy scale in this figure one can again see that the crystal is very hard, and that the HAS experiments appear to have difficulty in measuring phonons with energies near (and above) 40 meV [85]. Thus, in this work only the RW could be measured, and it is in good agreement with the calculations based on bulk model parameters.

2. *$KMnF_3$ and High-T_C Superconductors*

The perovskites form an interesting class of compounds of the general form ABX_3, where A and B are metals and X is usually oxygen but can also be another nonmetal such as fluorine [91]. In the normal geometry the A's are located at the corners of a cube, B is at the center, and the X's lie in the faces of the cube forming an octahedral environment around B. These materials can exhibit piezoelectric and ferroelectric behavior, and in general they undergo several phase transitions with temperature [91]. The difficulty for surface studies with these materials is in obtaining a good surface, which for ionic insulators requires that the material have a cleavage plane. In addition, the large number of atoms per unit cell translates into many optical modes which may lie close together in energy.

The Göttingen group was able to carry out both diffractive and inelastic

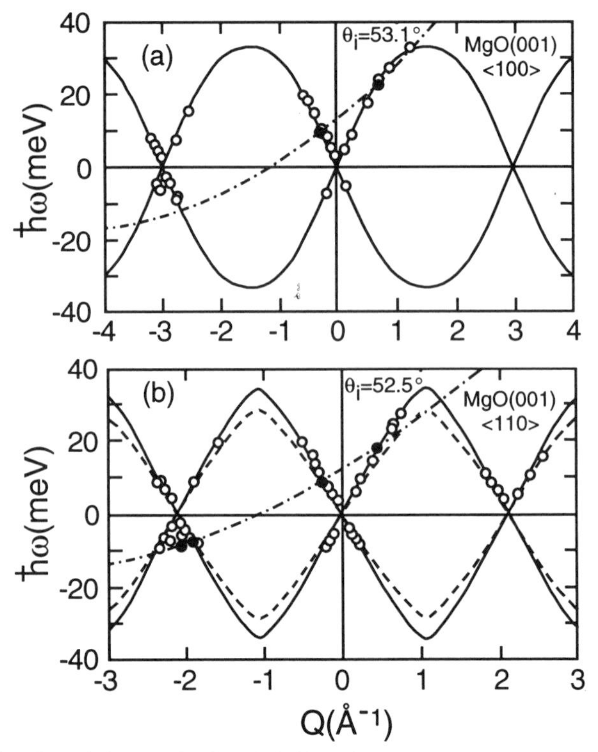

Figure 27. Extended zone plot for MgO(001) showing the data obtained from a number of TOF spectra. The solid curve is the calculated Rayleigh wave dispersion, while the dashed curve in the ⟨110⟩ direction is the S_7 shear horizontal mode which lies below the sagittal plane modes for this crystal in this direction. The dot–dashed line is a scan curve at the angles indicated. (Reproduced from Fig. 3 of Ref. 82, with permission.)

helium atom scattering experiments on $KMnF_3$ [92] for which slab dynamics calculations had been completed [93]. An additional complication was found for this material in that it appears to cleave in two (001)-type planes at the same time, a KF plane and an MnF_2 plane, and both contribute to the scattering signal. From the diffractive scattering this group was able to show that there is a surface phase transition which occurs at about 191 K, slightly higher than the corresponding phase transition which is known to occur in the bulk at about 188 K [92]. When a phase transition occurs, one expects the vibrational mode strongly associated with the displacive motion of the atomic rearrangements to become "soft"; that is, the frequency of the vibration should decrease toward zero as the change in lattice geometry takes place. This can be seen, in part, in Fig. 28, where the energy transfer peak labeled with an arrow can be seen to shift toward zero. This mode corre-

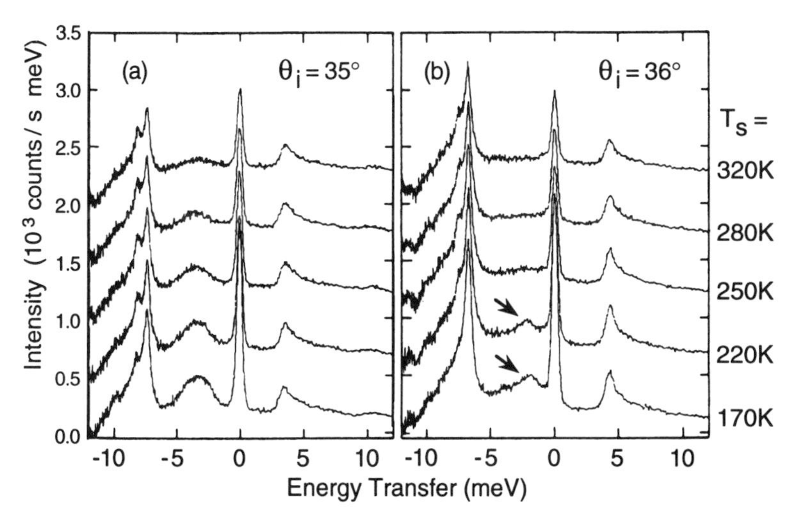

Figure 28. Time-of-flight spectra for $KMnF_3$ transformed to energy transfer distributions for several temperatures at incident angles 35° and 36°. The low-energy mode (arrows) seen on the creation side (negative energy transfer) at 36° is very close to the zone boundary and appears to be associated with the surface phase transition that occurs at 191 K. At 35° the low-energy peak corresponds to a mode further removed from the zone boundary, which persists up to high temperatures. (Reproduced from Fig. 18 of Ref. 92, with permission.)

sponds to momentum transfer near the zone boundary in the ⟨110⟩ direction. The broad peak in the figure appears to be made up of two components, one of which keeps the signal from vanishing at the phase transition temperature and complicates the interpretation. However, despite the complexity of the spectra for this crystal, the authors find general agreement between the experiment and the Shell Model calculations.

The Göttingen group has also carried out scattering experiments on the high-T_C superconductor $Bi_2Sr_2CaCu_2O_8$ for which relatively large crystals are available and which can be cleaved *in situ* to obtain the BiO(001) surface [94]. Calculations for this material using the Shell Model approach are in the process of being carried out, although there have already been some model calculations for others of these high-T_C materials [95]. To an even greater extent than for $KMnF_3$, the vibrational spectra become cluttered with peaks so that the interpretation of the data is still tentative.

C. Multiphonon Excitations

As indicated in Section III.B, the multiphonon scattering can also be investigated and interpreted in terms of the atom–surface potential. Most of this work has been carried out with alkali halides, but some experiments have also been done using metals. Figure 29 shows the typical behavior seen for

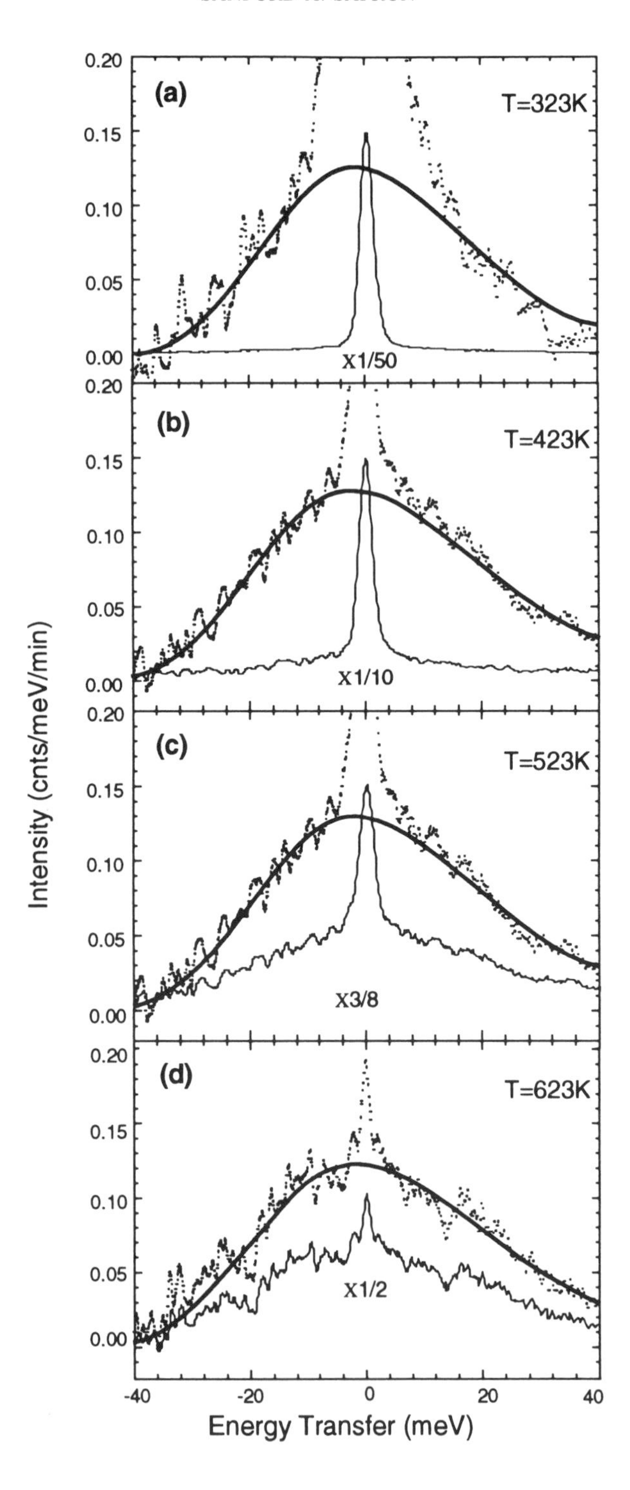
(a)
T=323K
X1/50
(b)
T=423K
X1/10
(c)
T=523K
X3/8
(d)
T=623K
X1/2
0.20
0.15
0.10
0.05
0.00
Intensity (cnts/meV/min)
-40
-20
0
20
40
Energy Transfer (meV)

the specular beam intensity as the surface temperature is raised, in this example for NaCl(001) [96]. The fits to the multiphonon "foot" shown in the figure have been obtained from the differential reflection coefficient, Eq. (18), with the form of the transition rate in Eq. (20) and the approximate form factor $|\tau_{fi}|^2$ given by Eq. (21).

By choosing values for the three parameters β, ω_D (or $\Theta_D = \hbar\omega_D/k_B$, the Debye temperature), and Q_C, one can fit the multiphonon background at all scattering angles and surface temperatures without further adjustment after normalizing the scattering intensity to one point, normally the specular intensity at one temperature. This is also illustrated in Fig. 30 for LiF at 723 K [51]. One should recognize in the figure that the theoretical curves do not include *any* elastic or single phonon contributions to the scattering intensity. For LiF one finds the Debye temperature to be 519 K, the stiffness parameter $\beta = 9.5$ Å^{-1}, and the cutoff parameter $Q_C = 8.5$ Å^{-1}. Other alkali halides tend to have lower Debye temperatures (softer crystals), smaller β values (less repulsive, or more polarizable, ions), and roughly comparable cutoff parameters, although only a few have been examined very carefully. For metals like aluminum the cutoff is typically much smaller, ~ 1 Å^{-1}, indicating that the interaction range is larger as is expected for a surface with less-localized electrons; and the β value is smaller, ~ 2.5 Å^{-1}, suggestive of a less steeply repulsive atom–surface potential [41]. It would be helpful for a wider range of materials to be investigated to explore more fully the significance of these potential parameters.

VI. CLEAN CRYSTALLINE SURFACES: SAMPLING OF MATERIALS

The ionic insulators discussed in some detail in the previous section have closed shell electronic configurations similar to the noble gases and electronic distributions which are localized around the electronic core. The principal interactions are Coulombic, although their polarizabilities appear to influence greatly the response of the electronic distribution to surface lattice vibrations. For other materials, particularly metals and some layered compounds, the conduction and valence electrons are best thought of as somewhat delocalized if not entirely free. These electrons are what the helium atoms scatter from, and their states of motion are significantly modulated by the vibrations of the atomic cores. Thus, for these materials HAS is very

Figure 29. Time-of-flight spectra for a series of temperatures at the specular angle (45°) for NaCl(001), transformed to energy transfer distributions. The large peak at zero energy transfer is the coherent elastic scattering (specular beam) which diminishes in intensity with temperature according to the Debye–Waller factor. The solid curves fitting the multiphonon "foot" of the elastic peaks are calculated from the Manson theory as described in the text. The incident helium wavevector for these experiments was 9.25 Å^{-1}, corresponding to an incident energy of 44 meV. (Reproduced from Fig. 1 of Ref. 96, with permission.)

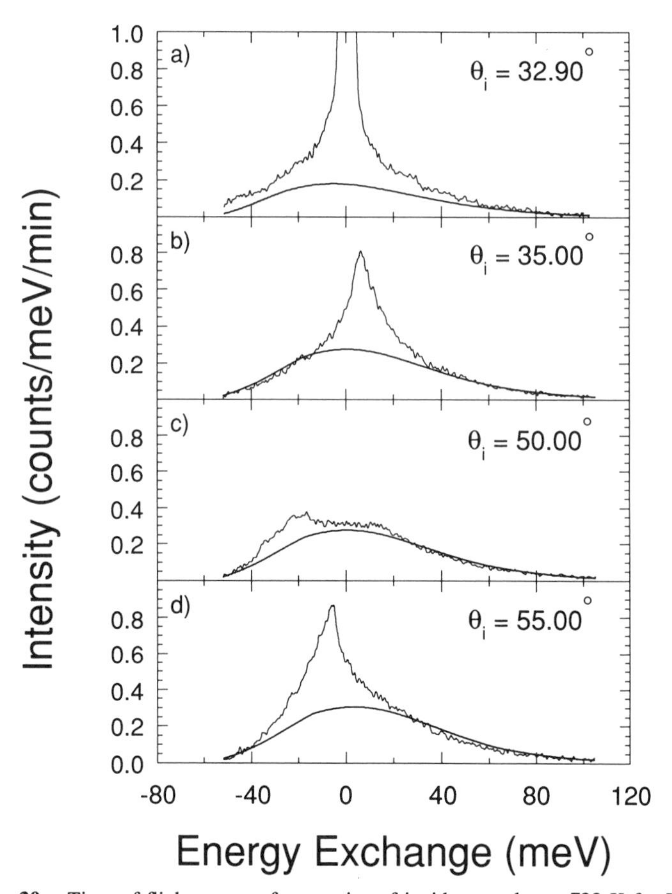

Figure 30. Time-of-flight spectra for a series of incident angles at 723 K for LiF(001), transformed to energy transfer distributions. The incident helium wavevector was 10.5 $Å^{-1}$ and energy 57.5 meV. The large peak at zero energy transfer at 32.90° is the coherent elastic scattering from Bragg diffractive scattering at this angle. The solid curves fitting the broad multiphonon distributions measured at these angles are calculated from the Manson theory as described in the text, which does not include any contributions from single phonons. (Reproduced from Ref. 51.)

sensitive to electron–phonon interactions (i.e., in more chemical language, to vibronic coupling) and can provide considerable insight into the proper description of the electronic structure at the surface. This sensitivity stands in contrast to the scattering of electrons and neutrons, which occurs predominantly from the high-electron-density cores and nuclei, respectively, and which therefore is not very responsive to the electron–phonon interactions. In this section we present some brief highlights of results that have been obtained by HAS to illustrate this point.

A. Metallic Surfaces

As examples of metallic surfaces, we present some experimental results from two very different families of metals: (1) the noble metals Cu, Ag, and Au, which have fcc lattices and are rather good conductors, and (2) the refractory metals Nb, Mo, and W, which have bcc lattices and are not especially good conductors.

1. Cu, Ag, and Au

The close-packed (111) surfaces of the noble metals were among the first surfaces studied by HAS [97]. Diffractive helium atom scattering shows that these surfaces have hexagonal symmetry and are rather flat with very small corrugation in the electronic density sampled by the helium atoms. They are also found to be relaxed by only a few percent and unreconstructed with the exception of Au(111), which exhibits a (23×1) structure [98].

Figure 31, taken from the Göttingen group, displays the dispersion curves from Cu(111) [99]. The data show that the surface vibrations consist of a Rayleigh wave and a longitudinal resonance which can be measured over virtually the entire surface Brillouin zone. The intriguing feature in this is that the intensity of the scattering associated with the longitudinal vibrational mode is greater than that for the Rayleigh mode. As discussed in Section

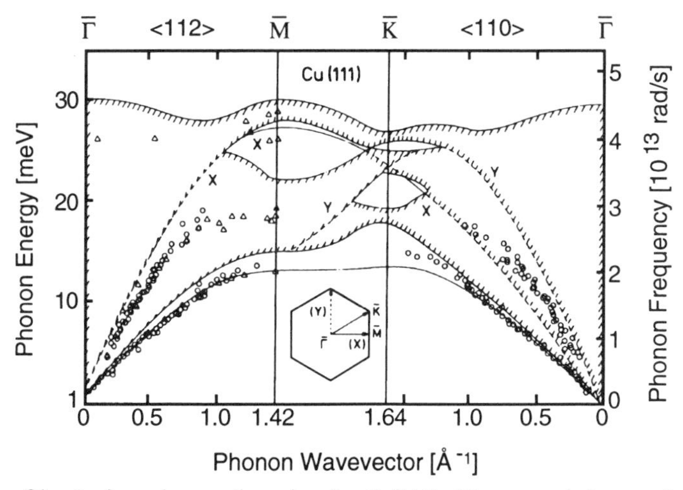

Figure 31. Surface phonon dispersion for Cu(111). The open circles are from HAS experiments, and the open triangles are from EELS experiments. The surface modes shown as solid lines and bulk band boundaries are based on a simple force constant model. The X and Y designations indicate the polarizations of the corresponding modes as identified in the reduced zone diagram in the inset. (Reproduced from Fig. 3 in Ref. 99, with permission.)

III.A, this is unexpected because the strongest interaction between the probe and the surface vibrations generally exists with modes having perpendicular polarization rather than longitudinal polarization. At the surface both of these vibrational modes are elliptically polarized sagittal-plane modes (as briefly discussed in Section II.B); that is, their respective polarization vectors have both surface normal and surface parallel components. While the relative proportion of each varies across the SBZ, generally the RW has the greater perpendicular component. Similar anomalously intense longitudinal phonon scattering signals are found for the other noble metals, including their (001) surfaces [100, 101]. However, this is probably a general property of metals in that the same phenomenon has also been observed in platinum(111) [50] and more recently in rhodium(111) [102], among others.

To explain these results in a Born–von Kármán treatment for the noble metals required fitting many levels of force constants, including nearest neighbor, next nearest neighbor, and so on, out to interactions several neighbors removed. In comparison with the bulk force constants these showed a softening of the surface of up to about 70% [103]. However, the parameters were not consistent with EELS measurements of the higher-energy modes which could be fit by a force constant scheme without such drastic softening of the surface force constants, nor were they consistent with the small surface relaxation observed for these metals [104].

To fit all the data, dispersion curves and intensities, the simple harmonic force constant picture had to be abandoned and replaced by models which deal more directly with the electronic charge distribution. The *Multipole Expansion Model* treats the electronic degrees of freedom through multipole expansion of the charge distribution about symmetry points in the unit cell [105]. This has now been refined into the *Pseudocharge Model* which incorporates six expansion points per cell at the midpoints between neighboring ion cores [99, 106]. The "extra" electron density which comes from expansion terms up through quadrupolar is responsible for the smoothing of the surface and, basically, arises from the decrease in the bonding between atoms at the surface in comparison with the atoms in the bulk. What is vital for explaining the HAS data are that the pseudocharge density couples strongly to the *longitudinal* vibrations of the metallic atoms such that even small longitudinal displacements can result in large displacements of charge and that the resulting charge displacement occurs primarily normal to the surface. This enhances the sensitivity of the helium atoms to the longitudinal vibrational mode (see Section III.A). At the same time, this electron–phonon interaction goes virtually undetected by the electron probes in EELS experiments because they scatter from a different electronic distribution. However, it is probable that this kind of low-energy interaction has great relevance for understanding adsorbate–substrate interactions in surface chemistry.

2. *Nb, Mo, and W*

An interesting contrast in behavior is found between the 5B element niobium [107] and the neighboring 6B elements molybdenum and tungsten [108]. The Nb(001) surface [107] is known to be stable with respect to reconstruction as are the hydrogen-saturated surfaces of W(001) and Mo(001) [108, 109]. However, LEED and x-ray data show that these *clean* surfaces of W and Mo undergo reconstruction which converts the (1×1) structures into a $c(2 \times 2)$ phase for tungsten at about 280 K and into a more complex $c(2 \times 2)$ phase modulated by a periodic lattice distortion for molybdenum at about 250 K [108]. The diffractive helium atom experiments, moreover, reveal a shoulder on the specular diffraction beam which indicates that the surface distribution of conduction electrons is not ordered simply in a (1×1) bulk-like distribution even in the high-temperature regime, and they suggest that the driving force for the reconstruction manifests itself well above the transition temperatures [110].

As one expects, then, the clean Nb(001) surface and the hydrogen-saturated surfaces of Mo(001) and W(001) have "normal" surface phonon dispersion curves, shown in Fig. 32 for Nb(001) [107], in that they consist of an RW and a longitudinal resonance which lie just below the corresponding bulk phonon bands over an extended range in temperature. These dispersion curves, in fact, have the same pattern as those discussed above for the noble metals. The similarity of these surfaces is evidently due to the ability of the hydrogen atoms to withdraw electrons from the 6B W and Mo atoms so that their electronic configurations at the Fermi level are very similar to that of the 5B niobium. The clean Mo(001) and W(001) surfaces, on the other hand, exhibit pronounced phonon anomalies with temperature. This can be seen in Fig. 33, where one observes drastic phonon softening in the longitudinal resonance for W(001) [110]. (From the figure it is not obvious that the soft modes are longitudinally polarized; rather, this feature comes out of the calculations [110].)

The difference in behavior between the clean surfaces and the hydrogen-saturated ones arises from a "Kohn anomaly" in the electron–phonon interaction [111, 112] made possible by the presence of filled electronic states at the Fermi level of the clean surfaces, which are otherwise depleted by hydrogen. In this case the electron–phonon interaction is the coupling between the longitudinal vibrational motion of the positively charged metal atom cores with the electronic charge density in the conduction band, perturbing the distribution of the electrons. Normally, the response of the "electron gas" is to screen the perturbing potential, limiting its effective range. However, when the topography of the Fermi surface satisfies certain conditions [111], a logarithmic singularity can appear in the slope of the screen-

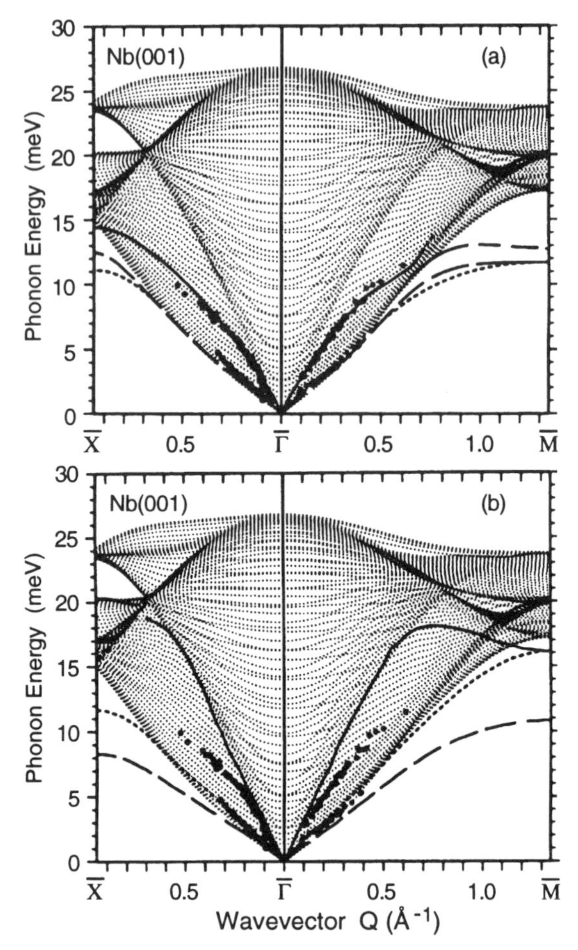

Figure 32. Surface phonon dispersion for Nb(001). The data are the solid points which were taken at 900 K. Panels *a* and *b* correspond to slab dynamics calculations with two different force constant models; the calculation in panel *b* uses the force constants from the bulk phonon fits. The solid lines represent the surface phonons and resonances polarized mainly longitudinally (or parallel), the lines with long dashes represent phonons polarized mainly perpendicularly, and those with short dashes are shear horizontal. (Reproduced from Fig. 6 of Ref. 107, with permission.)

ing factor at $q = 2k_F$, where k_F is the electron wavevector at the Fermi level, that leads to a long-range spatial perturbation of the electronic density, namely a charge density wave. For W(001) and Mo(001) it appears that the two-dimensional Fermi surface has the proper electronic states, which give rise to the charge density wave at $Q \approx 2K_F$. The periodicity of this wave does not need to be related to the surface reciprocal lattice vector; for W(001)

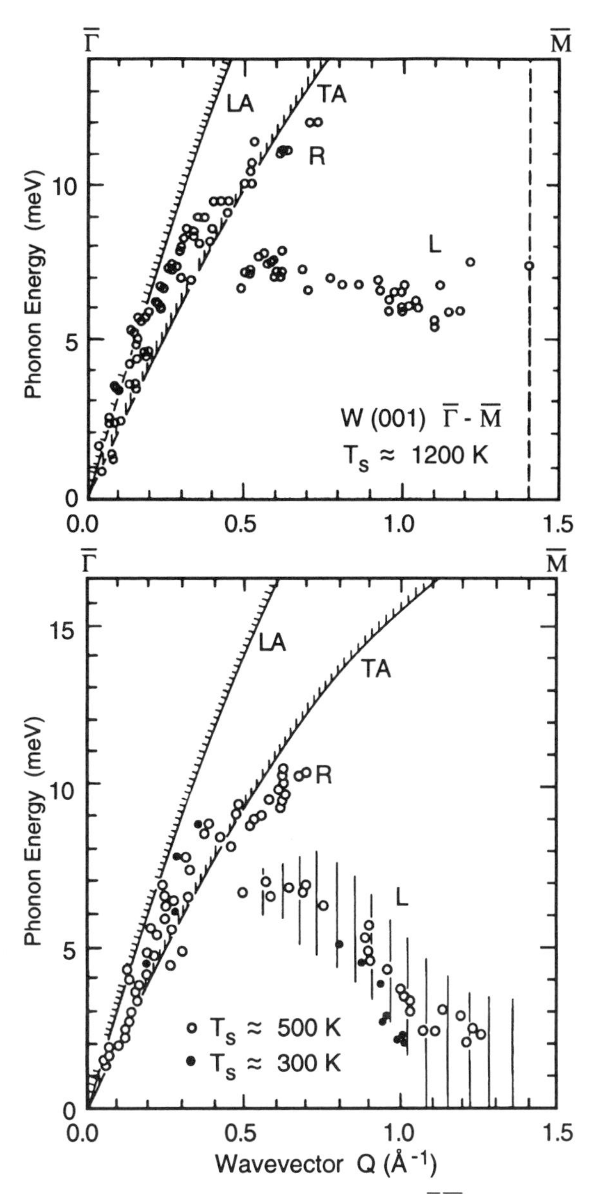

Figure 33. Surface phonon dispersion for W(001) in the $\overline{\Gamma}\,\overline{M}$ portion of the SBZ showing the measured Rayleigh wave (R) and longitudinal (L) modes. The data in the upper panel were obtained at 1200 K, while in the lower panel the data shown by open circles were obtained at 500 K and those represented by closed circles were obtained at 300 K. The edges of the transverse acoustic (TA) and longitudinal acoustic (LA) bulk bands are given by the hatched lines. The vertical lines in the lower panel denote the widths in the energy transfer distributions of these points. (Reproduced from Figs. 10 and 13 of Ref. 110, with permission.)

the charge density wave is incommensurate, but for Mo(001) it appears to be close to three-sevenths of the surface reciprocal lattice vector [108].

The modification of the electronic distribution through the charge density wave leads in turn to a softening of this phonon mode and thereby to the reconstruction of the surface. Thus, the forces which lead to the displacement of the atoms into the stable $c(2 \times 2)$ structure by the nominal transition temperature begin to be mobilized well above the transition temperature.

As in the case of the noble metals above, it is the sensitivity of the helium atom scattering to the distribution of electrons in the conduction energy states which has led to the detailed explanation of the surface properties of these metals. Intriguingly, for the W(110) and Mo(110) surfaces, it is still uncertain why saturation with hydrogen leads to just the opposite consequences, namely "giant" phonon anomalies, whereas the clean surfaces have normal dispersion curves [113].

B. Layered Materials

A review of HAS experiments on the layered materials graphite, GaSe, $TaSe_2$, and TaS_2 has been given by Skofronick and Toennies [54], and a summary of theoretical aspects of this work has been given by Benedek et al. [114]. One might think that the strongly anisotropic character of the bonding in these materials, strong intralayer forces and relatively weak interactions between layers, would lead to similar results for surface and bulk behavior as probed by HAS and by neutron scattering, respectively. However, only the graphite surface appears to behave according to predictions from the bulk properties [54].

The layered dichalcogenides have a number of interesting properties, including strong electron–phonon interactions that result in charge density waves and phase transitions. The mechanism is basically the same as for W(001) and Mo(001) described above. For the 2H polytype of $TaSe_2$ which was investigated by the Göttingen group [54, 115], there is hysteresis in the formation of phases depending on whether the sample is being heated or cooled. The normal phase exists above 122 K. On cooling below 88 K a commensurate (3×3) charge density wave phase is formed. In between, an incommensurate charge density wave phase is formed in which the wavevectors do not exactly match. On warming from the commensurate phase, at about 92 K a "striped" incommensurate phase is formed in which one of the three wavevectors remains commensurate while the other two are not. At 113 K the fully incommensurate phase is formed which goes over to the normal phase at 122 K. These results are consistently found in neutron scattering, electron microscopy, and diffractive helium scattering [54].

TOF measurements of the dispersion in the normal phase agree by and large with a force constant model based on the neutron scattering data [114]. However, at temperatures below the normal-incommensurate phase transition, the measurements show a softening in the RW at a different Q than is found in the neutron studies for the softening in the bulk longitudinal phonons. This suggests either that different electronic states at the Fermi level are involved for the surface than for the bulk or that hybridization of the RW with a temperature-dependent, surface longitudinal resonance is occurring. At the surface, both modes have sagittal plane polarization and hybridization can take place as the longitudinal mode softens with decreasing temperature and crosses the Rayleigh wave. One can see this crossing in Fig. 34 for the (3 × 3) commensurate phase at 60 K.

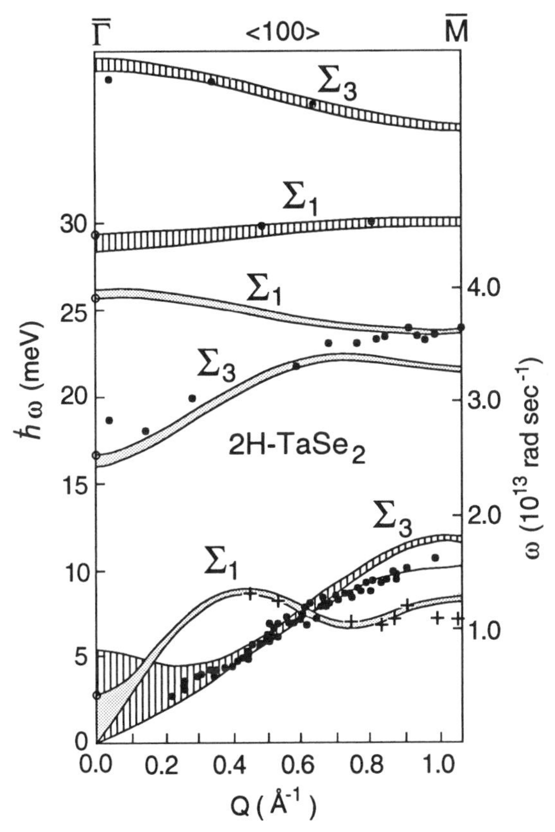

Figure 34. Surface phonon dispersion for $2H\text{-}TaSe_2$. The HAS data are shown as solid circles except for weak points which appear in the TOF spectra as hybridized longitudinal modes that are shown as crosses. All the data were obtained at 60 K, well into the low-temperature phase. The calculated striped and shaded regions, corresponding to transverse and longitudinal polarizations respectively, are the slab-adapted bulk phonon bands, while the solid line is a calculation for the Rayleigh wave based on the Dispersive Linear Chain Model (shown schematically in Fig. 35). The open circles at $Q = 0$ are from Raman scattering experiments. (This figure has been corrected from Fig. 23 in Ref. 54.)

The *Dispersive Linear Chain* (DLC) model was developed to account for the dispersion in these layered systems [114, 116]. As is shown schematically in Fig. 35, the wavevector-dependent force constants f_1, f_2, and f_3 couple the tantalum to the selenium atoms, the *inter*layer selenium atoms, and the *intra*layer selenium atoms, respectively. In addition, the two other force constants, d_1 and d_2, couple selenium atoms in neighboring subunits and tantalum atoms in neighboring subunits, respectively. Thus, the f's very nearly represent the surface normal forces and the d's represent the surface parallel forces, allowing one to sort out the polarization effects. The data analysis for $TaSe_2$ with the DLC model shows that there must be the equivalent of charge density accumulation between the Ta atoms in the tantalum plane associated with a charge density wave in the Ta conduction electrons, due to the perturbation of the surface. It has been pointed out that the fact that the anomaly occurs at different wavevectors for the bulk and surface suggests that one should view the stability of this material and its phase changes as a competition between different symmetries, both between the lattice and electron gas symmetries and between the surface and bulk symmetries [114].

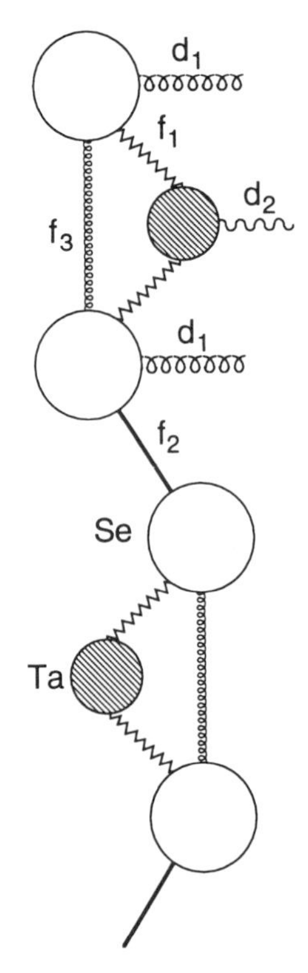

Figure 35. Schematic drawing of the Dispersive Linear Chain Model applied to 2H-$TaSe_2$. The force constants f_1 and f_3 describe the Ta–Se and Se–Se forces within a $TaSe_2$ subunit, and f_2 gives the Se–Se force between layers. d_1 and d_2 are the spring constants between Se atoms and between Ta atoms, respectively, for neighboring subunits in the same layer.

VII. EPITAXIAL OVERLAYERS, ADSORBATES, AND FILMS

Because of its sensitivity to surface features and the gentleness of its interaction with the surface, HAS is an ideal probe for studies of interfaces, including the structures formed by adsorbate deposition, the dynamics of the interactions of adsorbates with substrates, and the dynamics of the formation of overlayers and films. In the Florida State University (FSU) laboratory we have focused on ionic insulator growth and more recently on organic films.

A. Alkali Halide/Alkali Halide Systems

As discussed in Section V.A, the alkali halide family members are all rather similar chemically, differing mainly by size and polarizability. Hence, they constitute an ideal environment for probing the effects of lattice mismatch on the formation of epitaxial adlayers. Two systems have been investigated in great detail for their dynamical behavior during growth: (1) KBr/NaCl(001), where the lattice spacings are 4.67 Å and 3.99 Å, respectively [64], or ~17% mismatch [117], and (2) KBr/RbCl(001), where the lattice spacings are nearly the same, 4.67 Å and 4.65 Å, respectively [64], or <0.5% mismatch [118, 119]. (Note that these lengths are the surface lattice spacings; the bulk fcc lattice constants are $\sqrt{2}$ larger. See Fig. 13.) In addition, since the cleaved bulk surfaces of each of these materials has already been examined by HAS, the properties of the grown layers can be directly compared with the terminated bulk surface.

1. KBr/NaCl(001)

The growth of KBr onto NaCl was carried out by heating a piece of KBr crystal to about 400°C to form a molecular beam by sublimation. The NaCl(001) substrate which had been prepared in the usual fashion by cleaving was cooled to about 220 K in the scattering chamber. Under these conditions the layer-by-layer growth was followed by monitoring the oscillations in the helium specular beam intensity (see Section IV). A beam flag was deployed as shown schematically in Fig. 10 to stop the deposition at any point. Diffractive scattering experiments showed that a superstructure corrugation developed after two monolayers of KBr had been deposited, which persisted until about six monolayers had been deposited. The length scale of this superstructure is 28 Å, which corresponds to seven NaCl lattice spacings and six KBr lattice spacings. After about six monolayers of KBr had been deposited, only diffraction peaks from KBr(001) in the ⟨100⟩ orientation, the same azimuth as for the substrate NaCl, could be seen.

TOF spectra were taken after the growth of films of thicknesses two, three, four, and seven monolayers in order to compare the surface phonon

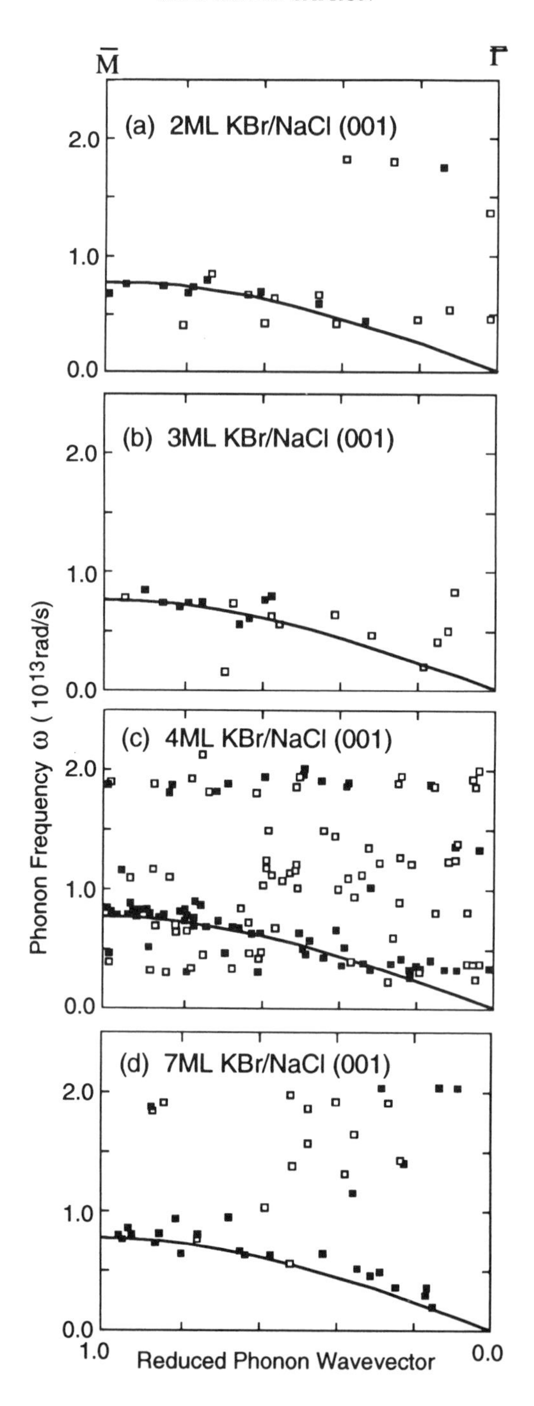

M̄
Γ̄
(a) 2ML KBr/NaCl (001)
(b) 3ML KBr/NaCl (001)
(c) 4ML KBr/NaCl (001)
(d) 7ML KBr/NaCl (001)
Phonon Frequency ω (10¹³rad/s)
Reduced Phonon Wavevector
2.0
1.0
0.0
1.0
0.0

dispersion with that of the cleaved KBr(001) surface. Several interesting features can be seen in Fig. 36. First, the frequencies of the RWs do not go to zero at the zone center (long-wavelength phonons) as one might expect for an acoustic vibrational motion. Then, even though there is a fair amount of scatter in the points due to signal levels considerably smaller than for the cleaved surface (Fig. 17), one can see that the surface optical mode S_2 is already developing at two monolayers and that by four monolayers the surface dispersion curves look very much like the cleaved KBr(001) surface. This can be seen better in Fig. 37 where the these two surfaces are directly compared. (There is much less extensive data for the seven monolayer film, but the optical mode and crossing resonance curves are still well-demarcated.) Finally, for the four-monolayer case, there appears to be another mode which lies below the RW that is dispersionless across the zone, which is not evident for fewer layers or for the seven-monolayer thick film.

Luo and coworkers [120] have proposed a simple harmonic spring model to fit the acoustic mode data. It employs one tangential and one radial force constant between the K^+ and Br^- ions, another pair of force constants for the repulsion between the two like ions (the same constants for each pair of ions), and a third pair of constants for the coupling of each K^+ and each Br^- to the NaCl substrate which is considered to be rigid. This latter approximation may not be too unreasonable here since the overlap between the vibrational bands of these materials is poor; the corresponding dispersion curves of NaCl lie at much higher frequencies than those of KBr (see Fig. 24). To fit the data, they find that the coupling between the adlayer ions and the substrate has to be much weaker than that between the ions in the adlayer; that is, the *intra*layer forces need to be much stronger than the *inter*layer forces. This is perhaps not surprising since the lattice spacing and frequency mismatches effectively decouple the two materials. The calculations also predict that the Rayleigh frequencies should not go to zero at the $\bar{\Gamma}$ point, but rather to a finite value. (This point will be discussed further in Section VII.B.) The decoupling of the KBr and NaCl also explains why the vibrational dispersion of the KBr film resembles that of the cleaved KBr even when the film is only a few layers thick (except for the finite Rayleigh frequency at the zone center). The origin of the dispersionless mode lying

Figure 36. Surface phonon dispersion of KBr films deposited onto NaCl(001) for two, three, four, and seven monolayers (ML). The solid squares correspond to strong peaks in the TOF spectra, whereas the open squares are calculated from weaker ones. The solid line is a sine curve matched to the RW phonon frequency at the $\bar{M}$ point for cleaved KBr(001), as in Fig. 17. Note that the frequencies at $\bar{\Gamma}$ do not go to zero. Also note that for the 4ML data (panel *c*), there appear to be a few good points (solid squares) which lie below the RW frequencies. (Reproduced from J. Duan, Ph.D. Dissertation, Florida State University, 1992.)

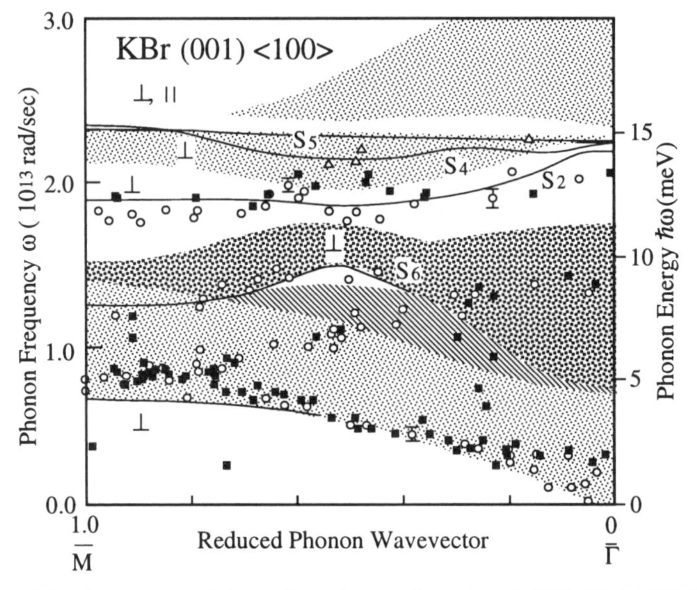

Figure 37. Comparison of the surface phonon dispersion of 4ML KBr/NaCl(001) with cleaved KBr(001) in the $\bar{\Gamma}\bar{M}$ region of the SBZ. The open circles and triangles are from cleaved KBr, the closed squares are from the 4ML film. The calculated bulk bands and surface modes are as in Fig. 17. (Reproduced from Fig. 5 of Ref. 117, with permission of Elsevier Science Publishers.)

below the RW for the four-monolayer film, however, is still not established, but there are suggestions that it results from the structure of the interface [121].

Recently, a large-scale calculation has begun on this system which employs more realistic pair potentials between all the ions, overlayer, and substrate [122]. Figure 38 shows the resulting "superstructure" obtained for a two-monolayer film in very preliminary work. This calculation suggests that the Br^- ions tend to be displaced outward in the top layer with respect to the K^+ ions.

2. *KBr/RbCl(001)*

Elastic diffractive scattering from the surface formed by the deposition of monolayer coverages of KBr onto RbCl(001) are sharply contrasting with KBr/NaCl(001) for the same growth conditions [52, 118, 119]. The Bragg peaks are located at the same positions as for the clean substrate, although the intensities shift to reflect the different surface corrugation. After about seven monolayers have been deposited, the relative heights of the Bragg peaks are essentially those of clean KBr(001). The diffraction pattern also

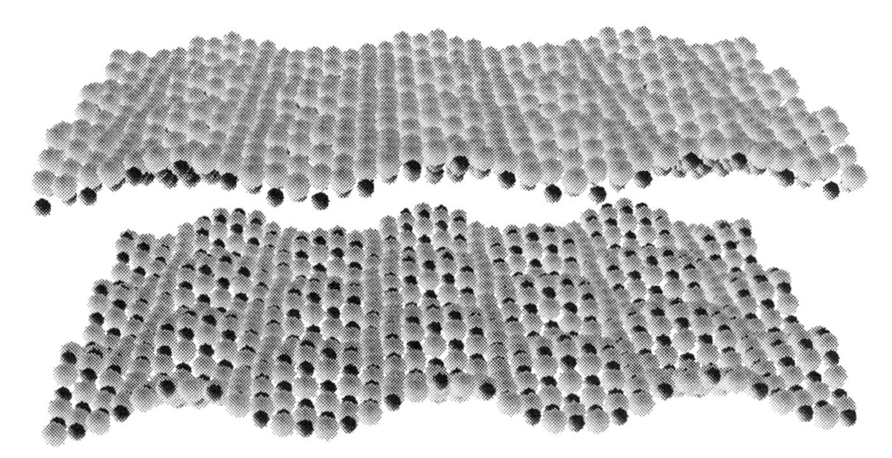

Figure 38. Calculated superlattice structure for 2ML KBr/NaCl(001). The unit cell in the calculation is 21 × 21 NaCl's and 18 × 18 KBr's, corresponding to the lattice mismatch for these materials. Only the two KBr layers are shown in the figure with their vertical separation exaggerated by a factor of 10. The viewing orientation of the layers is with the ⟨100⟩ direction into the plane of the paper. The darker atoms represent K^+ and the lighter ones represent Br^-. (This figure was provided by J. Baker.)

indicates that no superstructure is formed; evidently the lattices of these materials are similar enough that they fit together quite well.

There are some differences in the vibrational energies for a one-monolayer film of KBr compared to a cleaved KBr crystal, as can be seen in Fig. 39 [118]. The RW points lie a bit higher for the film, the S_2 optical modes lie a bit lower, and the crossing resonance is not well-developed. However, differing from the KBr/NaCl system, the RW here seems headed to zero at the zone center rather than to a finite limit.

It is difficult to treat the KBr/NaCl system by ''realistic'' models such as the Shell Model which works well for these ionic insulators because the unit cell is so large; from the superstructure found in the diffraction the real cell size is (7 × 7) NaCl's or (6 × 6) KBr's. For KBr/RbCl the unit cell is (1 × 1) and size poses no special computational difficulty. Schröder, and Bonart have undertaken the adaptation of the Shell Model to this overlayer/substrate system [119].

Although the parameters of the Shell Model can be interpreted in terms of ''real'' quantities such as polarizabilities, there is some arbitrariness in choosing values to fit the experimental data. For example, the interactions that need to be described at the interface of a monolayer of KBr on RbCl include those associated with the four compounds KBr, KCl, RbCl, and RbBr. However, the parameters in the literature for the K^+–K^+ interaction

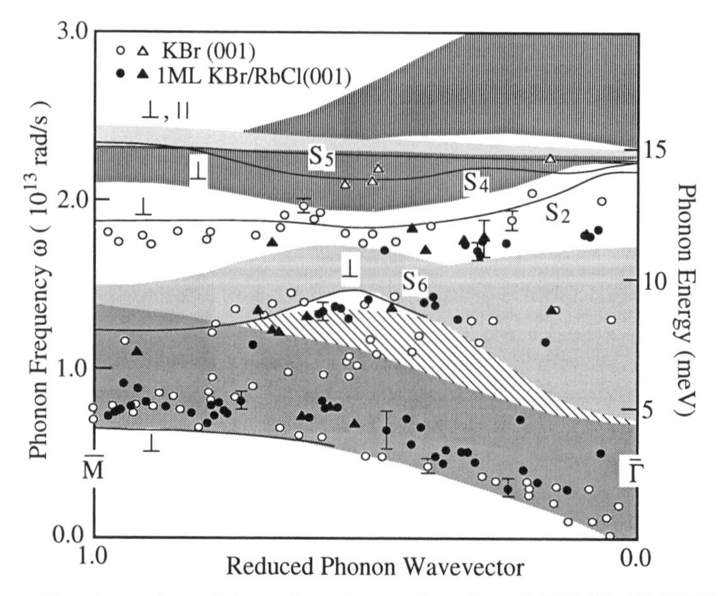

Figure 39. Comparison of the surface phonon dispersion of 1ML KBr/RbCl(001) with cleaved KBr(001) in the $\overline{\Gamma}\overline{M}$ region of the SBZ. The open circles and triangles are from cleaved KBr, and the closed circles and triangles are from the 1ML film. The calculated bulk bands and surface modes are as in Fig. 17. (Reproduced from Fig. 11 of Ref. 118, with permission.)

from fits to bulk KBr are not the same as for those from fits to bulk KCl, and similarly for the other ions. Nor are there parameters available from bulk fits which account for nearest-neighbor K^+–Rb^+ and Br^-–Cl^- interactions. Schröder and Bonart have set up criteria for choosing the parameters in a consistent fashion, which requires matching certain experimental quantities, and then have applied the Shell Model to this system. Data from this laboratory for the three-monolayer film are shown in Fig. 40, as an example, and the fit is rather impressive [119].

3. *KCN/KBr(001)*

KCN has an fcc lattice with rocksalt structure above the phase transition temperature of 168 K and cleaves to form (001) surfaces. In diffractive helium scattering this surface appears disordered, probably because the CN^- groups can orient in several equivalent orientations, and TOF experiments reveal primarily multiphonon inelastic contributions which are reasonably well fit by the Manson theory [52, 123]. (See Section III.B.)

The lattice spacing for KCN(001) is very close to that of KBr, 4.62 Å and 4.67 Å, respectively [64]. In work that is ongoing in this laboratory, a monolayer of KCN has been deposited onto KBr(001) for diffractive and

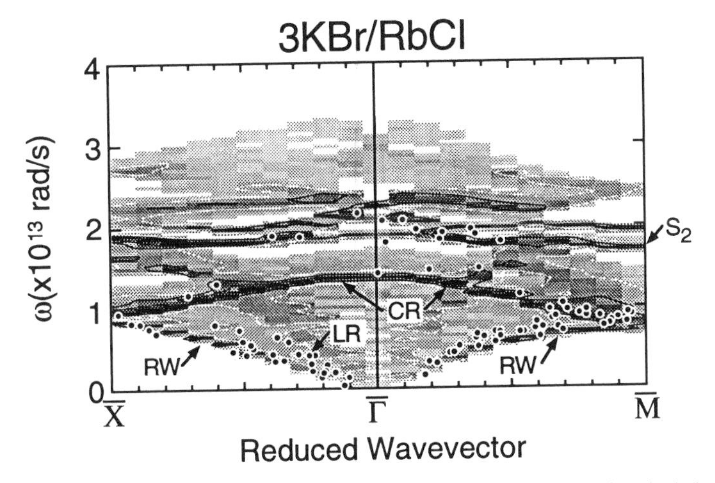

Figure 40. Surface phonon dispersion for 3ML KBr/RbCl(001). The shaded regions correspond to the surface projected density of states, with the darker shades representing higher state densities. The Rayleigh wave, longitudinal resonance, crossing resonance, and optical modes are indicated by RW, LR, CR, and S_2, respectively. (Reproduced from Ref. 119.)

inelastic scattering studies [124]. The single-monolayer film does give specular and small Bragg helium diffractive peaks which are not seen for cleaved KCN that suggest some long-range order on this surface. The inelastic scattering work so far has yielded single-phonon peaks in the TOF spectra which appear to correspond to a Rayleigh wave that lies slightly higher in energy than the RW for KBr(001) and to an optical mode. However, additional experimental work and model calculations are needed before these intriguing results can be fully understood [124].

B. Adsorbate/Metal Systems

There have been a number of adsorbate/metal systems examined by HAS in the past few years. The dynamics of such systems is of particular interest because of the implications for chemical catalysis and because of the growing reliance on the development of novel materials for specific technological purposes. A few systems are presented here as examples of the current state of experimental achievements.

1. Na/Cu(001)

One of the best-studied systems of metals with overlayers is this one [125]. The vibrational frequencies of the sodium overlayer have been measured layer-by-layer for films from 2 to 20 monolayers thick. The interesting feature is that the more layers, the lower the frequency of the vibration in the long-wavelength region, $\overline{\Gamma}$, similar in a general way to what was described

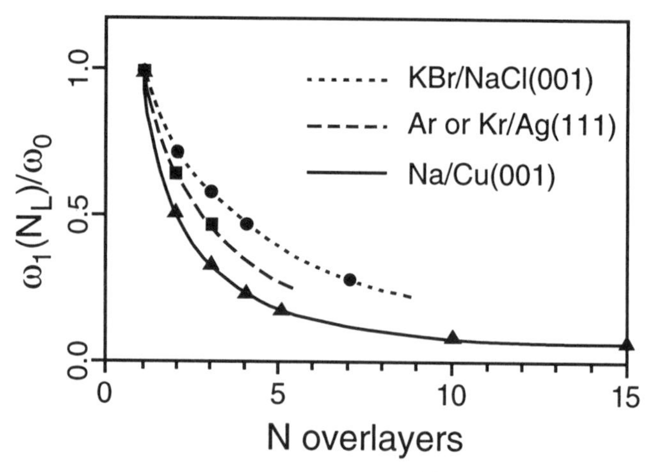

Figure 41. The lowest measured frequencies at the $\overline{\Gamma}$ point as a function of the number of overlayers, N, for KBr/NaCl(001), Ar or Kr/Ag(111), and Na/Cu(001). The frequencies have been normalized to the measured or estimated frequency for $N = 1$. (Reproduced from Fig. 1 of Ref. 128, with permission.)

above for KBr/NaCl. A graph showing a comparison of the vibrational frequencies at the $\overline{\Gamma}$ point for Na/Cu(001), KBr/NaCl(001), and the physisorbed system Ar or Kr/Ag(111) [126, 127] is shown in Fig. 41 [128]. The decrease in frequency with the number of overlayers N can be written as proportional to $N^{-\gamma}$ in all cases with $\gamma = 1$ for Na/Cu, 0.5 for KBr/NaCl, and 0.85 for Ar and Kr/Ag. Luo et al. [128] developed a simple model to explain this behavior, which consists of a linear chain that uses only two force constants, one between layers in the film and the other between the first layer of the film and the rigid substrate; that is, an *intra*film force constant k and an *inter*film force constant f. The result of this analysis is that when $f = 2k$ (i.e., when the interlayer force is about twice as strong as the intralayer force), then $\gamma = 1$ and incidentally the frequencies satisfy the same relation as do the frequencies of organ pipes with pipe length replaced by film thickness. (Hence, the Göttingen group refers to these vibrations as the "organ pipe" modes.) When $k >> f$ (i.e., when the intralayer forces are much greater than the interlayer forces), then $\gamma = 0.5$. As discussed above for KBr/NaCl, the mismatch between these materials seems to ensure that the film vibrates essentially independently of the substrate. At $\overline{\Gamma}$ the whole KBr film vibrates in phase (see Section II.A) with the frequency of a harmonic oscillator, $\omega_L = (f/m_L)^{1/2}$. Since the mass of the film is just proportional to the number of layers, one should expect $\gamma = 0.5$. For the physisorbed noble gas/Ag system, the interaction forces lie somewhere in between.

2. *Ar or Kr/Ag(111) and Kr/Pt(111)*

The close-packed metal surfaces, as discussed in Section VI, are very smooth because the highest-energy electrons, the conduction electrons, effectively spread out between the nuclei and inner shell cores. For physisorbed systems like these, usually incommensurate structures are formed because there is only a very weak potential coupled to the corrugation. The behavior at the $\overline{\Gamma}$ point for Ar and Kr/Ag(111) was discussed briefly above. For the dispersion across the SBZ, the situation is to some extent as in Section II.A (Fig. 4) for the optical mode dispersion in the linear diatomic crystal. When the coupling between unit cells is much weaker than that within the unit cell, the optical mode is flat or dispersionless. In this case the unit cell is the noble gas–silver surface. The analysis of the model of Luo et al. suggests that although the noble gas–surface force is weak, it is probably stronger than the single-monolayer Ar–Ar or Kr–Kr interaction for motion normal to the surface. However, as the number of noble gas layers increases, the interaction between the surface of the noble gas film and the silver should become comparatively weaker. This can be seen in the dispersion across the SBZ as shown in Fig. 42 for 1, 2, 3, and 25 layers [127]. One can clearly

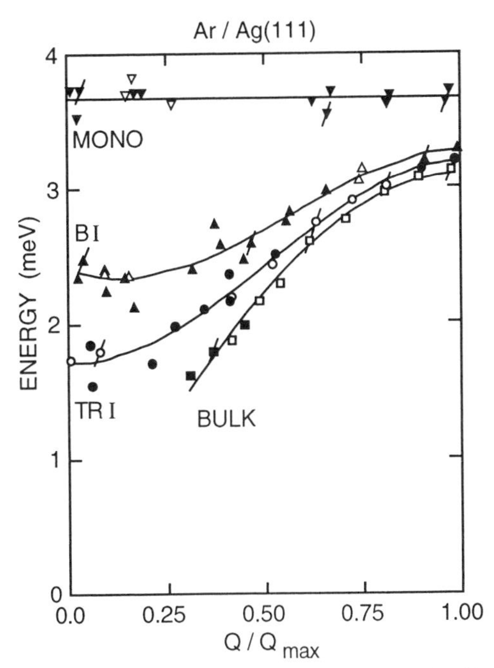

Figure 42. Surface phonon dispersion curves for Ar/Ag(111) for films of one monolayer (MONO), two monolayers (BI), three monolayers (TRI), and 25 monolayers (BULK). The progression shows the increase in dispersion as the film's surface becomes increasingly like the surface on a terminated bulk. (Reproduced from Fig. 3 of Ref. 127, with permission.)

see the formation of a bulk material in going from the monolayer Ar film with Einstein-like behavior (i.e., a dispersionless mode) to the fully developed dispersion of the argon Rayleigh wave at 25 monolayers.

The frequencies of dispersionless modes, such as in the above figure, normally lie below those of the RW of the clean metal at the zone boundary. However, since the Rayleigh wave goes to zero at the zone center, the two

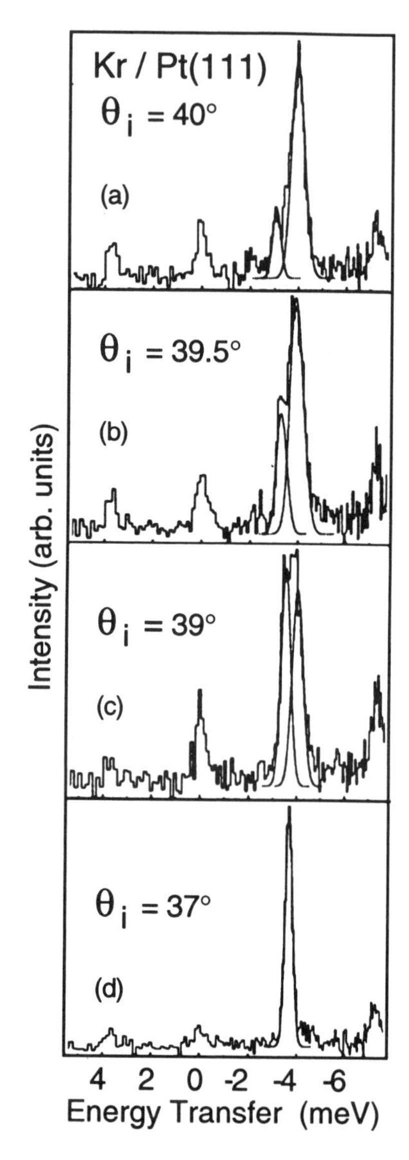

Figure 43. TOF spectra transformed to energy transfer distributions for a Kr monolayer on Pt(111) at four different angles. With decreasing incident angle, larger phonon wavevectors are probed. One can see that the two phonons which lie very close in energy in panel *a* merge together in panels *b* and *c* with increasing wavevector, and that only one phonon remains in the spectrum after a certain wavevector is reached (panel *d*). (This figure has been reproduced from Fig. 2 of Ref. 129, with permission.)

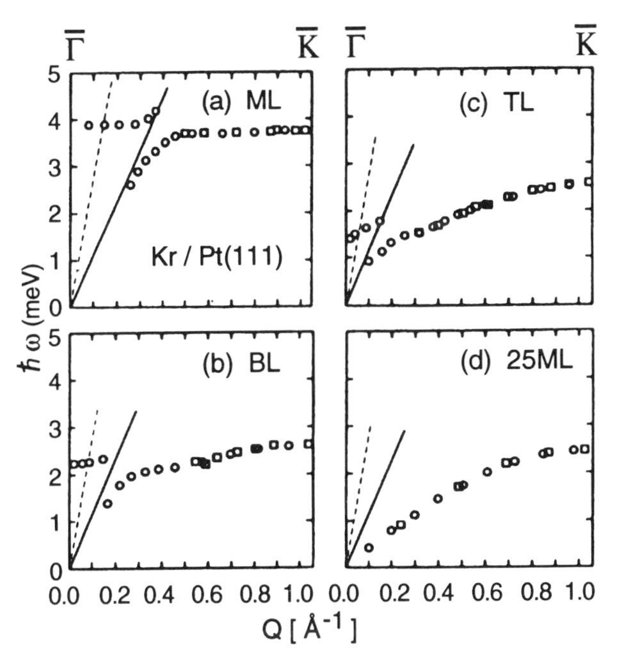

Figure 44. Surface phonon dispersion curves for Kr/Pt(111) for (*a*) one monolayer, (*b*) two monolayers, (*c*) three monolayers, and (*d*) 25 monolayers, showing the avoided crossing of the adsorbate and metal RW modes due to hybridization. The solid line is the clean Pt(111) RW dispersion curve, and the dashed line is the longitudinal bulk band edge. (Reproduced from Fig. 3 of Ref. 129, with permission.)

curves must cross. Accordingly, mode hybridization can occur in the intersection region since these vibrations have the same symmetry, if the coupling is strong enough. This effect is demonstrated very cleanly by Kern et al. [129] as shown in Figs. 43 and 44 for Kr on Pt(111). In the TOF spectra shown in Fig. 43, one can see the TOF peak broaden, split, and then sharpen dramatically as the Q values approach, reach, and then pass the intersection point as the incident angles are changed. Fig. 44 shows the mapped-out dispersion curves across the SBZ for 1, 2, 3, and 25 monolayers [129]. The coupling between the adsorbate surface modes and the substrate modes clearly decreases as the film thickness increases. The curve crossing illustrates an important coupling mechanism that surely must play a role in adsorbate–substrate interactions.

3. *Frustrated Translations: Benzene and CO on Rh(111)*

As with the weakly interacting noble gas/metal systems in the previous section, *isolated* adsorbates are always expected to have dispersionless vibrational modes with sagittal plane polarization, such as very-low-energy "frustrated" translational modes. The vibrations reflect the interaction with the substrate, although they do not appear to be particularly sensitive to differences in the adsorbate binding at different types of lattice sites [130]. In the past few years the importance of such low-frequency modes in determining the thermodynamic stability of adsorbed layers has begun to be understood, along with the realization that HAS is probably the best technique to measure them [131]. Again, this is possible because of the gentleness of the helium–target interaction and because of the extreme sensitivity of this probe to isolated surface defects.

As an example, Fig. 45 shows the TOF spectrum, converted to energy transfer, for isolated CO on Rh(111) where one can see the shift downward in phonon energy with isotopic mass change [130]. Since the vibration is of the whole adsorbate molecule, the frequency should vary as $m^{-1/2}$ and exchanging $^{13}C^{18}O$ for $^{12}C^{16}O$ should result in the fractional decrease in energy $(28/31)^{1/2} = 0.95$, in agreement with the 5.75-meV and 5.44-meV peak

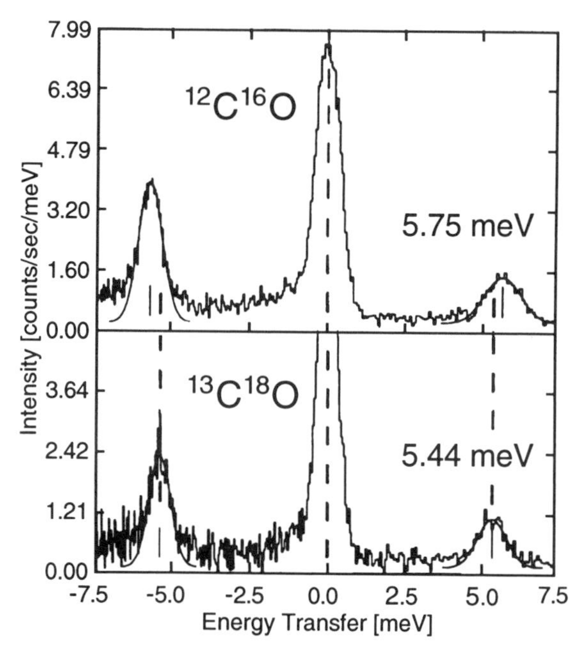

Figure 45. TOF spectrum transformed to energy transfer distribution for CO/Rh(111). The top panel shows the single-phonon creation and annihilation peaks for the frustrated translational motion of CO at 5.75 meV along with a diffuse elastic peak at zero energy transfer. In the lower panel, the shift in energy to 5.44 meV due to the heavier mass of the $^{13}C^{18}O$ isotope is clearly discernible (dashed vertical line). (Reproduced from Fig. 3 of Ref. 130, with permission.)

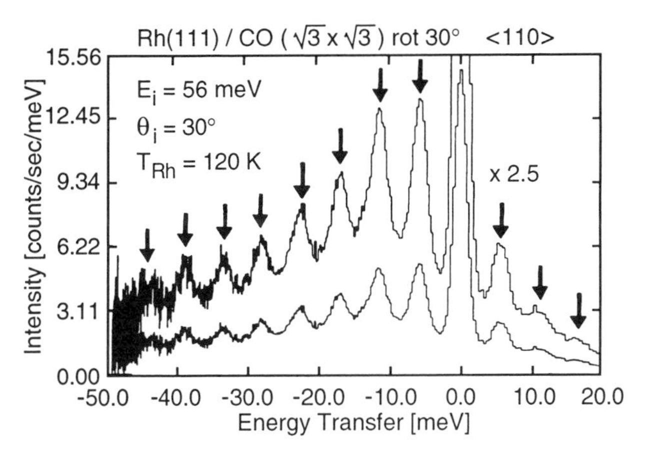

Figure 46. TOF spectrum transformed to energy transfer distribution for CO/Rh(111). For greater coverage of CO than in Fig. 45, the $(\sqrt{3} \times \sqrt{3})R30°$ structure is formed, which allows strong overtone excitations to occur. (Reproduced from Fig. 5 of Ref. 130, with permission.)

positions in the figure. As the CO coverage on the surface increases, a $(\sqrt{3} \times \sqrt{3})R30°$ structure is formed in which the frustrated translational mode decreases in energy from 5.75 meV to 5.6 meV, but more interestingly a number of overtone vibrational frequencies are seen, up to 8 as in Fig. 46 [130]. Although overtones have also been found in the noble gas/Ag(111) system [126, 127], the reason for so many harmonics here is not clear.

The benzene/Rh(111) system also exhibits very-low-energy vibrational modes, but the ones accessible to HAS TOF measurements here are identified by extended Hückel calculations as polarized *parallel* to the surface [130]. In addition, they are found to vary somewhat in energy depending on the coverage. The dispersion for a $(2\sqrt{3} \times 3)$ structure is shown in Fig. 47. One can see that the 13-meV frustrated translational mode is essentially flat (again, as in Fig. 4), indicating very little lateral coupling between benzene molecules, which is consistent with the relatively large spacing between benzenes (~5.6 Å) for this surface structure. The benzene dispersion curve also appears to have an avoided crossing with the RW of the clean Rh(111) surface analogous to that of Kr/Pt(111) in Fig. 44*a*.

C. Organic Films

Large organic molecules which may be employed in self-assembling monolayers (SAMs) or Langmuir–Blodgett (LB) films have become popular with a significant segment of the surface science community because of the many potential uses that can be envisioned for them. These materials have

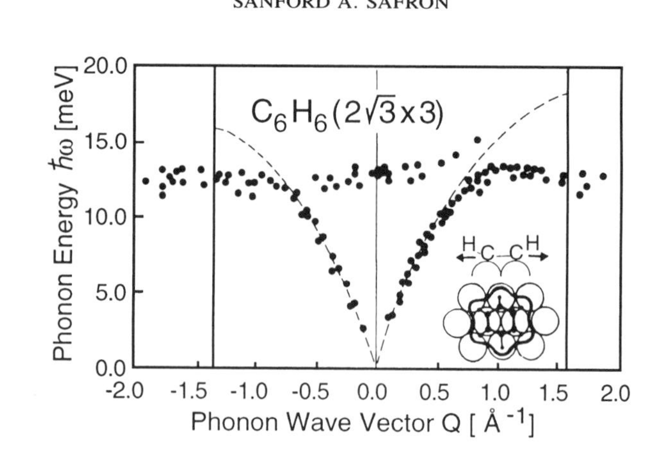

Figure 47. Surface phonon dispersion for C_6H_6/Rh(111). The phonon dispersion curve of the benzene frustrated translational mode appears to have an avoided crossing with the RW of the Rh(111) surface, similar to that of Kr/Pt(111) in Fig. 44*a*. The orientation of the benzene on the surface (top and side views) is shown in the inset. (Reproduced from Fig. 8 of Ref. 130, with permission.)

very many degrees of freedom, some with very low vibrational frequencies. For HAS, surfaces with these adsorbates can satisfy the Weare criterion [Eq. (19)] only with difficulty, and therefore may be incapable of being investigated under single-phonon scattering conditions.

For example, Wöll and Vogel [132] scattered helium atoms from an LB film from arachidic acid and methyl stearate in a molar ratio of 9 to 1 at a film temperature of 130 K. The resulting TOF spectrum yielded a very broad distribution of energy exchanges as shown in Fig. 48 without any distinct peaks to be seen. Manson and Skofronick, however, were able to extend the multiphonon scattering model discussed in Section III.B into the classical regime to fit this measured energy transfer distribution [133]. In this limit, the form factor can be represented by a hard-wall scattering matrix element so that the shape of the distribution depends only on the surface temperature and the mass of the target M. To obtain the rather good fit shown in Fig. 48, it was necessary to take $M = 15$ amu, the mass of a methyl group, in agreement with the expected structure of this LB surface which has methyl groups sticking up from the film. Analysis of the Debye–Waller exponent indicates that the average number of phonons exchanged in the helium atom collisions with this surface is ~ 15, far from the Weare criterion of Eq. (19), so that the interpretation of the results appears to be consistent.

Experiments have begun in the FSU laboratory which will make use of this kind of analysis for SAMs and physisorbed hydrocarbons. The ability to interpret these measurements with this model gives one encouragement that useful information can be obtained about the dynamics of rather complex systems through relatively easy HAS experiments.

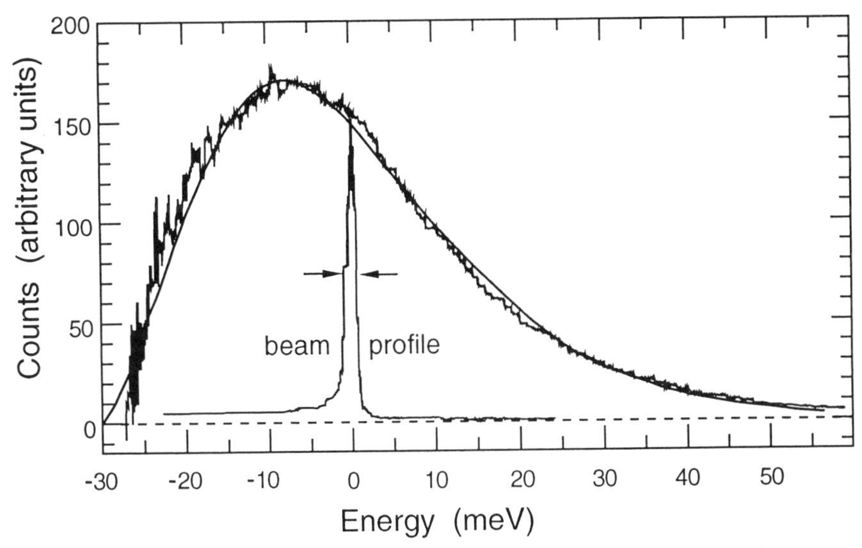

Figure 48. TOF spectrum transformed to energy transfer distribution for the Langmuir–Blodgett film described in the text at 130 K. The jagged line corresponds to the helium scattering data, and the solid line is calculated from the Manson theory as described in the text. The inset shows the initial energy spread in the helium atom beam. (Reproduced from Fig. 1 of Ref. 133, with permission.)

VIII. CONCLUDING REMARKS

The vibrational motion of atoms is intimately connected with the distribution of their bonding electrons. For alkali halides with localized charge distributions, neutron scattering studies found that the phonon dispersion, particularly for the optical modes, could not be adequately described by models which ignored polarizability effects. This led to the introduction of the Shell Model. HAS investigations of the surface vibrations of these materials have shown that the localized electronic distributions persist essentially unmodified at the surface except, possibly, for changes to the larger ions that are induced by surface relaxation. The Shell Model appears to be adequate also for the surface vibrational behavior of binary oxides and for more complicated materials with perovskite structures. However, because the optical modes are more sensitive to the electronic distributions, a greater effort needs to be made to acquire more data for comparison with model calculations.

For metals and layered materials which have delocalized electrons, the Born–von Kármán treatment appears to be able to reproduce the bulk vibrational mode dispersion and the high-energy surface phonons measured by EELS with a few harmonic force constants. However, this treatment breaks down for HAS experiments in reproducing the scattering intensities associated with different phonon modes and in interpreting the data that shows the occurrence of charge density waves which can lead to surface reconstruction. It is only by taking proper account of the electron–phonon coupling through the Pseudocharge Model for metals and Dispersive Linear Chain Model for layered materials that all the experimental results can be understood. (The Bond Charge Model, a cousin to both the Shell Model and the Pseudocharge Model, has been successful in describing the surface phonon dispersion of semiconductors.)

For surface chemistry the results obtained for clean crystalline surfaces are just the preamble. The recent experiments described above on layer-by-layer growth and the vibrational characterization of physi- and chemisorbed molecules and molecular films have an obvious and direct connection to issues ranging from heterogeneous catalysis to the development of novel materials. So do other experiments with high-resolution helium atom scattering which have not been included in this chapter. Two of these with the greatest potential for impacting this field are measurements of the phonons localized to surface steps and studies as a function of surface temperature of the quasielastic scattering of helium atoms associated with adsorbate diffusion and surface phase changes (e.g., pre-melting). In order to design films which are to be grown for specific properties, one needs to predict how molecules move on the substrate surface and arrange themselves into the film. Similarly, to control a chemical reaction at an interface requires understanding (a) the process of diffusion to the reactive site (which is very likely to be at an extended defect like a step edge) and (b) the dynamics of the adsorbate/substrate system at the site. Future HAS experiments will undoubtedly focus on these areas.

An additional useful development that has been motivated by HAS experiments is the adaptation of the theory of atom–surface scattering to treat multiphonon excitations. It would appear to be particularly valuable for the characterization of the vibrational properties of organic films such as Langmuir–Blodgett films and self-assembling monolayers which are currently of great interest.

It has often been remarked that surfaces are difficult to study because every technique in the repertoire of surface science touches on a different aspect of what we attribute to the surface, so that understanding the whole requires assembling all the jigsaw-puzzle-like pieces of information together. High-resolution helium atom scattering is very good at probing features, definitely associated with surface properties, that are very difficult to access in any other way and in just the past few years has joined several large pieces together. The future is likely to see a continuation of this constructive activity.

Acknowledgments

The author gratefully acknowledges the close and fruitful collaboration with Professor Jim Skofronick in the work described in this chapter from the HAS laboratory at the Florida State University. He also acknowledges the invaluable contributions of students and research associates, past and present, who have worked with us and is particularly indebted to Jeff Baker and Juan Hernández for help in preparing some of the figures for this manuscript. In addition, he would also like to thank his host Professor J. P. Toennies of the Max Planck Institut für Strömungsforschung, Göttingen, Germany, where part of this chapter was written. The helium atom scattering experiments at FSU are supported by the U.S. Department of Energy through grant FG05-85ER45208 and by NATO through grant 891059.

References

1. M. Born and T. von Kármán, *Z. Phys.* **13,** 297 (1912); M. Born and T. von Kármán, *Z. Phys.* **14,** 65 (1913).
2. M. Born and K. Huang, *Dynamical Theory of Crystal Lattices* (Oxford University Press, New York, 1954).
3. P. Brüesch, *Phonons: Theory and Experiment I* (Springer-Verlag, New York, 1982).
4. A. K. Ghatak and L. S. Kothari, *An Introduction to Lattice Dynamics* (Addison-Wesley, London, 1971).
5. B. N. Brockhouse and A. T. Stewart, *Rev. Modern Phys.* **30,** 236 (1958).
6. A. D. B. Woods, W. Cochran, and B. N. Brockhouse, *Phys. Rev.* **119,** 980 (1960); A. D. B. Woods, B. N. Brockhouse, R. A. Cowley, and W. Cochran, *Phys. Rev.* **131,** 1025 (1963); R. A. Cowley, W. Cochran, B. N. Brockhouse, and A. D. B. Woods, *Phys. Rev.* **131,** 1030 (1963).
7. B. Dorner, *Coherent Inelastic Neutron Scattering in Lattice Dynamics* (Springer-Verlag, Heidelberg, 1982).
8. For example, see a text in solid state physics such as N. W. Ashcroft and N. D. Mermin, *Solid State Physics* (W. B. Saunders, Philadelphia, 1976).
9. For example, see H. Bilz and W. Kress, *Phonon Dispersion Relations in Insulators* (Springer-Verlag, New York, 1979).
10. N. Cabrera, V. Celli, F. O. Goodman, and J. R. Manson, *Surf. Sci.* **19,** 67 (1970); J. R. Manson and V. Celli, *Surf. Sci.* **24,** 495 (1971).
11. G. Armand and J. R. Manson, *Surf. Sci.* **80,** 532 (1979).
12. A. C. Levi, *Il Nuovo Cimento* **54B,** 357 (1979).
13. S. S. Fisher and J. R. Bledsoe, *J. Vac. Sci. Technol.* **9,** 814 (1972).
14. B. R. Mason and B. R. Williams, Proceedings of the 2nd International Conference on Solid Surfaces, Japan, *J. Appl. Phys. Suppl.* **2,** part 2, 1974.
15. S. C. Yerkes and D. R. Miller, *J. Vac. Sci. Technol.* **17,** 126 (1980).
16. G. Brusdeylins, R. B. Doak, and J. P. Toennies, *Phys. Rev. Lett.* **44,** 1417 (1980).
17. G. Brusdeylins, R. B. Doak, and J. P. Toennies, *Phys. Rev. B*, **27,** 3662 (1983) and references cited therein.
18. R. S. Grace, W. M. Pope, D. L. Johnson, and J. G. Skofronick, *Phys. Rev. A* **14,** 1006 (1976).
19. J. P. Toennies and K. Winkelmann, *J. Chem. Phys.* **66,** 3965 (1977).
20. For example, see J. P. Toennies, *J. Phys. Condens. Matter* **5,** A25 (1993).

21. For example, see K. Kern and G. Comsa, *Adv. Chem. Phys.* **76,** 211 (1989).
22. For example, see J. P. Toennies, in *Surface Phonons*, W. Kress and F. W. de Wette (eds.) (Springer-Verlag, Heidelberg, 1991), Chapter 5.
23. For example, see R. B. Doak, in *Atomic and Molecular Beam Methods*, Vol. 2, G. Scoles (ed.) (Oxford University Press, New York, 1992), Chapter 14.
24. For example, see G. Benedek and J. P. Toennies, *Surf. Sci.* **299/300,** 587 (1994).
25. N. Esbjerg and J. K. Nørskov, *Phys. Rev. Lett.* **45,** 807 (1980).
26. H. Ibach and D. L. Mills, *Electron Energy Loss Spectroscopy and Surface Vibrations* (Academic Press, San Francisco, 1982).
27. W. P. Woodruff and T. Delchar, *Modern Techniques of Surface Science* (Cambridge University Press, Cambridge, 1986).
28. W. Kress, F. W. de Wette, A. D. Kulkarni, and U. Schröder, *Phys. Rev. B* **35,** 5783 (1987).
29. W. Cochran, *The Dynamics of Atoms in Crystals* (Crane, Russak and Co., New York, 1973).
30. L. L. Foldy and T. A. Whitten, Jr., *Solid State Commun.* **37,** 709 (1981).
31. B. Segall and L. L. Foldy, *Solid State Commun.* **47,** 593 (1983).
32. L. L. Foldy and B. Segall, *Phys. Rev. B* **25,** 1260 (1982).
33. F. W. de Wette, in *Surface Phonons*, W. Kress and F. W. de Wette (eds.) (Springer-Verlag, Heidelberg, 1991), Chapter 4.
34. G. Benedek and L. Miglio, in *Ab Initio Calculations of Phonon Spectra*, J. Devreese, V. E. van Doren, and P. E. van Camp (eds.) (Plenum Press, New York, 1982).
35. G. Benedek and L. Miglio, in *Surface Phonons*, W. Kress and F. W. de Wette (eds.) (Springer-Verlag, Heidelberg, 1991), Chapter 3.
36. T. S. Chen, F. W. de Wette, and G. P. Alldredge, *Phys. Rev. B* **15,** 1167 (1977).
37. F. W. de Wette, W. Kress, and U. Schröder, *Phys. Rev. B* **32,** 4143 (1985).
38. A. A. Maradudin, E. W. Montroll, G. H. Weiss, and I. P. Ipatova, *Theory of Lattice Vibrations in the Harmonic Approximation* (Academic Press, New York, 1971), *Solid State Physics*, Supplement 3.
39. For a recent review, see V. Celli, in *Surface Phonons*, W. Kress and F. W. de Wette (eds.) (Springer-Verlag, Heidelberg, 1991), Chapter 6.
40. For example, see G. Boato and P. Cantini, in *Advances in Electronics and Electron Physics*, Vol. 60, P. W. Hawkes (ed.) (Academic Press, New York, 1983), pp. 95–160.
41. F. Hoffmann, J. P. Toennies, and J. R. Manson, *J. Chem. Phys.* **101,** 10155 (1994).
42. For example, see D. Eichenauer, U. Harten, J. P. Toennies, and V. Celli, *J. Chem. Phys.* **86,** 3693 (1987).
43. A bulk phonon contribution to the time-of-flight spectrum has recently been identified by subtracting out all the other contributions. See Ref. 41.
44. J. R. Manson, *Phys. Rev. B* **43,** 6924 (1991).
45. J. H. Weare, *J. Chem. Phys.* **61,** 2900 (1974).
46. V. Bortolani, A. Franchini, N. Garcia, F. Nizzoli, and G. Santoro, *Phys. Rev. B* **28,** 7358 (1983).
47. V. Celli, G. Benedek, U. Harten, J. P. Toennies, R. B. Doak, and V. Bortolani, *Surf. Sci.* **143,** L376 (1984).

48. See also, R. David, K. Kern, P. Zeppenfeld, and G. Comsa, *Rev. Sci. Instrum.* **57,** 2771 (1986).
49. G. Chern, J. G. Skofronick, W. P. Brug, and S. A. Safron, *Phys. Rev. B* **39,** 12828 (1989).
50. See for example, K. Kern, R. David, R. L. Palmer, G. Comsa, and T. S. Rahman, *Phys. Rev. B* **33,** 4334 (1986).
51. G. G. Bishop, Ph.D. Dissertation, Florida State University, 1994.
52. E. S. Gillman, Ph.D. Dissertation, Florida State University, 1994.
53. G. Witte, private communication.
54. J. G. Skofronick and J. P. Toennies, in *Surface Properties of Layered Structures* (Kluwer Academic Publishers, The Netherlands, 1992), pp. 151–218.
55. For example, see B. Poelsema and G. Comsa, *Scattering of Thermal Energy Atoms*, Springer Tracts in Modern Physics (Springer, New York, 1989).
56. I. Estermann and O. Stern, *Z. Phys.* **61,** 95 (1930).
57. E. W. Kellermann, *Philos. Trans. R. Soc. London Ser. A* **238,** 513 (1940).
58. G. Bracco, M. D'Avanzo, C. Salvo, R. Tatarek, S. Terreni, and F. Tommasini, *Surf. Sci.* **189/190,** 684 (1987).
59. G. Benedek, G. P. Brevio, L. Miglio, and V. R. Velasco, *Phys. Rev. B* **26,** 497 (1982).
60. W. Kress, F. W. de Wette, A. D. Kulkarni, and U. Schröder, *Phys. Rev. B* **35,** 2467 (1987).
61. D. Evans, V. Celli, G. Benedek, J. P. Toennies, and R. B. Doak, *Phys. Rev. Lett.* **50,** 1854 (1983).
62. H. Hoinkes, *Rev. Modern. Phys.* **52,** 933 (1980); G. Vidali, M. W. Cole, and J. R. Klein, *Phys. Rev. B* **28,** 3064 (1983).
63. P. Cantini, G. P. Felcher, and R. Tatarek, *Phys. Rev. Lett.* **37,** 606 (1976); P. Cantini, R. Tatarek, and G. P. Felcher, *Surf. Sci.* **63,** 104 (1977).
64. R. W. G. Wyckoff, *Crystal Structures*, Vol. 1 (Wiley, New York, 1964).
65. S. A. Safron, G. Chern, W. P. Brug, J. G. Skofronick, and G. Benedek, *Phys. Rev. B* **41,** 10146 (1990).
66. G. G. Bishop, J. Duan, E. S. Gillman, S. A. Safron, and J. G. Skofronick, *J. Vac. Sci. Technol. A* **11,** 2008 (1993).
67. U. Schröder and D. Bonart, private communication.
68. S. A. Safron, W. P. Brug, G. G. Bishop, J. Duan, G. Chern, and J. G. Skofronick, *J. Electron. Spectrosc. Rel. Phenom.* **54/55,** 343 (1990).
69. F. W. de Wette, A. D. Kulkarni, U. Schröder, and W. Kress, *Phys. Rev. B* **35,** 2476 (1987).
70. G. Benedek, L. Miglio, G. Brusdeylins, J. G. Skofronick, and J. P. Toennies, *Phys. Rev. B* **35,** 6593 (1987).
71. G. Chern, J. G. Skofronick, W. P. Brug, and S. A. Safron, *Phys. Rev. B* **39,** 12838 (1989).
72. J. G. Skofronick, W. P. Brug, G. Chern, J. Duan and S. A. Safron, *J. Vac. Sci. Technol. A* **8,** 2632 (1990).
73. S. A. Safron, W. P. Brug, G. G. Bishop, G. Chern, M. E. Derrick, J. Duan, M. E. Deweese, and J. G. Skofronick, *J. Vac. Sci. Technol. A* **9,** 1657 (1991).
74. G. Benedek, R. Gerlach, A. Glebov, G. Lange, W. Silvestri, J. G. Skofronick, and J. P. Toennies, *Il Vuoto.*, in press.

75. G. Benedek, G. Brusdeylins, R. B. Doak, J. G. Skofronick and J. P. Toennies, *Phys. Rev. B* **28,** 2104 (1983).
76. W. Y. Leung, J. Z. Larese, and D. R. Frankl, *Surf. Sci.* **143,** L398 (1984).
77. J. Duan, W. P. Brug, G. G. Bishop, G. Chern, S. A. Safron, and J. G. Skofronick, *Surf. Sci.* **251/252,** 782 (1991).
78. W. P. Brug, G. Chern, J. Duan, G. G. Bishop, S. A. Safron, and J. G. Skofronick, *J. Vac. Sci. Technol. A* **10,** 2222 (1992).
79. J. P. Toennies, G. Witte, A. M. Shikin and K. H. Rieder, *J. Electron. Spectrosc. Rel. Phenom.*, **64/65,** 677 (1993).
80. J. Braun, P. Senet, J. P. Toennies, and G. Witte, to be published.
81. G. Brusdeylins, R. B. Doak, J. G. Skofronick, and J. P. Toennies, *Surf. Sci.* **128,** 191 (1983).
82. J. Cui, D. R. Jung, and D. R. Frankl, *Phys. Rev. B* **42,** 9701 (1990).
83. D. R. Jung, J. Cui, and D. R. Frankl, *J. Vac. Sci. Technol. A* **9,** 1589 (1991).
84. B. H. Brandow, *Adv. Phys.* **26,** 651 (1977).
85. However, G. Lange, C. Wöll, and co-workers have recently measured surface phonon modes with HAS for diamond which lie at about 70 meV; private communication.
86. C. Oshima, *Modern Phys. Lett. B* **5,** 381 (1991).
87. P. W. Palmberg, R. E. DeWames, L. A. Vredevoe, and T. Wolfram, *Appl. Phys.* **40,** 1148 (1969).
88. F. P. Netzer and M. Prutton, *J. Phys. C* **8,** 2401 (1975).
89. G. G. Bishop, J. Baker, E. S. Gillman, J. J. Hernández, S. A. Safron, and J. G. Skofronick, *J. Vac. Sci. Technol. A* **13,** 1416 (1995).
90. A. Swan, M. Marynowski, W. Frenzen, M. El-Batanouny, and K. M. Martini, *Phys. Rev. Lett.* **71,** 1250 (1993).
91. F. S. Galasso, *Perovskites and High T_C Superconductors* (Gordon and Breach Science Publishers, New York, 1990).
92. J. P. Toennies and R. Vollmer, *Phys. Rev. B* **44,** 9833 (1991).
93. R. Reiger, J. Prade, U. Schröder, F. W. de Wette, and W. Kress, *J. Electron. Spectrosc. Rel. Phenom.* **44,** 403 (1987); R. Reiger, J. Prade, U. Schröder, F. W. de Wette, A. D. Kulkarni, and W. Kress, *Phys. Rev. B* **39,** 7938 (1989).
94. U. Palzer, D. Schmicker, F. W. de Wette, U. Schröder, and J. P. Toennies, to be published.
95. For example, for $YBa_2Cu_3O_7$ see W. Kress, U. Schröder, J. Prade, A. D. Kulkarni, and F. W. de Wette, *Phys. Rev. B* **38,** 2906 (1988).
96. G. G. Bishop, W. P. Brug, G. Chern, J. Duan, S. A. Safron, J. G. Skofronick, and J. R. Manson, *Phys. Rev. B* **47,** 3966 (1993).
97. U. Harten, J. P. Toennies, and C. Wöll, *Faraday Discuss. Chem. Soc.* **80,** 137 (1985).
98. U. Harten, A. M. Lahee, J. P. Toennies, and C. Wöll, *Phys. Rev. Lett.* **54,** 2619 (1985).
99. C. Kaden, P. Ruggerone, J. P. Toennies, G. Zhang, and G. Benedek, *Phys. Rev. B* **46,** 13509 (1992).
100. G. Benedek, J. Ellis, N. S. Luo, A. Reichmuth, P. Ruggerone, and J. P. Toennies, *Phys. Rev. B* **48,** 4917 (1993).

101. N. Bunjes, N. S. Luo, P. Ruggerone, J. P. Toennies, and G. Witte, *Phys. Rev. B* **50,** 8897 (1994).
102. G. Witte, J. P. Toennies, and C. Wöll, *Surf. Sci.* **232,** 228 (1995).
103. For example, see V. Bortolani, G. Santoro, U. Harten, and J. P. Toennies, *Surf. Sci.* **148,** 82 (1984).
104. For example, see J. S. Nelson, M. S. Daw, and E. C. Sowa, *Phys. Rev. B* **40,** 1465 (1988).
105. For example, see C. S. Jayanthi, H. Bilz, W. Kress, and G. Benedek, *Phys. Rev. Lett.* **59,** 795 (1991).
106. N. S. Luo, P. Ruggerone, J. P. Toennies, and G. Benedek, *Phys. Scripta* **T49,** 584 (1993).
107. E. Hulpke, M. Hüppauff, D. M. Smilgies, A. D. Kulkarni, and F. W. de Wette, *Phys. Rev. B* **45,** 1820 (1992).
108. E. Hulpke, *J. Electron. Spectrosc. Rel. Phenom.* **54/55,** 299 (1990).
109. H.-J. Ernst, E. Hulpke, J. P. Toennies, and C. Wöll, *Surf. Sci.* **262,** 159 (1992).
110. H.-J. Ernst, E. Hulpke, and J. P. Toennies, *Phys. Rev. B* **46,** 16081 (1992).
111. See, for example, J. M. Ziman, *Principles of the Theory of Solids* (Cambridge University Press, Cambridge, 1972).
112. See also, U. Harten, J. P. Toennies, C. Wöll, and G. Zhang, *Phys. Rev. Lett.* **55,** 2308 (1985).
113. E. Hulpke and J. Lüdecke, *Surf. Sci.* **287/288,** 837 (1993).
114. G. Benedek, F. Hofmann, P. Ruggerone, G. Onida, and L. Miglio, *Surf. Sci. Rep.* **20,** 1 (1994).
115. G. Benedek, G. Brusdeylins, G. Heimlich, L. Miglio, J. G. Skofronick, J. P. Toennies, and R. Vollmer, *Phys. Rev. Lett.* **60,** 1037 (1988).
116. L. Miglio and L. Colombo, *Phys. Rev. B* **37,** 3025 (1988).
117. J. Duan, G. G. Bishop, E. S. Gillman, G. Chern, S. A. Safron, and J. G. Skofronick, *Surf. Sci.* **272,** 220 (1992).
118. S. A. Safron, J. Duan, G. G. Bishop, E. S. Gillman, and J. G. Skofronick, *J. Phys. Chem.* **97,** 1749 (1993).
119. E. S. Gillman, J. Baker, J. J. Hernández, G. G. Bishop, J. A. Li, S. A. Safron, J. G. Skofronick, U. Schröder, and D. Bonart, *Phys. Rev. B*, in press.
120. S. A. Safron, G. G. Bishop, J. Duan, E. S. Gillman, J. G. Skofronick, N. S. Luo, and P. Ruggerone, *J. Phys. Chem.* **97,** 2270 (1993).
121. N. S. Luo, private communication.
122. J. Baker and P.-A. Lindgård, private communication.
123. S. M. Weera, J. R. Manson, J. Baker, E. S. Gillman, J. J. Hernández, G. G. Bishop, S. A. Safron, and J. G. Skofronick, *Phys. Rev. B*, **52,** 14185 (1995).
124. J. Baker, private communication.
125. G. Benedek, J. Ellis, A. Reichmuth, P. Ruggerone, H. Schief, and J. P. Toennies, *Phys. Rev. Lett.* **69,** 2951 (1992).
126. K. D. Gibson and S. J. Sibener, *Phys. Rev. Lett.* **55,** 1514 (1985).
127. K. D. Gibson, S. J. Sibener, B. M. Hall, D. L. Mills, and J. E. Black, *J. Chem. Phys.* **83,** 4256 (1985).

128. N. S. Luo, P. Ruggerone, J. P. Toennies, G. Benedek, and V. Celli, *J. Electron. Spectrosc. Rel. Phenom.* **64/65,** 755 (1993).

129. K. Kern, P. Zeppenfeld, R. David, and G. Comsa, *Phys. Rev. B* **35,** 886 (1986).

130. G. Witte, H. Range, J. P. Toennies, and C. Wöll, *J. Electron. Spectrosc. Rel. Phenom.* **64/65,** 715 (1993).

131. G. Witte, H. Range, J. P. Toennies, and C. Wöll, *Phys. Rev. Lett.* **71,** 1063 (1993).

132. V. Vogel and C. Wöll, *J. Chem. Phys.* **84,** 5200 (1986).

133. J. R. Manson and J. G. Skofronick, *Phys. Rev. B* **47,** 12890 (1993).

ORDERING AND PHASE TRANSITIONS IN ADSORBED MONOLAYERS OF DIATOMIC MOLECULES

DOMINIK MARX

Max-Planck-Institut für Festkörperforschung, Stuttgart, Germany

HORST WIECHERT

Institut für Physik, Johannes Gutenberg-Universität, Mainz, Germany

CONTENTS

Advances in Chemical Physics, *Volume XCV*, Edited by I. Prigogine and Stuart A. Rice.
ISBN 0-471-15430-X

I. PRELIMINARY REMARKS

A. Introduction

The systematic investigation of superstructures of diatomic molecules on relatively smooth substrates and the transitions between the phases started about two decades ago with pioneering diffraction and calorimetric experiments of various kinds. In the meantime the field has matured, and the main tools of the early days (i.e., experiment and theory) have been supplemented with powerful simulation techniques, and it is certainly true that simulations helped a great deal in understanding the physics of these systems. In this review we focus on the homo- and heteronuclear diatomic molecules N_2 and CO physisorbed on the two smooth substrates graphite and boron nitride. In terms of the phase diagrams we cover (a) the density range from low coverages up to compressed monolayers including the onset of bilayer structures and (b) the temperature range below about 100 K.

Why do we concentrate on these particular systems? First of all, the isoelectronic diatomic molecules N_2 and CO have many molecular and bulk properties in common (see Section I.B), which is a signature of their quite similar van der Waals and electrostatic interactions. However, there are slight but crucial quantitative and qualitative differences which affect in a delicate way the phase diagrams in two dimensions. This situation is different in the case of O_2 monolayers with magnetic interactions due to uncompensated electron spins resulting in a triplet ground state with nonzero total spin ($S = 1$). These interactions are crucial to fully understand the phase behavior (see, e.g., Refs. 145, 221, 226, 247, 256, 331, 353, and 390 for experiments, Refs. 25, 96, 170, 207, and 210 for related simulations, and Refs. 106 and 220 for reviews, as well as Fig. 1 of Ref. 221 for a phase diagram of O_2 adlayers on graphite). The molecular hydrogen isotopes H_2, D_2, and HD behave under most circumstances just like the light spherical rare gas atoms, and most studies concentrate on aspects related to positional ordering (see, e.g., Refs. 82–84, 120, 121, 123, 131, 132, 158, 222, 243, 253, 278, 320, 365, and 377–379), where manifestations of quantum behavior can often be observed. Thus, the "diatomic nature" with its characteristic possibility of showing phase transitions entirely due to the ordering of internal degrees of freedom, such as orientational and/or head–tail ordering, is not "active," see, e.g., Refs. 269–271. A possible exception might be the behavior of H_2 and D_2 on graphite at very low temperatures as explored by nuclear magnetic resonance (NMR) techniques [180], but a firm conclusion is yet to be reached (see Refs. 21, 141, 157, and 218 for theoretical approaches related to this issue).

Second, after 20 years of research effort the phase diagrams of N_2 and CO physisorbed on graphite belong to the best-known truly molecular phase diagrams in two dimensions. Due to the wealth of accumulated data there is now a broad consensus about these phase diagrams reached so that many of the relevant questions are settled and are ready to be reviewed, but a comprehensive review is nevertheless still lacking; concerning unresolved problems we refer to Section VII. Parts of what we cover here as well as closely related topics have been discussed in the reviews and books on the theory of interactions of gases with solid surfaces [325], films on solid surfaces [87], ordering in two dimensions [205, 321, 333], calorimetric studies of a variety of phase transitions in two dimensions including the discussion of experimental techniques and substrates [220], neutron scattering from adsorbates [322, 350, 384], orientational ordering in N_2, CO, and other monolayers on graphite [106, 108, 109], Monte Carlo simulations of phase transitions in adsorbed layers based on lattice gas models [36], experimental results for a variety of adsorbed films [88, 347], and rare gas films on graphite [319], but we stress that this list is certainly incomplete.

Finally, both graphite and boron nitride have smooth and inert surfaces that interact with the adsorbed molecules only weakly; that is, these surfaces are not reconstructed by the adlayer and do not significantly perturb the properties of the adsorbed molecules. Other substrates such as lamellar halides [236] and MgO [79–81] have also been used for physisorption studies, but the information gained so far is much less comprehensive. The gas–surface interaction potential of alkali metals is weaker [68], but there are certain metals, alkali halides, and alkaline earth halides with surfaces which are not as smooth as those of graphite and/or modify the molecule–surface interactions much more than in the case of the archetypal physisorbates presented here (see Ref. 363 for a recent overview of relevant adsorption potentials). In addition, the isoelectronic hexagonal substrates graphite and boron nitride have in some sense closely related bulk and surface structures so that adlayers are expected to behave similarly in many respects. On the other hand, boron nitride is an insulator so that dielectric measurements which probe directly the ordering of charge distributions are feasible and offer a new source of information which can be carried over to graphite.

In this review we exclusively cover pure adlayers, and thus we exclude phenomena related to mixing of N_2 or CO with any other species. After the exploration by theory and simulation [142, 245] the impact of spherical impurities or vacancies on the orientational ordering and the associated ground states were investigated experimentally [391, 392] in mixtures of CO and N_2 with Ar (see Ref. 246 for a review and also see Refs. 70, 228, and 255). Random field behavior in two dimensions was first seen when CO on graphite was diluted with small amounts of N_2 molecules [380, 381] (see Refs. 288 and 289 for related simulation work). These topics, however, are more appropriate for reviews on disordered systems.

We stress that this chapter is also an attempt to present a review covering experimental, theoretical, and simulation results at the same time on an equal footing. Such a presentation seems to us to be very useful and actually quite natural because the knowledge accumulated for these diatomic monolayer adsorbates has emerged over the years from a fruitful interplay of these three disciplines. As a necessary consequence, experimentalists might not find all expected experimental details, and theoreticians might search desperately for the description of the theoretical techniques and the precise definition of potential models and so on. However, we regard as the main purpose of this chapter the compilation and review of results on the physics of these four adsorbate systems and not an attempt to provide introductions into the various methodologies and technical details, for which we will in general refer to the cited literature instead. We were guided by the idea to collect exhaustively and organize the available wealth of literature, and in this spirit we include an extensive list of references for further information. Thus, the style of the chapter is mainly encyclopedic with the aim to cover as much literature as possible, rather than focusing on a necessarily biased selection of a few papers.

The material is presented separately for each system and cross-referenced where necessary. For each system we have attempted to arrange the information by grouping it according to the main physical phenomena that are manifest in the phase diagram—that is, submonolayer coverages and melting, tricritical points, orientational and head–tail ordering, and, finally, compressed adlayers. Within these subsections, which are in some sense arbitrary and overlapping, the literature is presented in an essentially chronological order, which is sometimes rearranged for the sake of simplicity. Before we can discuss the monolayer properties of each system we felt it necessary to compile important molecular and bulk properties concerning N_2 and CO in Section I.B. Section I.C is devoted to the description of the substrates treated, as well as to the problems associated with defining the coverage of an adlayer on a substrate based on experimental measurements.

One central aspect of the present review is the orientational ordering of the molecular axis and the transitions between such ordered phases. In view of this we present in Section II.A the properties (i.e., structure of the phases and approximate phase diagram) of a very fruitful schematic model: electrostatic quadrupoles fixed on a triangular lattice. A further simplification of this model (which was extensively used in the quest for the order of the ideal herringbone transition), the anisotropic–planar–rotor model, as well as some of its generalizations, are introduced in Section II.B. The section on orientational order is closed in Section II.C by a discussion of a classification of some orientational transitions based on symmetry arguments and renormalization group results.

As an outcome of our review work we present as completely as possible phase diagrams of the four systems (see Figs. 8, 47, 49, and 66), which are obtained as a patchwork from several sources. We guide the reader through these phase diagrams in the first subsection for each system, and further detail is found in the corresponding subsections that follow. However, we give a warning right at the beginning: Even these phase diagrams are still subject to some uncertainty and possible inconsistency. We stress that the absolute uncertainty of reported coverages can be as large as 25% depending on the precise definition and the underlying technique (see Section I.C for a discussion of this aspect). Second, many phase boundaries are still speculative or controversial, the nature of the phase transition is often not known beyond doubt, and the structure of some phases is also subject to uncertainties. We do not recommend the use of these phase diagrams without reading the respective parts of the specialized subsections and taking them, as well as the original literature, into account.

Unfortunately, not much is known for N_2 and CO on graphite or boron nitride with respect to the nature of their commensurate–incommensurate transitions, and the order and mechanism of their melting transition, although various speculations and claims can be found in the literature. Therefore, and in view of the excellent reviews already available, we did not include detailed discussions of the theoretical implications of these topics. Many aspects related to the vast subject of commensurate–incommensurate phase transitions can be found in Refs. 10, 205, 260, and 364 from a theoretical perspective, whereas experimental results for Kr on graphite are described in Refs. 40, 323, and 329, for H_2, HD, and D_2 on graphite in Refs. 83, 84, 120–123, 243, 377, and 379, and for Kr and Xe on Pt(111) in Refs. 167 and 168. The possible peculiarities of the melting transition, when restricted to two dimensions, are covered, for example, in Refs. 129, 173, 205, 252, and 332.

Before closing this section, we draw the reader's attention to review articles of related fields which might be relevant to the material presented. Standard references on computer simulations in general are Refs. 3, 35, 74, and 143 and the series devoted to the Monte Carlo method 29, 33, 37. Finite-size scaling near phase transitions is discussed, for example, in Refs. 11 and 115 and an overview of its use in evaluating simulation data can be obtained from Refs. 28, 34, 38, and 293. Other relevant topics related to phase transitions are covered in the Domb–Green/Domb–Lebowitz series on "Phase Transitions and Critical Phenomena": Vol. 1 (*Two-Dimensional Ferroelectric Models*), Vol. 8 (*Critical Behavior at Surfaces*), Vol. 9 (*Theory of Tricritical Points*), and Vol. 12 (*Wetting Phenomena*). The origin of intermolecular interactions and their various theoretical formulations and representations are discussed in depth in Ref. 133. We do not cover questions related to (a) the physisorption potentials such as the determination of interaction potentials of gas atoms with single-crystal surfaces from gas–surface diffraction experiments [148] and (b) the modeling of such interactions in general [49, 50, 325]. We also refer to the respective sections of Refs. 18 and 298, and for specific systems we refer to Refs. 200, 324, 326, and 363, where also extensive tables of potential parameters can be found.

The experimental methods employed for studying properties of adsorbed films on graphite and boron nitride are to a large extent those of three-dimensional solid-state physics and surface physics. The basic properties of adlayers—that is, the two-dimensional equation of state (a relation between the variables temperature, coverage or density, and two-dimensional spreading pressure)—can be determined from adsorption isotherms at various temperatures [351]. Ellipsometry has turned out to be an extremely valuable tool for studying adsorption isotherms and film growth [147]. Heat capacity measurements are probably the most sensitive way to identify phase transitions and to map out phase boundaries of physisorbed systems. Several different heat capacity techniques have been developed: quasiadiabatic heat pulse calorimetry (see, e.g., Ref. [46]), comparative calorimetry [220], ac calorimetry [65, 334, 387], and scanning ratio calorimetry [204]. Dielectric studies of adsorbates on boron nitride are scarce [186, 223, 376], and we refer to Refs. 44, 124, and 224 for a treatment of the theoretical aspects of the dielectric behavior. Diffraction measurements with low-energy electrons (LEED), with x-rays, and with neutrons are the common techniques to investigate the structures of physisorbed systems, but also atomic and molecular beam diffraction (e.g., thermal helium scattering) can be applied [167]. Several books and review articles describe in detail the experimental methods and the underlying theory: LEED [75, 146, 150, 163, 281], x-rays [7, 85, 372], and neutrons [8, 201, 322, 350]. Explicit formulae for diffraction profiles of two-dimensional crystals and particularly of adsorbates on powdered graphite substrates have been given, for example, in Refs. 172, 309, 322, and 329. Inelastic neutron-diffraction measurements have also been used to explore the dynamics of physisorbed films [193, 322, 346, 384]. Quasielastic neutron scattering (QENS) is well-suited for measuring the diffusivity and mobility of molecular adsorbates with strong incoherent scattering cross sections [14, 26, 27]. Pulsed NMR techniques have been applied to study the in-plane orientational ordering of $^{15}N_2$ molecules on graphite [335]. The out-of-plane tilt angle of the N_2 on graphite system has been determined by nuclear resonance photon scattering (NRPS) [241], and that of CO on graphite has been determined by electron energy loss spectroscopy (EELS) [161].

Due to the weak binding of physisorbed particles to substrates, the application of low temperatures is indispensable to stabilize mono- and multilayer films. Principles of low-temperature technologies can be found in Refs. 290, 296, and 375. A general introduction to the physics of solid surfaces and physisorption is provided in Ref. 393, experimental details are contained in Ref. 383, and both experimental and theoretical aspects are discussed in Ref. 202.

B. Molecular and Bulk Properties of N_2 and CO

The two diatomic molecules N_2 and CO have similarities in many characteristics ranging from the molecular level to the bulk behavior. Among these properties are size, molecular mass, quadrupole moment, thermodynamic properties, and so on. In both cases the quadrupole moment plays a crucial role in determining major parts of the phase diagrams and the corresponding phase transitions in both three and two dimensions. However, CO carries a small dipole moment which causes subtle changes in the behavior in the bulk solid, and even more pronounced effects in the case of two-dimensional adlayers. Some molecular, bulk liquid, and bulk solid properties of N_2 and CO are presented in Table I for quick reference, where the sources of the reported data are also given. A much more detailed review concentrating on N_2 is provided in Ref. 312.

The charge density contours of the two isoelectronic molecules N_2 and CO are depicted in Fig. 1*a* and Fig. 1*b*, respectively. The outermost contour corresponds to a charge density of 0.002 atomic units and encircles approximately 95% of the total charge density of the molecules. The outermost contour can be viewed as a relative measure of shape and size of the molecules [9], indicating that the gross behavior is quite similar for N_2 and CO. The general shape of the molecules is prolate ellipsoidal associated with a pronounced molecular quadrupole moment which results in anisotropic (i.e., orientation-dependent) interactions. However, the molecular center of mass is displaced by about 0.2 Å from the geometric center of the outermost density contour in the case of CO, whereas both coincide, of course, for N_2 by symmetry. Also the charge density gradients are quite different for the carbon and oxygen ends of CO. Thus, in general the van der Waals interaction of these ends with other molecules and/or with corrugated surfaces is different. The most important low-order electrostatic multiple moments are the quadrupole moment in both cases, and in addition the dipole moment

TABLE I
Comparison of Molecular and Bulk Phase Properties of N_2 and CO

Property	N_2	CO	Unit
MOLECULAR PROPERTIES			
Dipole moment	0	0.374	10^{-30} Cm
	0	0.112	D
Quadrupole moment[a]	−5.07 (±6.34)	−8.34 (±9.37)	10^{-40} Cm^2
Bond length	1.10	1.13	Å
Charge contour (at 0.002 au)	3.34 × 4.38	3.36 × 4.48	Å
LIQUID-STATE PROPERTIES			
Triple-point temperature	63.150	68.127	K
Triple-point pressure	0.1246	0.1540	10^5 Pa
Melting enthalpy	720.9	835.6	J/mol
Melting entropy	11.4	12.3	J/K mol
Boiling temperature	77.36	81.638	K
Boiling enthalpy	5577	6040	J/mol
Boiling entropy	72.2	73.8	J/K mol
Critical temperature	126.26	132.85	K
Critical volume	89.19	92.17	10^{-6}m^3/mol
Critical pressure	33.991	34.935	10^5 Pa
SOLID-STATE PROPERTIES			
High-temperature β-phase	$P6_3/mmc$	$P6_3/mmc$	
Low-temperature α-phase	$Pa3$	$P2_13$	
Nearest-neighbor distance (0 K)	3.99	3.99	Å
Cubic lattice constant (α-phase)	5.644	5.647	Å
Hexagonal lattice constant[b] (β-phase) a_0	4.039	4.12	Å
	4.050	4.140	Å
Hexagonal lattice constant[b] (β-phase) c_0	6.67	6.80	Å
	6.604	6.723	Å
$\alpha \rightarrow \beta$ orientational ordering temperature	35.61	61.55	K
$\alpha \rightarrow \beta$ orientational ordering enthalpy	228.9	633.0	J/mol
$\alpha \rightarrow \beta$ orientational ordering entropy	6.4	10.3	J/K mol
Residual entropy	0	3.3	J/K mol

[a]The scatter of the data for the quadrupole moments are considerable so that we report the recommended value from Ref. 330 including a measure of the scatter of the data in parentheses.

[b]At 46 K for N_2 and 63 K for CO.

Source: Compiled from Table II of Ref. 41, Table 1 of Ref. 155, Table 1 of Ref. 389, Table 1 of Ref. 287, and Refs. 312, 330, and 348.

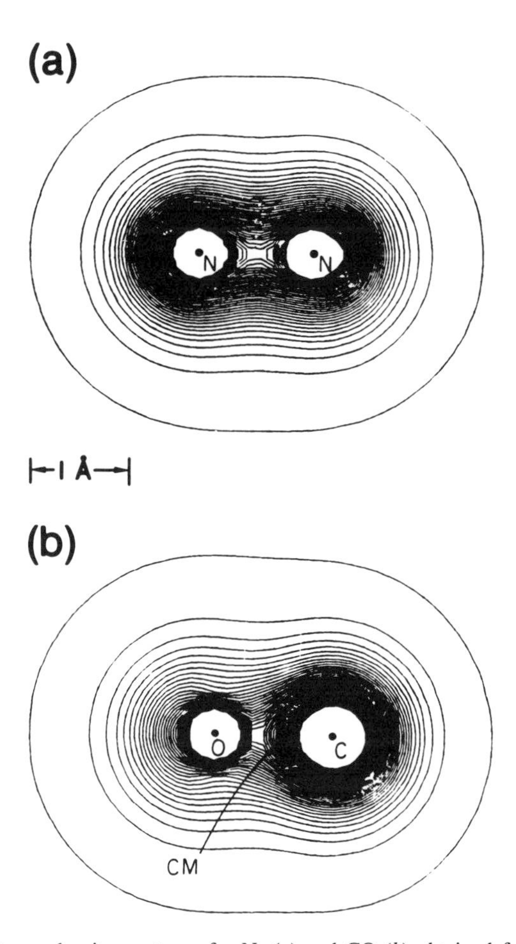

Figure 1. Charge density contours for N_2 (*a*) and CO (*b*) obtained from self-consistent field wavefunctions. The outermost contour corresponds to 0.002 atomic units and contains approximately 95% of the total electronic charge density. The center-of-mass of the CO molecule is marked by a dot and is displaced by about 0.2 Å from the center of the outermost charge contour shown. (Adapted from Fig. 1 of Ref. 108.)

for CO. Although the latter is very small (about 0.112 D) in comparison to other molecules, its presence has to be taken into account in order to fully understand the structures, the molecular dynamics, and also the phase transitions of CO adlayers. The electric dipole moment was determined to point from the negatively charged carbon atom to the positively charged oxygen atom [299], which is opposite to the sign one might expect for the dipole moment on the basis of simple electronegativity differences. Higher-order multipoles might in addition be necessary for a theoretical description of various properties on a fully quantitative level, which is especially true for CO (see Table I of Ref. 291 for a list also including an estimate for the 32-pole moment). The quadrupole moment of CO is significantly larger than that of N_2, the ratio varying somewhere between 1.1 to 1.9 depending on the choice of the experimental data [330]. The only small head–tail asymmetry in the case of CO will cause a behavior similar to that of N_2 over a wide range of temperatures, whereas a distinctly different behavior is expected at low temperatures, where the small dipolar interactions might become relevant.

The similarity of many molecular properties carries over to the bulk behavior in the liquid and solid states. The critical temperatures and volumes differ by only about 5%. Very similar values of the Lennard-Jones well-depth and length parameters which describe the liquid phase are obtained from fitting experimental data (see, e.g., Ref. 348 or Table 1 in Ref. 389). Both substances display similar crystal structures at low pressures and temperatures. The high-temperature/low-pressure β-solid has a hexagonal lattice, whereas the low-temperature α-phase has cubic symmetry. For both N_2 and CO the orientations of the molecular axes are disordered in the β-solid, the so-called plastic phase, whereas they are ordered in the α-phase such that the intermolecular quadrupolar interactions are optimized within the face-centered-cubic center-of-mass lattice structure. The cubic unit cell with identical lattice constants contains four molecules with specified orientations of all molecular axes (see Fig. 1 of Ref. 159 for a clear representation of this $P2_13$ structure of CO). The quadrupolar energy has been estimated to contribute [100, 175] about 14% to the total cohesion energy of α-N_2, and 19% in the case of α-CO, whereas the small dipole moment contributes only negligibly to the interaction energy [175]. The molecules do rotate around the (111) directions in the orientationally disordered β-phase. The larger molecular quadrupole moment for CO leads to an extension of the orientationally ordered α-phase up to a much higher temperature ($T_{\alpha\beta}$ = 61.55 K) as compared to that of the bulk N_2 solid ($T_{\alpha\beta}$ = 35.61 K). In addition, the asymmetry of the heteronuclear CO molecule results in an increased translation–rotation coupling in bulk CO solids [117, 118, 159].

Judging from purely energetic arguments the dipole moment of CO is expected to induce a head–tail ordered structure at very low temperatures in thermal equilibrium. A reasonable candidate structure for such a hypothetical fully ordered structure within the α-phase would be antiferroelectric with space group $P2_13(T^4)$ (see, e.g., Refs. 117 and 159). Based on a mean-field treatment of the dipole–dipole interactions the transition temperature to the fully ordered solid was estimated [230] to be smaller than 5 K. However, the nonvanishing residual entropy of 4.6 J/K mol reported in Ref. 76 has been interpreted as the signature of head–tail disorder; the residual entropy is defined as the difference between the calorimetric and the statistical entropies. A vanishing residual entropy is established for bulk N_2 in the α-phase, indicating complete order in the ground state. Later measurements of the heat capacity [52, 127, 317] of solid CO found no anomaly signaling a phase transition down to 0.8 K. The residual entropy was reevaluated [127] to amount to about 3.3 J/K mol, which is definitively less than the value $R \ln 2 \approx 5.76$ J/K mol that would be expected in the case of ideal statistical head–tail disorder.

A heat capacity study [5] revealed a small anomaly at about 18 K in bulk CO when the α-solid was annealed for several hours. This together with thermal relaxation phenomena was tentatively interpreted as a glass transition [5] taking place over the full range of stability of the α-phase from about 62 K down to 0 K. It was believed to arise from the freezing of the molecular head–tail reorientation as discussed in Ref. 86. According to this interpretation the residual entropy should be understood as a kinetic effect; note that the entropy removed below the glass anomaly amounts to only 0.04 J/K mol (see Ref. 154). This should be contrasted to the alternative scenario [230] of a cooperative phase transition in the thermodynamic sense, which would occur at a well-defined transition temperature in the stability range of the α-phase and which would lead to a long-range head–tail ordered ground state in the bulk solid phase.

Evidence in favor of dynamic disorder in α-CO was also obtained from nuclear quadrupole resonance studies [199, 369] and from dielectric measurements [251], which also indicate short-range antiferroelectric order [251]. The rate of molecular reorientation leading to head–tail flips was found to become vanishingly small at low temperatures: A head–tail reorientation time of about 5×10^{15} hours was estimated [369] at a temperature of 10 K. Thus, the extremely slow kinetics at the low temperatures which are energetically necessary to allow for head–tail ordering seems to prevent head–tail ordering in the bulk CO phase.

In the two-dimensional physisorbed environment the nearest-neighbor distance increases from 3.99 Å in the bulk α-phase to 4.26 Å in the commensurate monolayer on graphite. This change reduces the quadrupole–quadrupole and dipole–dipole interaction energies to about 72% and 82% of the bulk values, respectively. Additional features complicating the extrapolation from bulk to expected monolayer behavior are the different number of nearest neighbors and the geometry of the coordination shells, as well as the surface interactions with the substrate. How the similarities and differences between bulk N_2 and CO carry over to the two-dimensional physisorbed situation is the focus of Sections III and IV and Sections V and VI, respectively.

C. Substrates and Coverage

The most extensively used substrate for studying physisorbed films is graphite. It has hexagonal symmetry and belongs to the space group $P6_3mc$. The structure is depicted in Fig. 2*a* and consists of stacked planar sheets, in which the carbon atoms are essentially covalently bound by σ-like electrons into a honeycomb lattice. The lattice parameters are a = 2.459 Å and c = 6.708 Å [13] (or a = 2.456 Å and c = 6.696 Å [386]). This gives a nearest neighbor distance of 1.42 Å (1.418 Å) of the carbon atoms in the basal planes; the values in parentheses are obtained from the data of Ref. [386]. The interlayer distance is 3.354 Å (3.348 Å). The stacking sequence of the layers is of the type ABAB; however, stacking faults are often ob-

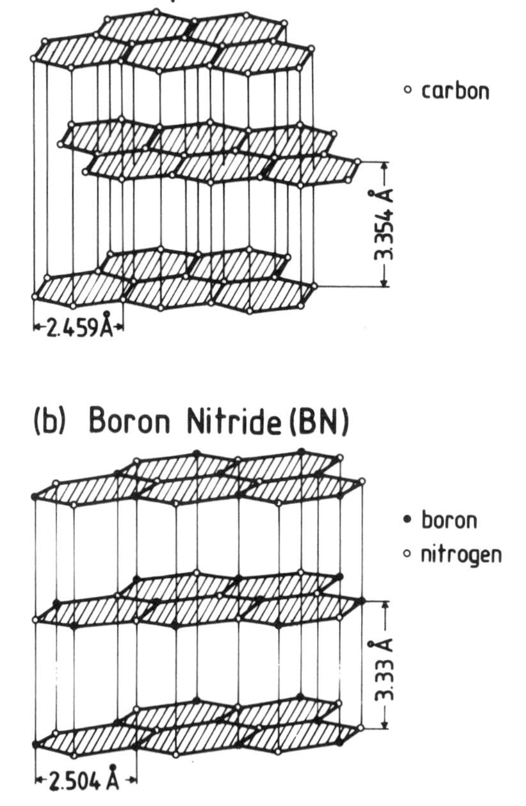

Figure 2. Schematic sketch of the graphite (*a*) and boron nitride (*b*) crystal structures including the lattice parameters.

served. The relatively large c/a ratio of 2.728 (2.726) and the corresponding large distance between the layers indicate that the bonds between the layers are much weaker than within the basal planes. This feature is responsible for the fact that graphite can easily be cleaved perpendicular to the c-axis. It also implies that it can be used as lubricant. The lattice constants mentioned above lead to the following positions of the principal low-angle Bragg peaks: the (002) reflection at about $Q = 1.873$ Å^{-1} ($Q = 1.877$ Å^{-1}), (100) at about $Q = 2.950$ Å^{-1} ($Q = 2.954$ Å^{-1}), and (101) at about $Q = 3.096$ Å^{-1} ($Q = 3.099$ Å^{-1}). Apart from the electrons associated with the σ-like bonds between carbon atoms in the basal planes, there are π-like electrons which are delocalized (see, e.g., Ref. 66 for a recent analysis of the bonding situation in graphite). The band structure of graphite is that of a semimetal. The π-like electrons are delocalized in the basal plane and form a quasi-two-dimensional system. This explains the electrical and thermal conductivity, the metallic luster, and the high absorption coefficient of visible light of graphite which leads to its black color. The transport properties are strongly affected by the anisotropy of the structure. Therefore the electrical conductivity of single-crystal graphite is almost by a factor of 10^4 larger parallel to the basal planes than that perpendicular to the planes, and for exfoliated graphite the difference still amounts to a factor of 10^2 [356, 357]. This is also the case for the thermal conductivity [94, 355].

For physisorption studies either natural single crystals of graphite or highly oriented pyrolytic graphite (HOPG) can be used. The highest quality of pyrolytic graphite with a mosaic spread of 0.4°, designated as ZYA graphite, can be obtained from Union Carbide (Union Carbide Coatings Service Corp., Cleveland, Ohio 44101-4924, USA). The average size of perfect domains between faults such as steps and grain boundaries is on the order of 1 μm. This size is usually called the coherence length. Experimental techniques which can be applied to characterize adsorbates on the nearly ideal surfaces are relatively rare. They comprise LEED, x-rays [149, 323], ellipsometry [147], and in one case also calorimetry [57].

In the overwhelming part of all investigations, however, exfoliated graphite is employed which is available in several varieties. Exfoliation refers to peeling apart or separating of layers. Grafoil (trademark of Union Carbide) and Papyex (trademark of Le Carbone–Lorraine, 92095 Paris La Defense 2, France) are such exfoliated graphite products which can be commercially obtained. They are prepared by intercalating graphite with sulfuric acid or $FeCl_3$, then heating it rapidly so that the pressure of the escaping gas drives the layers apart, and finally rolling it into thin sheets (of thicknesses between 0.2 and 1 mm), which partially orients the basal planes parallel to the sheets [87, 322]. The specific adsorption area ranges between 20 and 25 m^2/g. The orientations of the graphite crystallites are purely random in the sheet planes as in pyrolytic graphite. About 70% of the crystallites are a uniform powder, and 30% have a preferred orientation of the c-axis angle with respect to the normal to the sheets (mosaic spread) of about 30° full width at half-maximum. The coherence length of Grafoil is about 150 ± 50 Å and that of Papyex 250 ± 50 Å.

Graphite foam (product of Union Carbide) is uncompressed exfoliated graphite obtained at an intermediate stage of the production of Grafoil. The specific area is about 22 m^2/g. It can be regarded as a uniform powder with a coherence length of 900 ± 100 Å [39]. The density of foam is low, about 0.2 g/cm^3, and thus relatively large sample cells for adsorption studies are necessary.

ZYX graphite (product of Union Carbide) is a high-quality graphite substrate, which is exfoliated from pyrolytic graphite crystals of monochromator grade. The specific area is one order of magnitude smaller, about 1–2 m^2/g, than that of the other exfoliated graphite substrates. The percentage of crystals with preferred orientation is approximately 100%, and the mosaic spread is about 10° [39, 329]. For diffraction experiments with neutrons or x-rays, ZYX graphite has the advantage of having a rather large coherence length of 2000 ± 200 Å [39].

Graphitized carbon blacks have the highest specific surface areas of up to 200 m^2/g. While the substrates reviewed above are distinguished by their high surface homogeneity, carbon blacks are characterized by a pronounced surface heterogeneity and are of little use for the investigation of phase transitions of physisorbates. Their main usage dates back to the early days of physisorption studies.

Exfoliated graphite is cleaned by baking it under high vacuum, about 1×10^{-6} mbar, for several hours at 800–1000°C. This procedure removes strongly bonded chemical impurities introduced during the manufacturing process. The substance is then usually loaded into the sample cell in a glove box filled with helium gas. Subsequent pumping at room temperature is sufficient to remove residual contaminants. Studies on single crystals and highly oriented pyrolytic crystals require ultrahigh-vacuum (UHV) techniques. Fresh surfaces are prepared by cleaving in air with sticky tape. Mild baking in the UHV system cleans the surface finally.

An isoelectronic substrate closely related to graphite is boron nitride (BN). Though known for some time, it has only recently attracted attention. Like graphite, BN has hexagonal symmetry and is described by the space group $P6_3/mmc$ (see Fig. 2*b* for a sketch of the structure). The basal planes consist of honeycomb lattices, where the lattice sites are alternatingly occupied by boron and nitrogen atoms. The basal planes are arranged such that, unlike in the case of graphite, the atoms of one layer fall vertically above those of the layer below. The intraplane B–N distance is 1.446 Å, and the lattice parameters are $a = 2.50399$ Å and $c = 6.6612$ Å at 35°C [386]. The distance between the layers is 3.33 Å and the c/a ratio amounts to 2.66. Differences to graphite result from the fact that the π like electrons in BN are not delocalized as in graphite, but are bound to the nitrogen atoms. Consequently, BN is an insulator, possesses a diamagnetic susceptibility which is much smaller than that of graphite, and is white in color. These properties make BN suitable for dielectric [186, 223, 376], NMR [105], and optical [219] investigations. The depth of the potential well for adsorbate films on BN is shallower than that for the same adsorbates on graphite [234], and also the corrugation of the substrate potential is lower on BN than on graphite for classical rare gases and for N_2 [234]. Boron nitride is commercially available in the form of powders with different grain sizes (1–5 μm) from several companies (e.g., from Union Carbide Coatings Service Corp. or Elektroschmelzwerk Kempten GmbH, 87437 Kempten, Germany).

The cleaning procedure of BN is analogous to that of graphite. Recently it was observed [318] that washing BN powder in methanol prior to outgassing under vacuum at 900°C is an effective way of producing a good substrate. Such a treatment yielded high-quality adsorption isotherms.

Some of the properties of the substrates discussed are collected in Table II, and more details and references can be found in Refs. 39, 220, and 322.

One of the most important variables for describing the physics of adsorbed films is the coverage. To determine its absolute value to high precision is extremely difficult, whereas relative values can be measured much more accurately. The most frequently used method is to record an adsorption isotherm. It consists of introducing known amounts of gas into the sample cell filled with the substrate and measuring the resulting equilibrium vapor pressures at constant temperature. In the early days of physisorption studies

TABLE II
Characteristic Properties of Various Substrates

Substrate	Specific Area (m^2/g)	Percentage of Isotropic Powder	Percentage with Preferred Orientation	Mosaic Spread FWHM (Degrees)	Coherence Length (Å)	Density (g/cm^3)
ZYA graphite	2×10^{-4}	—	100	0.4	~ 10,000	2.26
Grafoil	22 ± 2	70	30	30	150 ± 50	1.0
Papyex	22 ± 2	70	30	30	250 ± 50	1.1
Foam	22 ± 2	100	—	—	900 ± 100	0.1–0.2
ZYX graphite	1–2	—	100	10	1800 ± 200	0.6
Carbon blacks	200	100	—	—	<50	<0.01
Boron nitride	5–15	100	—	—	Unknown	2.25

on exfoliated graphite the monolayer capacity was extracted from a smooth step occurring in an adsorption isotherm near monolayer completion and analyzing it by the so-called BET equation, named after Brunauer, Emmett, and Teller [51]. This method, however, is imprecise because there is no standard convention leading to a well-defined reference point. The density of a completed monolayer is not a fixed quantity, since it has a small but perceptible temperature dependence [46]. In addition, considerable compression of the first layer often takes place as the layer is completed and atoms or molecules start to populate the second layer.

Another, more precise method uses the fact that some adsorbates form commensurate structures on the surface of a substrate. An overlayer structure is designated as commensurate when its lattice spacing is a rational fraction of the lattice constant of the underlying substrate. Sometimes such structures are synonymously called *registered* or *epitaxial,* the latter term being, however, mostly used in connection with the growth of bulk material with the same structure as the substrate. Whether a commensurate layer is formed depends on many parameters, such as the depth and symmetry of the substrate potential wells, the spacing of the sites, the sizes of the adsorbed atoms or molecules and the interaction energies between them, the coverage, and temperature. Kr, N_2, and CO layers are known to condense into a commensurate $(\sqrt{3} \times \sqrt{3})\ R30°$ phase in the submonolayer regime (for the nomenclature see Ref. 95). This structure possesses a nearest-neighbor distance of $\sqrt{3}$ times the lattice constant a of graphite (i.e., 4.259 Å) and is rotated 30° with respect to the graphite lattice. In this structure the potential well in the center of every third graphite hexagon is occupied by one molecule, and consequently the areal density of this phase is 0.06366 $Å^{-2}$. Near monolayer completion Kr, N_2, and CO undergo a phase transition from the commensurate to an incommensurate phase. This transition is indicated by a small substep in an adsorption isotherm, which can be used as reference point for the calibration of the coverage. The amount of gas needed to form this substep is often defined as the most perfect commensurate phase and denoted as coverage unity $n = 1$. In diffraction work the commensurate–incommensurate transition can be identified directly; however, because of the presence of surface heterogeneities the volumetric coverage scale may differ from the crystallographic scale. In volumetric measurements, only the total amount of adsorbed gas can be determined, and it is often assumed that the first molecules dosed onto a substrate occupy the irregular and strongly binding heterogeneous sites. This may in some cases be as much as 10–20%. Being disordered, this part of the adsorbed film does not contribute to diffraction signals, which results in a difference between the coverage scales gained by both methods. Other sources of errors for determining the exact coverage of the perfect commensurate phase are interstitials and vacancies which may lead to discrepancies on the order of 3–5%.

Another reference point often being used is a substep occurring in adsorption isotherms recorded with N_2 at the two-dimensional fluid to commensurate solid coexistence region. Early measurements determined the top of this substep at 70 K [237], whereas later measurements changed it to 74 K [394] and 77 K [65], because it is easier to obtain reliable vapor pressure data at higher temperatures. It is estimated [394] that the difference in coverage scales at these temperatures is about 2%, but at coverages beyond $n = 1$ discrepancies of 24% between different measurements were found [394]. Such coverage differences ranging up to 30% have also been observed for Kr [39] and Xe [136] on ZYX graphite in the region between one and two complete monolayers. They were attributed to alternate-site adsorption, which means adsorption onto sites not belonging to the homogeneous basal planes of graphite. In the multilayer coverage regime, one also has to pay attention to capillary condensation.

Because of these uncertainties in the coverage scale, it is often confusing to compare results for the same system obtained by different research groups. In addition, the authors often do not define precisely the exact parameters on which their coverage scales are based. Therefore it turned out to be a difficult task to construct phase diagrams of CO and even more so for N_2 on graphite for this review, which were made as complete and reliable as possible by combining various results published in the literature.

A more consistent basis for a precise coverage scale might be heat capacity measurements in the vicinity of the order–disorder transition of the commensurate $\sqrt{3}$ phase of the helium [46, 98] and hydrogen [120, 121, 379] isotopes. The heat capacity anomalies display pronounced maxima in both magnitude and temperature at this transition, and a minimum is found in the low-temperature heat capacity. The coverage at which this effect occurs can be taken to be the perfect commensurate density of 0.06366 Å^{-2}. Of course, the presence of interstitials and thermally activated vacancies can also lead to some uncertainties. However, neutron diffraction measurements have revealed [83, 123] that they only amount to 2–4% depending on the adsorbate (H_2 or D_2) studied. This coverage scale turned out to be in good agreement with LEED data [83, 84] on graphite single crystals. Recent carefully performed comparative measurements using the same substrate sample based on this method on the one hand, and on the above-mentioned adsorption isotherm measurements with N_2 at 77 K on the other, found consistency between both methods within 5% [381]. Thus we conclude that the coverage error below monolayer completion is no more than 5%, whereas it can be considerably higher in the second and higher layer range.

II. ORIENTATIONAL ORDERING ON A TRIANGULAR LATTICE

A. Ground States and Mean-Field Theory

The pioneering theoretical approach to orientational ordering of linear molecules on a triangular lattice is presented in Ref. 21 and in much more detail in Ref. 141. These studies aimed to describe possible orientational ordering phenomena for H_2 and D_2 in the $J = 1$ rotational state (i.e., ortho-H_2 and para-D_2) in the $(\sqrt{3} \times \sqrt{3})R30°$ commensurate monolayer phase physisorbed on graphite (for the nomenclature see Ref. 95). Although originally devised for quantum mechanical hydrogen molecules, this investigation turned out to be extremely stimulating and even guiding in the search for and explanation of orientational ordering phenomena of physisorbed N_2 and also CO; a short discussion of the classical limit is presented in Ref. 141. Four possible orientationally ordered phases were proposed depending on the reduced temperature and a reduced energy scale characterizing the strength of the substrate potential. Based on the success of this type of modeling, several related classical models were later investigated to study (a) the effects caused by vacancies or isotropic impurities [70, 142, 228, 245, 246, 255] and (b) additional multipolar contributions [71].

In the model of Refs. 21 and 141, the centers of mass of the molecules are fixed at their sites on an ideal complete $\sqrt{3}$ lattice. The orientations of the molecules are determined by the electrostatic quadrupole–quadrupole interaction between molecules and by the crystal field of the substrate. Thus, the model consists of point quadrupoles on an ideal and rigid triangular lattice. In order for this model to be useful at the low temperatures necessary to describe orientational ordering of hydrogen molecules, it is necessary to include quantum effects in the treatment. Because of the large rotational constant of H_2 and D_2, it is sufficient at low temperatures to truncate the full Hilbert space to the $J = 1$ manifold with its three states [178, 320]. Thus J is assumed to still be a good quantum number even for molecules adsorbed on the surface, and higher-order contributions in J are neglected. The electrostatic quadrupolar interaction is found to be dominant in the solid at low pressures [178, 320]. The quadrupolar pair interactions restricted to only nearest-neighbor molecules can be expressed in the form

$$\mathcal{H}_{3D-EQQ} = \frac{20\pi}{9} (70\pi)^{1/2} \Gamma_0 \sum_{\langle i,j \rangle} \sum_{m,n} C(224; mn) \times Y_2^m(\omega_i) Y_2^n(\omega_j) Y_4^{m+n*}(\Omega_{ij}) \tag{2.1}$$

where $C(224;\, mn)$ is a Clebsch–Gordan coefficient, $Y_L^m(\omega)$ is a spherical harmonic of rank L and ω_i and Ω_{ij} denote, respectively, the spherical angles of the ith molecular axis and the line joining molecules i and j; all angles are measured relative to some fixed set of axes. The coupling constant

$$\Gamma_0 = \frac{6e^2Q^2}{25\,R_0^5} \tag{2.2}$$

describes the electrostatic quadrupole–quadrupole interactions between nearest-neighbor molecules with a quadrupole moment eQ at a distance R_0 apart.

The intermolecular Hamiltonian (2.1) is supplemented with a one-particle external out-of-plane crystal-field term representing the interaction of each molecule with the underlying substrate

$$\mathcal{H}_c = -\tfrac{5}{2}\,V_c \sum_i \left(\cos^2\theta_i - \tfrac{1}{3}\right) \tag{2.3}$$

where V_c is the crystal-field splitting for $J = 1$ molecules and the out-of-plane tilt angle θ_i is measured with respect to a z axis normal to the surface. The quadrupolar coupling parameter Γ_0 in (2.1) can be used to set the energy scale so that the phase behavior of the combined Hamiltonian (2.1) + (2.3) can be formulated in terms of the reduced temperature $T^* = T/\Gamma_0$ and the reduced crystal-field $V_c^* = V_c/\Gamma_0$.

The free energy of the anisotropic-rotor Hamiltonian (2.1) + (2.3) was investigated within the mean-field approximation. To this end, the many-body density matrix had to be factorized into a product of single-particle density matrices, thus neglecting the correlations of the fluctuations. In addition, a population of $J > 1$ levels is, of course, not possible by construction. The crystal-field term (2.3) simply tends to align all molecules parallel to the surface when negative ($V_c < 0$), whereas a perpendicular orientation is preferred for the other case ($V_c > 0$). The quadrupolar interaction, however, classically causes molecules to orient perpendicular to each other, resulting in the typical T-shape configuration with one molecule lying parallel and the other perpendicular to the line joining the pair. The competition of these interactions with the "geometric frustration" imposed by the triangular lattice in conjunction with thermal excitations results in rather complex cooperative phenomena and in interesting orientational ordering effects.

The mean-field phase diagram of the anisotropic-rotor model (2.1) + (2.3) as explored by a combination of numerical and analytical methods is presented in Fig. 3. The corresponding orientationally ordered phases are sketched in Fig. 4, and they are shown in Fig. 5 in a more realistic repre-

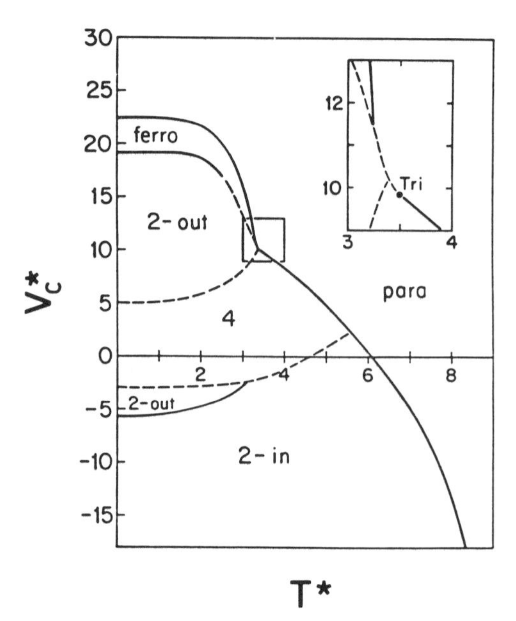

Figure 3. Mean-field phase diagram for the orientational ordering quantum $J = 1$ point quadrupoles in three dimensions on a triangular lattice with an axially symmetric out-of-plane crystal field (2.1) + (2.3); $V_c^* = V_c/\Gamma_0$ denotes the reduced crystal field and $T^* = T/\Gamma_0$ the reduced temperature, where Γ_0 is the nearest-neighbor quadrupolar coupling constant (2.2). Dashed (solid) lines represent first (second)-order transitions, the dot (Tri) in the inset marks a tricritical point, "para" denotes the pararotational (orientationally disordered) phase, and the structures of the ordered phases are shown schematically in Fig. 4. (Adapted from Fig. 2 of Ref. 141.)

sentation including the underlying graphite lattice. In the orientationally disordered or so-called pararotational phase at high temperatures, the behavior is dictated by the simple crystal-field term resulting in an orientation depending on the sign and magnitude of V_c; the pararotational phase does have a temperature dependent gap within the mean-field theory.

For large negative crystal field at low temperatures the stable structure is the ideal two-sublattice in-plane herringbone phase—that is, the "2-in" phase in Fig. 4*a*. The two sublattices which can be oriented in three different ways relative to the triangular lattice lead to six equivalent ground states. The excitation spectrum of this phase in general has a gap. In this phase the molecular wave functions are localized in the substrate plane, and classically the molecular axes are parallel to the surface; see Appendix A of Ref. 141 for an interpretation of the order parameters. Thus, the orientational degeneracy of the pararotational phase is broken by the quadrupolar interactions. A closely related structure was already proposed based on atomistic Lennard-

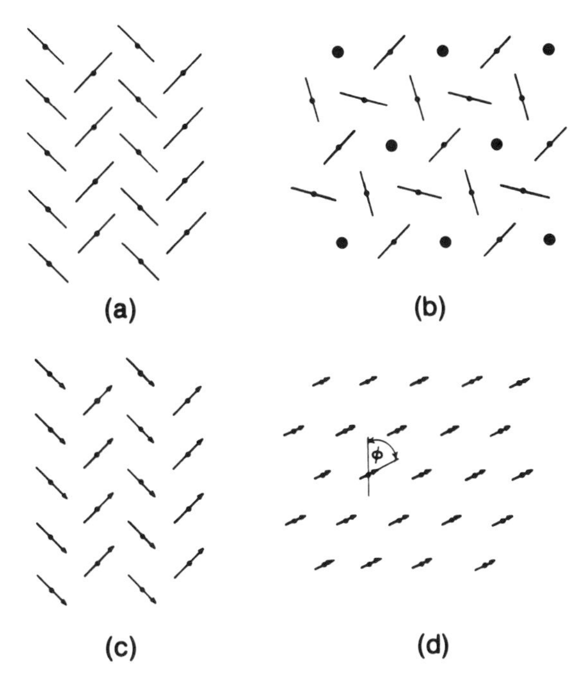

Figure 4. Schematic sketch of the four orientationally ordered classical ground-state phases occurring in the phase diagram (Fig. 3) of quadrupoles on a triangular lattice. (*a*) "2-in" or two-sublattice in-plane herringbone phase, (*b*) "4" or four-sublattice pinwheel phase where the large dots indicate the molecules perpendicular to the surface, (*c*) "2-out" or two-sublattice out-of-plane herringbone phase, (*d*) ferrorotational or XY-like gapless phase where all molecules are free to rotate uniformly by a constant phase angle ϕ. The arrows indicate a systematic out-of-plane tilt of the molecular axes. A representation of the in-plane herringbone phase (*a*) and the pinwheel phase (*b*) which includes the underlying honeycomb graphite lattice is given in Fig. 5. (Adapted from Fig. 2 of Ref. 108 and from Fig. 3 of Ref. 141.)

Jones potentials to describe the anisotropic N_2–N_2 and N_2–surface interactions in addition to the electrostatic quadrupolar effects, although four sublattices were necessary to describe it fully [125].

Just above the in-plane herringbone phase there is a small region (see Fig. 3) in which the molecules begin to tilt out of the plane and form a two-sublattice out-of-plane herringbone phase—that is, the "2-out" phase in Fig. 4*c*. Note that the order parameters describing various angles of the orientationally ordered phases assume their ideal value only in the ground state at $T = 0$, whereas some of these angles such as, for example, the tilting and sublattice rotation angles of the out-of-plane herringbone phase change systematically as a function of temperature.

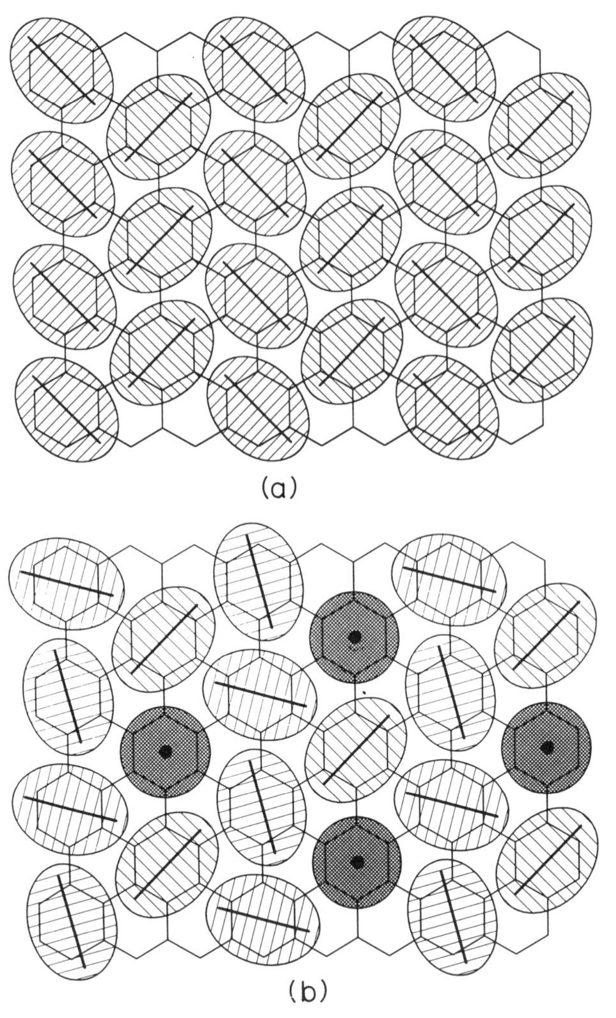

Figure 5. Schematic picture of (*a*) the ideal (2 × 1) two-sublattice in-plane herringbone structure and (*b*) the ideal four-sublattice pinwheel structure within the ($\sqrt{3} \times \sqrt{3}$)$R30°$ lattice which is commensurate on the "honeycomb" lattice of the (0001) basal plane of graphite. The principal axes of the ellipses correspond to the 95% electronic charge density contour given in Table I.

The "2-out" phase has a first-order phase boundary to the four-sublattice pinwheel or "4" phase for $V_c < 0$ (see Fig. 4*b*). In this phase one subset of molecules, the "pin molecules" which define their own sublattice, stands perpendicular on the surface, and the six nearest neighbors belonging to the other three sublattices are localized within the surface plane. This structure has eight degenerate ground states because the central pin site of the structure can be at any one of the four sublattices, and the rotation of the wheel molecules can be either right-handed or left-handed. The stability of the pinwheel phase with respect to quantum zero-point effects was also investigated, and the four sublattice state is found to be only weakly modified by those. It is speculated that the other phases with an excitation gap—that is, the in-plane and out-of-plane herringbone phases—should also not be destroyed by such fluctuations.

Another theoretically more interesting phase occurring at the largest crystal field still producing orientational order is the one sublattice ferrorotational phase (see Fig. 4*d*). Because there is only a single characterizing phase angle on which the energy does not depend, all molecules can be rotated without energy cost. This yields a gapless excitation spectrum, and this phase bears the same continuous symmetry as the XY model. A more sophisticated investigation should take into account the underlying sixfold, crystal field [164] and its Kosterlitz–Thouless aspects [164, 252].

Finally, taking into account only the electrostatic quadrupole–quadrupole interactions, the ratio of the values of the critical temperature for the quantum case and that for the classical case were also derived. The ratio of the classical transition temperature to that of its ($J = 1$) quantum analogue is found to be $\frac{5}{2}$ without the crystal-field effects. This result could be derived for a quite general quadratic Hamiltonian of the type (2.1). However, it was assumed that the transition is continuous, and the reported ratio no longer holds for a first-order transition. It is suggested [141] that this result may be relevant to the N_2 system.

This classification of the phases expected for orientational ordering of quadrupoles on a triangular lattice has important bearings on the expected experimental diffraction patterns [107, 108]. Structures (*a*)–(*c*) of Fig. 4, which have a repeat distance twice that of the lattice sites, can be distinguished in diffraction due to the different symmetries of the quadrupole arrangement within the unit cell. The in-plane herringbone phase (*a*) has two perpendicular glide planes, the pinwheel phase (*b*) has no glide planes, and the out-of-plane herringbone phase (*c*) has a single glide plane. This leads to systematic absence of reflections in the herringbone structures, which was crucial to determine the orientational structure of various phases occurring for N_2 and CO on graphite (see Section III and Section V). This limitation of the phases to be searched for in experiments to only a few well-characterized structures was the invaluable contribution of Refs. 21 and 141. The microscopic modeling of the interactions in the Hamiltonian (2.1) + (2.3), however, is certainly oversimplified and hence more qualitative than quantitative [108]; see Ref. 388 for an attempt to analyze orientational ordering in the triangular incommensurate phase (see Section III.E) based on a parameterization of the anisotropic-rotor model [21, 141]. Even further approximations have led to a very useful model, the anisotropic–planar-rotor model [244, 246, 274], which served over many years as the main workhorse to determine numerically the order of the ideal herringbone transition (see Section III.D).

In addition to the characterization of the possible phases, the mean-field predictions for the order of all occurring phase transformations were obtained [21, 141]. Those are shown in Fig. 3 as dashed and solid lines, which represent first- and second-order transitions, respectively. The critical exponents describing the continuous transition are, of course, the classical Landau exponents, and the transition temperatures are expected to be overestimated by roughly a factor of two in the present case. Fluctuations may drive several of the transitions from second order in the mean-field approximation to first order; that is, they may result in fluctuation-driven first-order transitions. Based on renormalization group results for the Heisenberg model with cubic anisotropy together with a conjecture, this was actually expected [141] to occur for the transition from the pararotational phase to the in-plane herringbone phase. The transition to the pinwheel phase and the tricritical point, on the other hand, should remain continuous transitions characterized by the corresponding nonclassical critical exponents (see also the discussion in Section II.C).

B. Anisotropic–Planar–Rotor Model and Some Generalizations

The quest to simplify further the three-dimensional quadrupolar Hamiltonian (2.1) + (2.3) led to a model where the substrate interactions are only treated in two limiting cases within classical statistical mechanics [274]. Again, the molecules are fixed with their centers of mass on a two-dimensional triangular lattice and interact exclusively via electrostatic quadrupole–quadrupole interactions. The large negative crystal-field case corresponding to $V_c \to -\infty$ in (2.3) was mimicked by constraining the molecules to rotate only in the plane parallel to the surface. This should be a good approximation to describe N_2 on graphite at sufficiently low temperatures as known from previous more realistic calculations [125, 326]. The other extreme case corresponding to the weak-field limit for the adsorbent–adsorbate interaction was realized by allowing the molecules to rotate in three-dimensional space without any external potential; that is, $V_c = 0$. In the first case, Monte Carlo simulations [274] revealed a transition to the herringbone superstructure necessarily of the in-plane type. The three-dimensional model without rotational constraints, however, led to pinwheel ordering at low temperatures. Both observations are in qualitative accord with the mean-field expectations [141]. This monolayer model was generalized to ideal bilayer models with both first and second nearest-neighbor interactions [275]. In the more realistic model the molecules in the first layer were restricted to rotate only in the surface plane and the molecular rotations in the second layer were unconstrained, whereas both layers were unconstrained in another variant of the model.

The case of vanishing external field—that is, $V_c = 0$ in (2.1) + (2.3)—was investigated in more detail in Ref. 174, where the quadrupoles on a triangular lattice are allowed to rotate in three dimensions. According to mean-field theory (see Fig. 3), one of the ground states, which has four-sublattice pinwheel order, transforms via a first-order transition to an intermediate in-plane herringbone two-sublattice structure, which finally has a continuous transition to the orientationally disordered high-temperature phase. Based on Monte Carlo simulations with nonperiodic boundary conditions with at most about 2000 quadrupoles and less than 9000 Monte Carlo steps, the mean-field prediction of multiple phase transitions is corroborated, but defect formation prevents the observation of the intermediate phase on a macroscopic scale [174]. Specifically, isolated pinwheels are believed to be a common localized defect appearing in domains of the intermediate phase [174]. The small system size in conjunction with the nonperiodic boundary conditions and no systematic finite-size scaling estimates certainly warrant further investigations of that special case.

The fruitful two-dimensional simplification [274] of the original model to describe orientational ordering of linear molecules on surfaces was furthermore pursued in Ref. 244 including an additional approximation. The Hamiltonian describing quadrupole–quadrupole interactions of molecules confined to rotate only in the surface plane can be written as [244]

$$\mathcal{H}_{2D-EQQ} = J \sum_{\langle i,j \rangle} \cos(2\varphi_i - 2\varphi_j) + K \sum_{\langle i,j \rangle} \cos(2\varphi_i + 2\varphi_j - 4\Theta_{ij}) \tag{2.4}$$

where φ_i denotes the angle of the ith molecules, and Θ_{ij} is the angle of the line joining the sites of molecules i and j; all angles are measured relative to some space-fixed axis (see Fig. 6). The six degenerate ground states of the model resulting from two sublattices, each with three possible glide line orientations, are sketched in Fig. 6. Only nearest-neighbor interactions are included in the summation over molecule pairs. Full electrostatic quadrupole–quadrupole interactions correspond to the special choice $K/J = 35/3$ for the coupling constants. The so-called anisotropic–planar–rotor model

$$\mathcal{H}_{APR} = K \sum_{\langle i,j \rangle} \cos(2\varphi_i + 2\varphi_j - 4\Theta_{ij}) \tag{2.5}$$

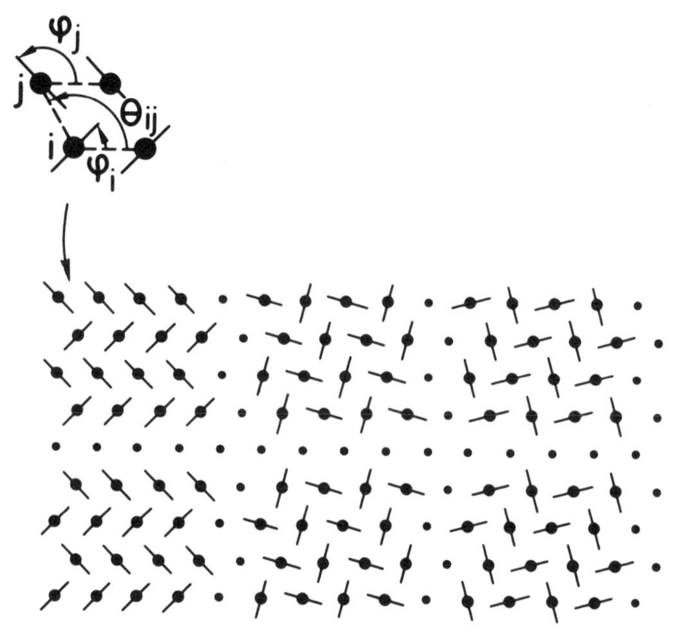

Figure 6. Schematic picture of the six $(2\sqrt{3} \times \sqrt{3})R30°$ herringbone ground states of the anisotropic–planar–rotor model on a triangular lattice (2.5). Small and large dots represent empty and occupied sites, respectively. (Adapted from Fig. 1 of Ref. 60.)

is obtained after neglecting the XY-like first term; that is, $J = 0$ in (2.4). Thus, only the anisotropic contribution to the full electrostatic quadrupole–quadrupole interactions is included in the anisotropic–planar–rotor Hamiltonian (2.5), whereas the more than ten times smaller isotropic part $\sim J$ is neglected in a first approximation [246]. This model served over the years as the workhorse to study and determine the nature of the ideal herringbone transition by computer simulations and analytical theory (see Section III.D for a discussion). Its phase behavior is simple: In addition to the high-temperature orientationally disordered phase there is only one ordered phase of the herringbone type, which has a sixfold degeneracy of the ground state (see Fig. 6). A more thorough discussion of this model including its dynamics and the influence of vacancies is given in Ref. 246.

Generalizations [71] of the bare anisotropic–planar–rotor model (2.5) include other multipolar interactions such as dipolar and octopolar terms with and without in-plane crystal-field modulations. Several such combinations were analyzed in the mean-field approximation, Landau theory, and spin-wave expansion [71]. The quadrupole–quadrupole model written in the form

$$\mathcal{H}_2 = \alpha_2 \sum_{\langle i,j \rangle} \cos(2\phi_i - 2\phi_j) + \beta_2 \sum_{\langle i,j \rangle} \cos(2\phi_i - 2\Theta_{ij}) \times \cos(2\phi_j - 2\Theta_{ij}) \tag{2.6}$$

reduces again to the Hamiltonian describing electrostatic interactions between quadrupoles for $\beta_2/\alpha_2 = -35/16$ and $\alpha_2 < 0$. Furthermore, the anisotropic–planar–rotor Hamiltonian approximation (2.5) is obtained [71] for the special choice $\alpha_2 = -\beta_2/2$ with the coupling constant $K = \beta_2/2$. The "2 model" (2.6) yields a rich phase diagram when the ratio β_2/α_2 is varied (see Figs. 2 and 3 of Ref. 71). In addition to the standard two-sublattice in-plane herringbone ground state with setting angles of $\pm 45°$, there is also a two-sublattice in-plane herringbone phase with setting angles of 0° and 90°, a ferromagnetic XY-like phase, a three sublattice orientational ($\sqrt{3} \times \sqrt{3}$) phase, and even a modulated incommensurate phase. The special case of (2.5) with its single phase transition from the disordered to the standard herringbone phase is recovered in Fig. 2*b* of Ref. 71 at $\beta_2/\alpha_2 = 2$. Introducing a sixfold-modulated crystal field results in a herringbone setting angle which depends on the amplitude of that perturbation. However, physically reasonable values for this contribution result only in relatively small deviations from 45°. Impurities were included in the calculations in Ref. 70.

The interactions between point dipoles in a plane can be formulated as

$$\mathcal{H}_1 = \alpha_1 \sum_{\langle i,j \rangle} \cos(\varphi_i - \varphi_j) + \beta_1 \sum_{\langle i,j \rangle} \cos(\varphi_i - \Theta_{ij}) \cos(\varphi_j - \Theta_{ij}) \tag{2.7}$$

where $\beta_1/\alpha_1 = -3$ and $\alpha_1 > 0$ yields electrostatic interactions between dipoles [71]. The mixed quadrupolar–dipolar model [71] is defined as

$$\mathcal{H}_{2-1} = \mathcal{H}_2 + \mathcal{H}_1 \tag{2.8}$$

where the number of independent parameters was reduced based on the relations $\alpha_1 = \lambda\alpha_2$ and $\beta_1 = \lambda\beta_2$. Thus, λ is a measure of the relative strength of dipolar versus quadrupolar interactions. Note that the combined Hamiltonian (2.8) does not take into account coupling terms between the quadrupolar and dipolar interactions, and that also the dipolar interactions are only effective between nearest neighbors. Specifically this quadrupolar–dipolar "2–1 model" (2.8) is relevant for the ordering of heteronuclear molecules on surfaces, such as CO on graphite. A tetracritical value

$$\lambda_c = \frac{(\frac{3}{2}\beta_2 - 1)}{(\frac{1}{2}\beta_2 + 1)} \tag{2.9}$$

depending on the energy scale β_2 is found at which the crossover between the critical behavior associated with quadrupolar and dipolar ordering occurs. A typical behavior of the setting angle φ^* is shown as the solid line in Fig. 7 for a certain choice of the parameters in (2.8) corresponding to stronger quadrupolar interactions compared to the dipolar ones, that is, $\lambda < 1$. Herringbone ordering with a constant setting angle of $\varphi^* = 45°$ sets in below the quadrupolar ordering transition temperature T_{c2}. In the combined quadrupolar–dipolar regime $T < T_{c1}$ ($< T_{c2}$), however, it is observed that the setting angle gets larger than 45° and shows a nonmonotonic behavior as a function of temperature. Inclusion of a sixfold-modulated in-plane crystal-field anisotropy

$$\mathcal{H}_{2-1,3} = \mathcal{H}_2 + \mathcal{H}_1 + V_3 \sum_i \cos 6\varphi_i \tag{2.10}$$

destroys the critical behavior associated with T_{c1} and leads to an increase of the setting angle over the entire stability interval $T < T_{c2}$ of the herringbone phase; see the dashed line in Fig. 7 for the special choice $V_3 = 0.1$ in (2.10).

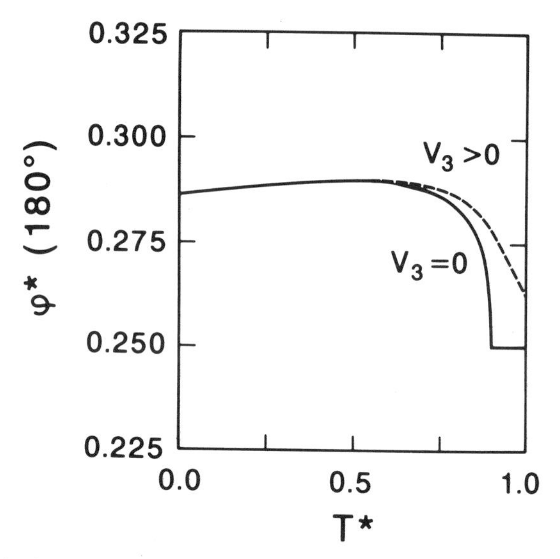

Figure 7. Setting angle of the herringbone phase of the quadrupolar–dipolar model (2.8) and the same model including a sixfold-modulated in-plane crystal-field (2.10) as a function of the reduced temperature $T^* = T/T_{c1}$ at $\lambda = 0.9$ for $\beta_2 = 2$, where $\lambda_c = 1$; note that α_2 only sets the temperature scale. Solid line: no crystal field. Dashed line: $V_3 = 0.1$. (Adapted from Fig. 7 of Ref. 71.)

A similar analysis is carried out for an octopolar term instead of the dipolar interactions, where analogous effects are observed [71].

A discretized model for quadrupoles with short-range interactions and discrete states on a triangular lattice was introduced and investigated by Monte Carlo simulations in Ref. 367. The interaction energies depend on the orientation of the molecules relative to each other and relative to the orientation of the bond between them. The three-state model has a particularly rich ground-state behavior including ferromagnetic, herringbone, antiferromagnetic, and so-called star structures, as well as mixed and possibly glassy phases.

The quantum mechanical generalization of the anisotropic–planar–rotor Hamiltonian (2.5) is devised and investigated by quasiharmonic, quasiclassical, and path-integral Monte Carlo methods in Refs. 213 and 218. Here, thermal fluctuations compete with quantum fluctuations which adds a qualitatively new dimension to the scarce one-dimensional phase diagram of the classical anisotropic–planar–rotor model (2.5). The phase behavior of this model is much richer, and the phenomenon of reentrant orientational quantum "melting" is observed in a certain regime of the phase diagram.

C. Symmetry Classification

The modern theory of critical phenomena suggests that the wealth of continuous transitions stemming from seemingly different systems and models can be classified according to only a few universality classes concerning static properties. Such concepts are reviewed in Ref. 307 with a special emphasis on applications and examples in surface physics. Of special relevance are the ferromagnetic Ising model, the three- and four-state Potts models, and the XY and Heisenberg models each with cubic anisotropy [95]. The central idea is that all systems whose Landau–Ginzburg–Wilson Hamiltonian has the same form have the same kind of phase transition. Ideally such a Hamiltonian with its parameterization is constructed by a coarse graining procedure starting from a microscopic model once the relevant order parameter(s) in the problem of interest are identified [95, 235]; see also Ref. 343 for a related approach. More often, however, such a Hamiltonian is postulated based on symmetry and physical arguments.

The paramagnetic-to-ferromagnetic transition of the two-dimensional Heisenberg model with cubic anisotropy is of special relevance for orientational ordering of linear molecules on surfaces [141, 259, 308]. The fourth-order term of its Landau–Ginzburg–Wilson Hamiltonian, which is invariant only under the symmetry operations of the cube, becomes dominant under renormalization. Depending on the sign of the quartic term the three-component order parameter points either to the six faces or to the eight corners of a cube, which lead to the notion of face- and corner-cubic symmetry. Thus, the face- and corner-cubic anisotropic systems have a different number of degenerate ground states—that is, six and eight, respectively.

Based on the approximate Kadanoff variational renormalization group scheme of the more general $n = 3$ cubic model, it is found that the Heisenberg model with face-cubic anisotropy necessarily leads for the paramagnetic-to-ferromagnetic transition, which is of interest here, to a first-order transition [259, 308]; see Ref. 297 for an early review of the renormalization group calculations of the relevant Potts and cubic models. Nevertheless, large fluctuations are expected to occur near the transition because very similar models are known to have continuous fixed points. Thus, when the Hamiltonian is propagated through its coupling parameter space during the renormalization group transformations, it spends much time in the vicinity of critical points and thereby accumulates large fluctuations [308]. Such a behavior is indeed observed [30] for the two-dimensional five- and six-state Potts models, where the strong fluctuations found at their first-order transitions may be caused by the "vicinity" of the $q = 4$ fixed point [308].

Similarly it is concluded that the transition of interest for corner-cubic anisotropy can either be first order or continuous. In the latter case the critical behavior has to be that of the two-dimensional Ising model up to logarithmic corrections [308]. However, the tricritical and critical behaviors are identical up to logarithmic corrections; in other words, no distinct tricritical behavior is expected to occur [308].

It can be shown that the Landau–Ginzburg–Wilson Hamiltonian constructed from the herringbone order parameter leads to the Hamiltonian of the Heisenberg model with face-cubic anisotropy, whereas the pinwheel order parameter leads to its corner-cubic-anisotropic version [141, 308]. Thus, the phase transition from the orientationally disordered phase to the (2×1) in-plane herringbone structure should be first order by symmetry, whereas the one to the (2×2) pinwheel phase might be first order or continuous with two-dimensional Ising exponents. Similar arguments in conjunction with several analogies led to the conclusion that the herringbone transition should have the same thermodynamic singularities including its first-order nature also in a floating-solid (incommensurate) phase (see Ref. 60 and Section III.E).

It has to be pointed out, however, that the arguments presented in this section for orientational ordering are not fully derived based on explicit microscopic modeling. On the contrary, they are only obtained after several renormalizations, simplifications, and invoking similarity assumptions [345]. For example, it is *a priori* not clear if the decoupling of the orientations of the underlying lattice from the molecular orientations as implicit in the mapping onto the $n = 3$ cubic model is a valid approximation (see Ref. 345 and Section III.D.2 for some further discussions).

III. N_2 ON GRAPHITE

A. Phase Diagram

The temperature versus coverage phase diagram of N_2 physisorbed on graphite as deduced from various experiments is presented in Fig. 8; we refer to Sections III.B–III.E for a more detailed discussion. Above 85 K, only a homogeneous fluid phase (F) is stable for submonolayer coverages. The centers of mass of the molecules show positional ordering in the commensurate $(\sqrt{3} \times \sqrt{3})R30°$ solid phase with disordered molecular orientations (CD) in coexistence with the fluid phase in the coverage range 0.2–0.8 for temperatures below about 50 K. The triangular lattice of the ideal $(\sqrt{3} \times \sqrt{3})R30°$ complete commensurate monolayer structure is represented by the centers of the adsorbed molecules in Fig. 5*a*. This superstructure consists of an equilateral triangular two-dimensional array where the superlattice spacing of 4.259 Å is $\sqrt{3}$ times larger than the lattice constant 2.459 Å of the underlying (0001) graphite basal plane surface. In addition, the superlattice is rotated by 30° with respect to the graphite lattice. The molecules in this structure occupy the potential wells in the center of every third graphite hexagon, which results in an areal density of 0.06366 molecules/Å^2. The coverage scale of the phase diagram shown in Fig. 8 is given in units of this density (see also Section I.C).

As the temperature increases, the registered solid–fluid coexistence region (CD + F) considerably narrows down to a strip that ends at a special point near 85 K and at a coverage of about 1.2 [65, 237]. The special point is believed to be a tricritical point in the universality class of the two-dimensional three-state Potts model [65, 237]. According to the incipient-triple-point model interpretation there is only two-phase coexistence between the $\sqrt{3}$ solid and a fluid phase without liquid–vapor coexistence for any temperature and density [65, 231]. In other words, no genuine critical or triple point can be found in this region of the phase diagram because the existence of the liquid phase has been suppressed by a more stable registered solid. The solid–fluid coexistence region only terminates at the aforementioned tricritical point. Center-of-mass melting to the fluid phase is believed to be of first order up to the complete monolayer, and continuous beyond. The fluid phase close to the melting line has some in-plane density modulations imposed by the corrugation of the underlying graphite lattice.

An orientational transition of the molecular axes is observed near 28 K

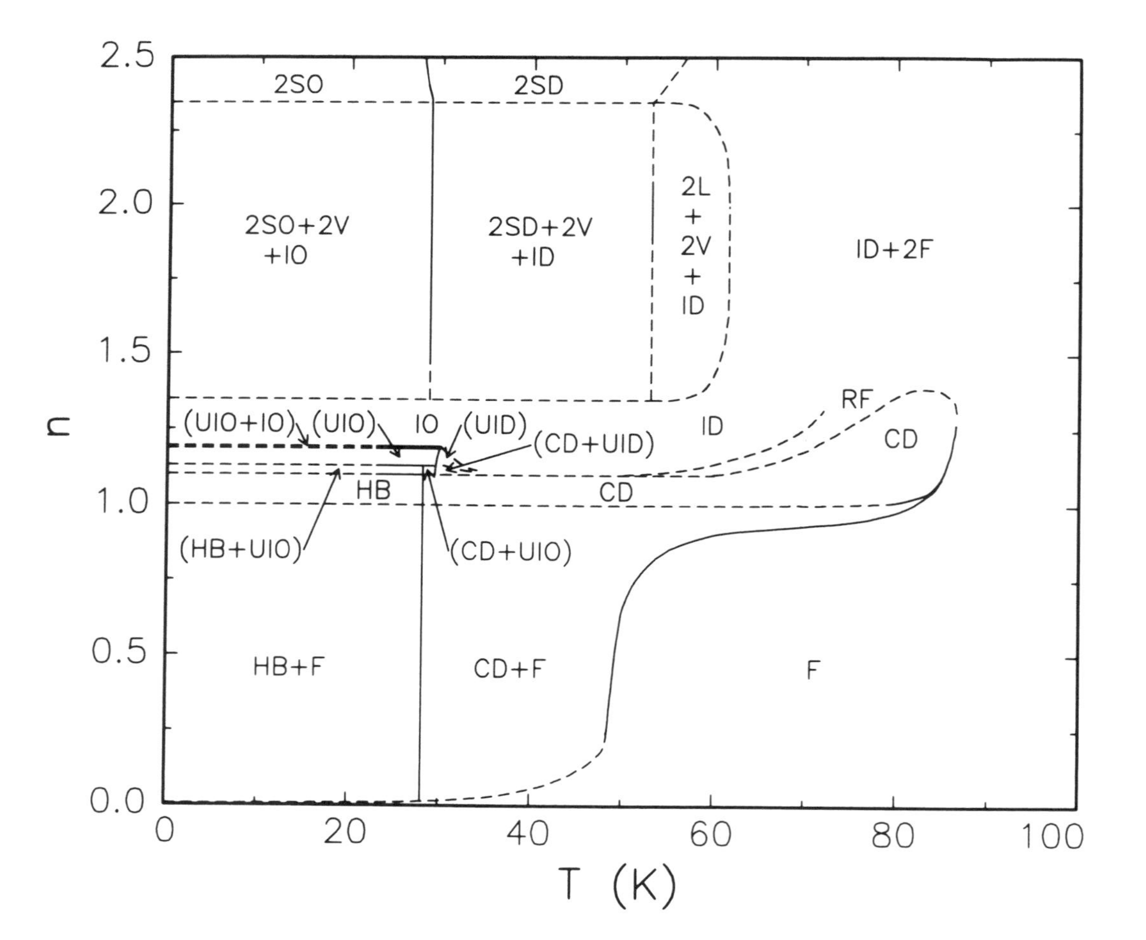

n
T (K)
2.5
2.0
1.5
1.0
0.5
0.0
0
20
40
60
80
100
2SO
2SD
2SO+2V
+IO
2SD+2V
+ID
2L
+
2V
+
ID
ID+2F
(UIO+IO)
(UIO)
IO
(UID)
(CD+UID)
ID
RF
CD
HB
CD
(HB+UIO)
(CD+UIO)
HB+F
CD+F
F

Figure 8. Phase diagram of N_2 physisorbed on graphite with the following phases: fluid without any positional or orientational order (F), reentrant fluid (RF), commensurate orientationally disordered $(\sqrt{3} \times \sqrt{3})R30°$ solid (CD), commensurate herringbone ordered $(2\sqrt{3} \times \sqrt{3})R30°$ solid (HB), uniaxial incommensurate orientationally ordered (UIO) and disordered (UID) solid, triangular incommensurate orientationally ordered (IO) and disordered (ID) solid, second-layer liquid (2L), second-layer vapor (2V), second-layer fluid (2F), bilayer orientationally ordered (2SO) and disordered (2SD) solid; the parentheses are included for clarity only. The solid lines are based on experimental results, whereas the dashed lines are speculative; part of those speculations are based on analogies to the phase diagram of CO on graphite in Fig. 49. Note that the transformation IO $\rightarrow$ ID is not known in detail. Coverage unity corresponds to a coverage of N_2 forming a complete $(\sqrt{3} \times \sqrt{3})$ commensurate monolayer. (The phase diagram is based on Fig. 1 of Ref. 72, Fig. 2 of Ref. 90, Fig. 1 of Ref. 237, Fig. 15 of Ref. 93, Fig. 2 of Ref. 106, Fig. 10 of Ref. 65, Fig. 2 of Ref. 394, Fig. 2 of Ref. 395, and Fig. 10 of Ref. 156.)

independently of coverage up to the full monolayer and slightly beyond [65, 394]. The type of low-temperature orientational ordering [92, 93, 99, 370] is the (2×1) in-plane herringbone orientationally ordered superstructure (HB) within the positionally ordered commensurate $(\sqrt{3} \times \sqrt{3})R30°$ solid phase as sketched in Fig. 4*a* and Fig. 5*a*; this is the "2-in" phase in the nomenclature of Ref. 141 possessing two sublattices with different orientations where all molecular axes are located in the surface plane on the average. The ideal herringbone orientational order–disorder transition (HB $\rightarrow$ CD) is believed to be a fluctuation-driven weak first-order transition [217, 273].

The commensurate $\sqrt{3}$ solids (HB and CD) still exist slightly above the complete monolayer coverage. Further compression of N_2 on graphite leads first to a uniaxial incommensurate phase [90, 93, 370] in coexistence with the commensurate phase (HB + UIO) slightly above monolayer completion. In this regime both phases are orientationally ordered at low temperatures in terms of a herringbone structure. The commensurate phase disorders (HB $\rightarrow$ CD) at about 28 K, and the uniaxial incommensurate phase disorders near 29 K (UIO $\rightarrow$ UID) so that there is coexistence between a commensurate disordered and a uniaxial incommensurate ordered phase (CD + UIO) [394]. Further compression results in the pure uniaxial incommensurate phase (UIO), which disorders orientationally around 29–30 K to UID. There are indications from simulations that the uniaxial incommensurate phase is nonuniform with striped domain walls [182, 283]. Next there is a coexistence region of the orientationally ordered uniaxial incommensurate phase with an orientationally ordered uniform triangular incommensurate phase (UIO + IO) [93, 242, 370, 388], which finally exists as a pure phase in the first layer (IO); the transformation to the orientationally disordered triangular incommensurate phase (ID) at higher temperatures is not known, but there are indications for a gradual loss of the orientational order [242, 394]. The center-of-mass structure seems to be close to a perfect equilateral triangular lattice. The orientational ordering of this triangular incommensurate phase (IO) is either of the pinwheel type (see Fig. 4*b*) or an out-of-plane herringbone-like ordering similar to Fig. 4*c*; see the discussion in Section III.E. At least the triangular incommensurate monolayer (IO) coexists already with molecules promoted to the second layer. This renders a precise experimental determination of absolute coverages difficult in the compressed regime, and it results in phase boundaries which might vary between different experimental techniques and research groups; see the detailed discussions in Section I.C and Section III.E. We mention here also that an alternative interpretation [370, 371] of the high coverage phase diagram including an amorphous phase [371] at low temperatures was put forward.

There are some early speculations [106] and more recent indications [112, 156] that a reentrant fluid phase (RF) exists for N_2 on graphite, which is squeezed in between the commensurate (CD) and triangular incommensurate (ID) disordered phases just above monolayer completion and temperatures of about 50–80 K (see Fig. 46 in Section III.E). But it is not fully clear how the strip of the reentrant fluid extends down to low temperatures and where this fluid undergoes solidification.

At coverages exceeding about 1.3 monolayers, an orientationally ordered bilayer solid phase (2SO) is in coexistence with the orientationally ordered incommensurate phase (IO) in the first layer and with a second-layer vapor phase [395]. An orientational disordering transition to the 2SD + 2V + ID coexistence region takes place at about 28–29 K. Compression beyond about 2.35 monolayers leads to a pure bilayer phase that is orientationally ordered at low temperatures (2SO) and disorders in the temperature range from about 29 K to 27 K to an orientationally disordered bilayer phase (2SD) [395]; even higher coverages are included in Fig. 2 of Ref. 395. We note, however, that it is not fully clear if the phases denoted by 2SO and 2SD are bilayer phases in the strict sense (where both first and second layers form a single structural entity) or a second-layer structure (where a second structurally independent layer sits on top of an essentially unperturbed first layer). The bilayer phase in the orientationally disordered coexistence region 2SD + 2V + ID melts to a second-layer liquid phase (2L) [72], which is in coexistence with the second-layer vapor (2V) on top of the orientationally disordered incommensurate phase in the first layer (ID). Finally, there has to be a two-dimensional critical point where the second-layer vapor (2V) and liquid (2L) phases become indistinguishable and form a second-layer fluid on the first-layer solid (ID + 2F).

B. Submonolayers and Melting

The differential heats of adsorption of N_2 on a number of carbon blacks were determined calorimetrically in Ref. 15. It was found that these heats of adsorption undergo a large variation as successive fractions of the surface are covered. This was confirmed in isosteric measurements of the heats of adsorption of N_2 on carbon blacks [166]. Searching for cleaner substrates than the carbon blacks the adsorption of N_2 on carbon dust was also studied [306]. Low-pressure volumetric isotherms [300] of N_2 on graphitized carbon black P-33 around 80 K and 90 K clearly indicated the formation of the first layers of physisorbed N_2 on graphite (see the short-dashed line in Fig. 9). It is found that the monolayer can be described approximately in terms of a two-dimensional gas for pressures corresponding to less than a coverage of 0.5 monolayers. Above this coverage the adsorption isotherm has the form

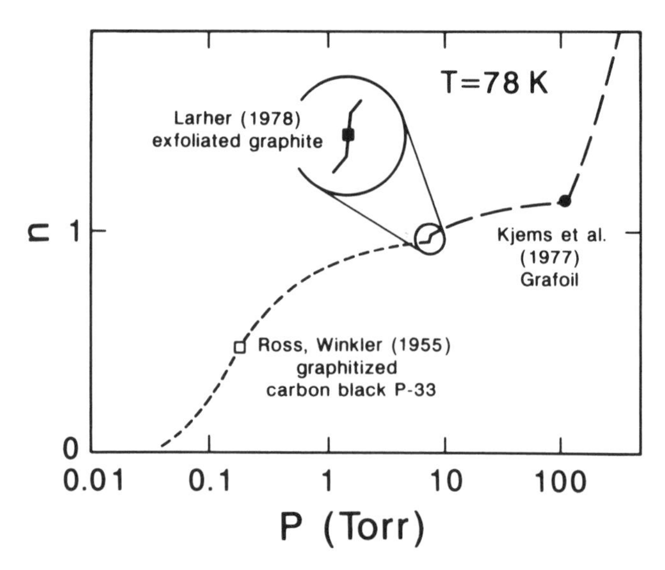

Figure 9. Adsorption isotherm [172, 189, 300] for N_2 on graphite at 78 K showing the adsorption of the first layer (short dashes), the fluid to commensurate solid transition (solid line and blowup), and second-layer adsorption and/or commensurate to incommensurate transitions (long dashes) in terms of the coverage n measured in units of the complete $\sqrt{3}$ monolayer. (Adapted from Fig. 1 of Ref. 91.)

of a Langmuir equation, which was interpreted as the adsorption of a mobile film on localized sites, where the sites are superimposed on the substrate by the packing of the adsorbate. The surface area covered by N_2 molecules on graphite was determined in Ref. 285 and compared to different models. Heats of adsorption of N_2 on the carbon black Graphon were determined in Ref. 16.

In a pioneering neutron scattering study [171] to explore the structure of quasi-two-dimensional adsorbed films, a low-temperature phase with long-range order was detected for N_2 on Grafoil; see Ref. 266 for the justification of the background subtraction procedure. The diffraction pattern could be understood in terms of a registered triangular phase with a lattice constant exceeding that of the graphite basal plane by a factor of $\sqrt{3}$. In this $(\sqrt{3} \times \sqrt{3})R30°$ phase, one-third of the hexagon sites of the honeycomb graphite lattice are occupied. The experiments [171, 172] were done in the coverage range of roughly 0.3 to 1.75 monolayers and from 10 to 90 K. The high-temperature disordered phase was considered to consist of N_2 molecules still constrained to epitaxy by the periodic potential of the graphite substrate; the diffraction profiles from adsorbed phases became indistinguishable from those of the bulk liquid above approximately 90 K. It was not possible to distinguish between an amorphous solid or a fluid-like disordered phase; the latter interpretation, however, was favored based on an analogy to ^{4}He monolayers on graphite. For coverages below about 0.8 the $(\sqrt{3} \times \sqrt{3})$ structure was found to persist up to and to disappear near 50 K, and this transition to the disordered phase was believed to be a first-order transition. The size of the $\sqrt{3}$ solid crystallites was estimated to be on the order of 100–1000 molecules.

As the coverage was increased to complete monolayer coverages the transition temperature increased beyond 70 K and reached a maximum at about 85 K. The diffraction profile at 0.87 monolayers and 70 K could only be satisfactorily fitted after decomposing it into contributions stemming from two coexisting phases—that is, the registered solid and the disordered high-temperature phase. These findings were essentially confirmed by an adiabatic heat capacity study with concomitant vapor pressure measurements [73]. In addition, the anomalies identifying the melting transition were found to stay constant near 49 K up to 0.9 monolayers (see Fig. 10). For coverages larger than 0.93, higher-order melting was proposed because no phase coexistence was observed [73], whereas the 48 K transition from the registered solid to the disordered phase was assigned to be a first-order change subject to finite-size broadening and shifting [73]. More specifically, it is argued that the low value of the melting temperature for partial monolayers is due to the large fraction of edge to interior molecules, which have different binding characteristics.

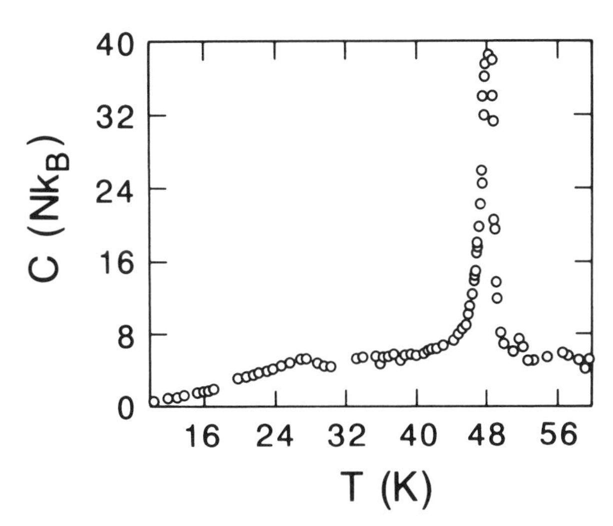

Figure 10. Heat capacity trace of N_2 on Grafoil at a coverage of about 0.61 monolayers. The anomalies near 28 K and 48 K stem from the orientational and melting transitions, respectively. (Adapted from Fig. 5 of Ref. 73.)

The characteristics of the riser regions in earlier vapor pressure isotherm measurements and the heat capacity traces around 75 K were, however, interpreted as being suggestive of a second-order phase transition to the registered solid in the submonolayer regime [54]. N_2 adsorption on graphitized carbon black was studied [135, 303] using isothermal calorimetry at 77 K. In accordance with Ref. 54, a small feature near the first step was detected in the gravimetric adsorption isotherm. This jump close to monolayer completion was interpreted as a phase transition occurring in the two-dimensional monolayer from a supercritical fluid to a localized state. But it was assigned to be a first-order transition [303] rather than a second-order one [54]. The molar enthalpy of adsorption extrapolated to zero coverage was determined [135] to be −9.7 kJ/mol at 77 K.

The observations by neutron scattering and calorimetric studies were initially interpreted in terms of the topology of standard three-dimensional triple-point phase diagrams [73, 171, 172]; see Fig. 11 for a proposal [73]. The melting transition below 0.8 monolayers was assigned to occur at a

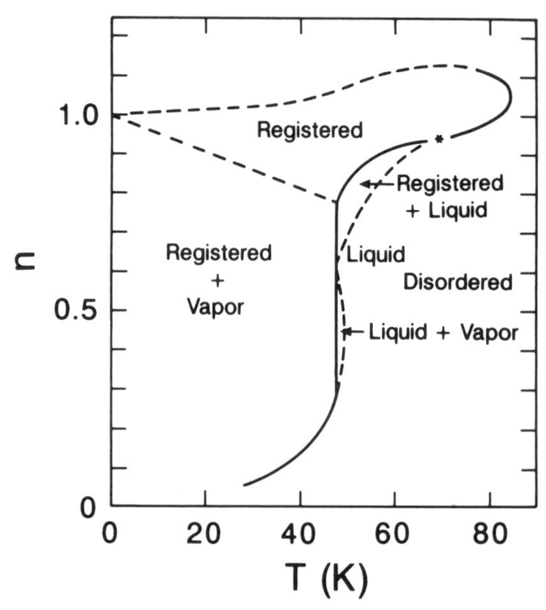

Figure 11. Proposed early low-coverage melting phase diagram of N_2 on graphite according to the triple-point model in analogy to three-dimensional phase diagrams. The hypothesized tricritical point (asterisk) marks the high-temperature end point of the coexistence region between registered solid and liquid. Note that the herringbone orientational transition around 28 K was already discovered (see Fig. 10), but was not included in the phase diagram. Dashed lines are speculative phase boundaries. (Adapted from Fig. 12 of Ref. 73.)

solid–liquid–vapor triple temperature of 49 K with the critical temperature being only 1 K or 2 K higher than the triple temperature [73]. In other words, the liquid–vapor coexistence region was thought to be too narrow to be observed in the presence of smearing. The increase of the melting temperature for higher coverages beyond 70 K was attributed to finite size [116], substrate inhomogeneity, and energy heterogeneity effects [89, 97]. According to this scheme, a registered solid–liquid coexistence region extending from 49 K to 65 K and at coverages between 0.6 and unity was postulated [73]. In addition, it was speculated [73] that the film has a tricritical point at the high-temperature end of the registered solid–liquid coexistence region around 50–65 K, because the 85 K transition from solid to the disordered phase was assigned to be continuous. The quasivertical risers observed in volumetric isotherms [189] (see the blowup in Fig. 9) gave strong evidence for a first-order fluid to commensurate solid transition up to temperatures as high as 82 K and led to speculations about a possible continuation as a second-order transition above 82–83 K.

A tricritical-point phase diagram strongly distorted by finite-size effects and inhomogeneities of various kinds as phenomenologically modeled by a temperature smearing ΔT was later proposed [276] in the framework of the three-state Potts lattice gas model to account for the peculiarities of the N_2 melting behavior; see also Ref. 20 for an earlier phase diagram without triple-point setup to describe Kr on graphite. This phase diagram (see Fig. 12) was based on real-space renormalization group calculations for a Potts lattice gas model on a triangular lattice derived from pairwise Lennard-Jones N_2–N_2 interactions and a periodic substrate potential of a given corrugation; the large diversity of phase diagram topologies covered by the particular lattice gas model is depicted in Figs. 6 and 7 of Ref. 277. The tricritical point for an ideal infinite system with no inhomogeneities (i.e., for the triangular lattice gas) is found to occur near 0.75 monolayer coverage and at about 50 K for the chosen parameterization aimed to mimic N_2 on graphite. The registered solid–fluid coexistence region extends below 50 K over almost the entire submonolayer range. This coexistence region is separated from the registered solid at large coverage above 0.75, as well as from the fluid phase below 0.75 monolayers by a boundary at 50 K that is independent of coverage in the range between 0.25 and 0.85 monolayers (see the solid line in Fig. 12 for a schematic picture). No two-phase coexistence occurs above 50 K, and the solid and fluid phases are separated by a continuous transition line that increases gradually in coverage from 0.75 with increasing temperature. Adding in a temperature smearing ΔT, it is found that the tricritical point is pushed to higher temperatures leading to a cusped phase boundary (see the dashed lines in Fig. 12). This yields a narrow strip of solid–fluid coexistence between 50 K and the increased tricritical tempera-

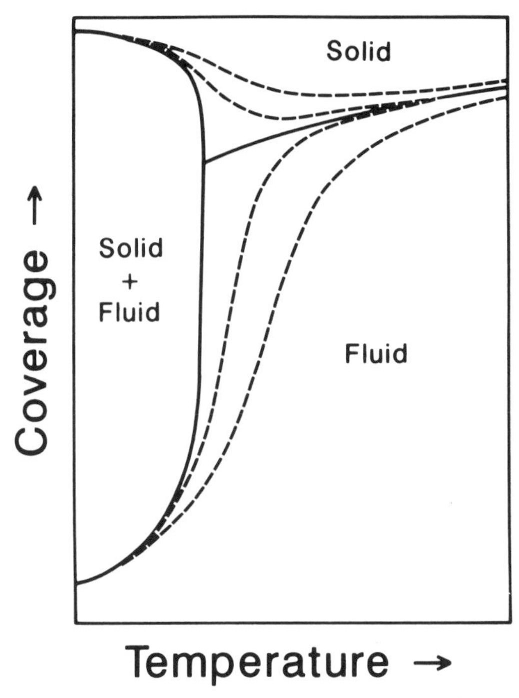

Figure 12. Proposed schematic low-coverage melting phase diagram of N_2 on graphite according to the tricritical point model. Full lines correspond to the ideal thermodynamic system; that is, $\Delta T = 0$. Dashed curves were obtained with two different temperature smearings $\Delta T > 0$. (Adapted from Fig. 3 of Ref. 276.)

ture. This topology reproduces approximately the at that time observed [73, 171, 172] phase boundary that separates the fluid phase from the coexistence region of solid and fluid phases at low coverages.

A third interpretation was put forward by analogy to the behavior of Kr on graphite which could be explained in terms of the incipient triple-point model where a liquid phase does not exist due to a solid phase much stabilized by surface effects [55]. Subsequent LEED intensity isotherms [91] on graphite crystals in the range from 45 K to 52 K were interpreted in favor of this model having a first-order transition from the commensurate solid directly to a supercritical fluid. Hysteresis was observed [91] when cycling up and down in pressure for all isotherm measurements of the fluid to commensurate solid transition. There is no evidence for a liquid–gas coexistence region down to a resolution of 0.5 K along the temperature axis, and a shift of the submonolayer phase boundary with coverage is found [91], which rules out the triple-point phase diagram in analogy to three dimensions. The insensitivity of the submonolayer phase boundary line on substrates with different domain sizes such as Grafoil, graphite foam, and single crystals makes the size-effect smeared tricritical phase diagram [276] unlikely.

Detailed ac heat capacity (and vapor pressure) experiments [65, 231] up to about 70 K were clearly inconsistent with the earlier explanations in terms of conventional triple-point and smeared tricritical-point phase diagrams [73, 171, 172, 276]. According to the incipient triple-point model [55], which was qualitatively supported in Ref. 91, the registered ($\sqrt{3} \times \sqrt{3}$) solid is significantly stabilized by the graphite physisorption potential. This leads to a coexistence of solid and fluid phases up to temperatures much higher than the "expected" liquid–vapor critical temperature (see Refs. 65, 231, and 261 for an explanation in terms of free energies.) As a consequence the existence of the liquid phase and, hence, of the two-dimensional triple and critical points is suppressed, and the commensurate solid disorders directly to the "supercritical fluid phase," bypassing the liquid phase. The resulting part of the phase diagram for N_2 on graphite is presented in Fig. 13.

The heat capacity data leading to the incipient triple point interpretation (see Fig. 1 of Ref. 231 or Fig. 5 of Ref. 65) were obtained from nitrogen films on graphite foam using ac calorimetry. This substrate has single crystal domains of linear dimensions of about 1000 Å being a factor of 5–10 larger than those of Grafoil (see Section I.C). At low coverages (0.287–0.600) the data show a single, sawtooth-like peak which drops sharply on the high-temperature wing; the temperature at maximum changes only slightly from 47.91 K to 48.99 K at the limits of this coverage interval. A less sensitive

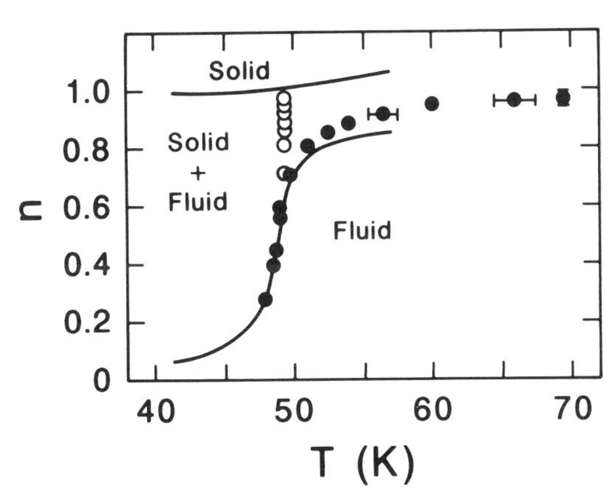

Figure 13. Proposed low-coverage melting phase diagram of N_2 on graphite according to the incipient triple-point model (see also Fig. 14). Filled circles indicate the positions of the heat capacity peaks at low coverages and the shoulders at higher coverages, and the open circles indicate the peak maxima at higher coverages; the filled circle at roughly 70 K stems from a vapor pressure scan at constant temperature. Solid curves represent model calculations. (Adapted from Fig. 2 of Ref. 231.)

adiabatic heat capacity study [156] yielded a slight increase from 45.5 K up to 49.4 K in the range from 0.169 to 0.666 monolayers and 53.2 K at a coverage of 0.820 monolayers. At coverages in the range from 0.721 to 0.955 monolayers, however, the shape changes and a shoulder instead of the sharp drop develops at the high-temperature side. The peak maxima occur at roughtly the same temperature of 49.3 K; the shoulder, however, moves rapidly with increasing coverage to higher temperatures. At a coverage of 0.975 the height of the 49 K peak decreases, and instead the shoulder is transformed to a genuine peak at 66 K. No more sign of a peak near 49 K is found for 1.010 monolayers. These heat capacity data are depicted in Fig. 13, where the full circles indicate the position of the peak at low coverages and the shoulder at higher coverages, and the open circles (which all occur at essentially the same temperature of 49.3 K) indicate the peak maxima at higher coverages. It is argued that the registered solid is in coexistence with a fluid in the region everywhere to the left of the solid circles in Fig. 13 below complete monolayer coverage; that is, the open circles are not part of an actual phase boundary. The evidence for such a coexistence below and also above 49 K at coverages exceeding roughly 0.8 monolayers comes from an analysis of the total heat capacity; see Refs. [65, 231]. The open circles in Fig. 13 are assigned to the rapid but continuous conversion of solid to fluid at fixed coverage due to the rapid rise of the fluid boundary of the coexistence region. The solid lines are obtained from simple model calculations and account qualitatively for the observed features of the phase diagram. The only minor inconsistency [231] is the increasing phase boundary of the solid to solid–fluid coexistence transition near 50 K and is probably due to crude modeling. This was changed in a more sophisticated approach [261] (see the solid lines in Fig. 14), which can also be generalized to include incommensurate phases and the respective phase diagrams [262].

The originally proposed triple-point topology was ruled out based on the absence of the required (in this case) critical point at some higher temperature [231], the lack of liquid–vapor coexistence [92, 231], and much too weak anomalies [231] along the "triple line." The tricritical scenario was disfavored [65, 231] because there is in fact two-phase coexistence found [189] to extend much above 49 K up to at least 82 K [189], and the solid phase does not show any signs of critical behavior [231]. In contrast, convincing evidence for a tricritical point (including the experimental assignment of its universality class) at the end of the solid–fluid coexistence strip at the much higher temperature of about 85 K could be given [65, 237] (see Section III.C). In addition, the phenomena observed for N_2 on graphite are very similar to those of Kr on graphite (and even show no change of the 49 K anomaly with coverage) which could successfully be explained based on the incipient triple-point model [55].

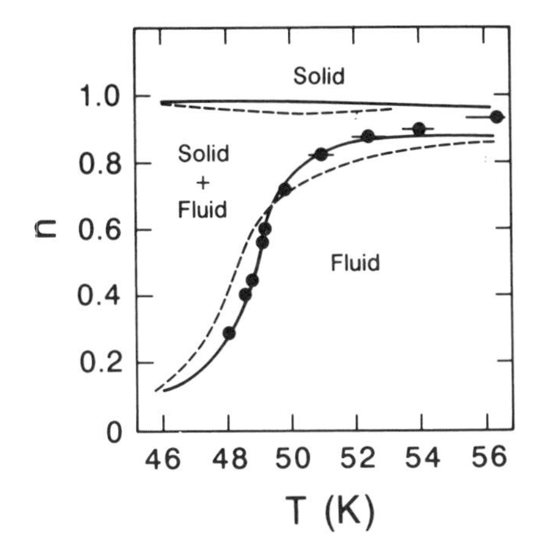

Figure 14. Proposed low-coverage melting phase diagram of N_2 on graphite according to an improved incipient triple-point model [261] (solid lines) and the density functional treatment [305] (dashed lines). Dots indicate the location of the coexistence boundary from heat capacity experiments. (Adapted from Fig. 3 of Ref. 261.)

A different approach to calculate the qualitative features of the phase diagram of N_2 on graphite was based on density functional theory [305]. In this particular method [110] the adsorbate is represented by a density distribution, and the free energy is written as a functional of this function to second order in the deviation from the uniform fluid phase, and the properties of the latter reference state are obtained from Monte Carlo data. The phase diagram is obtained by minimizing this functional with respect to its parameters, and the coexistence boundaries are obtained from Maxwell constructions. The N_2–N_2 interactions are modeled by isotropic Lennard-Jones interactions between the molecular centers of mass, and the surface potential is the first-order Fourier component [326]; see the relations (3.6), (3.7), and (3.8) with $\gamma_i \equiv 0$ given in Section III.D.1. It is found that the topology of the phase diagram changes qualitatively upon variation of the potential parameters; see the series of phase diagrams in Fig. 2 of Ref. 305, so that the phase diagram cannot be predicted from the calculation. However, a reasonable choice of parameters yields a final phase diagram, shown as dashed lines in Fig. 14, which captures the submonolayer phase diagram of N_2 on graphite quite well.

Heats of adsorption and molar entropies of N_2 on Grafoil were measured around 80 K and in the range from 0.011 to 1.2 monolayers [286]. The transition from a fluid to a registered two-dimensional solid phase was confirmed. The heat of adsorption of a single N_2 molecule on graphite was obtained [286] by extrapolation to the zero coverage limit. It amounts to 10.4 ± 0.1 kJ/mol at 79.3 K. Based on entropy estimates it is concluded that the N_2 molecules cannot be freely moving and rotating on the graphite surface in the low coverage limit. The heat capacity of N_2 on graphite was measured in the range from 3 K to 80 K at five submonolayer coverages [152] and molar entropies were obtained. The derived isosteric heat of adsorption at zero coverage is estimated to be 9.8 kJ/mol at 80 K. A value of 10.1 ± 0.2 kJ/mol was obtained from volumetric adsorption measurements [41]. An analysis of the second virial coefficients obtained from the same experiments yielded Lennard-Jones parameter sets for the N_2–N_2 and N_2–graphite interactions [41, 42].

In a molecular dynamics study [341] with realistic potentials (see Refs. 203, 340, 342, and 352 and Section III.D.1 for details), N_2 monolayers on graphite were investigated in the fluid phase close to 75 K in the range from about 0.2 up to full monolayer coverage. It is explicitly shown that the submonolayer fluid has a slight preference of the adsorption sites over the carbon atoms, whereas this preference increases dramatically near the complete monolayer. It is also demonstrated by decomposing orientational site–site correlation functions depending on the interparticle separation into a harmonic expansion that the in-plane orientational structure of the fluid is slightly affected by the intermolecular quadrupole interactions. The average height of the monolayer increases only slightly with coverage in the investigated range, whereas the width of the layer increases significantly. The dynamical out-of-plane tilt is compared to experimental results [241] in Fig. 15. It is also found that molecules standing perpendicular on the surface become more abundant on approaching the solid–fluid boundary. Finally, the isosteric enthalpy of adsorption is determined to be 9.29 kJ/mol in the limit of zero coverage [341], and a refit of the adiabatic calorimetry data of Ref. 286 (see above) to the same functional form yields 9.32 ± 0.2 kJ/mol at 79.3 K.

In an attempt to explore theoretically the melting transition of N_2 on graphite, a patch of molecules corresponding to a surface coverage of 0.5 monolayers was investigated by molecular dynamics [165]. Most of the

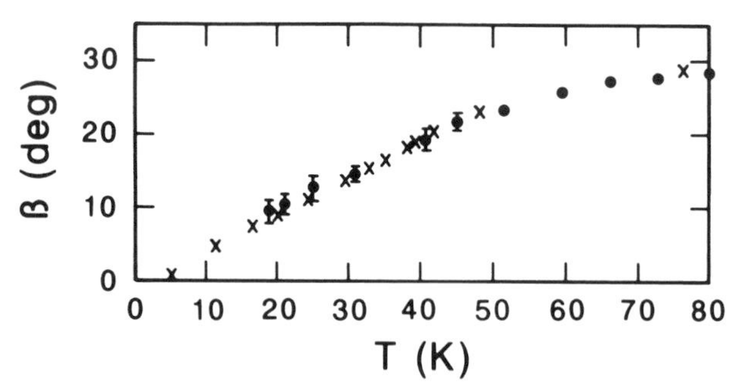

Figure 15. Out-of-plane tilt angle as a function of temperature for N_2 on graphite. Circles: nuclear resonance photon scattering of a compressed monolayer of coverage 1.05 ± 0.02 [241]. Crosses: molecular dynamics simulations of a complete monolayer [341]. (Adapted from Fig. 2 of Ref. 241.)

simulations were performed with 140 molecules in a rectangular parallelepiped with some runs extended to 210 molecules while keeping the coverage fixed; the simulation technique is in line with that reported in Refs. 203, 340, 342, and 352 (see also Section III.D.1 for discussions). The initial geometry was an infinite (periodic) strip where the randomly oriented molecules occupy ideal lattice sites of the commensurate $\sqrt{3}$ phase. The N_2–N_2 intermolecular potential was the so-called X1 model [248] consisting of a Lennard-Jones potential

$$u_{NN}(r) = 4\epsilon_{NN}\left\{\left(\frac{\sigma_{NN}}{r}\right)^{12} - \left(\frac{\sigma_{NN}}{r}\right)^{6}\right\} \tag{3.1}$$

with $\epsilon_{NN} = 36.4$ K, $\sigma_{NN} = 3.318$ Å, and a distance r between two nitrogen atoms located on different molecules; the intramolecular N–N bond length is 1.098 Å. The electrostatic part of the X1 potential is represented by point charges $q = -0.405e$ on the atoms and a point charge of $0.810e$ on the molecular center of mass yielding a quadrupole moment of -3.913×10^{-40} Cm^2. The periodic molecule–surface potential was Fourier and Euler–MacLaurin-expanded [326, 327]; see (3.6), (3.7), and (3.8) with $\gamma_i \equiv 0$ given in Section III.D.1. The melting transition of the orientationally disordered $(\sqrt{3} \times \sqrt{3})R30°$ solid based on the isotropic gas–solid site–site potential, which is also widely used in other investigations [203, 340, 342, 352], can be located at about 40 K (see the squares in Fig. 16), which is about 10 K below the experimental melting temperature at this coverage (see above). The trajectory plot above this melting transition is shown in Fig. 17*a*. It is clearly visible that there is coexistence between a gas-like phase and a liquid-like phase in the form of the original solid strip, which is a stable situation on the 70-ps time scale of the molecular dynamics run; note that a fluid phase should homogeneously cover the surface.

In order to improve the melting temperature and possibly the character of the phase above melting, an anisotropic atom–surface potential was devised [165]. The strongly anisotropic electronic properties of graphite can be modeled by a directional dependence of the polarizability of the carbon atoms such that they are more polarizable parallel to the graphite basal plane than perpendicular to it. An appropriate functional form for the resulting anisotropic dispersion attraction [61, 362] between a graphite carbon atom and a physisorbed gas atom is given by

$$u(\mathbf{r}) = -\frac{B}{r^6}\left[1 + \gamma_A\left(1 - \frac{3}{2}\cos^2\theta\right)\right] \tag{3.2}$$

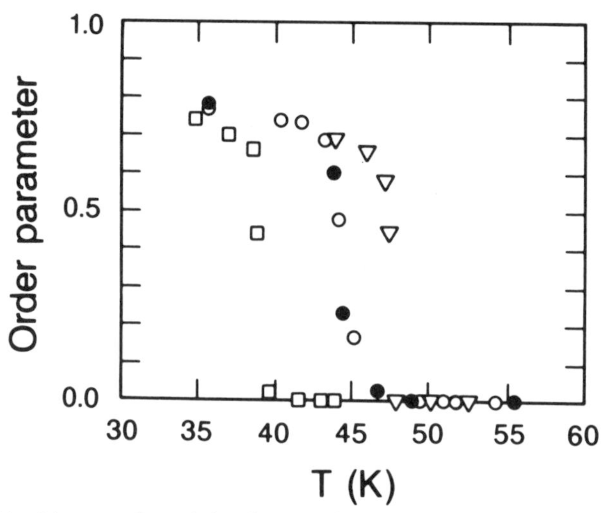

Figure 16. Measure of translational order of N_2 patches on graphite with respect to the ideal adsorption sites on the $(\sqrt{3} \times \sqrt{3})R30°$ lattice from molecular dynamics simulations. Anisotropic molecule–surface potential (3.3) with $\gamma_A = 0.4$ and $\gamma_R = -0.9$ (triangles), $\gamma_A = 0.4$ and $\gamma_R = -0.54$ (cirlces), and isotropic special case $\gamma_A = 0$ and $\gamma_R = 0$ (squares). Unfilled (filled) symbols refer to 140 (210) molecules. (Adapted from Fig. 1 of Ref. 165.)

where θ denotes the angle between the interatomic vector $\mathbf{r}$ of length r and the surface normal. The constants B and γ_A can be expressed in terms of anisotropic polarizabilities and atomic excitation energies, but $\gamma_A = 0.4$ is essentially independent of the gas atom for graphite and B is usually taken as an adjustable parameter. The repulsive interactions are included in a similar manner such that the result

$$u_{\mathrm{NC}}(\mathbf{r}) = 4\epsilon_{NC}\left\{\left(\frac{\sigma_{\mathrm{NC}}}{r}\right)^{12}\left[1 + \gamma_{\mathrm{R}}\left(1 - \frac{6}{5}\cos^2\theta\right)\right] - \left(\frac{\sigma_{\mathrm{NC}}}{r}\right)^{6}\left[1 + \gamma_A\left(1 - \frac{3}{2}\cos^2\theta\right)\right]\right\} \tag{3.3}$$

is an anisotropic Lennard-Jones-type potential to describe the nitrogen–graphite interactions, and $\gamma_R = \gamma_A = 0$ simplifies (3.3) to the original isotropic surface potential. The Fourier expansion coefficients can be given analytically similarly to Ref. 326. The overall molecule–surface energy is changed by less than 1% by using $\gamma_A = 0.4$ and $\gamma_R = -0.54$ from helium scattering data instead of the isotropic approximation $\gamma_R = \gamma_A = 0$, but the lateral barrier height (i.e., the surface corrugation) increases by about 65%.

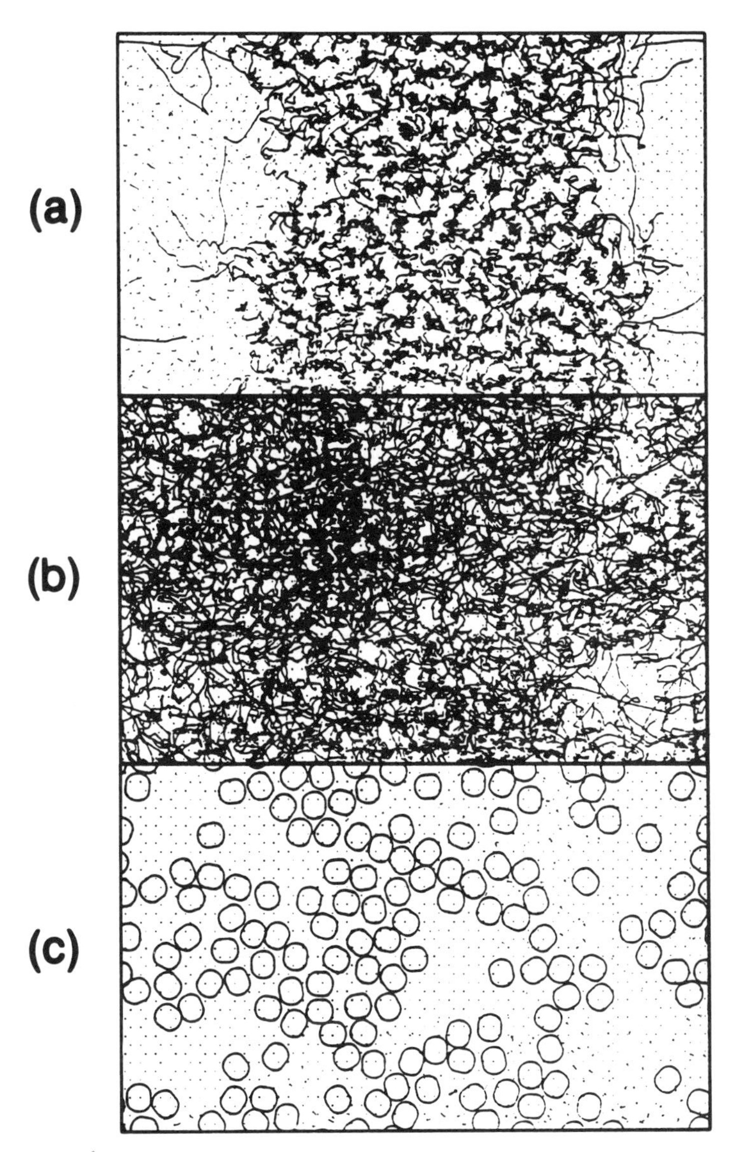

Figure 17. Molecular dynamics trajectories above the melting transition at a coverage of 0.5 monolayers. The center of mass of the N_2 molecules is projected on the graphite basal plane (*a*) for the isotropic molecule–surface potential (3.3) with $\gamma_A = \gamma_R = 0$ at 44 K and (*b*) for the anisotropic molecule–surface potential (3.3) with $\gamma_A = 0.4$ and $\gamma_R = -0.54$ at 55 K. (*c*) Top view of the final configuration of (*b*). (Adapted from Figs. 3 and 9 of Ref. 165.)

The minimum energy structure is again essentially an in-plane structure. With this anisotropic potential the melting temperature shifts to about 45 K (see the circles in Fig. 16) and to about 47–48 K when γ_R is arbitrarily changed to -0.9 (see the triangles in Fig. 16); note that there is not much change in the order parameter data when the particle number is increased from 140 to 210 molecules. The question of the character of the phase above the melting transition is unclear. The molecules seem to cover the surface area roughly homogeneously when heated to about 55 K with the $\gamma_R = -0.54$ potential (see Fig. 17*b*). Nevertheless, they have a tendency to cluster in small patches, and there is a clear evidence that the molecules still prefer the adsorption potential minima in the centers of the honeycomb hexagons (see Fig. 17*c*). A firm conclusion about the nature of this high-temperature-modulated fluid phase cannot be drawn from the reported simulations [165].

The dramatic increase of the melting temperature from about 49 K in the partial coverage regime to approximately 85 K near the complete monolayer coverage was systematically investigated by Monte Carlo methods using various boundary conditions [102, 184, 301]. The very realistic potential model is that of Ref. 181; see also Refs. 182 and 183 and Section III.E, together with a stronger modulation of the first-order corrugation term resulting from anisotropic dispersion interactions [61]. First of all a nearly circular island with free surface boundary conditions comprising 256 molecules, and systems of 16 and 64 molecules subject to periodic deformable boundary conditions were examined all in the *NPT* ensemble. The first case represents a finite patch physisorbed on graphite, whereas the second setup can be visualized as a large island. The *NPT* ensemble [279] allows for fluctuations of the lattice constants and shape and thus area of the simulation box, which allows for variations of the coverage; the particle number N, pressure P, and temperature T are constant in this ensemble. The mean coverage is an observable to be obtained as usual as a statistical average of its instantaneous values along the Monte Carlo trajectory. It is found (see Fig. 18) that the system, which prefers the registered $\sqrt{3} \times \sqrt{3}$ structure at low temperatures, melts at about 45 K as seen by the decrease of the average coverage. The simulation box of the *NPT* ensemble with periodic boundary conditions expands above the melting transition, and the heat capacity peak in Fig. 19*a* clearly signals this transition. The 256-molecule isolated island with free surface boundary conditions and the periodically replicated 16- and 64-molecule systems show the same behavior. This close comparison suggests that pure edge effects associated with the islands [73] cannot be a major contributor to melting properties of partial monolayers, since there are no edge molecules present in the *NPT* ensemble with periodic boundary conditions. In addition, the results for 16 and 64 molecules are the same, and there is no evidence for hysteresis effects as seen from the cooling run.

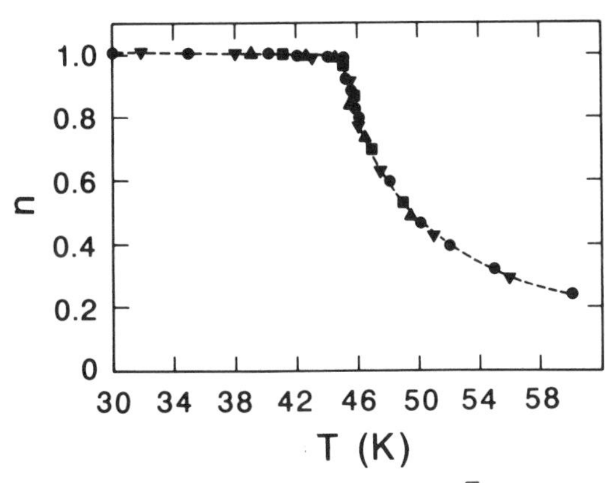

Figure 18. Surface coverage in units of the complete $\sqrt{3}$ monolayer for N_2 on graphite near the melting transition from Monte Carlo simulations. Triangles: 256 molecules forming an isolated patch subject to free surface boundary conditions. Circles (squares): 16 (64) molecules in the *NPT* ensemble with deformable periodic boundary conditions. Inverted triangles: same as circles but obtained from a cooling run starting in the fluid phase. (Adapted from Fig. 1 of Ref. 301.)

The phase above the melting transition is a modulated fluid [165], where the local densities near all graphite hexagonal centers are above the mean values. These density modulations extend up to about 10 K above the melting temperature and then disappear to yield an isotropic fluid. The behavior of the in-plane center-of-mass fluctuations of the molecules relative to their nearest graphite hexagon center $\langle(\Delta x)^2 + (\Delta y)^2\rangle^{1/2}$ increases dramatically above the transition and approaches the limiting value 0.917 for a uniform distribution, whereas the out-of-plane fluctuations $\langle(\Delta z)^2\rangle^{1/2}$ increase only very smoothly with temperature and in particular do not show any anomaly close to the transition (see Fig. 20*a*). The mean molecular orientations are in the substrate plane from 30 K to 65 K with increasing root-mean-square fluctuations from 15° to about 20°, and there is no evidence of bilayer promotion.

The complete monolayer is modeled by an *NVT* ensemble with again deformable periodic boundary conditions, where fixed particle number N and fixed area V results in a constant coverage, which was chosen to be unity [102, 301]. This models the situation where either the physical or the grain boundaries prohibit a thermal expansion of the complete monolayer in the graphite plane. Also in this case, a melting transition to a modulated fluid occurs, but the transition is now located at about 87 K; see the heat capacity anomaly in Fig. 19*b*. The transition does not show any noticeable hysteresis nor appreciable size effects, and the modulated fluid persists up

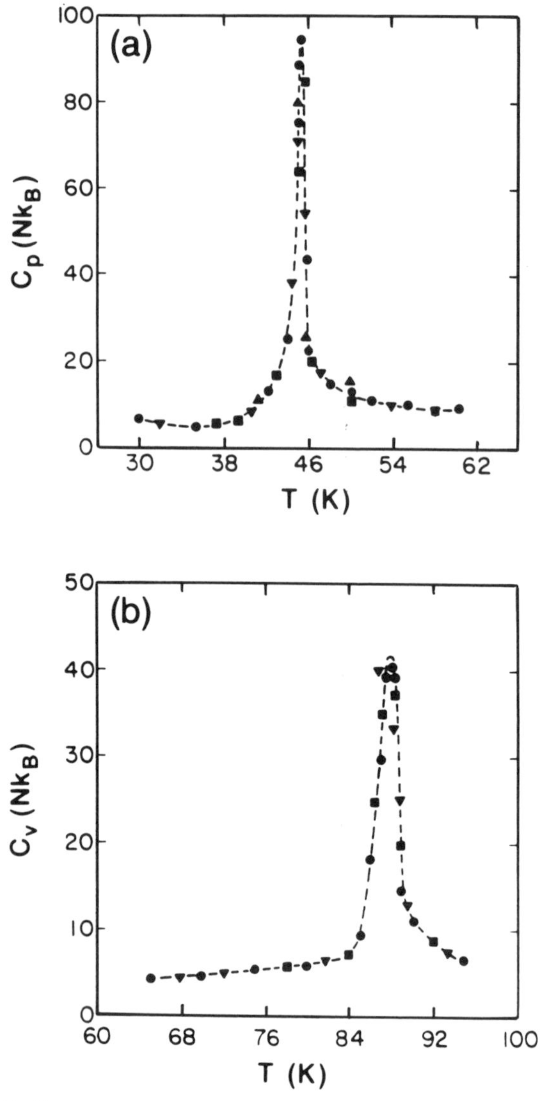

Figure 19. (*a*) Constant pressure heat capacity in units of Nk_B for N_2 on graphite near the melting transition from Monte Carlo simulations. Triangles: 256 molecules forming an isolated patch subject to free surface boundary conditions. Circles (squares): 16 (64) molecules in the *NPT* ensemble with deformable periodic boundary conditions. Inverted triangles: same as circles but obtained from a cooling run starting in the fluid phase. (*b*) Constant area heat capacity from Monte Carlo simulations. Circles (squares): 16 (64) molecules in the *NVT* ensemble with deformable periodic boundary conditions at coverage unity. Inverted triangles: same as circles but obtained from a cooling run starting in the fluid phase. (Adapted from Figs. 3 and 10 of Ref. 301.)

to approximately 93 K. The fluctuations of the molecular centers of mass in Fig. 20*b* show a qualitatively different behavior than those in Fig. 20*a*: The in-plane fluctuations increase smoothly and approach their asymptotic value, but the out-of-plane fluctuations develop a maximum near the melting transition. The equilibrium orientation of the molecules is still in-plane, and the root-mean-square out-of-plane orientational fluctuations are in the *NVT* ensemble at complete monolayer coverage higher than in the *NPT* ensemble and approach 35° for temperatures above roughly 85 K.

Thus, the difference in the melting behavior of partial and complete N_2 monolayers on graphite can be modeled and rationalized by different boundary conditions [102, 301] as already suggested [268] based on simulations of a model fluid; see also the recent study [250] for a systematic investigation of such effects. The near constant value of the melting temperature over a wide range of partial monolayer coverages can be understood in terms of island formation, and the rapid increase in the melting temperature near full coverage results from suppression of in-plane translational degrees of freedom. For partial monolayers there are many vacancies between islands that promote thermal fluctuations in the plane of the substrate which, in turn, facilitate self-diffusion, thermal expansion, and melting at a quite low temperature (about 49 K) compared to that of the complete monolayer (about 85 K). In the complete monolayer, however, all $\sqrt{3}$-lattice sites are occupied and thermal expansion in the plane is eliminated by the finite boundary imposed by grain boundaries or experimental constraints. Thus, only fluctuations normal to the plane can provide vacancies sufficient to initiate the melting transition. But the N_2–graphite interactions normal to the surface are nearly two orders of magnitude stronger than the in-plane N_2–N_2 interactions, roughly 1100 K/molecule versus 20 K/molecule [140], so that fluctuations perpendicular to the graphite plane and second-layer promotion are thermally activated only at considerably higher temperatures. This, in turn, results in a much higher melting temperature as the monolayer is completed, which exceeds even the bulk melting temperature of about 63 K (see Table I). However, if thermal expansion and fluctuations in the easy plane of the graphite surface are possible, melting is facilitated at much lower temperatures, which is possible for fractional monolayers consisting of registered islands of N_2 molecules. Finally, it has to be mentioned that all the discussed calculations [102, 301] predict registry at temperatures below the melting temperature, but only with a substrate corrugation potential with a deeper well [61] than that of the traditional Fourier expansion [324, 326]; see (3.6), (3.7), and (3.8) with $\gamma_i \equiv 0$ in Section III.D.1.

A subsequent investigation [185] with the same methodology [102, 301] concentrated on the dependence of the melting temperature on the size of the N_2 islands; the orientational ordering transition is discussed in Section

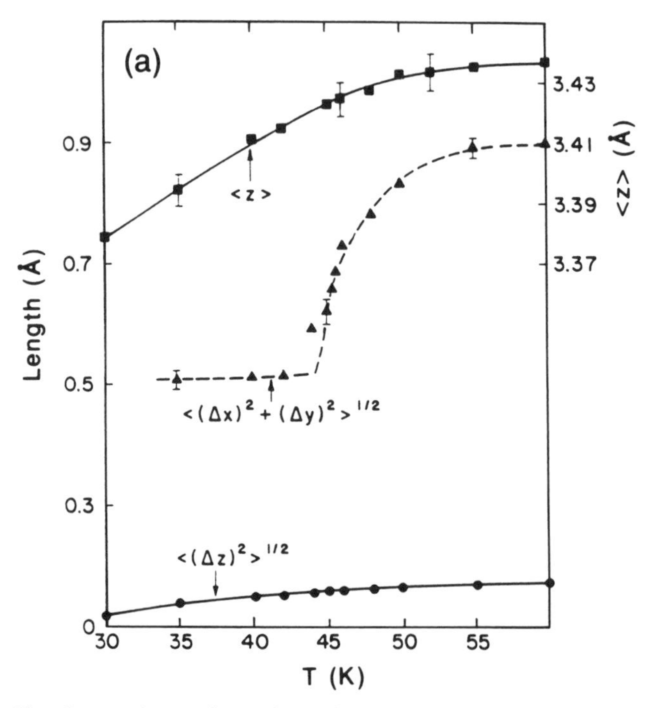

Figure 20. Center-of-mass fluctuations of N_2 on graphite with respect to the nearest hexagon center (registry point) near the melting transition obtained from Monte Carlo simulations. Left-hand scales: root-mean-square fluctuations perpendicular (circles) and parallel (triangles) to the surface plane. Right-hand scales: average center-of-mass position above the surface plane (squares). Fluctuations are obtained (*a*) in the *NPT* ensemble and (*b*) *NVT* ensemble at coverage unity, both with deformable periodic boundary conditions. (Adapted from Figs. 8 and 14 of Ref. 301.)

III.D.1. To this end, patches ranging up to 256 molecules subject to free surface boundary conditions were investigated by Monte Carlo simulations. The realistic potential model is the one of Ref. 181 (see Section III.E) and includes image charges and substrate-mediated dispersion interactions. It is found that the phenomenological melting temperature increases monotonically with increasing cluster size (see the dots in Fig. 21). However, the melting temperature seems to saturate at around 45 K, which is the value previously obtained for a model of an infinitely large island of N_2 on graphite [102, 301]. The calculations clearly show that the registered $\sqrt{3} \times \sqrt{3}$ structure is preserved for all cluster sizes until melting occurs, and that melting occurs directly from the $\sqrt{3}$ solid to the fluid phase with nearly unhindered self-diffusion.

The full coverage dependence of the melting temperature in the range

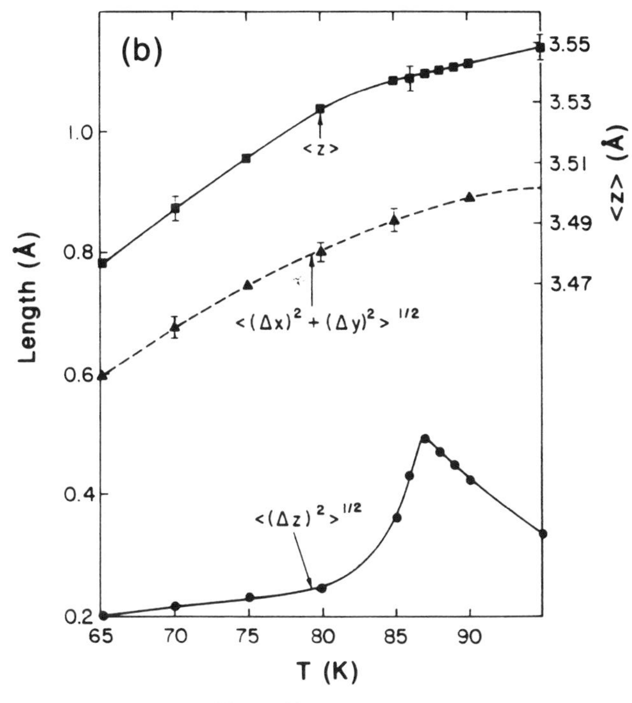

Figure 20. (*Continued*)

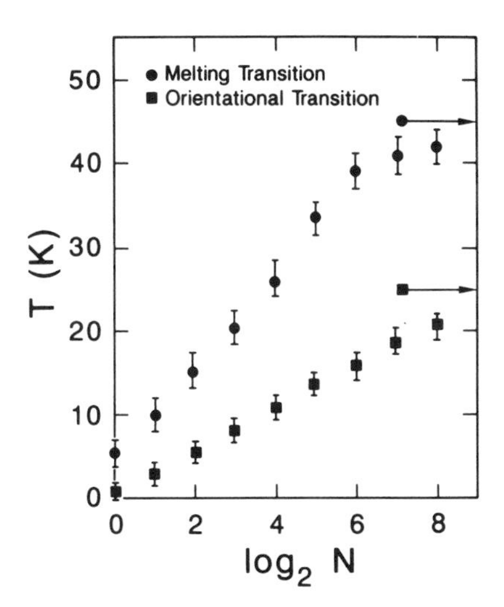

Figure 21. Orientational order–disorder (squares) and translational melting (circles) transition temperatures for N_2 monolayers on graphite as a function of the island size N from Monte Carlo simulations with free boundary conditions. The arrows with the corresponding symbols at the end mark the infinite system size estimates for these quantities within the same model. (Adapted from Fig. 1 of Ref. 185.)

from 0.2 monolayers up to the complete $\sqrt{3}$ monolayer was determined [103] using Monte Carlo simulations in the *NVT* ensemble with periodic boundary conditions and about 256 particles interacting with the same potential model as used in related previous studies [102, 181–183, 301]; see Section III.E. The important role of vacancies on the calculated melting temperature (circles) is evident from the phase diagram of Fig. 22, which is in striking agreement with experimental results [65, 73] up to monolayer completion; see, however, also Ref. 302. Inspection of instantaneous configurations in the range from 0.9 to 1.0 monolayers suggests that the vacancies tend to coalesce into relatively large voids, whereas the molecules register in the $\sqrt{3}$ arrangement. Near melting, large-scale fluctuations between a more homogeneous distribution of the vacancies over the system and situations with disordered and registered parts in coexistence are observed. At complete monolayer coverage, significant second-layer promotion sets in very near and above the melting temperature, but not below this temperature. This creates defects in the first monolayer and allows for melting. This mechanism is explicitly supported by the observation that artificially preventing second-layer promotion results in a higher melting temperature for the complete monolayer, whereas this constraint does not significantly affect melting at lower coverages. These results underline the idea of a defect melting mechanism for N_2 monolayers on graphite [102, 140, 301]. Based on a qualitative change of the heat capacity anomaly when increasing the coverage from less than 0.95 monolayers to the complete monolayer, it is

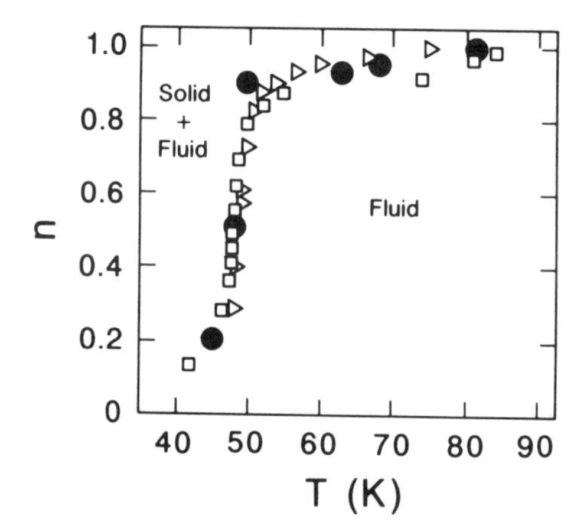

Figure 22. Submonolayer phase diagram near translational melting for N_2 on graphite; the coverage is measured in units of the complete $\sqrt{3}$ monolayer. Filled circles: Monte Carlo simulations in the *NVT* ensemble at the coverages 0.2, 0.5, 0.9, 0.93, 0.95, and 1.0. Squares: calorimetric measurements [73]. Triangles: calorimetric measurements [65]. (Adapted from Fig. 1 of Ref. 103.)

suggested that the transition dynamics change fundamentally as the vacancy concentration decreases [103].

C. Tricritical Point

The transition from the fluid phase to a registered monolayer solid is found by adsorption isotherm measurements [189] to be first order up to a temperature of about 82 K. Around this temperature the quasivertical risers become nonvertical and it is suggested that the order of the transition changes from first order to a continuous behavior near 82 K.

The strip of registered solid–fluid coexistence near monolayer completion narrows down as the temperature is increased and ends at a tricritical point (see Fig. 8). For temperatures higher than the tricritical one, the solid-to-fluid transition is a continuous transition, whereas it is of first order (i.e., with phase coexistence in the temperature-coverage plane) at lower temperatures [237]. The phase boundaries of this strip as shown in Fig. 23 in detail could be mapped using *in situ* vapor pressure isotherm measurements [237]; note that these solid–coexistence and coexistence–fluid transitions were (arbitrarily) defined to occur at reduced vapor pressures of 1.01 and 0.99 to account approximately for the found smearing effect of roughly $\pm 1\%$. These data allowed to obtain the coverage differences Δn of the solid–coexistence and coexistence–fluid transitions as a function of temperature and thus allowed for a fit

$$\Delta n = B(T_t - T)^{\beta_u} \tag{3.4}$$

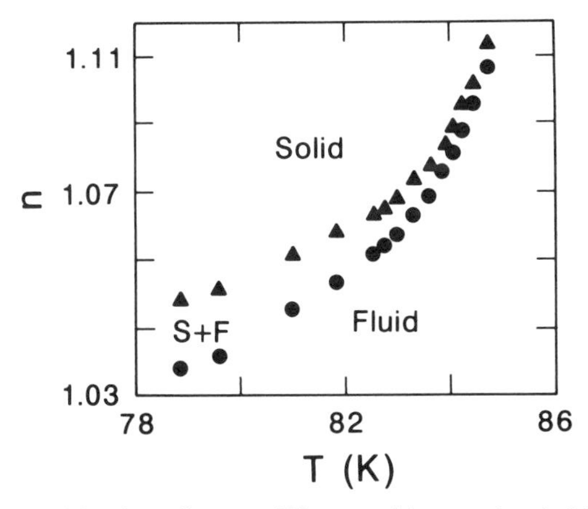

Figure 23. Detail of the phase diagram of N_2 on graphite near the tricritical point obtained from vapor pressure scans at fixed temperature. (Adapted from Fig. 1 of Ref. 237.)

appropriate [134, 194] to describe the tricritical divergence of the order parameter Δn at the tricritical point (T_t, n_t); note that the fit to such a law assumes that Δn is the order parameter obtained in the infinite system limit.

A reasonable model [307, 385] to describe the transition might be the three-state Potts model in two dimensions; the $(\sqrt{3} \times \sqrt{3})R30°$ superlattice has a triply degenerate ground state on the honeycomb lattice of the basal plane of graphite. In this case the "subsidiary" tricritical exponent β_u in (3.4), which describes the narrowing of the coexistence region [134, 194], adopts the value 0.5 [258]. Based on the vapor pressure data [65, 237], a fit of (3.4) (see Ref. 237 for details), yields a tricritical point at a temperature of 85.37 ± 0.06 K and a coverage of 1.19 ± 0.02 clearly beyond the complete monolayer with $\beta_u = 0.55 \pm 0.02$ in agreement with the theoretical value for the two-dimensional three-state Potts tricritical point. In spite of the cited very small error bars [237], the authors concede [65] that further studies have yet to confirm the existence and the location of the tricritical point. The good agreement of the experimental and theoretical tricritical exponent might be fortuitous because the commensurate solid–fluid coexistence boundary is nearly parallel to the experimental path of the isotherms (see Fig. 23), and hence the technique is not sensitive to the precise location of the tricritical temperature (see the remarks in Ref. 19 of Ref. 112). The reported tricritical temperature is consistent with (a) the earlier estimate [73] of about 85 K (where the transition was believed to be higher than first order) and (b) the outcome of the adsorption isotherm study [189] stating that the solid–fluid coexistence region extends at least up to 82 K. Based on a comparison of the data and their analysis near the tricritical point to Kr on graphite, it was later suggested [336] to reinterpret the N_2 data of Refs. 65 and 237 as being indicative of a tricritical point between 82 K and 84 K near complete monolayer coverage, in accord with the earlier volumetric study [189]. In any case, the tricritical temperature is very close to the maximum temperature at which the commensurate phase can still exist.

D. Commensurate Herringbone Ordering

1. Ground State and Orientational Order–Disorder Transition

The so-called "herringbone ordering" occurring at low coverages and temperatures is the most extensively studied orientational transition of N_2 on graphite including experiment, theory, and simulation.

A transition near 28 K at submonolayer coverages was discovered in a heat capacity study [73] (see the small low-temperature anomaly in Fig. 10) and interpreted in terms of a libration-hindered orientational transition of the molecular axes which occurs in bulk N_2 around 36 K between the fcc α solid and the hcp β phase (see Table I). It was speculated that the monolayer undergoes this collective transition without any structural change of the $\sqrt{3}$ lattice, and that it is driven by anisotropic interactions such as quadrupole interactions among N_2 molecules. The molecular axes were thought to be parallel to the surface plane, as argued in Ref. 326, with only a slight tilt out of this plane, and the herringbone arrangement was mentioned as one of two possible arrangements. Above the orientational transition the molecules start to rotate while staying pinned with their centers of mass on the $\sqrt{3}$-lattice sites.

Early theoretical speculations [326] concerning the energetics of dense, ordered N_2 monolayers on graphite were based on a model for the nonelectrostatic interactions of N_2 molecules with graphite in analogy to the description of rare gas physisorption [324, 325, 327]; the intermolecular interactions were represented by an atom–atom Lennard-Jones potential and the electrostatic quadrupole–quadrupole contribution. In these calculations the interaction between an adsorbed molecule and a substrate atom is considered to be a sum of atom–atom potentials, where each nitrogen atom of N_2 is considered to be a Lennard-Jones interaction site. Within this approach [324, 326] the atom–surface interactions are approximated by the sum of the pairwise interactions between each physisorbed atom and each atom of the unperturbed and rigid semi-infinite graphite block. This infinite periodic sum is expressed as a rapidly converging Fourier series in the position variables in the plane parallel to the surface. For a certain class of potentials, such as inverse power law interactions occurring in the Lennard-Jones potential [324] and also exponential potentials [17], the expansion coefficients can be obtained analytically, and they depend on the height of the interaction site above the surface; see (3.6)–(3.8) for explicit formulae for an anisotropic Lennard-Jones potential, and the special case of isotropic interactions discussed here is obtained from the given formulae for $\gamma_i \equiv 0$. In the spirit of the original atom–atom representation to model the molecular potential, the total molecule–substrate potential can be written as a sum over atom–substrate Fourier expansions (3.6). Specifically for N_2 on graphite it was shown that an adequate representation of the full interactions can be obtained after truncating this Fourier sum to first order in the orientations [326]. The zero-order contribution (3.7) is the energy averaged over all in-plane orientations (i.e., over the surface lattice at fixed distance and orientation relative to it) and depends only on the height of the atoms above the surface. The infinite sum over lattice planes defining the zero-order component can furthermore be approximated using the Euler–MacLaurin resummation [327]. A concise presentation of the essential formulae for practical applications can be found, for example, in Ref. 340.

Based on this type of modeling it is concluded [326] that the registered phase at very low temperature consists of molecules lying flat on the surface with fixed azimuthal angles. As the temperature is raised, finite tilt angles are thermally excited and free azimuthal rotation sets in, while the phase is still in registry with the graphite basal plane. Furthermore, it is stressed that an important consequence of the nonsphericity of the molecular shape is a strong coupling between tilt angle and absorbate–surface separation.

In a theoretical approach [125] to orientational superlattice ordering based on realistic microscopic modeling of the interactions between many N_2 molecules, a Lennard-Jones potential was used to describe the nitrogen–nitrogen and nitrogen–carbon interactions, and the molecular quadrupole was represented by point charges. Energy minimization resulted in a structure where the molecules are lying flat on the surface in a four-sublattice structure. It is, however, only a slightly distorted version of the ideal two-sublattice herringbone superlattice.

A prototype model for orientational ordering of electrostatic quadrupoles on a triangular net in a static crystal field was devised and systematically investigated based on mean-field, Landau, and spin-wave theory [21, 141]; see Section II.A for an extensive discussion.

In a simplified triangular lattice model for orientational ordering of N_2 on graphite, the molecule–surface interactions were approximated in two extreme limits and only quadrupolar intermolecular interactions were taken into account [274]; see Section II.B for details of the model. The finite temperature Monte Carlo study [274] uncovered in the case of the strictly two-dimensional model a phase transition (see Fig. 24) from the in-plane herringbone two-sublattice structure to an orientationally disordered high-temperature phase. Estimating the quadrupolar coupling constant, a transition temperature of 28 ± 5 K was obtained for a system of at most 12×12 molecules. The same model allowing for rotations of the molecular axes

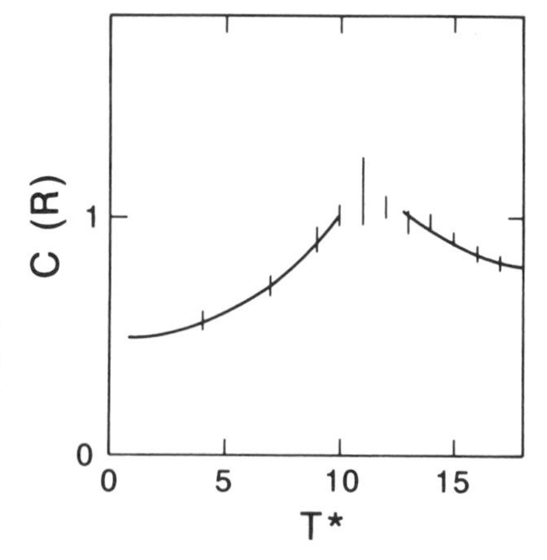

Figure 24. Specific heat including error bars from canonical Monte Carlo simulations of a two-dimensional triangular lattice of quadrupoles as a function of the reduced temperature $T^* = T/\Gamma_0$, where Γ_0 denotes the nearest-neighbor quadrupolar coupling constant (2.2). (Adapted from Fig. 2 of Ref. 274.)

in three dimensions with no substrate interactions showed a transition from a pinwheel-like low-temperature phase to the rotationally disordered high-temperature phase.

Resonance photon scattering from $^{15}N_2$ monolayers on Grafoil at 78 K has been used to determine the tilt of the molecular axes relative to the surface [240]. It was found that the molecules are preferentially aligned parallel to the graphite plane from near monolayer coverage down to about 0.8 monolayers, whereas no preferred orientation was obtained for compressed monolayers at a coverage of 1.5.

In the first study to assign experimentally the type of orientational ordering [99], a superlattice neutron diffraction peak corresponding to a doubling of the unit cell of the overlayer was observed below 30 K for 0.8, 1.0, and 1.2 monolayers (see Fig. 25), and a structural transition of the center-of-mass lattice was ruled out. Doubling and the herringbone structure was already predicted for N_2 on graphite based on theoretical considerations [21, 125] shortly before the experiment and also guided the interpretation of the

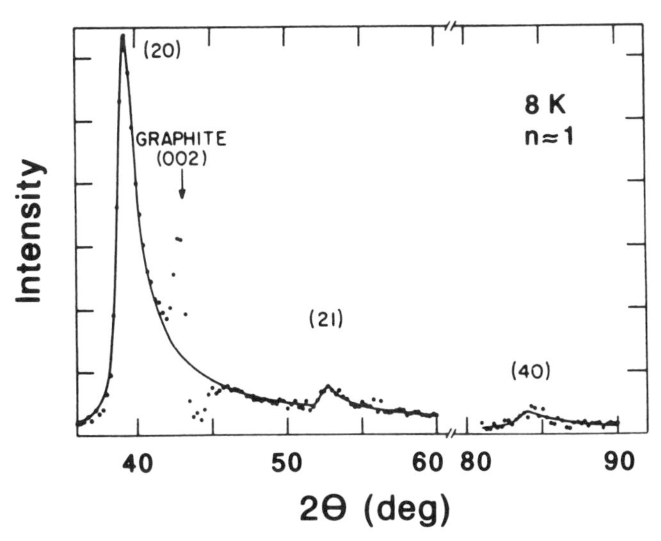

Figure 25. Neutron diffraction pattern from N_2 on Grafoil sheets at 8 K and at complete monolayer density as a function of the scattering angle. The background scattering of the substrate and the sample cell is subtracted from the data. The peak indexed (21) is the superlattice reflection from the (either herringbone or pinwheel) orientationally ordered ground state, the (20) and (40) peaks are due to the underlying $(\sqrt{3} \times \sqrt{3})R30°$ structure, and the residual scattering around the (002) graphite reflection stems from incomplete subtraction and/or interference effects between the overlayer and the graphite basal planes. Better-resolved complete monolayer diffraction patterns can be found in Figs. 2 and 3 of Ref. 370; see also Fig. 62. (Adapted from Fig. 1 of Ref. 99.)

experimental data. Concurrently the first finite temperature simulations of a simple but nevertheless sufficient Hamiltonian were undertaken [274]. However, the neutron diffraction data [99] could not be used to distinguish between the various proposed superstructures [21, 125, 141, 274], and the superlattice was indexed as (2×2).

The unequivocal assignment as a (2×1) orientationally ordered in-plane two-sublattice structure with the molecular centers on the commensurate $(\sqrt{3} \times \sqrt{3})R30°$ lattice sites (i.e., the $(2\sqrt{3} \times \sqrt{3})R30°$ superstructure) was given in a subsequent LEED study [92] on a graphite single-crystal surface. The experimental setup and procedure is described in great detail in Ref. 93, where also other parts of the phase diagram were explored; the coverage was near a complete $\sqrt{3}$ monolayer, remained constant between 20 K and at least 33 K, and may have changed slightly only at higher temperatures. Kinematic calculations [93] indicate that the out-of-plane tilt angles of the molecular axes have to be less than about 2° in order to be consistent with the observations. Other possible candidates [141] such as the two-sublattice out-of-plane herringbone arrangement (i.e., the "2-out" structure; see Fig. 4*c* and the four-sublattice pinwheel structure (i.e., the "4" phase; see Fig. 4*b*) could be ruled out based on the qualitative features of the LEED diffraction patterns; a detailed technical report on the expected Bragg diffraction pattern for pinwheel, in-plane, and out-of-plane herringbone structures including intensity calculations for LEED experiments is given in Ref. 107. The herringbone transition occurs near 27 K (see Fig. 26).

In spite of the strong drop of the LEED diffraction intensity in the range from 25 K to 30 K near the herringbone transition, it was nevertheless possible to observe substantial residual intensity above the transition [92, 93] (see Fig. 26). The intensities do not vanish above 30 K and persist until at least 40 K. This is different from the early neutron diffraction data [99] where the superlattice reflection vanished sharply above the transition. Also in the x-ray study [242] (see Fig. 26) the intensity drop at the transition was much sharper. This was attributed [108] to the fact that vibrations and librations perpendicular to the surface affect the LEED data much more than the x-ray data because of different measurement geometries. Thus, the x-ray data may better represent the true in-plane herringbone order parameter [108]. The LEED intensity, in addition, may not necessarily represent the square of the long-range order parameter [93], but it only gives a measure of the "amount of orientational order" in some loose sense [93]. The LEED tail was attributed [92, 93] to short-range angular order in the orientationally disordered phase, and/or to finite-size effects near the transition, and/or to substrate heterogeneity. The coherence length of the single-crystal graphite surface might extend to up to 10,000–1,000,000 Å, but the length over which the LEED apparatus can easily detect spatial correlations was esti-

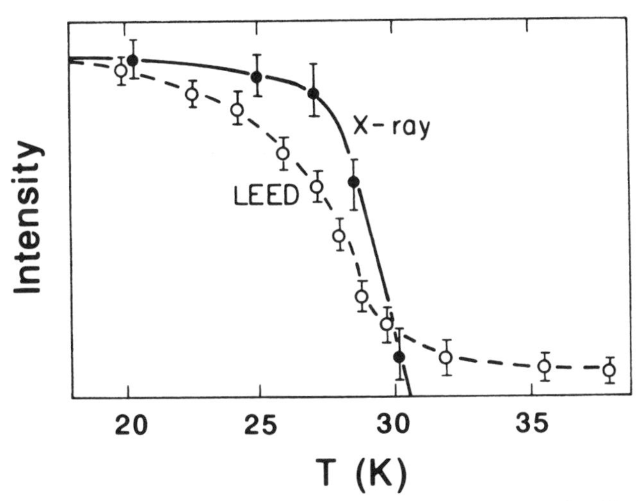

Figure 26. LEED [93] (open circles and dashed line) and x-ray [242] (filled circles and solid line) intensity as a function of temperature for commensurate N_2 on graphite. The intensity plotted on a linear scale is the herringbone superlattice peak intensity divided by the peak intensity of the $(\sqrt{3} \times \sqrt{3})R30°$ center-of-mass structure present in both the orientationally ordered and disordered phases; the data are arbitrarily normalized to the same intensity at 18 K. (Adapted from Fig. 6 of Ref. 108.)

mated to be in the range of roughly 80–200 Å, and may even extend up to the order of 1000 Å in favorable cases [92, 93].

The ground-state study based on realistic atomistic modeling [125] was considerably refined in Ref. 48. The nonelectrostatic part of the holding potential was constructed using the Fourier decomposition for Lennard-Jones potentials where the parameters were obtained from combination rules [324, 326]. The role of image forces arising from the permanent multipole moments of the physisorbed molecules was highlighted. The effects of static and dynamic screening of multiple fields are approximately included by surface charges on an effective graphite surface plane. Specifically the anisotropic McLachlan substrate-mediated dispersion interactions [49, 225] [see (3.5)] were included, which depend not only on the intermolecular separation but also upon the positions with respect to the surface. For the electrostatic contribution to the holding potential of one molecule, the hexadecapole moment of the isolated molecule is included in addition to its quadrupole moment. For the lateral interactions of the adsorbed molecules, the interactions of molecular quadrupoles and of the quadrupoles of their images with respect to the surface are included; in some calculations also the quadrupole–hexadecapole interactions and their image terms were added. Two parameterizations of the Lennard-Jones and quadrupolar intermolecular interactions were used according to Ref. 125 and to the X1 potential (3.1) [248]. The resulting laterally averaged molecule–graphite holding potential is very similar to the one obtained in Ref. 326 without any electrostatic contributions, which is caused by large cancelations. The molecular axis is found parallel to the surface near the minimum of the holding potential, and a tilt occurs for distances larger than 3.8 Å from the surface. The site of minimum energy is found to be at the center of the honeycomb cells of the graphite lattice when the lateral interactions are included. It is found that the quadrupolar orientational energy and the (nonelectrostatic) orientational energy due to molecular overlaps are comparable in this modeling. The ground-state superstructure is the in-plane herringbone two-sublattice structure in registry with the graphite lattice [48]. Limitations of this calculation are the poor knowledge of the molecular quadrupole moment, the crude estimation of the screening charges and the electrostatic response near the surface, and the multipole expansion of the electrostatics, which is also used at short distances [48].

An attempt to precisely localize the herringbone orientational transition in the submonolayer regime by heat capacity measurements based on ac calorimetry [232] yielded a temperature of 27.00 $\pm$ 0.03 K (see Ref. 65). This transition temperature stays essentially constant up to coverages slightly above the complete monolayer, whereas it rises for higher compressions [65]. In a more detailed study of the near-monolayer regime around the herringbone transition by the same group [394], a slightly different transition temperature of 28.1 K was inferred in the submonolayer range; this is due to another calibration of the temperature scale. Heat capacity traces [394] spanning the coverage range from the submonolayer regime up to incommensurate phases are depicted in Fig. 27; see, however, the remarks concerning the coverage scale [394] in Sections I.C and III.E. It was not possible to precisely locate the upper boundary of the coexistence region of fluid and commensurate solid. However, the fact that the herringbone orientational peak at 28.1 K remains unshifted up to 1.10 monolayers suggests that the pure commensurate solid is stable in a narrow coverage range of less than 0.1 monolayers [394]. The heat capacity anomaly is shifted to 29.4 K above 1.14 monolayers. The orientational transitions above monolayer completion are discussed in more detail in Section III.E. The latter slightly higher transition temperature [394] is also in better agreement with NMR spin-echo data [335] yielding an estimate of 28.1 $\pm$ 0.1 K for $^{15}N_2$ on exfoliated graphite (Papyex-N) at a coverage of 1.00 $\pm$ 0.02 monolayers.

The heat capacity peaks obtained by adiabatic calorimetry [156] using Grafoil GTA show maxima at slightly lower temperatures in the submonolayer regime at about 27.5 $\pm$ 0.1 K from 0.169 up to 1.000 monolayers. A value of 28.7 K was reported beyond the complete monolayer at a coverage

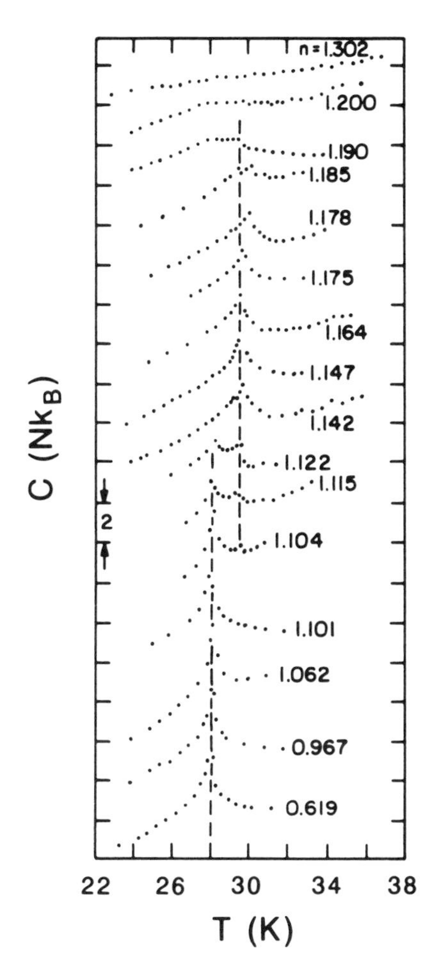

Figure 27. Reduced heat capacity of N_2 films adsorbed on graphite for various coverages n as given in the figure; the coverage is given in units of the complete $\sqrt{3}$-monolayer coverage. The vertical dashed lines at 28.1 K and 29.4 K mark the orientational transition temperatures in the commensurate and uniaxial incommensurate phases, respectively. (Adapted from Fig. 1 of Ref. 394.)

of 1.109; here a different kind of calorimeter vessel and Grafoil MAT was used. Based on extracted molar entropies and enthalpies of adsorption, it was explicitly demonstrated that the ground state of N_2 on graphite is a completely ordered structure [153, 156, 286]. The molar entropy change associated with the heat capacity anomalies near 28 K is determined to be approximately $R \ln 2$, which corresponds to the disordering of a two-sublattice structure [156]. The data also confirm that the adsorbed molecules form islands or clusters on the surface [156].

A molecular dynamics study based on an atomistic model with translations and rotations unconstrained in three dimensions was conducted together with heat capacity measurements [232]; a more complete account of the simulation part was given in Ref. 340, and essentially the same model was used in several other investigations [203, 342, 352]. This was the first attempt in the framework of finite-temperature simulations to use the microscopically detailed Fourier expansion technique of the molecule–surface interactions [324, 327]; see (3.6)–(3.8) with $\gamma_i \equiv 0$, which was previously explored in ground-state calculations, together with a site–site representation of the N_2–N_2 interactions. The simulations were done with 96 molecules at full monolayer coverage in terms of the $\sqrt{3}$ lattice. Around the heat capacity maximum at 34 K it is found that no sharp change of out-of-plane orientational and translational order in the $(\sqrt{3} \times \sqrt{3})R30°$ registered structure occurs, whereas the indicator for the in-plane herringbone-type order changes significantly as a function of temperature. The latter order parameter, however, is only a measure of long-range order as long as the ergodicity of the system is effectively broken—that is, as long as the system essentially resides in the initial kind of the six herringbone structures from Fig. 6.

These simulation results [232, 340] strongly suggest that the experimental heat capacity anomaly is essentially caused by losing the in-plane orientational long-range order in the herringbone structure. They also support the qualitative validity of the more simplified models where translations are frozen and rotations are strictly constrained to in-plane motion [56, 211, 212, 217, 244, 273, 274]. The root-mean-square amplitude of the perpendicular vibrations of the centers of mass of the physisorbed molecules amounts to less than approximately 0.5 Å even at 40 K, and no evidence for the promotion of molecules to the second layer is found [340]. However, the tilt angle relative to the surface plane and the center-of-mass height and thus the out-of-plane vibrations of the molecules are strongly coupled [340] such that molecules far away from the surface can more easily tilt perpendicular to the surface, which is actually observed; even at 20 K the molecules can undergo end-over-end flips.

The distribution function of the in-plane orientations computed in Ref. 340 has a sixfold-modulated contribution at all temperatures due to the crystal field caused by the neighboring molecules; the underlying graphite lattice imposes the same symmetry, but its effect is much smaller. The low-temperature behavior of this distribution in the herringbone phase is, of course, dominated by two peaks induced by the superstructure. The high-temperature phase shows only a small amount of residual herringbone ordering. This is attributed to the possibility that molecules stay perpendicular on the surface above the transition. These tilted molecules create effective vacancies in the two-dimensional structure which facilitates in-plane disordering [340]. However, the rapid decay of the herringbone order parameter above the transition is probably an artifact of its definition which is not symmetry-broken [232, 340].

It is suggested that the underestimation of the experimental calorimetric transition temperature [232] by 6 K should be attributed to a smaller effective quadrupole moment of N_2 in the monolayer than experimentally measured. However, a positive shift of the transition temperature by several Kelvin is easily induced by finite-size effects [217], whereas a negative shift of about 10% can be traced back to quantum effects on the librations [211, 212]. In addition, different representations in terms of potential models based on very similar data can again easily result in 10% effects on the transition temperature [165, 217].

The nature of the herringbone structure was furthermore elucidated by computer simulations [340]. The orientational order vanishes when the quadrupolar contribution to the intermolecular potential is switched off using otherwise identical conditions [340]. A patch of 151 molecules corresponding to 0.67 $\sqrt{3}$ monolayers was simulated in order to explore the effects of periodic boundary conditions and the submonolayer behavior [340]. The patch was found to be registered with a nearest-neighbor distance of 4.26 Å and herringbone-ordered below 33 K. Thus, the potential functions induce the correct registry of the $\sqrt{3}$ solid, and the periodic boundary conditions do not introduce major orientational correlations. In addition, the transition temperature does not change with coverage, which is also found in the submonolayer experiments. Finally, the lateral contributions to the molecule–surface potential were switched off, resulting in a smooth surface without lattice structure. The average nearest-neighbor distance is with 4.28 Å at 16 K only slightly larger than the registered solid [340]. Thus, the nearest-neighbor distance and commensurability or incommensurability of the superlattice are mostly determined by intermolecular N_2–N_2 interactions for N_2 monolayers on graphite. This means that incommensurate phases are not formed automatically by reducing the commensurating surface potential, but by increasing the surface pressure, which is effected in an experimental situation by putting on a partial second layer [340].

Stimulated by the different finite-size behavior from various experimental studies [65, 92, 93, 99, 232, 335], the influence of boundary conditions on the herringbone transition was investigated in more detail [104]. The model system of 1000 quadrupoles fixed on a triangular lattice with rotations confined to the lattice plane was investigated with Monte Carlo simulations. Three types of boundary conditions were applied. The infinite system was modeled using usual periodic boundary conditions applied to 50 × 20 molecules. The so-called weak boundary conditions create an infinitely long periodic strip with a width of 20 molecules and should model a N_2 patch on the graphite surface. The strong boundary conditions are obtained by fixing the molecules on the long side of the strip in the perfectly aligned herringbone structure and applying periodic boundary conditions only along the long axis. This should model edge effects and pinning of adsorbate molecules by surface defects. It is found that the weak boundary conditions reduce the transition temperature only slightly by less than 5%, whereas the strong boundary conditions wash out the peak in the heat capacity. However, the width of the strip is quite narrow, and the authors state explicitly that their results for the order parameters are still changing even for their longest runs of 22,000 Monte Carlo sweeps. A mean-field theory is presented which takes into account the different boundary conditions and which also reproduces qualitatively a smearing of the transition for the weak and strong boundary conditions.

An x-ray study [242] focused on the herringbone ordering at a coverage of 0.83 ± 10% monolayers. In agreement with earlier studies, this orientational ordering transition was found to occur in the range of 27–30 K, centered at about 28 K (see Figs. 26 and 28). It is a sharp transition and the ordering seems to be complete below about 15 K. The pinwheel structure could not be used to fit the observed diffraction patterns, whereas the in-plane herringbone structure did give a satisfactory description. However, this result was obtained with a relatively large value of the Debye–Waller factor based on a mean-square displacement of 0.1 Å^2, which is by a factor of 10 larger than the experimental value determined later by neutron diffraction for registered N_2 on graphite [370].

The out-of-plane orientation of the molecular axes as a function of temperature was determined by nuclear resonance photon scattering [241] at a

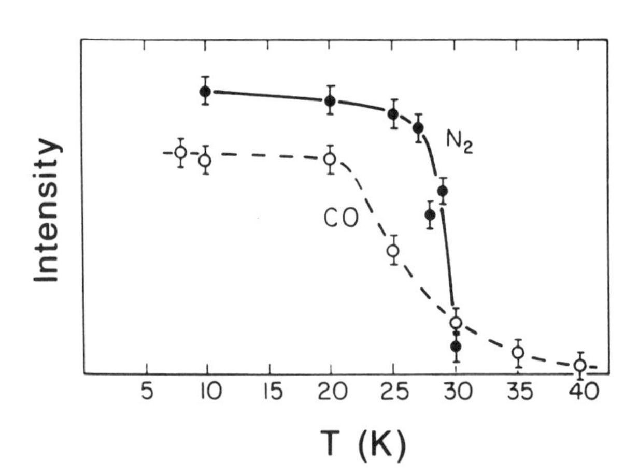

Figure 28. X-ray intensities as a function of temperature for commensurate N_2 (filled circles and solid line) and CO (open circles and dashed line) on graphite (Papyex) at a coverage of about 0.83 ± 10% monolayers. The intensity plotted on a linear scale is the herringbone superlattice peak intensity divided by the peak intensity of the $(\sqrt{3} \times \sqrt{3})R30°$ center-of-mass structure present in both the orientationally ordered and disordered phases. (Adapted from Fig. 4(a) of Ref. 242.)

coverage of 1.05 ± 0.02 $\sqrt{3}$ monolayers (see Fig. 15). It is found that the average out-of-plane tilt angle increases roughly linearly from about 10° at 20 K to 20° at 40 K, and it seems to reach a plateau value of 25–30° above 60 K. The obtained temperature dependence is in striking agreement with molecular dynamics simulations at coverage unity [341]. Note that nonzero tilt angles are a measure for the dynamical out-of-plane motion around an in-plane equilibrium structure and do not represent a static tilting of the molecular axis [161].

In the very detailed neutron scattering study [370] of the herringbone-ordered phase on Papyex at 1.0 monolayers and 10.6 K, it is found that the diffraction patterns could be well-fitted with an in-plane herringbone structure. However, the resolution for various angles determining the unit cell was limited to about 5–10°. In particular, in the fit model of the herringbone phase the out-of-plane tilt angle could be increased to 15° without discernible degradation of the fit, which would be much larger than the value expected from the experimental curve [241] shown in Fig. 15.

The dynamics of solid N_2 monolayers on graphite was analyzed by harmonic lattice dynamics and molecular dynamics methods [59] for a variety of potential models in the ground state as well as at 5 K and 17 K. The static structural properties of the two potential models used—that is, the N_2–N_2 interactions of Refs. 69 and 249 and the molecule–surface potential according to Ref. 326—are very similar for the herringbone commensurate phase even when the corrugation term is excluded from the calculations (see also Ref. 48). The corrugation does also have only a minor effect on the frequencies of the dispersion curves, whereas its shape changes considerably (compare parts *a* and *b* of Fig. 2 or 3 of Ref. 59). The two in-plane acoustic modes have zero frequency at the Γ point in the case of the flat surface, and this increases to only about 6 cm^{-1} when corrugation is included, which reflects the weakness of the restraining surface potential. By changing the amplitude of the corrugation, it is found that in-plane modes are virtually unchanged, whereas the out-of-plane frequencies are slightly higher in the flat case. It is confirmed that the out-of-plane motions are determined by the molecule–surface interactions, while the in-plane motions are dominated by the intermolecular potential. The two intermolecular models [69, 249] are also very similar concerning dynamical properties including their speeds of sound (see also the dispersion curves Figs. 2 and 3 of Ref. 59). The analysis of the molecular dynamics trajectories in terms of the dynamical structure factor (i.e., the Fourier transform of the van Hove correlation function) yielded well-resolved phonons for the out-of-plane motions, while the in-plane motions are strongly mixed. The phonon frequencies extracted from a one-phonon approximation to the finite temperature dynamic structure factor agree well with the ones obtained from the harmonic lattice dynamics calculations in the ground state.

The band structure of N_2 on graphite was also calculated and compared to spectra from angle-resolved ultraviolet photoelectron spectroscopy [310]. The qualitative features of the spectra are in agreement with those expected for a herringbone structure. In addition, it was made explicit that N_2 is physisorbed on graphite in contrast to being chemisorbed by showing that its electronic properties are not much changed in the adsorbed state relative to the gas phase.

The characteristics of clusters of herringbone-ordered domains (see Fig. 6 for the six possible domain structures) were studied in the vicinity of the herringbone transition by Monte Carlo methods with 128×128 (i.e., 16,384) molecules and up to 50,000 Monte Carlo steps [2]. The model is given by (2.4); that is, planar electrostatic point quadrupoles are pinned on a triangular lattice and interact with the full quadrupole–quadrupole term including the scalar contribution which is omitted in the anisotropic–planar-rotor model (2.5). Above the phase transition, clusters of only moderate size were found, whereas a large fraction of the area is covered by a single large cluster in the low-temperature phase. This pattern changes close to the transition point. By analysis of cluster histograms it is found that a certain surface area A is equally likely to be covered by a single cluster of size A, two clusters each of size $A/2$, four of size $A/4$, or any other suitable combination of clusters of different sizes. This was taken to be consistent with the association of the transition with slow critical-like fluctuations on large length scales, prominent domain walls, and continual transformations between different herringbone domains. The cluster boundaries become highly irregular, and the Haussdorff or fractal dimension decreases from the Euclidean value of two deep in the herringbone-order phase to about 1.6 near the transition. These features are along the lines of the characteristics expected for both fluctuation-driven weak first-order transitions [141, 217, 244, 246, 273, 308, 345] and second-order transitions [56].

The microscopic features of the herringbone transition were investigated in considerable detail with Monte Carlo simulations of 64 molecules in a rectangular simulation box with periodic but continuously deformable boundary conditions [183]. The very realistic potential model used for the simulations at full monolayer coverage is described in Ref. 181 (see also Section III.E), and averages were taken over 10,000 Monte Carlo steps. Different "transition temperatures" around 25 K for the fully three-dimensional model are reported based on different indicators of the herringbone ordering, such as heat capacity and orientational order parameters. This, however, is a manifestation of pronounced finite-size effects as various quantities approach their behavior in the limit of the infinitely large system with different size corrections for the same reduced distance to the transition point, which becomes especially obvious for quite small system sizes (see, e.g., Refs. 11, 28, 34, 38, 115, and 293).

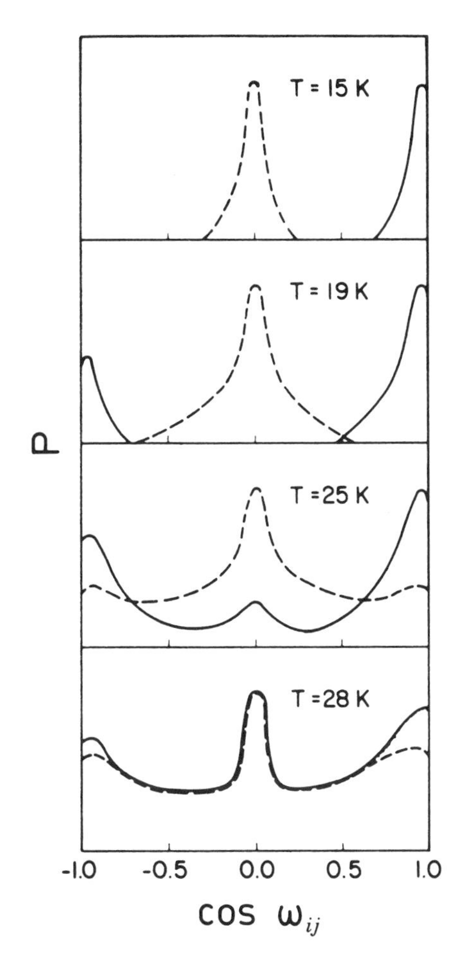

Figure 29. Distribution function $P(\omega_{ij})$ of the cosine of the relative angle ω_{ij} between two N_2 molecules i and j in the $(\sqrt{3} \times \sqrt{3})R30°$ commensurate phase on graphite obtained from Monte Carlo simulations for either first, second, and third nearest-neighbor pairs (ij). Solid line is for pairs (ij) which belong to the same sublattice, and dashed line is for the intersublattice pairs. The herringbone transition is located in this model at around 25 K. (Adapted from Fig. 4 of Ref. 183.)

An interesting microscopic behavior is uncovered [183] by the distribution function $P(\cos \omega_{ij})$ of Fig. 29, where ω_{ij} denotes the angle between the axes of two molecules i and j; here i and j are up to third nearest neighbors. Deep in the herringbone solid at 15 K the distribution for pairs (ij) belonging to the same sublattice (solid line) and the one for intersublattice pairs (dashed line) show peaks centered at zero and unity. This reflects pronounced herringbone order because it corresponds to parallel and perpendicular arrangements of the molecular axes, respectively. These peaks broaden around 19 K, and a new peak for intrasublattice pairs appears around -1, which shows that $\pm 180°$ reorientations (leading of course to identical configurations for the homonuclear diatomic molecule N_2) occur at that temperature far below the transition. The sublattice distinction starts to lose its meaning at around 25 K and disappears at around 28 K. Interestingly, the otherwise broad distribution shows at 28 K still a well-developed central peak centered at zero. This persistent feature reflects that there is considerable local quadrupolar order left even above the herringbone transition, which favors the typical perpendicular T-shaped intermolecular configurations.

The size dependence of the herringbone orientational transition was later studied as a function of the cluster size [185] with the same methodology as before in Ref. 183; the melting transition is discussed in Section III.B. The clusters were modeled as patches of up to 256 N_2 molecules on graphite with free surface boundary conditions using the same realistic potential model as previously [181] (see Section III.E), which also includes image charges and substrate-mediated dispersion interactions. The orientational disordering temperature increases strongly with increasing size of the patches and seems to reach for around 256 molecules the value obtained from simulations of 64 molecules experiencing periodic but continuously deformable boundary conditions [183] (see the squares in Fig. 21). However, no onset of a real saturation of the transition temperature as observed for the melting transition indicated by the circles can be identified. For all investigated island sizes the equilibrium orientations subject to out-of-plane fluctuations are in the substrate plane until melting occurs. The $\pm 180°$ flip rate is found to increase with temperature as expected.

Quantum effects on the herringbone ground state and on the orientational phase transition of N_2 on graphite were investigated [211, 212] using path-integral Monte Carlo methods specialized for quantum-mechanical rotations in two dimensions [209]. The path-integral Monte Carlo method (for reviews see, e.g., Refs. 22, 128, and 311) allows us to investigate large quantum-mechanical systems at finite temperatures without invoking quasiclassical, quasiharmonic, or mean-field-type approximations. The partition function of the quantum system in its path-integral representation is mapped onto an effective classical partition function with an additional dimension, the imaginary time axis or Trotter dimension [114]. The only additional approximation with its associated error in addition to those already present in classical simulations stems from the finite discretization along the imaginary time axis. This systematic error is well under control when systems of different discretization numbers are simulated [22, 128, 311]. In the model studied [211, 212] the molecules are restricted to rotate in the two dimensions of the surface plane, and their centers are fixed on a triangular lattice representing the adsorption sites of the commensurate $(\sqrt{3} \times \sqrt{3})R30°$ phase on graphite. The intermolecular interactions are represented by the X1 potential [248], and the periodic N_2–graphite interactions are modeled with the aid of the first-order Fourier expansion [324, 326, 327]. Thus, this model is a compromise between the very schematic anisotropic–planar–rotor Hamiltonian (2.5) and simulations treating all translational and rotational degrees of freedom. The intramolecular N_2 vibrations are frozen, the vibrational constant is as high as 3400 K for N_2, and the rotational constant used in the simulations is 2.9 K. The whole study is based on 900 quantum N_2 molecules and on discretizations of up to 500 at the lowest temperature, which corresponds computationally to simulations with 450,000 pseudoclassical molecules.

The orientational transition [211, 212] occurs in the classical model near 38 K as judged from the inflection point of the order parameter curve. Quantum fluctuations effectively soften the potential and decrease the transition temperature by 10%. In addition, the transition is broadened as can be judged from the heat capacity anomaly, which is found to decay to the required ground-state value zero at low temperatures. The ground-state saturation value of the order parameter is not unity as in the classical case, but

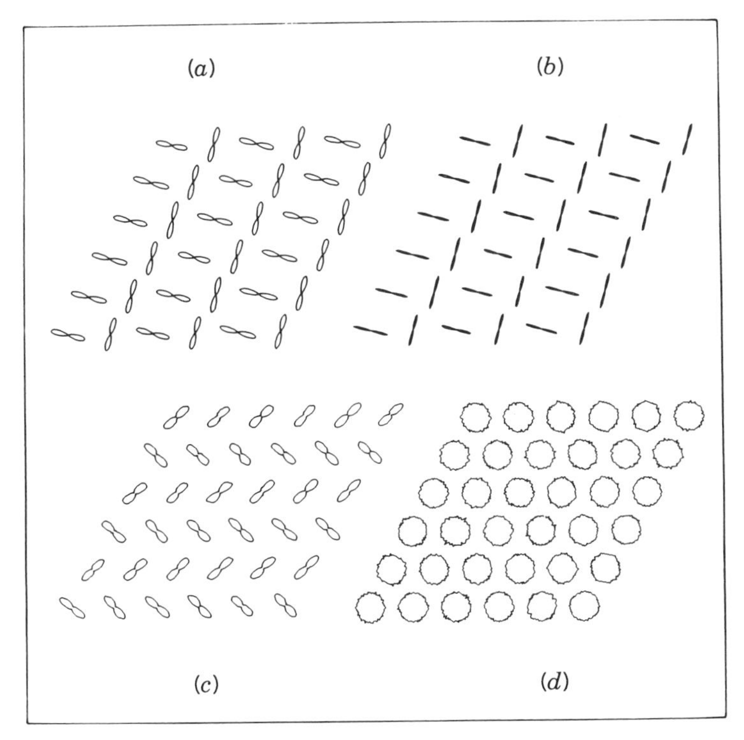

Figure 30. Angle distributions from path-integral Monte Carlo simulations of complete monolayer $(\sqrt{3} \times \sqrt{3})R30°$ N_2 on graphite. Plotted in direction of a given angle is the probability of finding this angle in the simulation; the maximum diameter is out of scale. (*a*) 5 K quantum simulation, (*b*) 5 K classical simulation, (*c*) 35 K quantum simulation, (*d*) 200 K classical simulation. (Adapted from Fig. 16 of Ref. 212.)

it is also depressed by about 10% because the operator of the order parameter and the Hamiltonian do not commute. Physically, this effect comes from the zero-point quantum librations of the N_2 molecules in their potential wells corresponding to the herringbone structure. In Fig. 30 the (quantum–statistical) angular distributions of a patch of 36 out of the 900 quantum molecules are shown. Plotted in the direction of a given angle is the probability of finding this angle in the simulation; the maximum diameter is out of scale and the same in all graphs. Classically the molecules are perfectly aligned along the herringbone axes at 5 K, and the resulting distribution (Fig. 30*b*) is extremely narrow. At the same temperature, quantum librations broaden this distribution considerably (see Fig. 30*a*) while preserving its general shape. The root-mean-square amplitude of these librations is calculated to be 14°; this value compares favorably to the 13° obtained from a quasiharmonic calculation using a very similar model [298]. At 35 K the distribution becomes broader (see Fig. 30*c*), and a considerable fraction of the molecules assumes an in-plane orientation perpendicular to the herringbone axes. Free rotation and correspondingly an isotropic distribution results far above the order–disorder phase transition in Fig. 30*d*; note that this high temperature was chosen only for technical convenience.

The range of validity of widely used quasiharmonic and quasiclassical approximations was checked against the benchmark data from the full quantum simulations [211, 212]. The quasiharmonic estimate of the zero-point energy of 30.4 K is in quantitative agreement with the value 30 ± 1 K from the quantum simulations with the full potential. This simple approximation describes the behavior quantitatively up to about 10–15 K. It is found that the second-order Feynman–Hibbs quasiclassical potential [114, 349], which is a quadratic expansion in the quantum fluctuation, describes the system quantitatively down to about 35 K. It starts to deviate systematically below approximately 30 K, breaks down completely below 20 K, and finally results in a disordered ground state. A systematic investigation of quantum effects on orientational ordering in two dimensions within the quantum generalization of the anisotropic–planar–rotor model (2.5) can be found in Refs. 213 and 218.

Several N_2–N_2 intermolecular potentials were examined with respect to their herringbone transition temperature under otherwise identical conditions based on canonical Monte Carlo simulations with 900 classical N_2 molecules [217]. The simulations were performed in strictly two dimensions and the centers of mass were fixed on a triangular lattice. No finite-size extrapolations whatsoever were performed so that the transition temperatures obtained from the heat capacity maxima are only of a qualitative nature. The N_2–graphite interaction was modeled by the first-order Fourier expansion technique [324, 326, 327] and is included only for the realistic atom–atom potentials, whereas this contribution is traditionally excluded in the case of the more schematic anisotropic–planar–rotor model (2.5). The latter model yields a transition temperature of about 25 K for a coupling constant of $K = 33$ K. Next, the X1 potential [248] yields a transition near 35 K followed by the model of Ref. 181 near 45 K. The empirical Lennard-Jones model of Ref. 340 and the *ab initio* potential in its site–site representation [23] both have the transition slightly above 50 K. The transitions occur at much lower temperatures closer to experiment if translations and rotations in full three dimensions are allowed for in the simulations. However, this series of data is an explicit demonstration that the reasonable transition temperature of the anisotropic–planar–rotor model obtained by using the bare electrostatic N_2 quadrupole moment is entirely fortuitous. It seems that (a) the neglected nonelectrostatic contributions to the overall anisotropy of the molecule and (b) the renormalization of the potential due to the neglected out-of-plane librations and center-of-mass oscillations at the adsorption sites effectively cancel each other. This highly idealized model is nevertheless an invaluable tool to investigate the order of the ideal herringbone transition both by simulations and analytical means (see Section III.D.2). In addition, the data show that commonly used empirical or *ab initio* potential models easily yield transition temperatures which deviate by as much as ± 15 K when employed under identical conditions. Thus, much of the reported close quantitative agreement between experiment and simulation seems to be due to the specific combination of the potential model used, finite-size effects from very small systems, and different simulation methods and boundary conditions.

The Brillouin zone center gap frequency in the phonon spectrum has been investigated for the commensurate herringbone N_2 monolayer physisorbed on the basal planes of graphite (Papyex) by inelastic neutron scattering at 4 K [137, 192]. On a smooth substrate with no corrugation, the acoustic phonons of a crystalline monolayer have zero frequency at the Γ point of the Brillouin zone. Such zero wavevector modes correspond to a uniform translation of the entire monolayer solid on the substrate. This is not so for a commensurate monolayer, where the lack of translational invariance due to the corrugation in the holding potential provides a finite restoring force to such a sliding. The resulting translational modes have a nonzero frequency at the zone center $\mathbf{q} = 0$ of the phonon dispersion curve, the zone center gap frequency. The magnitude of this frequency is related to the corrugation of the in-plane adsorption potential at the registry positions in the limit of small excitations. The neutron energy-loss spectra show modes near 0.4 THz (which corresponds to approximately 1.6 meV, 20 K, or 13 cm^{-1}) and 1.5 THz resulting from in-plane and out-of-plane bouncing motion, respectively. In an attempt to reproduce the neutron spectra, a model consisting of the X1 intermolecular N_2 potential [248] and the Fourier expansion of the atom–atom Lennard-Jones N_2–graphite interactions [326] was used in a harmonic ground-state lattice dynamics calculation of rigid N_2 molecules in the herringbone structure (see also Ref. 298 and Section III.E). The obtained spectra are qualitatively similar to the measured one, but the gap frequency is underestimated by a factor of two (see also Ref. 19 and Section III.E). An increase of this frequency was obtained by effectively increasing the corrugation using anisotropic nitrogen–carbon dispersion interactions [61, 362] with the model (3.3) of Ref. 165 (see Section III.B). The obtained value of 0.26 THz shows the right trend but still significantly understimates the experimental result.

The significant discrepancies of a factor of two between the measured [137, 192] and calculated [19, 137] zone-center frequency gap for commensurate herringbone monolayers was investigated in great detail with lattice dynamics calculations [138]. The N_2–N_2 interactions were modeled with the X1 potential [248] supplemented by substrate-mediated dispersion interactions [49, 225] and image charges [49, 298], and the Fourier expansion [324, 326] of the periodic potential was used. Several models for the holding potential were examined with respect to the produced phonon gap. They include in particular (a) the charge distortion model [254] where the aspherical atomic charge distribution of the carbon atoms in graphite is represented as local axially symmetric quadrupole moments at the carbon sites derived from organic molecules [361] or (b) using quadrupole parameters obtained for bulk graphite by x-ray diffraction [67]. This approach is similar in spirit to the modeling of anisotropic dispersion interactions [61, 362] (3.2) used earlier to obtain better agreement with experimental melting temperatures by increasing the substrate corrugation [165] (see (3.3) in Section III.B). This model [165], as well as the one which includes only substrate-mediated interactions as in Ref. 298, is also studied [138]. Only the commensurate herringbone structure with two molecules per unit cell is investigated. It is concluded that the interactions arising from aspherical charge distributions at the graphite atomic sites may suffice to remove the discrepancy with the experimental data [137, 192] concerning the theoretical underestimation of the zone-center gap. In particular the quadrupole moment derived from experiments on bulk graphite [67] leads to an underestimation of the gap frequency by only 25% without deteriorating the already obtained agreement [137] at higher frequencies. In addition, the corrugation is also increased in the sense required to bring the melting temperatures in agreement with experiment [165]. The parameterization of the quadrupole moment derived from three aromatic molecules with sp^2 bonding [361], however, results in a stable out-of-plane herringbone structure instead of the actual in-plane monolayer (see also the remarks in Ref. 284).

In a following study [140] the potential model introduced in Ref. 138

was used in an extensive investigation of structural and dynamical properties of commensurate N_2 monolayers on graphite including the herringbone and melting transitions. Constant-temperature molecular dynamics simulations with 224 molecules under periodic boundary conditions were carried out. Static properties were determined from runs of up to about 600 ps using a timestep of 0.0025 ps, and the dynamics was analyzed from an additional 200 ps. The resolved frequency domain ranges from approximately 0.05 THz up to about 4 THz, which corresponds to 20 ps and 0.125 ps total time and time interval in the evaluation of correlation functions, respectively.

The empirical X1 potential (3.1) [248] consisting of a Lennard-Jones term for the nonelectrostatic part and a three-point-charge model for the electrostatics is used [140] to describe the N_2–N_2 interactions. The substrate-mediated dispersion energy as formulated by McLachlan [49, 225] is included

$$E_{ij}^{\mathrm{McL}} = \frac{C_{s1}}{(r_{ij}r_{i'j})^3}\left[\frac{4}{3} - \frac{(r_{jz} + r_{iz} - z_{\mathrm{lay}})^2}{r_{i'j}^2} - \frac{(r_{jz} - r_{iz})^2}{r_{ij}^2}\right] - \frac{C_{s2}}{r_{i'j}^6} \tag{3.5}$$

with the particular choice of parameters $C_{s1} = 5.7898 \times 10^4$ KÅ^6 and $C_{s2} = 2.9468 \times 10^4$ KÅ^6; $z_{\mathrm{lay}} = 3.37$ Å is the distance between the basal planes of graphite, r_{iz} is the z coordinate of atom i, and i' is the image of atom i in a mirror plane parallel to the surface and shifted $z_{\mathrm{lay}}/2$ outward. The McLachlan energy contribution (3.5) is distributed over atom–atom pairs. The molecule–substrate interaction is based on the modified anisotropic Lennard-Jones potential [61] (3.3) with the original values [61] for the coefficients $\gamma_A = 0.4$ and $\gamma_R = -0.54$.

The separation of the N_2–graphite interaction into a laterally averaged term and corrugation amplitude

$$E_{\mathrm{surf}}^{\mathrm{disp}}(z, \mathbf{r}) = E_0^{\mathrm{disp}}(z) + \sum_g E_g^{\mathrm{disp}}(z) \exp[i\mathbf{g}\mathbf{r}] \tag{3.6}$$

is based on the Fourier decomposition [324, 326], where $\mathbf{r}$ is the lateral position vector of a nitrogen atom, z is its vertical distance from the graphite surface, and $\mathbf{g}$ are the two-dimensional reciprocal lattice vectors of the graphite basal planes. For the present case the laterally averaged contribution reads

$$E_0^{\mathrm{disp}}(z_\alpha) = \epsilon_{\mathrm{NC}}\frac{4\pi}{A}\left[\frac{2}{5}\frac{\sigma_{\mathrm{NC}}^{12}}{z_\alpha^{10}} - \frac{\sigma_{\mathrm{NC}}^6}{z_\alpha^4}\right] \tag{3.7}$$

where A is the area of the unit cell and z_α is the vertical distance from the atom to the αth basal plane in the graphite crystal. The Fourier coefficients are given by

$$E_g^{\mathrm{disp}}(z_\alpha) = \epsilon_{\mathrm{NC}} \frac{2\pi}{A} \sum_{g \neq 0} \left(\left\{ \sigma_{\mathrm{NC}}^{12} \times \left[\frac{1}{30} \left(\frac{g}{2z_\alpha} \right)^5 (1 + \gamma_R)\, K_5\, (g z_\alpha) - \frac{1}{150} \gamma_R\, z_\alpha^2 \left(\frac{g}{2z_\alpha} \right)^6 K_6(g z_\alpha) \right] - \sigma_{\mathrm{NC}}^{6} \left[2 \left(\frac{g}{2z_\alpha} \right)^2 (1 + \gamma_A) K_2(g z_\alpha) + \gamma_A z_\alpha^2 \left(\frac{g}{2z_\alpha} \right)^3 K_3(g z_\alpha) \right] \right\} \times \sum_n \exp\left[i\mathbf{g}(\mathbf{r} - \boldsymbol{\tau}_n) \right] \right) \tag{3.8}$$

where K_i is the modified Bessel function of order i, and $\boldsymbol{\tau}_n$ is the two-dimensional position vector of the carbon atoms in the unit cell of the basal planes of graphite; 15 layers α of basal planes were included in the evaluation of (3.7), and the corrugation is calculated from the topmost plane and for the first two shells of reciprocal lattice vectors.

In addition to these more traditional terms the corrugation increase due to an aspherical charge distribution around the sp^2 bonded carbon atoms in graphite is included in form of local quadrupole moments on the carbon sites [138, 254, 361]; see the discussion above in this section. The additional electrostatic energy of a point charge q on a nitrogen molecule and a quadrupole moment at a graphite carbon atom position is

$$E_{q\Theta} = \frac{1}{2}\, q\Theta_C \left[\frac{3r_z^2}{r^5} - \frac{1}{r^3} \right] \tag{3.9}$$

where $\mathbf{r}$ is the distance vector between the charge and the quadrupole, and $\Theta_C = +3.336 \times 10^{-40}\ \mathrm{Cm}^2$ obtained from x-ray data for the valence charge distribution for bulk graphite [67] was used. The Fourier representation of this contribution is given by

$$E_{\mathrm{surf}}^{\mathrm{el}}(z, \mathbf{r}) = \frac{\pi}{A}\, q\Theta_C \sum_n \sum_g g \exp[-gz] \exp\left[i\mathbf{g}(\mathbf{r} - \boldsymbol{\tau}_n) \right] \tag{3.10}$$

in addition to the term (3.8).

Based on this modeling, molecular dynamics simulations [140] for a complete $\sqrt{3}$ monolayer are carried out from 10 K to 80 K—that is, ranging from the harmonic herringbone solid through the orientationally disordered solid to the fluid. The herringbone transition occurs at around 22 K, below which enhanced $\pm 180°$ head-to-tail flips of the homonuclear molecules occur. The sixfold symmetry in the orientations persists up to about 50 K, above which nearly free rotation is found. The distribution of the out-of-plane tilt angle changes qualitatively on passing through the orientational disordering transition. This distribution can be described as a single Gaussian centered around the in-plane angle at low temperatures, but it develops in addition to this major contribution shoulders corresponding to perpendicular molecules when the orientational transition is approached. These belong to transient pinwheel defects in the solid. The orientationally disordered triangular solid melts around 73 K, and promotion of molecules into the second layer becomes only significant a few Kelvin below the melting point. Then with further increase in temperature, molecules are promoted to the second layer, vacancies are created in the monolayer, and the solid finally melts. Thus, the solid of nearly spherical N_2 molecules with their small aspect ratio melts via a defect mechanism by promoting molecules into the second layer which leaves enough free surface area for the other molecules to translationally disorder; see also the discussion of this mechanism [102, 103, 301] in Section III.B. This should be contrasted to the footprint reduction mechanism as found for more elongated molecules [139] where sufficient free surface area can be created by pure out-of-plane tilting of molecules. Such a mechanism is not effective for a nearly spherical molecule like N_2 (see the contour plot in Fig. 1). The aspect ratio being close to unity is also responsible for the orientational disordering prior to translational melting. The average perpendicular distance of the molecules from the surface increases smoothly in the range from 10 K to 70 K, and then increases drastically, caused by molecules in the second layer and desorption processes.

The long simulation times [140] up to 600 ps allowed the authors to estimate two-dimensional diffusion coefficients from the mean-square displacements. They amount to 1.8 and 2.8×10^{-5} cm^2/s at 75 K and 80 K, respectively. For the higher temperature, preliminary estimates of about 2.4 $\times 10^{-5}$ cm^2/s and 1.1×10^{-4} cm^2/s for the first- and second-layer diffusion constants are obtained. At low temperatures the layer is a nearly harmonic solid with a strong mixing of azimuthal motions and lateral translations and a weaker coupling with the polar motion and perpendicular vibrations. Just above the orientational disordering temperature at around 25 K the molecules perform librational motion between diffusive steps as expected in a hindered rotor regime. This picture changes at about 50 K where free diffusive rotational motion occurs. The temperature dependence of the zone-center frequency gap of the acoustic phonons is obtained from the dynamical structure factor (see Fig. 31). First of all there is no sign of a floating solid with a zero frequency gap, but there is a strong Bragg reflection. The decrease of the gap frequency with increasing temperature results from various factors, such as the anharmonicity of the corrugation and the reduction of the corrugation from a thermal expansion of the overlayer height which both lead

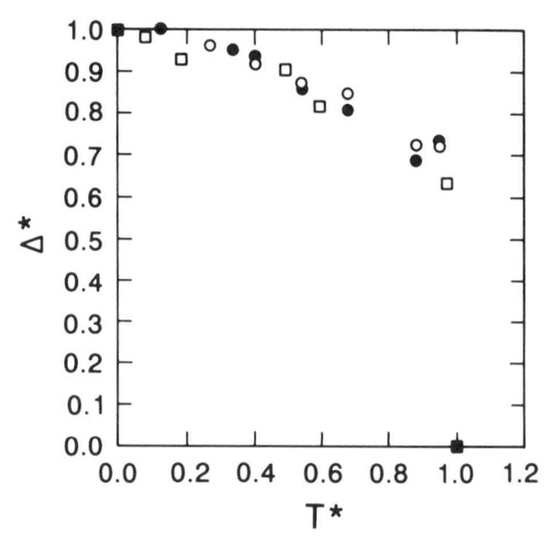

Figure 31. Relative Brillouin-zone-center gap frequencies for translational in-plane modes for N_2 on graphite as a function of temperature; the gap frequencies $\Delta^* = \Delta(T)/\Delta_0$ are normalized with the corresponding ground-state gap frequency Δ_0, and the temperature $T^* = T/T_m$ is normalized by the melting temperature T_m of the $\sqrt{3}$ solid. Circles (unfilled and filled circles refer to the x and y directions of the in-plane translations, respectively): constant temperature molecular dynamics simulations, $\Delta_0 = 0.30$ THz, and $T_m = 73$ K. Squares: inelastic neutron scattering data [192], $\Delta_0 = 0.40$ THz, and $T_m = 72$ K. (Adapted from Fig. 15 of Ref. 140.)

to a decreasing sharpness of the commensurate lattice. The effects of multiphonon processes are found to be quite small, though they increase with temperature, and the gap frequencies determined from two reciprocal lattice vectors were indistinguishable. Finally, the agreement of the temperature dependence of the relative Brillouin-zone-center gap frequency is well reproduced by the elaborated potential model of Ref. 140 as judged by comparing to the experimental curve [137, 192] (see Fig. 31); note that the results are normalized with the lattice dynamics ground-state gap frequency of 0.30 THz, which is 25% smaller than the corresponding experimental value of 0.40 THz or 19 K. In spite of this trend being well-reproduced, the authors [140] conclude that the specification of the adsorption-induced interactions for the nitrogen on graphite system (3.5)–(3.10) still seems to be seriously incomplete.

Some of the results obtained for the commensurate herringbone structure are discussed in Section III.E because they served many authors as starting and test cases for calculations of compressed monolayers and bilayers. In particular, we refer to the study [19] of the lattice dynamics of the herringbone structure summarized in Table III.

2. *Order of the Phase Transition*

The assignment of the order of the herringbone transition was discussed in the literature over many years. The LEED results [92, 93] (see Fig. 26) were consistent with an interpretation in terms of a first-order transition with pronounced rounding effects. However, these data could not rule out a continuous transition so that this study was not accurate enough to decide the order of the transition. In addition, there is the problem to relate the LEED superlattice spot intensities to the proper long-range order parameter of the system [93, 108] (see the presentation in Section III.D.1).

A heat capacity study [65, 232] of the herringbone transition yielded also evidence in favor of a discontinuous scenario. The heat capacity maxima along the temperature scan could not be fitted in terms of an algebraic singularity; that is, a critical point analysis seemed inappropriate to describe the available data. However, the quality of the data was still insufficient to allow for a more sophisticated analysis. It was concluded that finite-size effects were unimportant because the transition temperature and the widths of the peaks were insensitive to changes of the coverage [65, 232]. In addition, no sign of hysteresis occurring at first-order transitions was found in the heat capacity traces [65].

In the NMR study [335] a hysteresis of about 2.5% and a small discontinuity at 28.1 K in the temperature dependence of the order parameter was interpreted in terms of an evidence for a first-order herringbone transition. However, this assignment is based on NMR echo amplitudes which define only local order parameters. For the authors of the x-ray study [242] it seemed also probable that the herringbone orientational transition is a first-order transition.

In summary, the experimental results did not give a compelling answer to the question concerning the order of the herringbone transition, but they are rather only consistent with a first-order interpretation (see also the detailed discussion in Ref. 246).

The pioneering mean-field theoretical study [21, 141] (see Section II.A for a detailed discussion) predicted a second-order transition from the orientationally disordered to the herringbone phase. But it is of course well known that the underlying mean-field factorization of all higher correlation functions in terms of one-particle densities is an extremely severe simplification for models in low spatial dimensions. At the first place the resulting mean-field transition temperature is for the present case, as expected, a factor of two too large compared to the exact result within the model. In addition, it is possible that the artificially decoupled long-range fluctuations are actually relevant and turn a continuous scenario into a fluctuation-driven first-order transition according to Landau's classification. This scenario was already favored in Ref. 141 based on a comparison to a more general model. This assignment could be reinforced [308] after more renormalization group results concerning the underlying Heisenberg model with face-cubic anisotropy were obtained (see Section II.C). However, it is known that the underlying approximate real-space renormalization group approaches might give wrong results. The two-dimensional q-state Potts model is one example where simple real-space methods first gave the wrong order of the transition for $q > 4$.

The aim of the first exploratory Monte Carlo investigation [274] was to introduce a more simplified generic model for further studies of herringbone ordering (see Section II.B and Section III.D.1). The systems with 6×6 and 12×12 molecules were clearly too small and the statistical averaging insufficient to establish the order of the transition.

The first Monte Carlo simulation [244] (see also the broad discussion in Section 5.3 of Ref. 246) focusing on the order of the herringbone transition was based on the strictly two-dimensional anisotropic–planar–rotor model (2.5), which is similar to the planar model used in Ref. 274 except that the quadrupole interactions are treated only approximately (see Section II.B for further details). The instrument of diagnostics was the three-component herringbone order parameter [244, 246] defined as

$$\Phi_\alpha = \frac{1}{N} \sum_{i=1}^{N} \sin[2\varphi(\mathbf{R}_i) - 2\eta_\alpha] \exp[i\mathbf{Q}_\alpha \mathbf{R}_i] \tag{3.11}$$

with $\alpha = 1, 2, 3$ and

$$\begin{aligned} \mathbf{Q}_1 &= 2\pi(0, 2/\sqrt{3})/a, & \eta_1 &= 0 \\ \mathbf{Q}_2 &= 2\pi(-1, -1/\sqrt{3})/a, & \eta_2 &= 2\pi/3 \\ \mathbf{Q}_3 &= 2\pi(1, -1/\sqrt{3})/a, & \eta_3 &= 4\pi/3 \end{aligned} \tag{3.12}$$

Here, the angles $\varphi(\mathbf{R}_i)$ denote the orientations of the N_2 molecules on the positions $\mathbf{R}_i$ of the triangular lattice with lattice constant a as in the definition of the Hamiltonian (2.5) itself. The corresponding reciprocal lattice vectors $\mathbf{Q}_\alpha$ and phases η_α distinguish the three components $\alpha = 1, 2, 3$ of the order parameter.

The main outcome of the study [244] was that the herringbone transition within this model is of first order and occurs at $T_c = 25.575$ K, or at $T_c^* = 0.775$ when reduced by the appropriate numerical value $K = 33$ K of the coupling parameter in (2.5) based on the gas-phase quadrupole moment for N_2. Though the system sizes $L \times L$ were already quite large (with linear dimensions of $L = 20$, 40, 80, and 100), the statistical effort (500–5000 sweeps over the lattice) does not provide the basis for a firm conclusion distinguishing a weak first-order from a continuous transition. The assignment is mainly based on roughly 1.5% hysteresis loops of the order parameter [where the largest of the three components (3.11) was chosen to define *the* order parameter] and internal energy, but second-order transitions may also show such a behavior as a manifestation of finite-size effects or finite-averaging time effects. Hardly any finite-size effects were observed, and specifically the heat capacity as presented in Fig. 5.3.5 (for coverage unity denoted by $x = 1$) of Ref. 246 does not show a systematic size dependence of both height and position of the peak for $L > 20$. This is difficult to reconcile with the theoretical expectation that the maximum of the heat capacity for sufficiently large systems should increase for increasing system size as L^d or $L^{\alpha/\nu}$ for first- or second-order transitions, respectively; $d = 2$ is the space dimensionality and α and ν are the critical exponents of heat capacity and correlation length, respectively. The size-independent behavior found could only be rationalized in the marginal case of a logarithmic critical divergence with $\alpha = 0$, or if a Kosterlitz–Thouless-like phase intervenes similarly to an XY model in a symmetry breaking external field [164, 252].

Generalizing Kikuchi's cluster-variational method for continuous variables, an improved mean-field theory of the anisotropic–planar–rotor Hamiltonian (2.5) was published [62, 345]. The cluster-variational method [169] is a systematic expansion of the free energy in terms of clusters composed of n sites—that is, "spins." This approach goes beyond the simple mean-field theory concerning nonuniversal quantities such as transition temperatures, whereas the critical exponents remain, however, at the level of the classical Landau exponents. Fluctuations on the length scale of the largest cluster are taken into account exactly, instead of factorizing them as in the mean-field approximation, which is obtained for $n = 1$. The problem in the present case is that the series expansion in n-site clusters is not monotonically convergent due to frustration effects on the triangular lattice. As a consequence, the $n = 2$ Bethe approximation as the first step beyond the mean-field theory completely fails because it does not produce any phase transition at all (see the lucid discussion in Ref. 62). Only when the basic plaquette in the form of the nearest-neighbor triangle is taken as the smallest cluster (i.e. for $n = 3$) a second-order transition occurs with much better herringbone transition temperatures of 32.835 K ($T_c^* = 0.995$), 32.934 K (0.998), and 35.046 K (1.062) for the potential used in the mean-field study [141], the anisotropic–planar–rotor model (2.5), and the full quadrupole potential (2.4), respectively. However, in view of the quite large correlation length of the herringbone transition revealed later (see below), the fluctuations cannot be termed short-range, so that the good agreement between this analytical treatment and the simulations is somewhat fortuitous. In principle, the high accuracy of the cluster-variational method for studying first-order transitions is guaranteed only if the cluster size used is of the order of the correlation length, which would be impossible for the present problem. A systematic study of the further convergence of the series expansion in n-site clusters is extremely involved. In addition, it may well be that higher terms again worsen the low-order triangular result.

The simultaneously published finite-lattice real-space renormalization group calculation [345] is ideally suited to take into account the crucial long-range fluctuation effects. This approach goes beyond the symmetry arguments presented in Refs. 141 and 308 in that a more realistic two-dimensional microscopic model is explicitly investigated by renormalization group methods. The resulting fixed-point Hamiltonian corresponds to a discontinuity fixed point associated to a first-order transition. However, the studied Hamiltonian does not have the full continuous rotational degrees of freedom, but it is severely discretized in the sense that only ground-state properties are kept in the form of an anisotropic six-state model with full quadrupolar interactions. In addition, for technical reasons the Hamiltonian had to allow for vacancies and thus for the chemical potential as an additional variable. The transition temperature depends, as usual, on the rule used for the real-space block transformations, and a simple majority rule yields a transition temperature of 20.13 K ($T_c^* = 0.61$), whereas other rules gave even lower temperatures. The various transition temperatures obtained from these analytical treatments are collected in Fig. 32.

An important theoretical result [345] is that the obtained discontinuity fixed-point Hamiltonian is a generalization of and qualitatively different from the ones of the three-axes cubic or six-state Potts models in two dimensions. This difference could be traced back to preserving the coupling between the molecular orientations and the lattice bonds orientation of the original quadrupolar interactions. The renormalization group map used in Ref. 345 has two different discontinuity fixed points: One is the well-known six-state Potts and the other represents the quadrupolar Hamiltonian with its herringbone phase. Thus, the lattice bond asymmetry included in the quadrupolar Hamiltonian is not irrelevant. The symmetry arguments [141, 308] presented in Section II.C rely, however, on the mapping of the herringbone problem onto the three-axes cubic model.

An unorthodox approach viewing the herringbone transition as an orientational freezing transition in the framework of the density functional theory of freezing within the anisotropic–planar–rotor model (2.5) was explored in Ref. 316; for references related to this technique see the cited literature in Ref. 316. The promising feature of that technique is that long-range fluctuations are taken into account by summing over contributions from whole classes of infinitely large clusters; the simple mean-field theory can be recovered within a well-defined approximation. This summation,

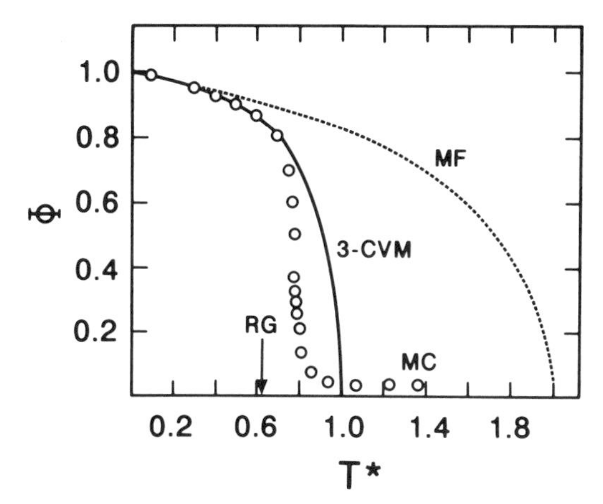

Figure 32. Herringbone order parameter for the anisotropic–planar–rotor model (2.5) as a function of the reduced temperature $T^* = T/K$. Circles: Monte Carlo results [244]. Dotted line: mean-field approximation [62, 141]. Solid line: triangular cluster-variational method [62]. Arrow: first-order transition temperature obtained from a real-space renormalization group treatment of a planar quadrupolar six-state model [345]. (Adapted from Fig. 2 of Ref. 345.)

however, can only be done approximately in an uncontrolled way. Despite the fact that much better results for the structure of the orientationally disordered phase could be obtained, the transition temperature of 63.36 K ($T_c^* = 1.92$) is only marginally shifted from $T_c^* = 2$ of the simplest mean-field theory. In addition, the herringbone transition is found to be continuous as in mean-field theory. This failure of the density functional approach might be caused by a too severe truncation of a certain sum over lattice vectors, by a failure of the standard second-order approximation of the density functional, and most probably by unsatisfactory closures for the integral equations [316]. Another aspect is that it was explicitly shown for the anisotropic–planar–rotor model that pronounced short-range angular correlations up to at least next-nearest neighbors are present well above the transition also in the infinitely large system [316]. The correlation functions Figs. 1*b*–*d* and 2*a* of Ref. 316 would correspond to roughly 45 K when rescaled with the experimental transition temperature. Because there are no substrate heterogeneity and finite-size effects present in the (approximate) theoretical study, it is suggested that the significant LEED tail [92, 93] above the transition (see Fig. 26) is at least to some extent caused by short-range angular correlations. In addition to this, there seems also to be a finite-size contribution to this phenomenon. The true long-range herringbone order parameter as shown in Fig. 9 of a Monte Carlo study [217] does also decay slowly in a system of linear dimension $L = 30 \approx 130$ Å, which is about the resolution length scale of the LEED apparatus [93]. The rapid decay with a quasivertical part is only seen for much larger systems [217].

An extensive Monte Carlo study of the herringbone transition within the approximations underlying the anisotropic–planar–rotor model (2.5) is reported in Ref. 56. This investigation differs in two respects from previous work. First, finite-size scaling is systematically used to uncover the nature of the phase transition. Second, the statistics is orders of magnitude better than previously (1–5 $\times$ 10^6 Monte Carlo sweeps), the systems are larger (with linear dimensions ranging from $L = 12$ up to $L = 120$), and, finally, advanced methods to evaluate statistical averages (the histogram reweighting technique; for a review see Ref. 337) were exploited. Binder's fourth-order cumulant [31, 32] for the energy [64]

$$V_L = 1 - \frac{\langle E^4 \rangle_L}{3\langle E^2 \rangle_L^2} \tag{3.13}$$

was used as a detector for the order of the phase transition; $\langle \bullet \rangle_L$ denotes canonical averages for a system of $N = L \times L$ molecules. This quantity approaches the trivial value 2/3 far away from any transition in both the low- and high-temperature phases. At any phase transition in a finite system it develops a minimum owing to the non-Gaussian character of the energy distribution. However, only in the case of second-order transitions, this minimum vanishes in the limit $L \to \infty$. At a first-order transition, on the other hand, it converges as L^{-d} to a nontrivial minimum value

$$\min V_L(T) = \frac{2}{3} - \frac{1}{3}\frac{(E_+^2 - E_-^2)^2}{(2E_+E_-)^2} + \mathcal{O}(k_B T_c^2 L^{-d}) \tag{3.14}$$

depending in leading order on the energies of the disordered E_+ and ordered E_- phases at a transition temperature of T_c in the infinite system. The finding that the minimum of the energy cumulant approached 2/3 for large systems was interpreted as strongly suggesting a continuous transition in Ref. 56. Other quantities such as the heat capacity and the orientational ordering susceptibility were also found to be strongly dependent on system size. First of all, a critical temperature of 25.08 K ($T_c^* = 0.76$) was determined from the systematic size dependence of both the peak positions of the heat capacity and the susceptibility. As a second step the critical exponents and the universality class were determined from the finite-size behavior of these quantities. The critical exponents α and ν (obtained by invoking the so-called hyperscaling relation $\nu d = 2 - \alpha$ between critical exponents and the space dimension d) are extracted from the maximum of the heat capacity and γ

(together with ν taken from the heat capacity data together with hyperscaling) from the order parameter susceptibility. The four critical exponents (α = 0.335 $\pm$ 0.002 (0.333), β = 0.03 $\pm$ 0.05 (0.111), γ = 1.6 $\pm$ 0.2 (1.444), and ν = 0.832 $\pm$ 0.002 (0.833), where the three-state Potts values are quoted in parentheses), are compared to the ones of the nearest-neighbor three-state Potts model in two dimensions [385]. It is finally concluded that the herringbone transition of the anisotropic–planar–rotor Hamiltonian (2.5) is continuous and belongs to this universality class.

The orientational correlation length as a function of temperature was determined in another Monte Carlo study [217, 273] using again the standard anisotropic–planar–rotor Hamiltonian (2.5). The decay of a correlation function, and more explicitly the behavior of the correlation length as the transition is approached, is one of the most direct distinguishing features between different transition types. Of course this is a very demanding task since one has to compute at each temperature a whole function instead of only simple moments of observables. Correlation functions along the three symmetry axes of the triangular lattice were defined as

$$\Gamma_\alpha(l) = \left\langle \frac{1}{N} \sum_{i=1}^{N} \cos[2\varphi(\mathbf{R}_i) + 2\varphi(\mathbf{R}_i + l\mathbf{a}_\alpha)] \right\rangle \tag{3.15}$$

where $\varphi(\mathbf{R}_i)$ is the molecular angle at lattice position $\mathbf{R}_i$, $\{\mathbf{a}_\alpha\}$ denotes lattice vectors ($|\mathbf{a}_\alpha| = a$) along these axes α, and l runs over the neighbors in half the two-dimensional Wigner–Seitz cell along these directions. It is well known that the leading order decay of such correlation functions is exponential

$$\Gamma_\alpha(l) \propto \exp[-l/\xi] \tag{3.16}$$

for large separations l, but because (3.16) is only asymptotically correct, there may be strong systematic corrections to this expression for short distances. In addition, one clearly has strong systematic corrections to (3.16) also for large l: The periodic boundary conditions artificially fold back the correlations because $\Gamma_\alpha(l) \equiv \Gamma_\alpha(L - l)$ has to be satisfied by construction. Both effects can be taken approximately into account either by restricting the range where a simple exponential (3.16) is fitted to the data or by fitting the correlation function to a symmetrized form

$$\Gamma_\alpha(l) \propto l^{-\lambda} \exp[-l/\xi] + (L - l)^{-\lambda} \exp[-(L - l)/\xi] \tag{3.17}$$

trying to incorporate correction terms. However, in both cases (3.16) and (3.17) various artifacts can easily be introduced yielding biased correlation lengths.

Such effects can at least be minimized by using a procedure [273, 314] defining ξ_l via the discrete logarithmic derivative of (3.16)

$$\delta_m \ln \Gamma_\alpha(l) := \ln \left[\frac{\Gamma_\alpha(l) - \Gamma_\alpha(\infty)}{\Gamma_\alpha(l + m) - \Gamma_\alpha(\infty)} \right] = \frac{m}{\xi_l} \tag{3.18}$$

where $\Gamma_\alpha(\infty) := \Gamma_\alpha(L/2 >> l >> \xi)$ denotes the constant asymptotic value of (3.15) which vanishes, of course, in the disordered phase. We do not have to distinguish the three correlation lengths measured along the different symmetry axes because they are identical by symmetry, and $m = 2$ takes care of the fact that $\Gamma_\alpha(l)$ oscillates with period two because of the antiferromagnetic-like ordering of the herringbone phase. The advantage of this approach is that no (possibly uncontrolled) fitting is involved and especially that the range where ξ_l approximates the true ξ can be assessed by inspection: Only if a plot of (3.18) versus l yields a plateau for a certain window of intermediate distances $l \in [l_{\min}, l_{\max}]$ may an effective correlation length ξ be extracted safely from such a plateau value $2/\xi_l$. Thus, only in the case of the observation of a plateau in (3.18) the linear dimension of the system and the quality of the data (sufficient statistics and equilibration on the relevant length scales) are sufficient for a meaningful computation of ξ.

Using system sizes ranging up to linear dimension of $L = 180$, and a statistical effort of up to 1,500,000 Monte Carlo sweeps over the whole lattice, the herringbone orientational correlation functions were obtained in the relevant temperature range (see Fig. 33 and also Figs. 12 and 13 of Ref. 217 for a few examples). The reduced temperature range covered was about one decade both above and below the transition, and the latter was approached up to 2% and 0.3% from above and below, respectively. The extracted correlation lengths (see Fig. 34) were analyzed in terms of two different scenarios concerning the order of the transition. As the transition temperature is reached ($T \rightarrow T_c^{\pm}$), a first-order transition would manifest itself in a behavior of the type

$$\xi = \xi_{\pm} + c_{\pm} (1 - T/T_c) + d_{\pm}(1 - T/T_c)^2 + \cdots \tag{3.19}$$

because at a first-order transition ξ is a regular function of temperature, and thus a Taylor expansion at T_c should exist. Here ξ_+ (ξ_-) denotes the finite correlation length at the transition T_c upon approaching it from above (below); $c_{\pm}$ and $d_{\pm}$ are unknown constants. A second-order transition would correspond to a critical divergence

$$\xi = \xi_0^{\pm} |1 - T/T_c|^{-\nu} \tag{3.20}$$

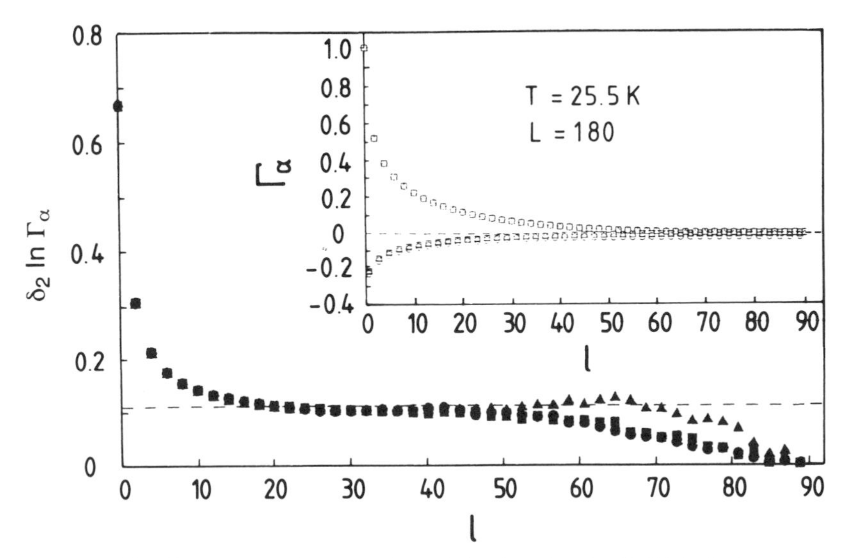

Figure 33. Herringbone orientational correlation functions Γ_α (3.15) in the inset and the logarithmic derivatives $\delta_2 \ln \Gamma_\alpha$ (3.18) as a function of distance l in units of the lattice constant $a = 4.26$ Å in the disordered phase of the anisotropic–planar–rotor model (2.5) from Monte Carlo simulations at $T = 25.5$ K and a linear system size of $L = 180$. The different symbols distinguish the three symmetry axes $\mathbf{a}_\alpha$ and the dashed line marks the plateau $2/\xi$. In the inset all three Γ_α fall on top of each other and the different symbols denote here the two oscillating parts of the antiferromagnetic-like ordering pattern. (Adapted from Fig. 1 of Ref. 273.)

of the correlation length as the transition is approached ($T \to T_c$). The critical exponent [385] $\nu = 5/6$ and the universal critical amplitude ratio [294] $\xi_0^+/\xi_0^- = 4.1$ are known for the nearest-neighbor three-state Potts class [385] in two dimensions, which was proposed in Ref. 56 as the universality class of the herringbone transition.

From the plot of the correlation length as a function of temperature in Fig. 34, it was judged [273, 217] that the first-order behavior [141, 244, 308] describes the available data better than the continuous three-state Potts scenario proposed in Ref. 56; the correlation length according to the Potts fit overestimates the calculated one by roughly a factor of two close to the transition. The simple linear fit—that is, (3.19) with $d_\pm = 0$—yields finite correlation length estimates of $\xi_+ \approx 23$ (roughly 100 Å) and $\xi_- \approx 12$ (50 Å) at the transition upon approach from above and below, respectively. A much better fit is, of course, obtained if the quadratic term $(1 - T/T_c)^2$ in the deviation from the transition temperature is included in (3.19). This leads to correlation length estimates of $\xi_+ \approx 37$ (150 Å) and $\xi_- \approx 11$ (50 Å).

In view of the conflicting studies [56, 244, 273], a more detailed finite-

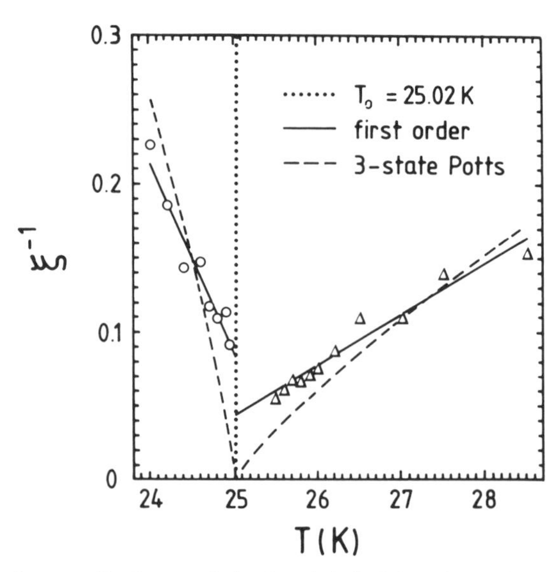

Figure 34. Inverse effective correlation length ξ^{-1} of the anisotropic–planar–rotor model (2.5) from Monte Carlo simulations in units of the lattice constant $a = 4.26$ Å as a function of temperature; the extrapolated transition temperature $T_c = 25.02$ K as obtained independently from energy cumulants is marked by a dotted line. Solid lines correspond to a fit assuming the simple linear dependence (3.19) expected close to the transition for a first-order transition, while dashed lines assume the critical behavior of the three-state Potts model in two dimensions (3.20). (Adapted from Fig. 2 of Ref. 273.)

size analysis of thermodynamic quantities near the herringbone transition of the anisotropic–planar–rotor model (2.5) was carried out in Ref. 217. The order parameter components (3.11)–(3.12) were defined as previously [244, 246], but contrary to the procedure used in Ref. 244 a total order parameter was introduced

$$\Phi = \left[\sum_{\alpha=1}^{3} \Phi_\alpha^2 \right]^{1/2} \tag{3.21}$$

as usually done for models with vector order parameters; note that at variance with the three-state Potts model the three components (3.11) constituting the scalar order parameter (3.21) are independent of each other. The largest linear dimension used was $L = 180$, and the averages of the observables were based on typically 1,000,000 Monte Carlo steps.

Contrary to Refs. 244 and 246 a strong size dependence of total energy and its heat capacity is observed [217], which is in accord with Ref. 56. Thus, the study [244] definitively relied on insufficient equilibration and statistics. Based on these size dependencies and a comparison to simulations of related models [63, 187], it was concluded that a Kosterlitz–Thouless-like scenario [252] can be ruled out because no strong size dependence of the specific heat should occur in this case [63, 187].

The analysis of the energy cumulant (3.13) essentially confirms the qualitative finding of Ref. 56; this was interpreted in Ref. 56 as strongly suggesting a continuous herringbone transition. However, it was shown based on various finite-size estimates that the same data can be described equally well using the finite-size dependencies for first- and second-order transitions [217]; the critical exponents of the proposed [56] three-state Potts model in two dimensions were used. The reason simply is that both sets of size dependencies are numerically very similar, and that the more accurate analysis necessary to discriminate was not possible with the available quality of the data. For a first-order behavior, it was concluded that the energy difference between the ordered and disordered phases must be extremely small, which is qualitatively consistent with the obtained energies for different system sizes [217].

The order parameter cumulant was also investigated in addition to the energy cumulant [217]. It was shown previously [368] (for a concise overview see Ref. 38) that Binder's fourth-order cumulant [31, 32]

$$U_L = 1 - \frac{\langle \Phi^4 \rangle_L}{3\langle \Phi^2 \rangle_L^2} \tag{3.22}$$

can also help in distinguishing different orders of phase transitions in addition to just locating second-order transitions; the order parameter Φ is defined according to (3.11), (3.12), and (3.21). In the case of second-order transitions, this quantity adopts a nontrivial universal value U^* at the critical point, irrespective of the system size in the scaling limit. At first-order transitions, U_L develops a characteristic minimum at a temperature that approaches the transition temperature as $\tilde{L}^{-d}$ to leading order; the linear dimension in this analysis is rescaled

$$\tilde{L}^d = \frac{L^d \Phi_-^2}{2k_B T_c \chi_-} \tag{3.23}$$

where $\Phi_\pm$ and $\chi_\pm$ denote the values of order parameter and its susceptibility as $T \rightarrow T_c^\pm$. This minimum, however, occurs only if $\tilde{L}$ exceeds an unknown minimum system size. In addition to the minimum, an effective crossing point of the cumulants occurs at a temperature T_{cross}. The value of the rescaled cumulant [i.e., $0 \leq g_L(T) \leq 1$]

$$g_L(T) := \frac{U_\infty(\infty) - U_L(T)}{U_\infty(\infty) - U_\infty(0)} \tag{3.24}$$

at this point is given in the Gaussian approximation by [368]

$$g_L\,(T_{\text{cross}}) \approx 1 - \frac{n}{2q} + \frac{2n}{q}(1 + q)\left(n\left[\frac{\chi_+}{\chi_-}\right]^2 - 2q\right)\tilde{L}^{-d} \tag{3.25}$$

where the order parameter dimension and the number of (degenerate) ordered phases are denoted by n and q, respectively.

The order parameter cumulant data [217] for different system sizes show a crossing point at about 0.65, and they do not display a minimum even for the largest system size $L = 180$. The universal value for the intersection point assuming the respective continuous three-state Potts scenario of Ref. 56 would be $U^* \approx 0.65$ (see Ref. 69 in Ref. 217). Assuming a first-order transition, the crossing point $U_\infty(T_{\text{cross}})$ is expected to occur at around 0.61 for the infinitely large system, and the first-order size correction estimated based on (3.24)–(3.25) yields a value of 0.67. Similarly to the outcome for the energy cumulant, it was concluded [217] that the behavior of the order parameter cumulant could be rationalized for both the first-order and the proposed [56] continuous transition scenarios.

It is concluded in Ref. 217 that an analysis of the order of the herringbone transition entirely based on thermodynamic quantities might be very misleading, because it was shown that most of the data can be rationalized in terms of both first- and second-order transitions. Thus, an analysis along these lines would require systems which are orders of magnitude larger than those available in Refs. 56 and 217, but only this would allow to reliably estimate the latent heat and the order parameter jump at the transition.

The orientational correlation length at the transition was used to rationalize this finding [217, 273]. A rough and qualitative comparison of the magnitude of ξ_+ with rigorous results for the q-state Potts model in two dimensions [43], which has a first-order transition for $q > 4$, shows that the herringbone transition should indeed be classified as a quite weak first-order transition: The Potts correlation lengths from above T_c for $q = 7$, 8, and 9 are 48.0, 23.9, and 14.9, respectively. This supports the speculations [308] about the large fluctuations near the herringbone transition, where the analogy to the situation as encountered in the q-state Potts model in two dimensions was invoked (see the discussion in Section II.C). Although it is known for the two-dimensional Potts model [385] that the transition is of first order for $q > 4$, one observes extremely large fluctuations near T_c in the case of Potts models with only a few more states than four so that the detection of the nature by simulations is very intricate [30, 197]. The reason is that renormalization trajectories may spend much time in the vicinity of this second-order fixed point and thus accumulate large fluctuations before finally flowing in the correct discontinuity fixed point. The authors of Ref. 345 actually confirm explicitly such a behavior for their anisotropic quadrupolar six-state model.

These effects might also be the reason [217] why a set of quite reasonable critical exponents could be fitted in Ref. 56 assuming a continuous transition and invoking finite-size scaling. However, as demonstrated in Ref. 368 for the case of the three-state Potts model in three dimensions, which is known to have a first-order transition [385], such a scaling might not yet be sufficient to unambiguously judge on the type of the transition for small systems. It was found [368] that the order parameter and its moments could nicely be scaled just as for critical phenomena, and the heat capacity maximum does not yet show the exponent d as a function of system size as expected for a first-order case. Instead an effective critical exponent, $\alpha_{\text{eff}}/\nu_{\text{eff}}$ can easily be fitted.

It is concluded [217] that an interpretation of the ideal herringbone transition within the anisotropic–planar–rotor model (2.5) as a weak first-order transition seems most probable, especially since previous assignments [56, 244] can be rationalized. This phase transition is fluctuation-driven in the sense of the Landau theory because the mean-field theory [141] yields a second-order transition. Assuming that defects of the $\sqrt{3}$ lattice and additional fluctuations due to full rotations and translations in three dimensions are not relevant and only renormalize the nonuniversal quantities, these assignments should be correct for other reasonable models and also for experiment [217].

The question of the (reduced) transition temperature for the herringbone transition of the anisotropic–planar–rotor Hamiltonian (2.5) seems much less controversial. The first estimate [244] $T_c^* = 0.775$ was already close to 0.76 obtained 10 years later by much more advanced techniques including size corrections [56]. This value was also confirmed (0.758) based on another finite-size extrapolation [217].

E. Compressed Monolayers

Before discussing compressed monolayers with their even richer behavior than the submonolayer regime we note that the experimental determination of the coverage gets more difficult due to possible second-layer adsorption [90, 93]. Thus, the coverages marking a phase boundary may differ between various experimental techniques and groups (see the discussion in Section I.C).

In the registered commensurate $\sqrt{3}$ phase of N_2 on graphite there is one N_2 molecule for every three carbon hexagons in the ideal case (i.e., without defects of any kind). Upon compression, it is possible to pack more molecules on the surface. An incommensurate phase is obtained if the ratio of the number of adsorbed molecules to the number of substrate sites is an irrational number (see Section I.C). There are different structural realizations possible. First, the molecules can lose commensurability in both directions on the surface. This leads in the case of N_2 on graphite to the triangular incommensurate phase; the alternative names honeycomb or hexagonal incommensurate phase are motivated by the topology of the domain walls. However, the molecules can also retain their commensurate spacing in one direction only. The resulting uniaxial incommensurate phase is sometimes also called "striped phase" because parallel domain walls are expected to occur in the ground state.

The question which type of incommensurability occurs for a specific physisorbate is of considerable theoretical interest, as documented in the reviews [10, 260, 364]. According to simple theory, the incommensurate phase just at its onset is believed to be composed of locally commensurate regions separated by periodic arrays of discommensurabilities, also called domain walls. The structure of the domain wall is thought to be controlled by the wall energy per unit length and the sign of the wall crossing energy. If the crossing energy is positive, then an uniaxial incommensurate phase with parallel domain walls is expected. At higher incommensurabilities this phase can transform to a triangular incommensurate phase with a hexagonal domain wall pattern. If the crossing energy is negative, the uniaxial compressed phase is not stable and the commensurate–incommensurate transition takes the adsorbate directly into the trigonal incommensurate phase. Also in this respect, N_2 on graphite is a rich system because it might display both types of scenarios depending on orientational ordering and thus temperature. There is also the possibility that a reentrant fluid phase is located in between the commensurate and incommensurate phases [77, 78]. In addition, the incommensurate superlattice might be rotated slightly with respect to the underlying substrate lattice [227, 263–265, 267] leading to epitaxial rotation.

A compressed solid phase in the range from 10 K up to 78 K was first observed by neutron scattering [171, 172] at coverages exceeding unity with a lattice constant of 4.04 Å, or about 5% less than the commensurate monolayer. It was, however, not clear if this so-called "dense solid" was still a compressed monolayer, or if the more dense packing of the film was linked to the formation of a second layer [172]; the latter was assumed in fitting some diffraction profiles. The melting of the "dense solid" to the high-temperature disordered phase, which occurs in the range from 78 K to 85 K, seemed to be a continuous rather than a discontinuous transition [172]; this was also concluded based on calorimetric data [73]. Beyond monolayer completion coexistence between the "dense solid" and the registered $\sqrt{3}$

phase was found by decomposing the respective neuton diffraction scans into a mixture of both contributions. The interpretation in terms of a second-layer growth was favored based on compressibility data [73] above monolayer completion.

More detailed LEED measurements [93] have shown that as the coverage is increased beyond the complete $\sqrt{3}$ monolayer a triangular incommensurate solid is formed in the temperature range from 31 K to 35 K—that is, above the herringbone transition at about 27 K. The lattice constant was found to be 5% smaller than that of the commensurate $\sqrt{3}$ phase. The epitaxial rotation angle of the adsorbate lattice with respect to the $(\sqrt{3} \times \sqrt{3})R30°$ graphite lattice could also be determined [93]. Invoking several assumptions, no epitaxial rotation was found up to a difference in lattice constants of 2%, and it increases to 0.5° at 5% compression. The order of this phase transition from the orientationally disordered $\sqrt{3}$ phase to the orientationally disordered triangular incommensurate solid could not be determined experimentally [93], although a continuous transition was assumed to determine the lattice constants and epitaxial rotation angles of the compressed phase. This was motivated by the theoretical expectation that this transition is of the continuous type described in Ref. 292.

Compressed phases were also studied by LEED [90, 93] below the herringbone ordering at a temperature of 15 K. Here, it was found that first an incommensurate solid which is only uniaxially compressed is formed [90] in the coverage range from 1.02 to 1.05 $\sqrt{3}$ monolayers. The existence of this phase has also been observed by synchrotron [257] and neutron diffraction [370] experiments. But contrary to the LEED study [93], evidence for such a uniaxial incommensurate phase was also obtained at temperatures exceeding the herringbone transition (i.e., above 30 K) in synchrotron x-ray [257] and heat capacity [394] investigations. The uniaxial character is thought to be induced by the anisotropy of the commensurate orientationally ordered herringbone phase [93, 257]. This structure was believed to be basically the same as that for the commensurate phase at low temperatures (i.e., a two-sublattice in-plane herringbone structure) except that the rectangular unit cell is compressed along the long side compared with the commensurate phase. This structure can exist in one of three separate domains depending on the angle of the glide line relative to the graphite lattice. The sublattice angles are assumed to be similar to the commensurate phase, and the precision of the measurements [93] cannot rule out a slightly oblique uniaxially incommensurate unit cell with an angle deviating by $\pm 1°$ from the ideal value of 90°. Again, the order of this phase transition could not be determined by experimental means; however, it is not expected [93] to have an intermediate fluid phase [77, 78] and might be a realization of the special continuous scenario described in Ref. 292.

Upon further compression at 15 K to approximately 8% more than the $\sqrt{3}$ monolayer, a LEED pattern is found which cannot be explained in terms of a superposition of three uniaxially compressed domains [93]. However, the pattern is consistent with an approximately uniformly compressed incommensurate triangular phase with roughly herringbone order. The orientational superlattice spots are much weaker than for the uniaxial incommensurate phase at lower coverages. It has to be considered that in the compression range above 1.05 $\sqrt{3}$ monolayers a significant out-of-plane tilting is expected to occur [93] which actually allows for the higher compression than in the uniaxial case.

A subsequent LEED study [388] of the triangular incommensurate phase around 1.10–1.12 $\sqrt{3}$ monolayers near 22 K gave evidence for a triangular incommensurate phase possessing a systematic out-of-plane tilt of the molecular axes in both sublattices; the lattice was found to be triangular within the experimental resolution of less than roughly 1%. The superstructure was assigned to be a "2-out" herringbone-like structure (see Fig. 2 of Ref. 388), similarly to an out-of-plane structure already proposed [141] as a possible orientationally ordered structure on commensurate lattices. The precise geometrical parameters could not be derived unambiguously [388]. The triangular incommensurate phase was also found in synchrotron [257], x-ray [242], and neutron scattering [370] studies. At higher coverages than the uniaxial incommensurate phase the coexistence of this and the triangular incommensurate phase is suggested from the LEED patterns [388], which is consistent with x-ray scattering data [257]. It is also suggested [388] that the uniaxial incommensurate structure whose density is near that of the triangular incommensurate phase almost completely loses its orientational long-range order at 22 K. The density of the triangular incommensurate phase is about 1.10–1.12 $\sqrt{3}$ monolayers at maximum compression.

The herringbone orientational transition was investigated in a model for a floating-solid (incommensurate) phase [60]. In this case no conventional true long-range order is expected, and melting transitions might follow the route of two-step continuous melting including essential singularities of thermodynamic functions along the lines of the Kosterlitz–Thouless–Halperin–Nelson–Young scenario (see Ref. 252). This loss of true long-range positional order might have consequences on phase transitions of internal degrees of freedom, such as molecular orientations. The unperturbed model of Ref. 60 designed to address this question in the context of the herringbone ordering of N_2 on graphite is a discretized version of the Heisenberg model with face-cubic anisotropy [259, 297], which has the same symmetry properties as the full quadrupolar Hamiltonian—that is, a herringbone ground state and a first-order transition in two dimensions (see also Section II.C).

Within this three-axis cubic model [259, 297], each site on the triangular lattice is occupied by an Ising spin and a three-state Potts variable. The strength of the interactions is ruled by J_{aniso}, which increases as the quadrupolar coupling constant [i.e., Γ_0 in (2.2) or K in (2.4) or (2.5)] increases. The effect of dislocations (which act exclusively on the Ising degrees of freedom), as well as their interactions in the Coulomb-gas approximation [252], is finally incorporated. Here the binding energy of the dislocations is ruled by the elastic constant K_{dis} of the underlying positional lattice, where, roughly speaking, large (small) K_{dis} corresponds to a stiff (floppy) lattice.

In parts based on analogies to the simpler case of a two-dimensional Ising antiferromagnet perturbed by dislocations, it is shown that the thermodynamic singularities associated with the orientational transition in the floating-solid phase, below the dislocation-unbinding temperature, are the same as in the commensurate phase. Specifically, it is concluded that in the presence of a thermal distribution of dislocations the transition is still of the face-cubic Heisenberg type—that is, a first-order transition. The very schematic phase diagram for the dislocation perturbed three-axis cubic model in the plane spanned by the relevant parameters J_{aniso}^{-1} and K_{dis}^{-1} is shown in Fig. 35. The physical parameters describing N_2 on graphite place this system in the region of a stiff lattice and a low quadrupolar coupling strength. Thus, the herringbone ordering transition of N_2 on graphite occurs in both commensurate and incommensurate lattices on a path crossing somewhere the

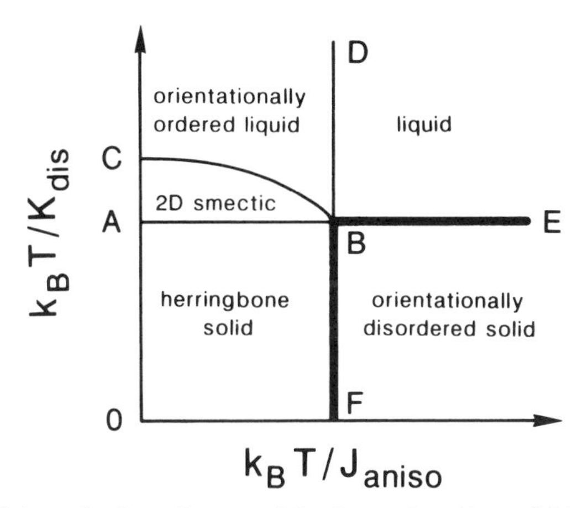

Figure 35. Schematic phase diagram of the three-axis cubic model including the effect of dislocations to model the herringbone transition in floating-solid (incommensurate) phases (see Section III.E for details). The orientational and melting transitions of real N_2 on graphite are expected to occur on a path crossing somewhere the bold solid lines BF and BE, respectively. (Adapted from Fig. 5 of Ref. 60.)

BF line, and the positional melting transition from the orientationally disordered solid to the fluid occurs across BE. It is interesting to note that for sufficiently floppy lattices and strong quadrupolar interactions it would be possible to have transitions across AB and CB to a phase with positional disorder, which is nevertheless orientationally ordered.

A heat capacity investigation [394] of compressed monolayers was essentially consistent with the complementary LEED studies [90, 93] and a concurrent x-ray investigation [242]. However, more detailed phase boundaries and stronger evidences as to the order of several transitions in the higher coverage regime could be reported [394]. The definition of the complete commensurate $\sqrt{3}$ monolayer (i.e., coverage unity) varies between Refs. 65 and 394 due to a different choice of the calibration (see the general discussion in Section I.C). The coverage scale in Ref. 394 is roughly 2% larger than that used in Refs. 65 and 232; that is, coverage unity in Ref. 394 corresponds to 1.02 monolayers in Ref. 65. In addition, there seems to be a shift in coverage of about -0.24 monolayers when converting the higher coverages reported in the last six entries in Table II of Ref. 65 to the scale reported in Ref. 394, where the highest coverage [65] of 1.370 seems to correspond to a coverage in between 1.104 and 1.122 in Ref. 394. The latter calibration [394] is believed to be more reliable by the authors of Ref. 394 (see their Ref. 14).

The submonolayer orientational ordering anomaly at 28.1 K remains unshifted up to 1.122 monolayers and is no more detectable above this coverage. However, there is a peak located at 29.4 K in a partly overlapping range from 1.104 up to much higher coverages, where it gets shifted to slightly higher temperatures. This anomaly is attributed to the orientational (herringbone) transition in the uniaxial incommensurate phase. The compression leads to reduced nearest-neighbor distances between the nitrogen molecules. This in turn leads to a higher transition temperature since the herringbone phase is stabilized by quadrupole–quadrupole interactions which are increased upon compression.

Orientational ordering transitions [394] clearly occur in the commensurate phase (possibly in coexistence with the fluid) up to 1.101 monolayers at 28.1 K, as well as in the uniaxial compressed phase from 1.142 to 1.178 monolayers at 29–30 K; based on adiabatic calorimetry [156] a heat capacity peak at 28.7 K was reported at a coverage of 1.109 which was interpreted as the transition to the uniaxially compressed incommensurate phase, whereas this peak is located near 27.5 K in the submonolayer regime. The ac heat capacity scans [394] in the triangular incommensurate phase at 1.200 and 1.302 monolayers show only very weak and broad features in the relevant temperature range from 22 K to 32 K. Thus, it is concluded [394] that their orientational order is only gradually lost as the temperature is increased similarly to what was suggested in the x-ray study [242]. Note that in accord with the synchrotron investigation [257] but at variance with the LEED studies [90, 93], the uniaxial incommensurate phase is believed [394] to exist also in an orientationally disordered form at temperatures above roughly 30 K with an orientational phase transition in between; note that the LEED data of Ref. 388 suggest that the uniaxial incommensurate phase loses its orientational long-range order at its highest density even at 22 K. The bilayer growth upon compression beyond the triangular incommensurate phase seems to begin around 1.30 monolayers with a coexistence regime [395]. The features of the orientational transitions together with the structural assignments of the LEED study [93] led to a determination of various coexistence regions and phase boundaries of the uniaxial and triangular incommensurate phases as shown in Fig. 36; the phase boundaries above 30 K are speculative, and it is possible that the uniaxial incommensurate orientationally disordered region is not a distinct phase [395].

The transition from the commensurate solid to the uniaxial incommensurate phase, as well as that from the uniaxial to the triangular orientationally

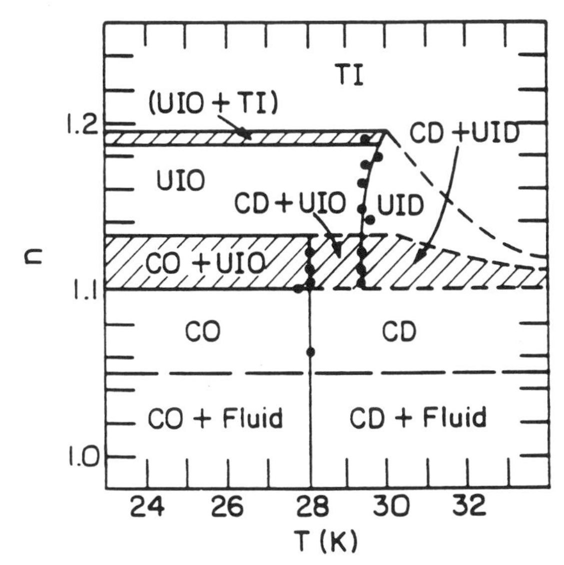

Figure 36. Compressed-monolayer and low-temperature phase diagram of N_2 on graphite (solid lines) based on heat capacity measurements. Filled circles represent positions of heat capacity peaks. Hatched regions denote two-phase coexistence, and dashed lines as well as boundaries separating UIO + TI, UID, and TI phases are speculative. The following abbreviations have been used: Commensurate $\sqrt{3}$ solid (C), incommensurate $\sqrt{3}$ solid (I), uniaxial (U), triangular (T), orientationally disordered (D), orientationally ordered (O). (Adapted from Fig. 2 of Ref. 394.)

ordered incommensurate phase, is assigned to be first-order transitions [394]. The first assignment is at variance with some theoretical expectations for the commensurate–incommensurate transition [10]. The orientational disordering within the triangular compressed phase is only a gradual loss of order without going through a proper phase transition [242, 394]. The orientational phase transition within the uniaxial incommensurate phase is believed [394] to be a first-order transition, an assignment based on a power law analysis of the heat capacity similar to that of Refs. 65 and 232. Such a first-order nature would be consistent with an analysis [60] (see above) devised for orientational disordering within a floating-solid (incommensurate) phase similar to the one presented for the herringbone transition on a commensurate lattice [308] (see Section II.C).

The structure of the triangular incommensurate phase at about 1.17 ± 10% monolayers was also investigated by x-ray diffraction of N_2 on Papyex [242]. In qualitative agreement with the early neutron scattering study [99], it is found that the orientational ordering occurs in this compressed phase (see Fig. 37) over a much broader range than in the submonolayer case (see Fig. 28 and Fig. 26 and also Section III.D). This range extends roughly from 35 K down to 10 K, where the ordering is still incomplete. The slight expansion of the incommensurate superlattice and the onset of orientational

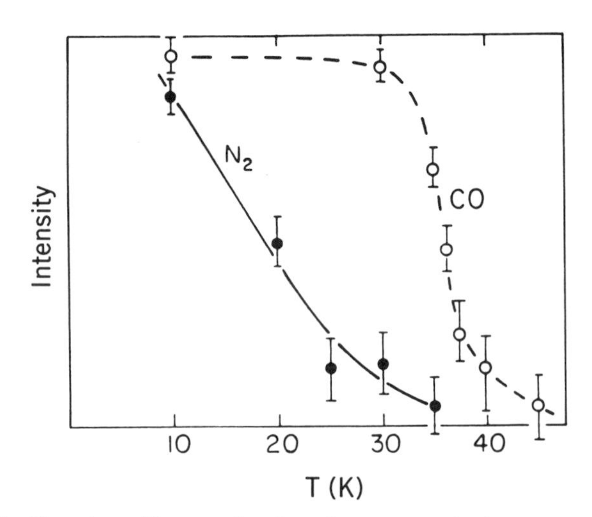

Figure 37. X-ray intensities as a function of temperature for incommensurate N_2 (filled circles and solid line) and CO (unfilled circles and dashed line) on graphite (Papyex) at a coverage of about 1.17 ± 10% monolayers. The intensity plotted on a linear scale is the superlattice peak intensity of the orientationally ordered structure divided by the peak intensity of the $(\sqrt{3} \times \sqrt{3})R30°$ center-of-mass structure present in both the orientationally ordered and disordered phases. (Adapted from Fig. 6 of Ref. 242.)

disorder occur simultaneously within the experimental uncertainty [242]. There is evidence that the center-of-mass lattice is slightly distorted from the ideal triangular incommensurate lattice, and the type of orientational order (in-plane herringbone, out-of-plane herringbone, or pinwheel) could not be inferred based on the available data [242].

Finally an extensive neutron diffraction study [370] is conducted in the compressed monolayer range at 11 K in order to investigate in more detail the type and geometry of the various ordered structures; the results for the herringbone commensurate phase are discussed in Section III.D. A uniaxial compression along the long axis of a rectangular unit cell of up to about 3.3% is observed. The out-of-plane tilting is at most 15°, and the angles between the sublattices and the unit cell axes are 40–50°. Thus, this phase is indeed best described as a uniaxially compressed two-sublattice in-plane herringbone phase [93, 370]. The neutron diffraction patterns [370] also confirm the finding [394] of a coexistence region between the uniaxial and triangular incommensurate phases. The pattern of the triangular incommensurate structure is best fitted as a slight oblique distortion from the ideal hexagonal symmetry with the same rotation of the two sublattices and an out-of-plane tilt of 10–20° (i.e., as a two-sublattice out-of-plane herringbone structure) [93, 370, 388]. The precise values of the lattice parameters deviate from the ones inferred by LEED [388]. A reinvestigation [371] of this diffraction pattern in terms of a pinwheel structure [141] was stimulated by the stability of the pinwheel phase for the fully compressed monolayer obtained from model calculations [298]. It is indeed found that a four-sublattice pinwheel structure with again a slightly oblique cell and the "wheel molecules" tilting out of the graphite plane by about 10° provides at least as good a fit to the diffraction profile as the two-sublattice out-of-plane herringbone structure [371].

Lattice distortions occurring in incommensurate monolayers might lead to ferroelastic temperature-driven transitions instead of simple orientational order–disorder transitions depending on the coupling strength between orientational and translational degrees of freedom [206]. In the latter case, the center-of-mass lattice structure is identical in the orientationally ordered and disordered phases, whereas the anisotropy in the intermolecular interactions leads to a lattice distortion in the orientationally ordered phase for the first type of transition. A strictly two-dimensional model based on Lennard-Jones interactions but without electrostatic interactions and corrugation potential was devised and investigated in some detail by various methods [162, 206, 344]. The molecules can rotate and move only in the surface plane. This model comes close to a description of the δ-phase of O_2 monolayers on graphite (see Refs. 220 and 221 and the literature cited in Section I.A) which are incommensurate to the substrate lattice and where the quadrupolar moment can be neglected (see also Refs. 207, 208, 210, 216, 239, and 315). Using constant pressure molecular dynamics [279], periodic boundary conditions, and 400 O_2 molecules, it is found that the phase transition from the orientationally ordered ferroelastic to the disordered paraelastic plastic phase is of first order and is accompanied by a lattice distortion from a distorted (isosceles) to an equilateral triangular lattice [344].

Using the Monte Carlo method a very realistic model for N_2 on graphite was investigated emphasizing the consequences of compressing the monolayer on orientational properties [282]. The model is very similar to what was used in Ref. 48 (see Section III.D.1); that is, it includes in addition to the static molecule–surface interactions [326] also surface-mediated interactions which modify the bare intermolecular potential. The system consists of a fixed number of 64 molecules which are free to move in the three dimensions of a periodic rectangular cell, and up to 40,000 Monte Carlo steps were used. Order parameters were obtained by projecting on a specific average reference structure obtained from Monte Carlo simulations well in the ordered phase at 10 K, instead of using an invariant such as the combination (3.11), (3.12), and (3.21); note that the projection does not take into account the presence of several degenerate ground states and might be misleading in the transition region. The 12×8 array of the honeycomb adsorption sites in the complete herringbone monolayer was uniaxially compressed to an 11×8 structure to yield a coverage of 1.09 monolayers. The herringbone transition in the complete monolayer case was found to occur around 28 K. The low-temperature incommensurate phase at 1.09 monolayers has a ground state very similar to the uncompressed case with molecules lying flat on the surface in a herringbone arrangement. However, the transition is broader than in the complete monolayer, and it is found that some molecules stand up and form pinwheels at 20 K, which is well below the disorder transition. Energy minimization calculations using the same potentials but only four molecules in the unit cell were carried out; only the surface area was constrained in these calculations. A uniaxially compressed herringbone structure was found at a coverage of 1.025 and 1.05 monolayers (see also Ref. 283), while the molecules begin to tilt out of the plane by less than 5° upon further compression to 1.09 and 1.13 monolayers. Finally a pinwheel structure is found for a compression to more than 1.16 monolayers.

The study of incommensurate N_2 monolayers was pursued [283] using systems which were larger by more than an order of magnitude than they were previously [282]. Only the N_2–graphite interactions were slightly modified, and the largest system consisted of 2704 molecules in a 104×26 periodic rectangular array of molecules. The key finding, which is only observable in sufficiently large systems, is that the physisorbate first prefers upon compression beyond the complete monolayer a striped domain wall uniaxial incommensurate structure over a uniformly compressed uniaxial incommensurate configuration in accord with theoretical expectations [10, 260, 364] derived from simple Frank–van der Merve (or Frenkel–Kontorova) type models [10]. Such a striped configuration for a 52 $\times$ 12 molecule system with a coverage of 1.026 monolayers is depicted in Fig. 38, where one can clearly see the perpendicular striped domain wall in the left portion of Fig. 38*a*, which separates two periodically connected parts of a standard commensurate herringbone structure, see the inset in Fig. 38*a*. At 10 K these domain walls are stable on the time scale of the simulation as can be seen, for example, from Fig. 38*b*. It was found that the number of striped domain walls grows with system size as theoretically expected based on simple models [10] and that the total width of the stripe amounts to about 35 Å. Heating the samples results in pinwheel defects in the domain walls and finally in orientational disordering around 30 K. Using again the projection definition of the herringbone order parameter from Ref. 282, it was found that the domain walls have enhanced orientational order relative to the commensurate regions and that the orientational order in the domain walls persists to higher temperatures than in the commensurate regions. In addition, the domain walls are still present well above the orientational transition. Simulations at a coverage of 1.013 monolayers confirmed the reported character of the domain walls, and an essentially uniformly compressed uniaxial incommensurate phase was finally obtained at 1.08 monolayers. No incommensurate two-sublattice out-of-plane herringbone phase (see Fig. 4*c*) could be stabilized in these studies [282, 283] even experimenting with a variety of potential models [284]; see, however, the remarks concerning out-of-plane herringbone ordering in Refs. 137, 181, and 182.

Uniaxially compressed phases and the commensurate reference structure were furthermore examined [342] by molecular dynamics simulations along the lines of the work reported in Refs. 232 and 340 (see Section III.D.1), except for small alterations in the potential models. The 96 molecules were put into a rectangular cell which was uniaxially compressed by 5% perpendicular to a glide line of the herringbone sublattice structure; that is, the center-of-mass lattice is contracted toward the glide line; this compression allows the same periodic boundary conditions to be effective for both adsorbate and graphite lattices. It should be noted, however, that even this does not ensure a simulation of the true equilibrium situation because every solid accommodates even in equilibrium a certain number of vacancies and interstitials. In simulations with a constant number of particles the net number of such defects is actually constrained to some constant value, which is not necessarily the correct equilibrium value [338, 339]. Two temperatures well below and above the orientational disordering transition at 15 K and

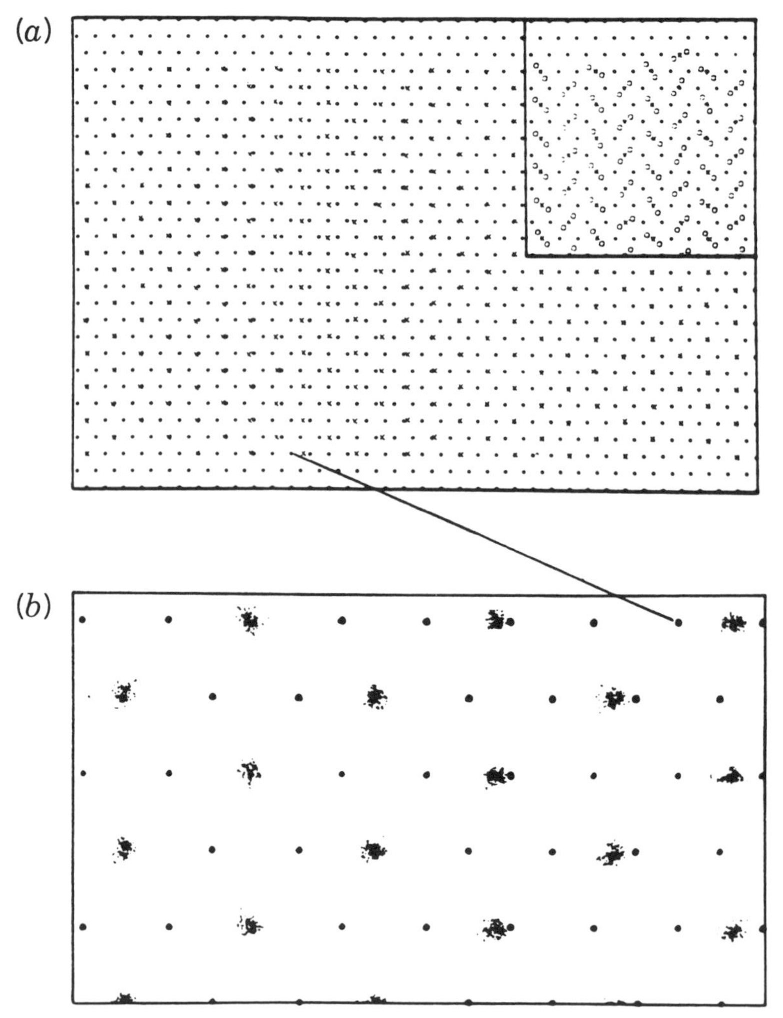

Figure 38. Striped domain wall in a model of uniaxially compressed N_2 monolayers on graphite (*a*) observed in Monte Carlo simulations of 52 × 12 molecules at 10 K and a coverage of 1.026 monolayers. Dots denote the centers of the honeycomb hexagons of the graphite basal plane, and crosses mark the mean positions of the molecular centers of mass. The inset of (*a*) shows the herringbone order in the commensurate region at the left and right boundaries of (*a*). The center-of-mass distribution in the region of the domain wall of (*a*) sampled from the Monte Carlo trajectory is magnified in (*b*). (Adapted from Fig. 1 of Ref. 283.)

40 K, respectively, were chosen. First of all it is found that the positional ordering of both the commensurate and incommensurate structures are preserved at 40 K. The uniaxial incommensurate phase floats over the graphite lattice along the long axis, whereas the commensurate phase is locked to the lattice because all molecules sit on the minimum adsorption sites in the centers of the honeycomb cells. The most probable height of the molecular layer above the surface was found to be constant at 3.3 ± 0.1 Å for the considered temperatures. The in-plane distribution of the molecular axes sharpens upon uniaxial compression of the monolayer. The orientation of the herringbone-ordered phase relative to the substrate is changed from the ideal value of 45° to $42.8 \pm 0.15°$ and $41.0 \pm 0.15°$ for the commensurate and uniaxial incommensurate solids, respectively. The average value of the out-of-plane tilt of the molecules is not much affected at the chosen compression in the low-temperature solid; that is, the molecules prefer to alter their in-plane ordering rather than tilting away from the surface. Thus, the orientational order–disorder transitions in both the commensurate and the particular uniaxial incommensurate phases seem to be very similar. Again, pinwheel-like defects are found to appear at the higher temperature in the solid phases. Actually such pinwheel arrangements seem to be a kind of natural defect structure to form the necessary grain boundaries between herringbone domains belonging to different domain structures. Noteworthy is that furthermore the system does not seem to be stable under a compression parallel to the glide line, and it prefers to reorder by forming different herringbone-ordered domains in structures contracted toward the local glide lines. In a subsequent simulation [358] extended to longer times and bilayers, it was found that such a sublattice reorientation of the initially parallel compressed phase can be completed.

Dynamical properties of the commensurate and uniaxial incommensurate phases according to the model of Refs. 232, 340, and 342 could also be explored by the molecular dynamics technique used [203, 352]. It is found that in-plane and out-of-plane motions can be analyzed separately for orientationally ordered N_2 on graphite [203]. The 40-ps simulations below the orientational ordering transition (see Ref. 342) show [203] that the amplitude of reorientation is small and the out-of-plane motion nearly harmonic in both phases, whereas the in-plane motion is more complex, because it is anharmonic and collective. The out-of-plane motion in the disordered phases is still harmonic, but more strongly damped, and the in-plane dynamics cannot be analyzed any more in terms of a cumulant expansion. Thus, there is little qualitative difference between the reorientational motion observed in the commensurate and uniaxially compressed solids. Only the out-of-plane motion is slightly less damped in the uniaxial phase, and the fluctuations from the planar configuration are more pronounced.

The phonon and libron dynamics in the orientationally ordered commensurate, disordered commensurate, and 5% uniaxially compressed solid phases was analyzed [352] in terms of the dynamic structure factor $S(\mathbf{q}, \omega)$ obtained at about 15 K, 25 K, and 40 K on a 20-ps time scale (see Ref. 342 for simulation details). The herringbone transition seems to occur just below 25 K, which is lower than the 33–34 K reported in Ref. 340, because a smaller quadrupole moment is used in Ref. 352. Note that the modeling of the molecule–surface interactions as a static external field [59, 203, 340, 342, 352] precludes any vibrations of the substrate carbon atoms, and thus any coupling between the phonons of the monolayer and the graphite. This study confirms many features pointed out in Ref. 59 (see Section III.D.1), such as the qualitative similarity of the commensurate and uniaxially compressed solids, a strong coupling of the translations leading to a considerable phonon dispersion, and a coupling between the translational and librational phonon modes in the orientationally ordered phase. The latter mixing is responsible for the structure of the frequency spectra for in-plane motion reported in Fig. 2 of Ref. 203. The lowest frequency mode in these four-peak spectra is believed to be due to the acoustic mode which couples to the higher-frequency librational modes. The out-of-phase librational and acoustic modes have very similar spectra for certain $\mathbf{q}$ values, which is a manifestation of the strong repulsive interactions between molecules in the same herringbone sublattice upon longitudinal compression of that sublattice. This leads to favoring [342] the uniaxial contraction toward the glide line relative to uniaxial compression parallel to it. The disordered phase cannot be described in terms of librations in the sense of small deviations from some reference configuration because the spectra are broad and rather featureless. The phonons in the uniaxial phase usually have higher frequencies than those in the commensurate phase owing to the increased surface packing fraction.

The stability of various compressed phases, especially the uniaxially compressed case, was investigated in great detail based on Monte Carlo and energy minimization schemes [181, 182]. The interatomic potential was a more accurate refit [101] of *ab initio* data [23] for the N_2–N_2 interactions in the well and long-range region supplemented with Gordon–Kim electron gas calculations [198] for the small separation regime and includes electrostatic multipole interactions. The van der Waals part of the molecule–surface potential is represented by its first-order Fourier decomposition [326], which is supplemented with (a) Coulomb interactions between the point charges representing the N_2 multipoles and their images induced in the graphite substrate and (b) McLachlan substrate-mediated dispersion interactions [49, 225] using the parameterization of Ref. 282. The supercell in which the energy minimizations or Monte Carlo simulations (using 16 molecules and typically 2000 Monte Carlo steps) are performed have deformable, periodic boundary conditions [279]; that is, the length of the two lattice vectors and their relative angle are independent variables to be optimized.

The energy versus coverage plot obtained from the static energy minimization [181, 182] is shown in Fig. 39. The solid curves are obtained for a flat surface—that is, without the first-order corrugation term modeling the lateral potential [326]. In the coverage interval from 1.11 to 1.32 monolayers the four-sublattice pinwheel structure seems to be stable, and above 1.32 monolayers the two-sublattice out-of-plane herringbone phase with a tilt angle of about $\pm$ 46° (see Fig. 4*c*) is judged to be stable. Removing the incommensurability condition (i.e., introducing a corrugation term) results in a stable absolute minimum for the in-plane herringbone phase at coverage unity (see the solid circle in Fig. 39). The dashed line extending from coverage unity to approximately 1.06 monolayers is a uniaxial incommensurate phase; all parameters of these various phases are reported in Table I of Ref. 181. This picture is exclusively based on static energy minimizations

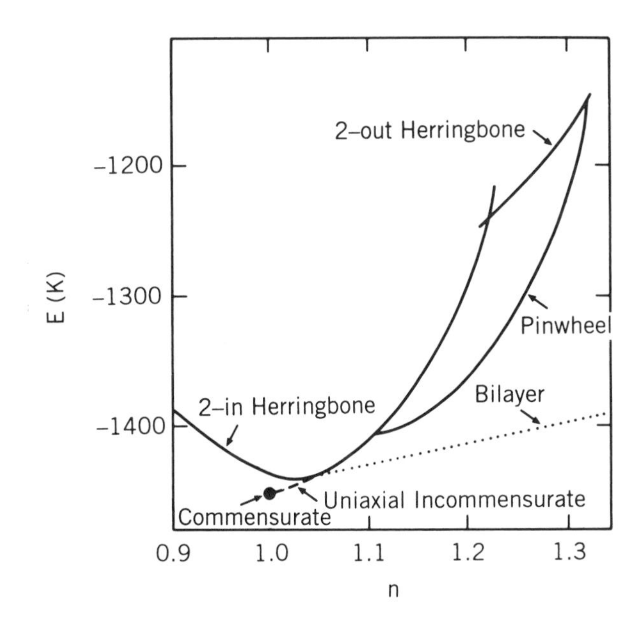

Figure 39. Total energy and corresponding phases for N_2 on graphite as a function of coverage from energy minimizations and Monte Carlo simulations. Solid lines correspond to energies for a flat surface without lateral energy variation, the filled circle marks the absolute minimum corresponding to the in-plane herringbone commensurate solid including surface corrugation, the dashed line corresponds to the uniaxially compressed phase, and the schematic dotted line stems from a qualitative double-tangent construction with respect to the stable bilayer at densities exceeding the frame. (Adapted from Figs. 3 and 4 of Ref. 181.)

of monolayer unit cells, and it changes when larger cells allowing for multilayers are used in the Monte Carlo simulations at 5 K. In particular, the creation of the bilayer being completed at a coverage of about 2.2 changes this scenario qualitatively: The double-tangent construction renders monolayer phases other than the commensurate and uniaxial incommensurate phases metastable (see the dotted line in Fig. 39 for a schematic and qualitative sketch of this line). Thus, there is no thermodynamically stable pinwheel monolayer phase within the underlying model, and no indication of a nearly triangular incommensurate herringbone phase is found. The uniaxially compressed phase is stable as a pure phase from coverage unity to 1.06 monolayers as in Fig. 39, followed by a region of coexistence with the bilayer up to 2.2 monolayers according to the schematic dotted line included in Fig. 39. It is claimed that the uniaxial phase evolves continuously from the commensurate phase as the density is increased above unity. The herringbone transition in the commensurate phase could be located at around 25 ± 2 K.

The second part of the paper [182] is exclusively devoted to the uniaxial incommensurate phase using the Monte Carlo method with variable cell at 5 K and the potentials as in Ref. 181. The most interesting result (Fig. 40) strengthens the finding [283] that the uniaxial phase is a striped incommensurate phase close to coverage unity, which seems to transform into a modulated phase for higher surface coverages. This figure shows the relative displacements of the centers of mass of the N_2 molecules from the lattice sites of the ideal $(\sqrt{3} \times \sqrt{3})R30°$ commensurate phase as a function of lattice position along the uniaxially compressed axis—that is, perpendicular to the glide lines of the herringbone sublattices. At coverage unity the molecular positions form a straight line which signals the ideal commensurate $\sqrt{3}$ phase. Two commensurate regions displaced by one hexagon spacing which are separated by a uniaxially compressed stripe with a width of about 48 Å is observed at 1.0135 monolayers. A well-defined stripe can be observed even at a coverage of only about 1.0067 monolayers, whence the calculations suggest that the striped uniaxial phase extends down to a coverage of unity. More stripes appear proportional to the compression of the monolayer similar to the observation in Ref. 283, and the orientational structure clearly is an in-plane arrangement.

A qualitative change occurs only at about 1.056 monolayers, where it is no more possible to unambiguously identify commensurate stripes [182]. There, the situation can be described as commensurate modulations in an otherwise uniform uniaxial incommensurate structure. Following an earlier suggestion [362], test calculations where the lateral substrate energy is increased by a factor of two compared to the original potential [326] were performed at several coverages. The results are qualitatively similar to what is shown in Fig. 40, only the boundaries between the domain walls seem to

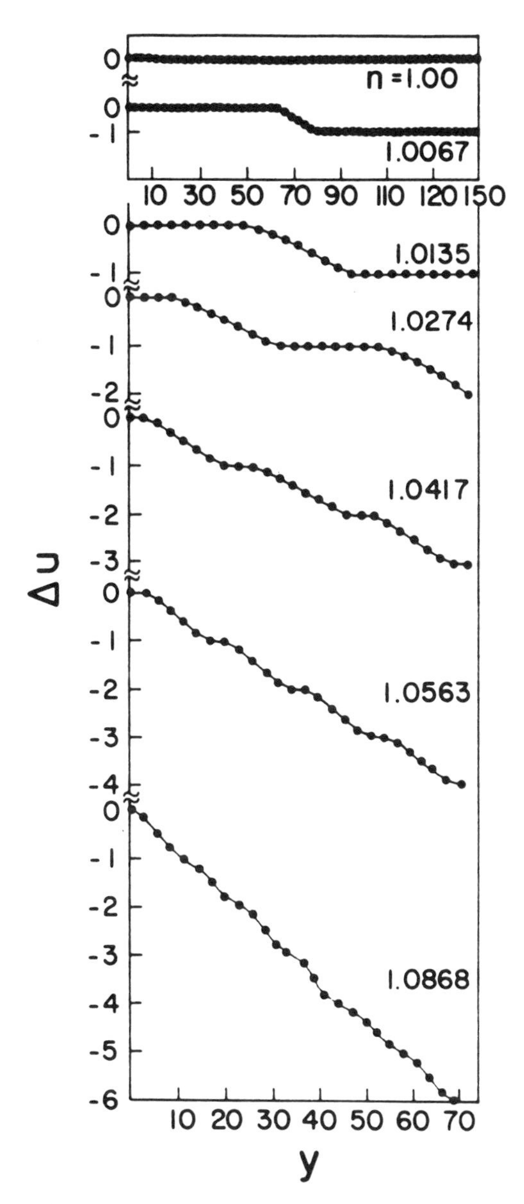

Figure 40. Relative displacements Δu of the centers of mass of the N_2 molecules from the lattice sites of the ideal $\sqrt{3} \times \sqrt{3})R30°$ commensurate lattice on graphite as a function of the lattice position y along the uniaxially compressed axis (i.e., perpendicular to the glide lines of the herringbone sublattices) from Monte Carlo simulations at 5 K. The coverages n in units of the complete monolayer are indicated near the corresponding curves. The lengths Δu and y are measured in units of the graphite lattice spacing $a = 2.46$ Å. Note that the curves for the coverages unity and 1.0067 are obtained in a box with twice the extension along y compared to all other curves, hence the different scale of their y axes. (Adapted from Fig. 1 of Ref. 182.)

be more sharply defined. Thus, the nonuniform striped uniaxial incommensurate phase of N_2 on graphite is within this model thermodynamically stable in the coverage range from unity to about 1.06 monolayers. The structure of the uniaxially compressed stripes is an in-plane herringbone arrangement, and the lattice constant parallel to the herringbone glide lines is not altered from the commensurate value 4.26 Å by the compression. The width of the uniaxial incommensurate stripes appears independent of coverage, and their number is proportional to the box length perpendicular to the glide lines (i.e., the y axis of Fig. 40) and to the deviation of the coverage from unity. The conclusions especially concerning the stability and nature of the "modulated uniaxial incommensurate" phase above about 1.06 monolayers have partly speculative character in view of the small particle numbers and the necessary use of deformable but periodic boundary conditions. Finally it is proposed that the system may support mass-density waves characterized by a propagation of the high-density uniaxial incommensurate stripes.

The compressed phases including second layers were investigated by constant temperature molecular dynamics [24, 328, 358–360] with realistic potentials and techniques very similar to the ones used in earlier studies of that series [232, 340, 342] (see also the presentation in Section III.D.1). The first study [328, 359] concentrated on the high-temperature regime at about 74 K and a coverage range from the complete monolayer up to a compression to 2.5 layers, using 96–240 particles depending on coverage. The density distribution [359] of the molecular centers of mass normal to the surface is presented in Fig. 41; in order to obtain fully converged thermodynamic averages, very long equilibration and sampling times are required, but the distributions shown might come reasonably close to such averages. There is a clear division of molecules of the system into first-, second-, and third-layer molecules. Promotion to the second and third layers seems to occur at 1.1 and 2.5 monolayers, respectively. The number of molecules in the first layer (see Fig. 42) increases first upon compression up to about 1.5 monolayers, but finally decreases again to a value which is only slightly higher than in the commensurate case. The reasons for that are not entirely clear, but some energetic arguments are presented [359]. A significant degree of commensurability is present at coverages close to those of a commensurate monolayer. This picture changes at higher coverages where interlayer transfers occur: Here the first layer appears not to be commensurate any more. In the first layer the molecules are preferentially located above the adsorption sites, whereas such a tendency is no more observed for molecules in the second layer [359]. For all coverages simulated, the first layer molecules have a preference to be coplanar to the graphite surface, but the number of molecules that are almost perpendicular to the surface is nonnegligible and increases as the film is squeezed. The second-layer molecules behave clearly

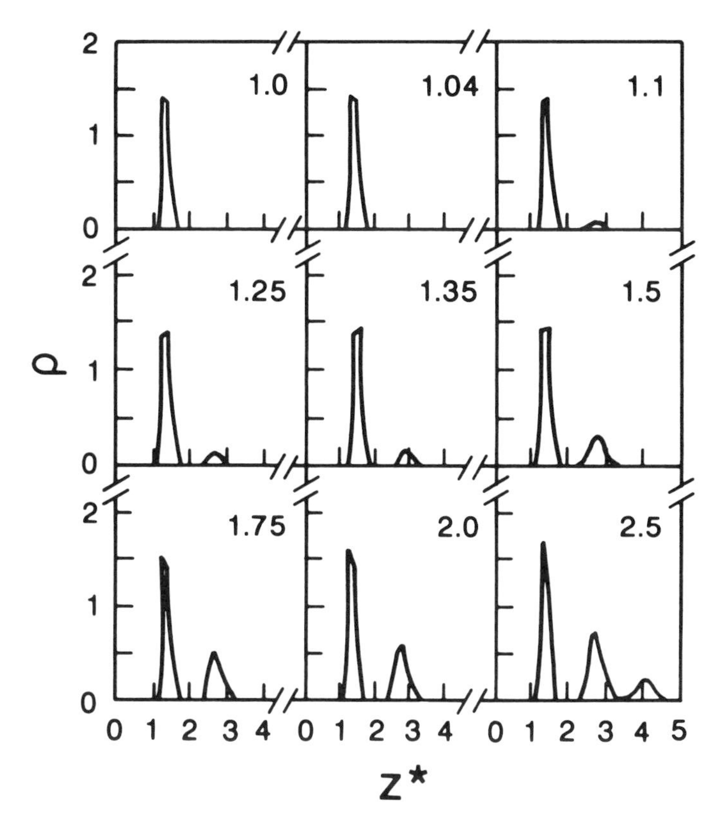

Figure 41. Density distribution of the center-of-mass distance of N_2 molecules perpendicular to the graphite basal plane from isothermal molecular dynamics simulations at 74 K for various coverages as indicated. The total coverages are given in the figure, and $z^* = z/a$. (Adapted from Fig. 3 of Ref. 359.)

fluid-like. Overall, the density of the first layer, the orientations of the first-layer molecules, and the density of the second layer are coupled in a complicated way.

The dynamical analysis of the isothermal molecular dynamics simulations [359] was presented in Ref. 360. The motion perpendicular to the surface for the first-layer molecules is for all coverages of the damped oscillatory type around the minimum of the total potential normal to the plane; the frequency increases slightly as the number of second-layer molecules increases. The translational velocity correlation function parallel to the surface stems from an apparently liquid-like phase of the first-layer molecules. Inspection of trajectory plots suggests that much of the observed first-layer motion is in directions parallel to the rows of molecules and can be inter-

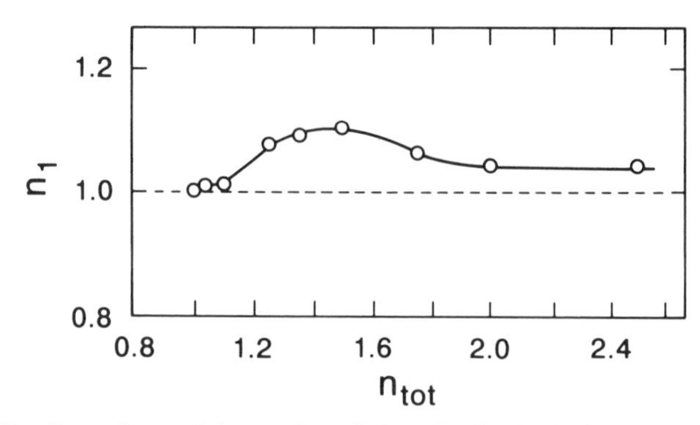

Figure 42. Dependence of the number of N_2 molecules in the first layer n_l on graphite as a function of the total coverage n_{tot} from isothermal molecular dynamics simulations at 74 K. All coverages are reported in units of the commensurate $\sqrt{3}$-monolayer coverage. (Adapted from Fig. 6 of Ref. 359.)

preted as hopping to adjacent vacant sites. Thus, the large diffusion constants are not signatures of a fluid phase, but of jump diffusion via defect sites in the solid. The barrier height to translation seems to play a minor role at the simulation temperature of 74 K; however, see the discussion of melting in Section III.B. The dynamics of the reorientational motion is of the hindered-librational type perpendicular to the plane, whereas the in-plane motion is more complex. An interpretation in terms of a random walk does not describe the situation because of indications for some dynamical coherence or memory effects [360].

Simulations of the triangular incommensurate monolayer [24] at coverage 1.116 reveal that this phase consists of a mixture of vertical and horizontal molecules considerably blurred by thermal librational motion (see Fig. 43). More perpendicular molecules are found when the density of the single monolayer is increased to 1.210. The situation of the triangular incommensurate phase at 1.116 monolayers remains qualitatively similar when a second layer of the same density is added on top of the first monolayer with only the fraction of perpendicular molecules being reduced; note that the question of the thermodynamic stability of the triangular incommensurate monolayer relative to second-layer promotion remains unresolved. Inspection of configurations of the triangular incommensurate monolayers at these two coverages shows that the perpendicular molecules form the "pin molecules" of local pinwheel defects and that the irregular defect patterns change on the timescale of about 10 ps at temperatures around 25 K and 40 K. This pinwheel orientation of the six neighbors of individual perpendicular defect molecules is also confirmed for the commensurate, uniaxial incommensurate, and bilayer systems.

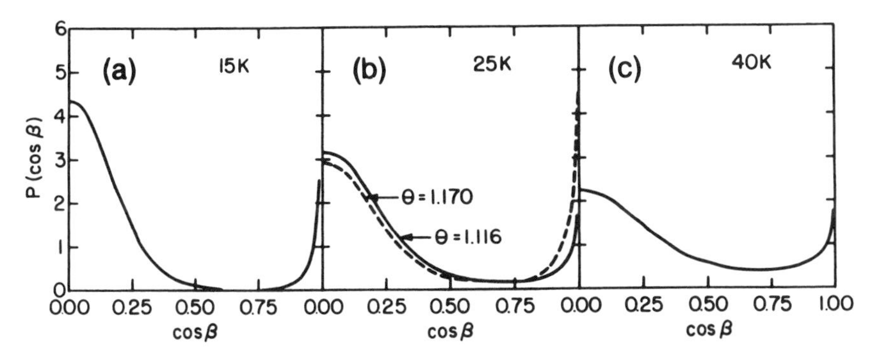

Figure 43. Distribution $P(\cos\ \beta)$ of the cosine of the out-of-plane tilt angle β for N_2 molecules on graphite in a triangular incommensurate phase obtained from molecular dynamics simulations at monolayer coverages of 1.116 and 1.170 marked by solid and dashed lines, respectively, at (*a*) 15 K, (*b*) 25 K, and (*c*) 40 K. (Adapted from Fig. 3 of Ref. 24.)

Quasiharmonic lattice dynamics calculations were performed for selected monolayer and bilayer ground-state structures [298]. This technique allows us to minimize the quasiharmonic Helmholtz free energy instead of just the total energy of the system, and it includes harmonic quantum mechanical zero-point vibration and libration effects; the zero-point energy of the low-temperature bulk N_2 solid reduces the cohesive energy by about 15% [298]. Several combinations of interaction models were systematically studied. Two N_2–N_2 pair potentials [181, 248] in a site–site representation were combined with the first order Fourier expansion of the substrate potential [324, 326]. Substrate-mediated interactions to model the effect of substrate screening of the electric multipoles were included in some calculations in the form of electrostatic images, and the dynamic response to fluctuating dipole moments were described by the spherically averaged McLachlan interaction. These calculations were performed for both the static and the harmonic lattice approximations. Concerning incommensurate superlattices it has to be noted that the ansatz [298] does not allow for long-range structures such as domain wall formation because there is no provision for modulations of the adlayer along the incommensurate axis by the periodic substrate potential. In addition, this approximation artificially enhances the periodic energy contributions for commensurate structures.

The $(\sqrt{3} \times \sqrt{3})R30°$ in-plane (2×1) herringbone monolayer was found to be the most stable configuration for all types of modeling [298]. The uniform uniaxial incommensurate monolayer has also in-plane herringbone orientational order even up to higher compressions, and a search for a stable out-of-plane herringbone monolayer of the type shown in Fig. 4*c* led to no viable candidates. However, pinwheel structures exist for certain potential models, but they are energetically very similar to the herringbone phases and might be favorable at higher densities. Lattice instabilities for the pinwheel monolayers were encountered when the electrostatic image terms were properly included and could be cured by neglecting these terms. In general the results are quite sensitive to the interaction models, but the inclusion of harmonic zero-point energies does not alter the stability sequence within the models. The ground-state harmonic zero-point root-mean-square librations are 18° in the commensurate herringbone monolayer, and 23° and 16° for the pin and wheel molecules, respectively, in the pinwheel monolayer at a coverage of about $1.12\sqrt{3}$ monolayers using the X1 potential [248] without image charges. These amplitudes are slightly larger than the value 15.6° obtained for the orientationally ordered $Pa3$ α-N_2 bulk structure using the X1 potential [248]. In summary, the modeling of Ref. 298 is consistent with the following succession of structures as the coverage is increased: commensurate herringbone monolayer, uniaxially compressed incommensurate herringbone monolayer, pinwheel monolayer, and only then bilayers of the pinwheel or uniaxially compressed incommensurate herringbone type.

An alternative to the classic site–site expansion [324, 326] for the anisotropic N_2–graphite surface potential was developed [18]. This potential combines the following three ingredients: a spherical expansion [6] in symmetry-adapted free-rotor functions of the molecule which represents the anisotropy of the potential explicitly, an anharmonic translational-displacement expansion, and a Fourier expansion which reflects the translational symmetry parallel to the surface. In order to accurately describe in this way the interaction between an N_2 molecule and the graphite substrate, it appears sufficient to include only one corrugated layer and 10 flat layers where only the lowest-order Fourier term is kept. Specifically for a single N_2 molecule sitting on the minimum energy adsorption site it is explicitly demonstrated that the in-plane anisotropy appears negligible with respect to the out-of-plane anisotropy. In the commensurate in-plane herringbone structure the out-of-plane crystal-field anisotropy is strongly determined by the N_2–graphite potential, whereas the in-plane anisotropy is dominated by the N_2–N_2 interaction. These calculations also indicate that anharmonic terms in the molecular displacement expansion are important and will influence the out-of-plane translational vibrations.

Phonons and librons in herringbone and pinwheel commensurate and incommensurate N_2 monolayers on graphite were investigated based on quantum-mechanical mean-field and time-dependent Hartree methods for the ground state [19]. The latter method includes on a systematic basis rotation–translation coupling which is neglected within the mean-field approximation, and it is able to treat motions with larger (but still finite) amplitudes around some equilibrium structure because higher-order anharmonicities are taken into account. Calculations for the ortho and para species of the N_2 molecules yield practically the same results for the ordered monolayers as expected. The potential consists of an *ab initio* N_2–N_2 potential [23] in its spherical representation and for comparison also in the site–site fitted form. The higher multipole moments are included exactly in the converged expansion in spherical harmonics without approximating them by the asymptotic multipole expansion. The molecule–graphite potential is modeled by the empirical Lennard-Jones atom–atom expression [324, 326], but its parameters σ_{NC} and ϵ_{NC} were also systematically varied. The $\mathbf{q} = 0$ phonon and libron frequencies and the full dispersion curves for commensurate and incommensurate herringbone monolayers as well as for an incommensurate pinwheel phase are collected in Table III and are shown in Fig. 44 for the most accurate model—that is, the time-dependent Hartree method with quartic anharmonicity in the displacements and the spherical expansion of the *ab initio* N_2–N_2 potential [23].

Anharmonicity effects are relatively small, but significant: the commensurate herringbone monolayer is lifted by 0.054 Å relative to the static

TABLE III

Lattice Frequencies [19] (in cm^{-1}) at the Γ Point of the Two-Dimensional Brillouin Zone for Orientationally Ordered N_2 Monolayer on Graphite from Time-Dependent Hartree Lattice Dynamics Including Up to Quartic Displacement Terms, the Spherical Expanded *Ab Initio* N_2–N_2 Potential [23], and an Empirical N_2–Graphite Potential [326][a]

Mode	C-HB	IC-HB	IC-PW
In-plane translations	6.9 (B_1)	0.0 (B_1)	0.0 (E_1)
	7.1 (B_2)	0.0 (B_2)	31.4 (B)
	35.6 (B_2)	51.4 (B_2)	38.8 (E_1)
	48.8 (B_1)	80.5 (B_1)	62.4 (E_1)
			73.5 (B)
Out-of-plane librations	40.6 (B_2)	19.5 (B_2)	24.3 (E_1)
	51.6 (B_1)	37.5 (B_1)	50.6 (B)
			52.1 (E_1)
Out-of-plane translations	55.9 (A_1)	53.0 (A_2)	51.7 (A)
	55.5 (A_2)	56.1 (A_1)	53.1 (E_2)
			54.5 (A)
In-plane librations	60.4 (A_1)	78.6 (A_1)	41.7 (E_2)
	69.9 (A_2)	88.4 (A_2)	68.8 (A)

[a]The full spectra are shown in Fig. 44. C-HB: commensurate in-plane herringbone structure (*p*2*gg*) with a nearest-neighbor distance of 4.26 Å. IC-HB: incommensurate in-plane herringbone structure (*p*2*gg*) with optimized lattice constants of 4.15 Å and 6.87 Å. IC-PW: incommensurate pinwheel structure (*p*6) with a nearest-neighbor distance of 4.075 Å.

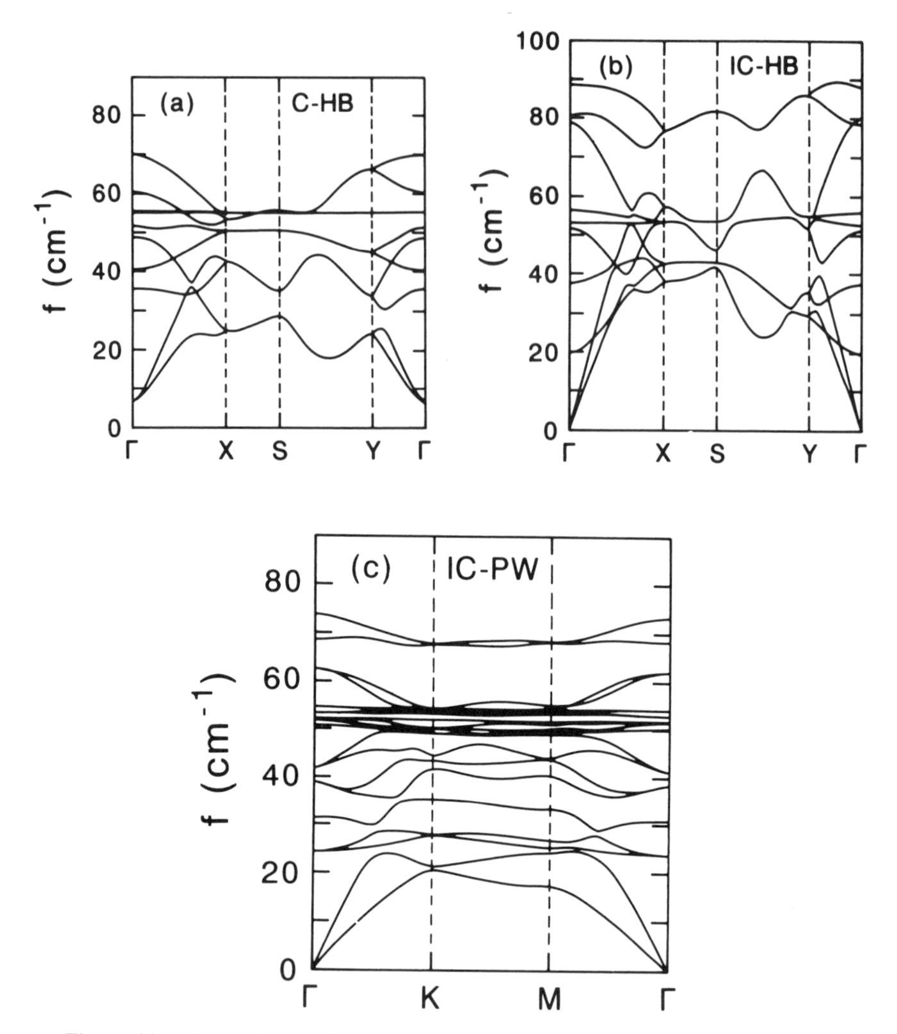

Figure 44. Phonon–libron dispersion curves in the two-dimensional Brillouin zone for orientationally ordered N_2 monolayers on graphite from time-dependent Hartree lattice dynamics calculations including up to quartic anharmonic displacement terms, the spherical expanded *ab initio* N_2–N_2 potential [23], and an empirical N_2–graphite potential [326]. (*a*) C-HB: commensurate in-plane herringbone structure [symmetry group $p2gg$ where $\Gamma = (0, 0)$, $X = (\pi/a, 0)$, $Y = (0, \pi/b)$, and $S = (\pi/a, \pi/b)$] with a nearest-neighbor distance of 4.26 Å. (*b*) IC-HB: incommensurate in-plane herringbone structure (symmetry group $p2gg$) with the uncorrugated N_2–graphite potential optimized lattice constants of 4.15 Å and 6.87 Å. (*c*) IC-PW: incommensurate pinwheel structure (symmetry group $p6$ where $\Gamma = (0, 0)$, $K = (2\pi/3a, 0)$, and $M = (\pi/2a, \pi/2a\sqrt{3})$) with the uncorrugated N_2–graphite potential and a nearest-neighbor distance of 4.075 Å. The Γ-point frequencies are reported in Table III. (Adapted from Figs. 2, 3, and 4 of Ref. 19.)

equilibrium height, and its vibrations are shifted by several wave numbers (see Tables II, IV, and V of Ref. 19 for details). The harmonic character [59] of the dispersion curves of Fig. 44 and the near separation between in-plane and out-of-plane motions [203] except near avoided crossings are confirmed with a much more elaborate methodology. The two lowest frequencies in Table III corresponding to acoustic phonons are not zero because of the corrugation in the commensurate phase and yield a phonon gap of about 7 cm^{-1}. The anharmonic shifts of the in-plane and out-of-plane phonons and librons are smaller than in the bulk phases; in particular, the phonon gap is not much affected by translational and librational anharmonicities. The frequency shifts found when the more accurate spherical expansion of the intermolecular potential was replaced by its site–site fit amount to at most a few wave numbers. The Lennard-Jones parameters characterizing the N_2–graphite holding potential were systematically varied for the commensurate herringbone monolayer mainly in order to probe its impact on the acoustic phonon gap. It is found that this gap depends on both parameters σ_{NC} and ϵ_{NC}, but it was again [137] not possible to find within physically reasonable bounds a value close to the experimental value of about 13 cm^{-1} for the Brillouin-zone-center frequency gap for in-plane vibrations with zero wave vector [137, 192] (see above in this section). Because the acoustic phonon gap is directly related to the substrate corrugation, it is concluded that an atom–atom model as generally used to describe the N_2–graphite interactions [324, 326] cannot represent the correct surface corrugation. However, only the bare Lennard-Jones potential without any substrate mediated corrections, such as image charges and anisotropic dispersion interactions [49, 61, 165, 225, 362], was taken into account (see also the work of Refs. 138 and 140 and its discussion in Section III.D.1).

For incommensurate monolayers it is found that the in-plane herringbone structure is more stable than the pinwheel structure at zero pressure [19]. The dispersion curve Fig. 44*b* is qualitatively very similar to the one for the commensurate herringbone monolayer (Fig. 44*a*), except for the vanishing phonon gap in the incommensurate case. Due to the compression the frequencies of the in-plane modes are, however, considerably higher with a more pronounced dispersion and slightly larger anharmonicity shifts. At higher pressures, but probably still before bilayer formation, the pinwheel structure seems to be more stable. Its dispersion curve in Fig. 44*c* is, as expected, very different from the compressed herringbone case, and the frequencies are much lower at the same coverage as for the incommensurate herringbone case Fig. 44(*b*).

Heat capacity scans from 3 K to about 100 K based on adiabatic calorimetry [156] revealed new features at a coverage of 1.109 $\sqrt{3}$ monolayers

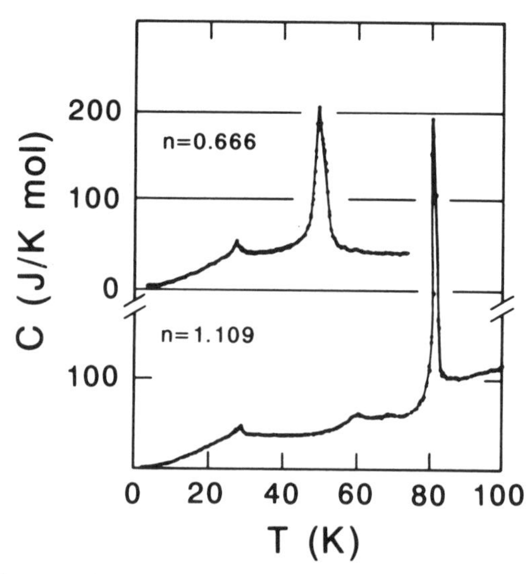

Figure 45. Heat capacity traces from adiabatic calorimetry for N_2 on graphite for coverages of 0.666 (above) and 1.109 (below) monolayers in units of the complete $\sqrt{3}$ monolayer. (Adapted from Fig. 2 of Ref. 156.)

as compared to lower coverages (see Fig. 45). In addition to the herringbone order–disorder peak which is shifted from 27.5 K at 0.666 to 28.7 K at 1.109 monolayers, and the melting peak which is shifted from 49.4 K to 81.9 K for the same coverages, there are two new anomalies found at 60.2 K and 68.6 K in the compressed regime. Though the origin of these peaks is not clear from these data, the existence of a reentrant fluid phase squeezing in between the uniaxially compressed incommensurate and commensurate phases could not be ruled out; an early reentrance speculation is included in the tentative phase diagram (Fig. 2 of Ref. 106), and reentrance was reported to occur for N_2 on graphite in Ref. 20 of Ref. 112. Note that clear evidence for such a reentrance was obtained in the case of CO on graphite in Ref. 112 (see Section V.F and the phase diagram in Fig. 51). The partly speculative phase diagram resulting from the heat capacity data of Ref. 156 is shown in Fig. 46, where the measurement path for the highest coverage is sketched as a dashed line.

Finally we only mention that, for example, calorimetric investigations [72, 395], LEED [91], high-energy electron diffraction [313], neutron diffraction [45, 371], high-resolution electron energy loss spectroscopy [304], and ellipsometric studies [111, 147, 366] were extended much beyond the fully compressed N_2 monolayer into multilayer and film regimes.

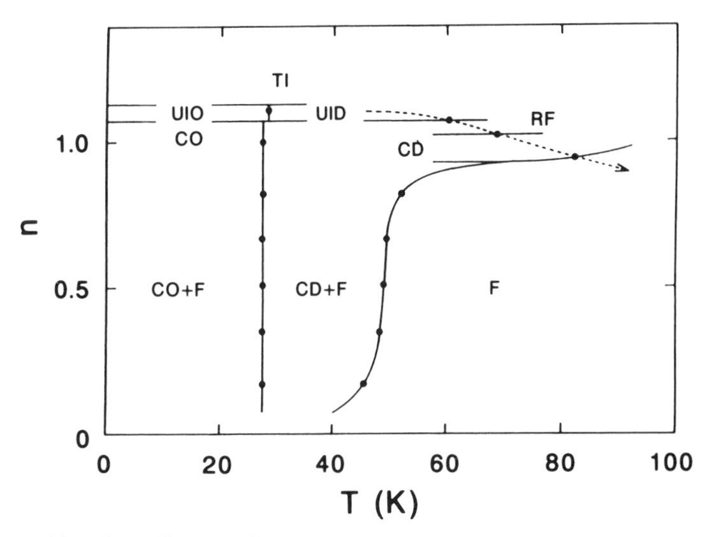

Figure 46. Phase diagram of N_2 on graphite based on adiabatic heat capacity data; coverage is reported in units of the complete $\sqrt{3}$ monolayer. Orientationally ordered commensurate phase (CO), orientationally disordered commensurate phase (CD), orientationally ordered uniaxially compressed incommensurate phase (UIO), orientationally disordered uniaxially compressed incommensurate phase (UID), triangular compressed incommensurate phase (TI), fluid phase (F), speculative reentrant fluid phase (RF). The measurement path for the highest coverage in Fig. 45 is sketched by the dashed line. (Adapted from Fig. 10 of Ref. 156.)

IV. N_2 ON BORON NITRIDE

The phase diagram of N_2 on boron nitride was studied by adsorption isotherm measurements [1] in the range from 51 K to 75 K. The assignment of the phases in the tentative phase diagram in Fig. 47 was inferred by comparing the characteristics of the thermodynamic data to those for CO, Kr, and N_2 on graphite. Isotherms above 65.8 K were found to be essentially featureless, signaling that no phase changes occur in this range, whereas the traces below that temperature show a sharp substep and a kink in the slope. The step is located near monolayer completion, whereas the kink is found at coverages slightly above the substep [1]. Correspondingly, the isothermal compressibility displays a very sharp and pronounced peak and a cusp at the respective pressures (see Fig. 48). In analogy to CO, Kr, and N_2 on graphite, the sharp isotherm substeps for N_2 on boron nitride are interpreted as corresponding to a coexistence region of the commensurate solid and fluid phases. The kinks at coverages slightly above the substeps have been identified as indicating the melting from the pure commensurate solid to a reentrant fluid phase. It is remarked that N_2 on boron nitride might represent a borderline

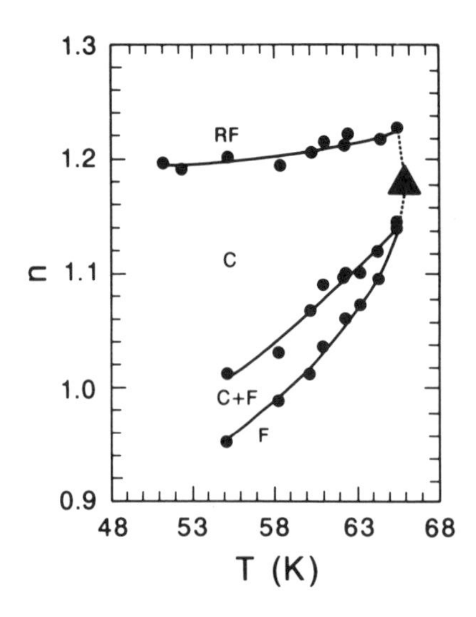

Figure 47. Phase diagram of N_2 on boron nitride based on adsorption isotherms; coverage is reported in units of the complete $\sqrt{3}$ monolayer obtained from the top of the fluid to commensurate solid isotherm substep at low temperatures less than 51 K. Commensurate solid phase (C), fluid phase (F), reentrant fluid phase (RF). The solid lines correspond to phase boundaries based on measured features, the dotted line is an expected phase boundary, and the triangle marks the tricritical point. Second-layer growth instead of a transition to an incommensurate solid phase is expected beyond the reentrant fluid phase in the temperature range studied. (Adapted from Fig. 4 of Ref. 1.)

case for commensuration, with substrate and adsorbate parameters being just marginally sufficient for forming a registered solid phase. The tricritical point at the end of the commensurate solid–fluid coexistence regime is located at a coverage of about 1.2 monolayers at 65.8 K, which is nearly 20 K lower than that for the same adsorbate on graphite (see Section III.C). This was attributed to be a likely consequence of the smaller corrugation and physisorption well depths on boron nitride as compared to graphite (see Section I.C).

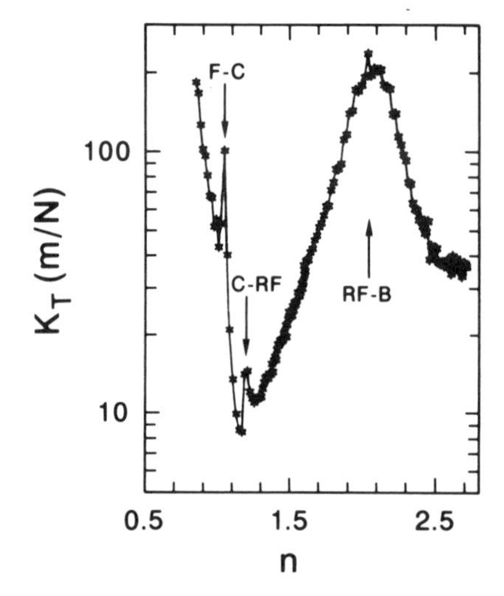

Figure 48. Semilogarithmic plot of the isothermal compressibility of N_2 on boron nitride at 60.8 K as a function of the coverage in units of the complete $\sqrt{3}$ monolayer. The peak sequence starting at low coverages is attributed to the fluid to commensurate solid F-C and commensurate solid to reentrant fluid C-RF transitions and finally to second-layer growth RF-B (instead of a transition from the reentrant fluid to an incommensurate solid phase). (Adapted from Fig. 5 of Ref. 1.)

The isothermal compressibility peak at highest coverage, however, is not attributed to the transition from the reentrant fluid to an incommensurate solid. Following the evolution of the second-layer growth from lower temperatures, this anomaly near two monolayers (see Fig. 48) is associated to the formation of a second layer on the substrate. It is also suggested [1] that such an interpretation instead of the transition to an incommensurate phase might also hold for the similar case of CO on graphite reported in Ref. 112 (see Section V.F), where a broad isothermal compressibility peak over a coverage interval of nearly one layer is observed. No evidence for a transition from the reentrant fluid to an incommensurate solid was obtained for temperatures in the studied range from 51 K to 75 K, so that this transition for N_2 on boron nitride should occur below 51 K.

V. CO ON GRAPHITE

A. Phase Diagram

The temperature versus coverage phase diagram of CO physisorbed on graphite as deduced from various experiments is presented in Fig. 49; we refer to Sections V.B–V.F for a more detailed discussion, and to the corresponding phase diagram of N_2 on graphite in Fig. 8 for comparison. Above about 95 K only a homogeneous fluid phase (F) is stable for submonolayer coverages [112]. The centers of mass of the molecules show positional ordering in the commensurate $(\sqrt{3} \times \sqrt{3})R30°$ solid phase with disordered molecular orientations (CD) in coexistence with the fluid phase (CD + F) in the coverage range 0.2–0.8 for temperatures below about 50 K.

As the temperature increases, this registered solid–fluid coexistence region (CD + F) considerably narrows down to a strip that ends at a special point at 93–94 K. This special point is believed to be a tricritical point in the universality class of the two-dimensional three-state Potts model [112]. Similarly to the incipient-triple-point picture for N_2 monolayers on graphite, there is only two-phase coexistence between the $\sqrt{3}$ solid and a fluid phase without liquid–vapor coexistence for any temperature and density; that is, no genuine first-layer critical or triple points involving a first-layer liquid phase exist. Center-of-mass melting to the fluid phase is believed to be of first order up to the complete monolayer and is believed to be continuous beyond the tricritical point near 93–94 K. Following the melting line much beyond this first tricritical point, a second tricritical point at about 85 K seems to exist at higher coverages [112] (see Fig. 51 for details); this assignment, however, is still tentative [112]. On the phase boundary in between these two tricritical points the commensurate solid phase (CD) melts continuously to the fluid phase, which shows a reentrant behavior (RF) beyond the high-coverage stability boundary of the commensurate solid; thus

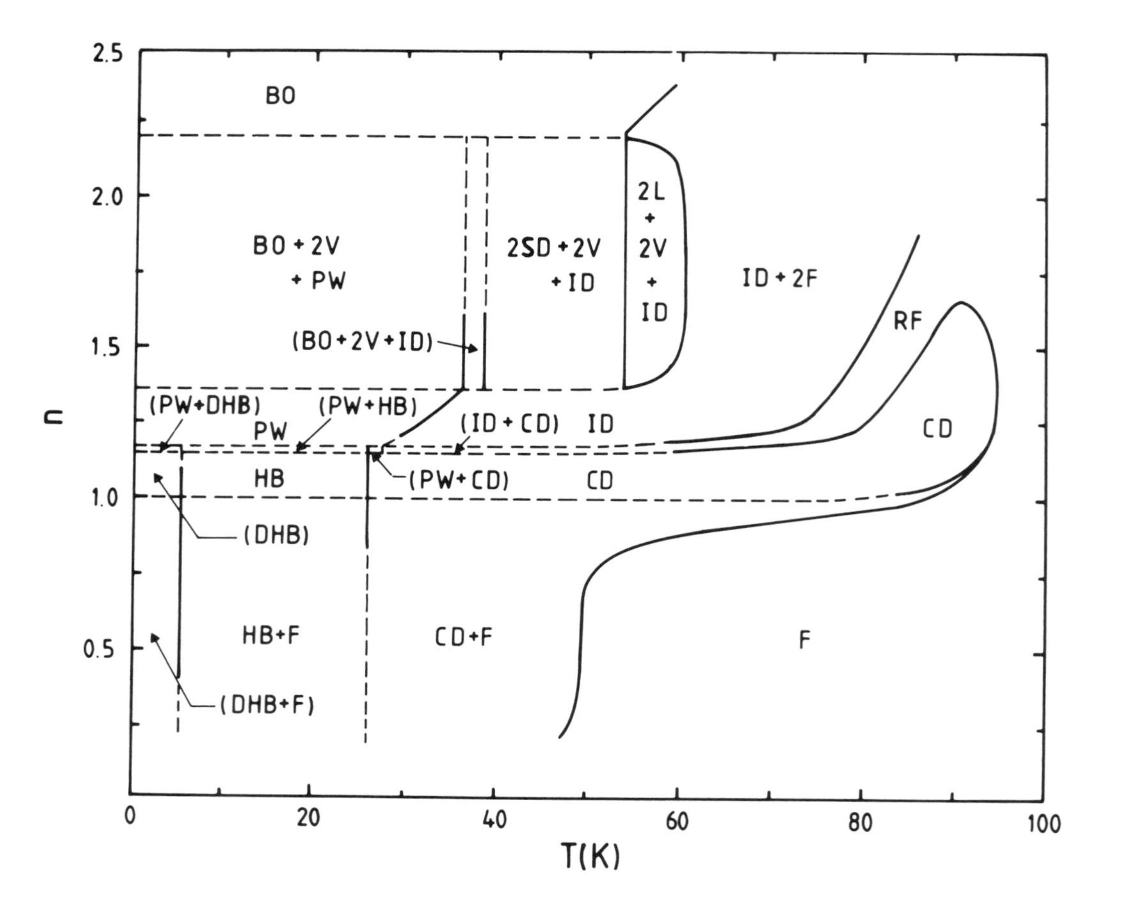

BO
BO + 2V
+ PW
(BO + 2V + ID)
2SD + 2V
+ ID
2L
+
2V
+
ID
ID + 2F
RF
CD
(PW+DHB)
PW
(PW+HB)
(ID + CD)
ID
HB
(PW+CD)
CD
(DHB)
HB+F
CD+F
F
(DHB+F)
u
2.5
2.0
1.5
1.0
0.5
0
20
40
60
80
100
T(K)

Figure 49. Phase diagram of CO physisorbed on graphite with the following phases: commensurate orientationally disordered $(\sqrt{3} \times \sqrt{3})R30°$ solid phase (CD), commensurate orientationally ordered herringbone phase (HB), head–tail (dipolar) ordered commensurate herringbone phase (DHB), fluid (F), reentrant fluid (RF), incommensurate orientationally disordered solid (ID), incommensurate orientationally ordered pinwheel phase (PW), second-layer orientationally disordered solid (2SD), second-layer liquid (2L), second-layer vapor (2V), second-layer fluid (2F), bilayer orientationally ordered solid (BO); the parentheses are included for clarity only. The solid lines are based on experimental results, whereas the dashed lines are speculative. Note that the phase boundary (ID + CD) → RF and the orientational disordering temperature of the BO phase are unknown. Coverage unity corresponds to a coverage of CO forming a complete $(\sqrt{3} \times \sqrt{3})$ commensurate monolayer. (The phase diagram is based on Fig. 1 of Ref. 112, Fig. 2 of Ref. 113, and Refs. 380 and 381.

the commensurate solid extends at most up to about 1.70 monolayers and 95 K. Such a high coverage implies that the commensurate monolayer beyond coverages of about 1.2 is in coexistence with a fluid layer on top [112]. Obviously, a high density of the second-layer fluid is needed to stabilize the commensurate phase of the first layer up to such high temperatures of 95 K. The strip of the reentrant fluid phase (RF) in between the commensurate (CD) and incommensurate (ID) orientationally disordered solids narrows with decreasing temperature. It can either hit a coexistence region of the commensurate and incommensurate orientationally disordered solids (ID + CD) or extend down to roughly 25–30 K, where orientational order in the form of the pinwheel and herringbone solids sets in [229].

A broad and weak orientational transition of the molecular axes is observed at about 26 K. The transition temperature is independent of coverage up to the full monolayer and beyond to about 1.2 monolayers; the nature of the transition is still unclear. The type of low-temperature orientational ordering is probably of the herringbone type (HB) within the positionally ordered commensurate $(\sqrt{3} \times \sqrt{3})R30°$ solid phase; the projection of the molecular axes on the surface plane is shown in Figs. 4*a* and 5*a*. There is evidence that the CO molecular axes are approximately parallel to the graphite surface with some out-of-plane tilting. A transition to a ferrielectric head–tail ordered herringbone phase (DHB) occurs at 5.18 K up to 1.08 monolayers. The head–tail transition (DHB → HB) [153] belongs to the universality class of the two-dimensional Ising model [214, 215, 380]. A probable candidate for the completely ordered ground state is sketched in Figs. 54*a* and Fig. 53*b*, and the molecular axes are probably slightly but systematically tilted out of the surface plane. This fully ordered phase seems to coexist with the fluid phase in the submonolayer regime (DHB + F) as the herringbone phase does (HB + F) at higher temperatures.

The commensurate $\sqrt{3}$ solid extends up to a coverage of approximately 1.15 monolayers. Further compression of CO on graphite leads first to a pinwheel incommensurate phase (PW) at low temperatures. This should be contrasted to N_2 on graphite, where a uniaxially compressed phase is believed to be the first stable incommensurate phase. However, in Ref. 156 such a uniaxial incommensurate phase is proposed beyond commensurability in the range from 1.04 to 1.05 monolayers also for CO on graphite and finally transforms to the pinwheel phase at a coverage of 1.05 (see Fig. 64), and independent support is provided in Ref. 176 in the range from 50 K to 62 K in the slightly compressed monolayer (see the discussion in Section V.F). There seem to be small coexistence regions between about 1.15 and 1.19 monolayers of the pinwheel and herringbone phases (PW + HB), as well as between the pinwheel and head–tail ordered herringbone phases (PW + DHB) below about 5.2 K. The pinwheel phase is stable as a pure phase up to about 35 K, and the highest possible compression is about 1.4 monolayers. The orientational disordering transition of the pinwheel phase to the incommensurate orientationally disordered solid (PW → ID) at about 30–35 K belongs to the two-dimensional Ising universality class [113].

Beyond about 1.4 monolayers, the first-layer pinwheel phase (PW) with a second-layer vapor phase (2V) on top coexists in the region denoted by BO + 2V + PW with a bilayer orientationally ordered phase (BO). This coexistence region extends to about 2.2 monolayers where the bilayer becomes stable as a pure phase; the orientational disordering of the pure bilayer upon increasing the temperature is not investigated. Warming up the adsorbate from the BO + 2V + PW coexistence regime results first in the disordering of the pinwheel phase at about 36 K into an orientationally disordered incommensurate phase (ID) in the coexistence region BO + 2V + ID. At about 39 K the bilayer orientationally ordered phase (BO) transforms into a second-layer orientationally disordered solid phase (2SD), which is in coexistence with a second-layer vapor (2V) on top of the first-layer ID phase. Finally, second-layer liquid (2L) and vapor (2V) phases coexist beyond the respective triple temperature of about 53 K on top of the disordered incommensurate solid (2L + 2V + ID) up to a critical temperature of approximately 60 K, which constitutes a second-layer two-dimensional Ising-like critical point [112]. Beyond this critical temperature there is a second-layer fluid (2F) on top of the orientationally disordered incommensurate phase (ID) in the first layer.

B. Submonolayers and Melting

Volumetric adsorption isotherm measurements [348] of CO on Grafoil and exfoliated graphite were conducted in the temperature range from about 71 K to 92 K. A transition from a fluid to a registered solid is detected in the form of quasivertical risers (see Fig. 50 for such isotherms on exfoliated graphite and Grafoil near 76 K). The difference of the curves is an explicit demonstration of the different domain size distributions in both substrates. The CO data could directly be compared to N_2 data, and a very similar behavior with almost equal areal densities of both adsorbates were found. The isosteric heat of CO and N_2 on graphite differ by less than 4% over the whole range of measured temperatures, and the law of corresponding states can be used to scale many thermodynamic parameters of the fluid to registered solid transition for both systems. By analogy it is concluded [348] that the observed transition of CO on graphite is that from a fluid phase to a registered $(\sqrt{3} \times \sqrt{3})R30°$ solid CO monolayer. The density of the monolayer solid is almost constant up to about 85 K, which indicates a negligible vacancy content in the registered phase.

In the calorimetric study [287] heats of adsorption, heat capacities, and

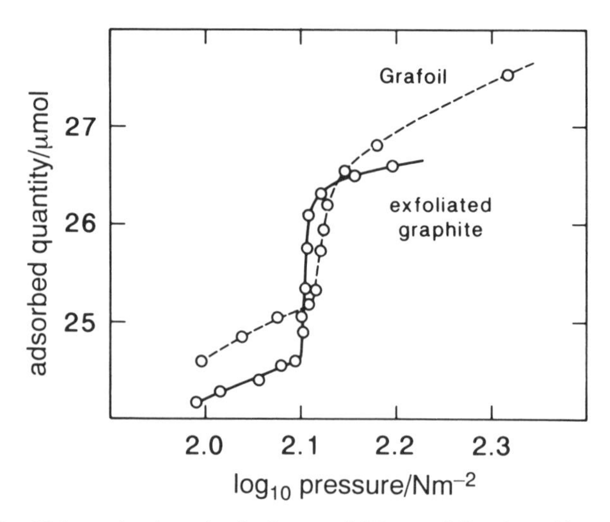

Figure 50. Volumetric adsorption isotherms of CO on exfoliated graphite (solid line) and Grafoil (dashed line) indicating the transition from the fluid phase to the registered $(\sqrt{3} \times \sqrt{3})R30°$ monolayer solid. (Adapted from Fig. 1 of Ref. 348.)

adsorption isotherms were obtained for CO on graphite up to 1.2 monolayers at various temperatures from 20 K to 119 K. The isosteric heat of adsorption at zero coverage was determined to be in the range from 10.9 $\pm$ 0.1 kJ/mol to 11.2 $\pm$ 0.1 kJ/mol for temperatures from 79.3 K to 119 K; these results are in agreement with earlier measurements and also with data for N_2 on graphite (see Table 2 in Ref. 287 for further details). It is shown that the ratio of the experimental heats of adsorption of CO and N_2 is consistent with the properties of the molecules. The coverage dependence of the integral heat of adsorption was also calculated based on an empirical potential [238] including molecule–surface interactions [287]. The kink in the low-temperature isotherms observed in Ref. 348 is also found and assigned to the fluid to commensurate solid transition. Vertical segments in the isotherms at 58.0 K and 60.4 K near half coverage were interpreted as being indicative for liquid–vapor coexistence, and a corresponding phase diagram is proposed that is strongly influenced by what was known at that time for N_2 on graphite.

LEED investigations [389] in the low-coverage regime showed that the commensurate $(\sqrt{3} \times \sqrt{3})R30°$ arrangement of the scattering centers extends up to about 45 K where desorption takes place. The LEED pattern observed at 45 K is essentially identical to that of commensurate orientationally disordered N_2 on graphite. The submonolayer melting transition at higher CO coverages near 0.83 monolayers occurs at about 80 K and is quite sharp as deduced from x-ray diffraction experiments [242]; it is speculated that the solid–fluid transition is of the first-order type. Upon compression to about 1.17 ± 10% the melting is found [242] to occur in the range of 80–90 K. Thus, there is a similar steep raise of the melting temperature in both CO and N_2 on graphite as the complete monolayer is approached from the submonolayer regime.

Low-coverage volumetric isotherm data [41, 42] were used to extract the isosteric heat of adsorption extrapolated to zero coverage, the gas–solid virial coefficient, and the two-dimensional second virial coefficient. It is found that fitting the two-dimensional virial coefficient obtained from the measurements in the vicinity of the surface in terms of Lennard-Jones intermolecular potentials reduces the well depth obtained from the bulk by about 20%. In addition, the effective quadrupole moment of CO needs to be significantly reduced [42] by as much as about 50%. These fitted parameters are believed to account for various substrate mediation effects in some effective way (see Refs. 41 and 42 for more details concerning these parameterizations). It is also concluded [41] that the asymmetric empirical parameterization of Ref. 238 should be replaced by models in which the similarity between the isoelectronic CO and N_2 molecules is exploited for the nonelectrostatic contributions as in Ref. 17.

The shape of the melting curve was mapped by ac heat capacity and vapor pressure isotherm measurements on exfoliated graphite foam [112]. The steep increase of the phase boundary between the commensurate solid–fluid coexistence region and the fluid phase as the complete monolayer coverage is approached from below (see Fig. 51) is very similar to the topology of the phase diagram of N_2 on graphite (see Fig. 13 or Fig. 22). Based on the different contributions to the molar entropy derived from calorimetric data [156], it is explicitly demonstrated that the in-plane rotation is quite free, whereas the in-plane translations are highly hindered for submonolayer CO on graphite at 80 K (i.e., in the fluid phase).

C. Tricritical Point

A critical point or possibly multicritical point is observed in volumetric adsorption measurements [348]. Based on the qualitative change in the behavior of the quasivertical risers as a function of temperature, it is suggested that this point is located at about 89 K for CO on graphite; an identical experiment for N_2 on graphite [189] yielded 82 K (see Section III.C).

The ac heat capacity and vapor pressure isotherm experiments of CO on exfoliated graphite foam [112] allowed the detailed investigation of the narrowing of the commensurate solid–fluid coexistence region as the tricritical point is approached; see Fig. 51 for this part of the phase diagram and Fig. 52 for the calorimetric data corresponding to the paths I to VII marked in the phase diagram. The vapor pressure isotherms indicate that the transition

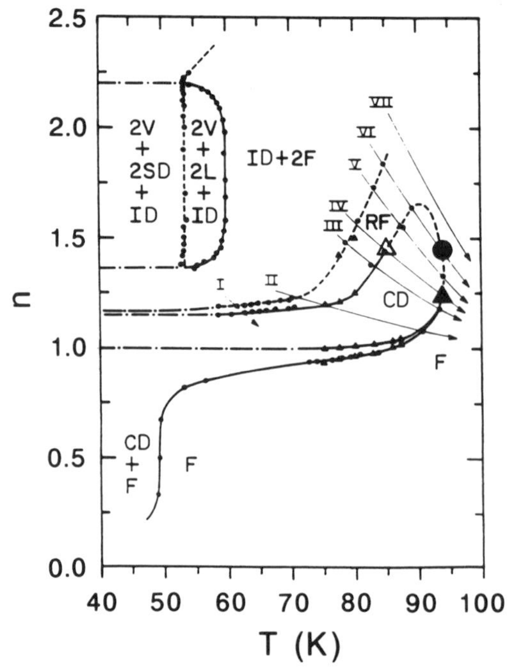

Figure 51. Experimental phase diagram of CO physisorbed on graphite with the phases: fluid (F), commensurate (CD) and incommensurate (ID) orientationally disordered solids, reentrant fluid (RF), second-layer fluid (2F), vapor (2V), liquid (2L), and orientationally disordered solid (2SD) phases. Filled circles and triangles represent phase boundary locations from heat capacity and vapor pressure measurements, respectively. Solid and dashed lines indicate phase boundaries believed to be associated with first-order and continuous transitions, respectively; dash–dotted lines correspond to speculated boundaries. The large filled triangle and the large filled circle mark the two-dimensional Potts tricritical and critical points, respectively; the tricritical point marked with an open triangle is tentative. Lines I–VII with arrows are experimental paths of the heat capacity scans shown in Fig. 52. Coverage unity corresponds to a coverage of CO forming a complete ($\sqrt{3} \times \sqrt{3}$) monolayer. (Adapted from Fig. 1 of Ref. 112.)

from the commensurate solid to the fluid is first order at least up to 87.2 K as evidenced from the extremely large value of the compressibility over a finite coverage range. This is a consequence of the solid–fluid coexistence region which results in a vertical step in the isotherm. The width and height of the commensurate solid-to-fluid melting peak in the heat capacity scan II in Fig. 52 is very similar to what is observed for N_2 on graphite [65] and is indicative for first-order melting. Both characteristics change dramatically when the coverage is increased from scan IV to scan VI, while scan V might correspond to an intermediate situation (see Fig. 52).

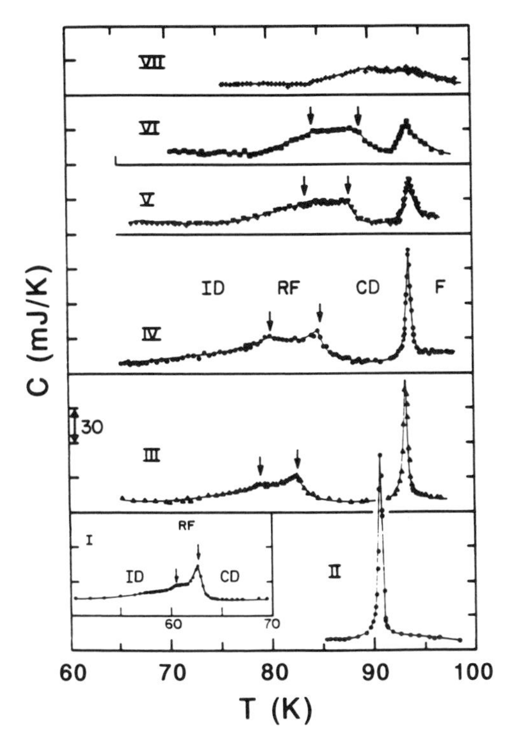

Figure 52. Heat capacity scans for CO on exfoliated graphite foam for the paths I–VII indicated in the phase diagram Fig. 51 as a function of temperature for seven initial low-temperature coverages 1.35 (I), 1.35 (II), 1.90 (III), 2.20 (IV), 2.60 (V), 2.90 (VI), and 3.51 (VII). For the symbols see Fig. 51. (Adapted from Fig. 2 of Ref. 112.)

This crossover in the behavior of the melting transition from the commensurate solid to the fluid phase is analyzed by evaluating [112] the heat capacities in this region of the phase diagram in terms of the relation

$$C = B_0 + B_1 T + B_\alpha |1 - T/T_c|^{-\alpha} \tag{5.1}$$

where the constant and linear terms represent the background contribution, which is associated with the regular lattice part of the heat capacity. This dependence describes the data in scans IV and VI reasonably well, and the power law seems to hold from $1 - T/T_c = 3 \times 10^{-3}$ to 3×10^{-2} (see Fig. 4 of Ref. 112). The exponents α were fitted to be 0.85 ± 0.08 and 0.32 ± 0.06 for the two scans IV and VI with T_c being 93.65 K and 93.90 K, respectively. The obtained exponents are consistent with the corresponding tricritical (0.833) and critical (0.33) heat capacity exponents of the two-dimensional three-state Potts model (see Ref. 112). The scans I–III appear to be indicative of first-order melting to the fluid phase, whereas scan V might show the crossover from tricritical to critical three-state Potts behavior; these data could not be fitted using (5.1). Thus, as the coverage is increased, the melting transition of the commensurate phase displays first-order, tricritical-point, and continuous behavior [112]. Similar to N_2 on graphite (see Section III.C), the highest temperature of the commensurate solid phase is reached very close to the tricritical point.

D. Commensurate Herringbone Ordering

The LEED pattern [389] for submonolayer CO on graphite below about 25–30 K is very similar to that observed for commensurate $(2\sqrt{3} \times \sqrt{3})R30°$ herringbone-ordered N_2 monolayers on graphite (see Fig. 3 in Ref. 389 for a direct comparison). The intensities and spot profiles, however, are quite different from those of N_2 on graphite, which is a first indication that the submonolayer orientational order–disorder transition of CO on graphite might be significantly different from that of N_2 on graphite (see also Ref. 108). A crude estimate [389] of the correlation length indicates that it may be as small as 200 Å at 22 K. The LEED superlattice spots for CO on graphite reach their saturation value below about 25 K, decrease to half that value at 30–35 K, and are decayed for temperatures above approximately 40 K (see Fig. 4 of Ref. 108).

Similarly to the case of head–tail ordering (see Section V.E), a systematic out-of-plane tilt of the molecular axes should break one of the two glide line symmetries which are both present in the in-plane herringbone phase. Since no corresponding extra spots were observed it is concluded based on LEED that no such long-range systematic tilt exists for herringbone-ordered CO on graphite in the studied temperature range [389]; the largest systematic out-of-plane tilt angle which would still be compatible with the resolution is estimated [389] to be at most 5°. Note that this glide line argument does not exclude the possibility of a random out-of-plane tilt discussed in Ref. 242 (see below), which would occur in a head–tail disordered situation where one end of the CO molecule prefers to be closer to the surface than the other.

The superlattice diffraction intensities and profiles in a simultaneous x-ray diffraction study [242] of CO on Papyex were measured at 10 ± 1 K and a coverage of about $0.83 \pm 10\%$ monolayers. The fits of the spectra in terms of several models such as the in-plane herringbone (a), pinwheel (b), and out-of-plane herringbone (c) structures of Fig. 4 was not unambiguous, and especially the (21) peak is relatively small. An acceptable agreement for otherwise reasonable fit parameters such as domain size, lattice parameter, structure, and temperature factors was obtained for the out-of-plane herringbone structure (c) of Fig. 4 with a very large tilt angle of $26 \pm 12°$ with respect to the surface. The experiment, however, cannot distinguish between random and systematic tilting, nor is it sensitive enough to detect a possible head–tail ordering. However, the amount of tilting is larger than one would expect from dynamical librations and could be taken as indicative of static tilting [161]. The in-plane herringbone arrangement (a), on the other hand, can also yield a reasonable fit of the experimental data if the range of the orientational order is reduced to about 200 Å as compared to 450 Å for the translational order of the triangular center-of-mass lattice. Thus, these x-ray results [242] do not have the sensitivity to resolve unambiguously the type of orientational order in submonolayer CO on graphite. However, they indicate together with the LEED data [389] that there is probably more orientational disorder present in the in-plane herringbone structure for CO than for N_2 on graphite. This disorder could be associated [242] with random static or systematic out-of-plane tilting and/or with orientational order of shorter range than the translational order in the underlying $(\sqrt{3} \times \sqrt{3})R30°$ structure.

The herringbone ordering transition deduced from x-ray diffraction [242] in the commensurate phase is much broader for CO as compared to N_2 on the same piece of graphite and also sets in at lower temperatures (see Fig. 28 for a direct comparison). The plotted x-ray intensity ratio [242], which is a measure of the amount of in-plane orientational order, starts to drop at about 20 K, decreases gradually, and is practically decayed only at about 40 K. In addition, the quadrupole ordering transition temperature is unexpectedly low in view of a much larger quadrupole moment in the case of CO; note that the corresponding $\alpha \rightarrow \beta$ orientational ordering temperature of bulk CO is much larger than that of bulk N_2 which reflects as expected the disparity of the quadrupole moments (see Table I). It is concluded [242] that pure point quadrupolar interactions are not adequate to model the orientational ordering of commensurate CO monolayers on graphite.

The pioneering finite temperature Monte Carlo study [282] of the orientational disordering of commensurate and incommensurate monolayers of CO on graphite was based on empirical potentials and 64 molecules in a rectangular periodic cell. The CO–surface interactions were modeled with the Fourier representation [324, 326], and the necessary Lennard-Jones parameters were obtained from a fit to the measured isosteric heats of adsorption [287]. The nonelectrostatic CO–CO intermolecular interactions were based on Buckingham-type potentials as parameterized in Refs. 238 and 287. The electrostatics was represented by a three-site point-charge distribution located on the molecular axis [282]. The chosen values yield a reasonable representation of the moments up to the hexadecapole; the dipole moment, however, is larger than the experimental value. These interactions were augmented with surface-mediated dispersion terms [49, 225]. Image charges were not included in these calculations because of some divergent behavior in the case of CO on graphite [282].

The commensurate initial herringbone configuration at coverage unity was stable at 10 K up to 25 K. Further heating the sample resulted in a smeared disordering with some sort of herringbone-like short-range order [282]. Translational diffusion sets in at about 55 K, and the systems appears to be fluid at 60 K. Using a hybrid potential for CO composed of an *ab initio* atom–atom potential [23] for N_2 in conjunction with the CO multipole moments [282] gave evidence that the herringbone phase is destabilized by the nonelectrostatic contribution. It is also suggested, based on simulations of free CO patches on graphite which contract to incommensurate structures without applying pressure, that the CO molecule as represented by the potential model [282] is probably too small. Other potentials [117] were also investigated with little success, so that the authors [282] conclude that a much better CO–CO potential is called for.

The angular distributions of CO molecules on highly oriented pyrolytic graphite composed of 1-μm microcrystallites are investigated by resonance electron energy loss spectroscopy (EELS) [161]. Resonance electron scattering is sensitive to the orientation of molecules on surfaces, so that this technique can be used to probe orientational ordering in adsorbates. Two types of disorder should be considered for CO on graphite. Dynamic tilting is caused by the thermal motion of the molecular axes around some symmetric equilibrium orientation, whereas static tilting is a consequence of asymmetries and is also present in the ground state at zero temperature. The EELS spectra show that the CO molecular vibrations and resonance excitation mechanism are essentially unchanged upon adsorption, which explicitly demonstrates the weak interactions of the physisorbed molecules with the substrate. The angular distributions of the lowest vibrational excitation give strong direct evidence that the CO molecules lie approximately parallel on the graphite surface in the commensurate herringbone phase at 25 K and at a coverage which is slightly less than unity [161]; there are also indications that the spectra cannot be explained by scattering from isolated oriented molecules but from molecular monolayer structures. Heating up the sample to 38 K at the same coverage yields angular distributions which give evidence that the molecules are largely constrained to rotate about an axis normal to the surface with significantly less out-of-plane rotation [161]. However, both data sets cannot be fully understood in terms of only dynamical librations, and the assumption of some static tilting might be a possible explanation. It is speculated that at higher temperatures the molecules might be able to reduce the amount of their static tilting by increasing their average distance to the surface; an increase of the distance between overlayer and surface as a function of temperature is observed in simulations of N_2 on graphite [301] [see, e.g., the average height $\langle z \rangle$ (squares) in Fig. 20*a*]. Thus, the increased dynamical libration contribution due to the higher temperature could be compensated by a reduction of the static tilting by lifting the overlayer slightly [161]. In conclusion, the EELS study [161] indicates that the CO molecular axis is approximately parallel to the graphite basal plane in the submonolayer regime, but there is evidence of some out-of-plane tilting.

The ac heat capacity traces [113] in the commensurate solid range at 0.85, 0.95, and 1.00 monolayers uncover weak and broad anomalies centered at about 26 K. Typically, they have a full width at half-maximum as large as 2.5 K and a height of only 0.4 k_B/molecule, whereas the analogous representative data [113, 232] for the herringbone transition of N_2 on graphite are 0.5 K and 5 k_B/molecule, respectively. These anomalies are identified as related to the orientational disordering from the herringbone phase of CO on graphite [113]. Based on the scaling of the orientational transitions within the bulk solids (see Section I.B and Table I), one would expect that CO should disorder at higher temperatures than N_2 on graphite. Guided by the behavior of the quadrupolar lattice model [141] (2.1) + (2.3) it is proposed [113] that the lower temperature found in experiments is related to the change of the magnitude of the substrate interaction V_c relative to the strength of the quadrupolar intermolecular interactions Γ_0. Assuming a negative crystal field of similar strength V_c for both CO and N_2 on graphite, the ratio $V_c^* = V_c/\Gamma_0$ is smaller for CO as compared to N_2 because of the larger CO quadrupole moment eQ [see (2.2)]. This promotes out-of-plane tilts of the molecular axes, which ultimately results in the appropriate parameter ranges in the stabilization of the out-of-plane herringbone and pinwheel phases (see the phase diagram, Fig. 3). Even without changing the character of the phases the value of V_c^* might be small enough to allow for substantial out-of-plane tilting, which might lead to frustration effects and random tilting inducing a gradual loss of the in-plane orientational order. This interpretation would be in line with some evidence [161, 242] for stronger out-of-plane tilts in the orientationally ordered submonolayer phase of CO as compared to N_2 on graphite. As a final consequence of this interpretation, however, the herringbone disordering of CO on graphite should not be viewed as a genuine phase transition in the thermodynamic sense.

The very gradual and continuous loss of orientational order of submonolayer CO on graphite is consistent with the LEED [108, 389] and x-ray [242] results (see Fig. 28). Such a qualitatively different behavior of CO and N_2 on graphite at the orientational transition is easily possible because the reduced crystal field—that is, $V_c^* = V_c/\Gamma_0$ in a model like (2.1) + (2.3)—is probably smaller for CO than for N_2 (see also Ref. 113). It has been concluded [282, 391] that a possible reason for the peculiar behavior of CO on graphite might be that its orientational ordering is not long-ranged due to the orientational strain induced by the anisotropy of the CO molecule. In mixtures of CO with Ar, a sharpening of the transition occurs [391], which might arise because the strain is reduced by random dilution with Ar. Strains due to a strong substrate-induced orientational field are also considered as a possible cause for the probably continuous melting behavior of submonolayer Ar adsorbed on graphite [4, 233, 396]. It has been suggested [4, 233, 396] that such a continuous transition may be a realization of a Kosterlitz–Thouless–Halperin–Nelson–Young-type melting process [252, 332]. According to this theory, melting in an ideal two-dimensional solid occurs in a two-step process, in which the solid first melts into a hexatic phase by the unbinding of dislocation pairs; analogously, in a two-dimensional superfluid the superfluid state is destroyed by the dissociation of vortex pairs. For solids with sixfold symmetry the hexatic phase is characterized by sixfold quasi-long-range bond-orientational order. This phase transforms at higher temperatures via a second transition by the unbinding of disclination pairs into an isotropic fluid phase with short-range bond-orientational order. The theory predicts that the specific heat displays no divergences but only essential singularities above the transition temperatures [252, 332]. In a two-dimensional solid exposed to an external field (e.g., a substrate corrugation potential), the melting transition is expected to be modified only slightly [164, 252, 332].

We mention these theoretical results here because the orientational disordering transition of CO on graphite seems to resemble a Kosterlitz–Thouless–Halperin–Nelson–Young-type melting process phenomenologically in several respects. There is some evidence in computer simulation studies of N_2 on graphite (see, e.g., Refs. 140 and 340) that transient pinwheel-like defects are created when passing through the herringbone transition. In addition, it was stressed that local pinwheel-like defects are natural boundaries between domains of different herringbone order [342, 358]. These effects may be enhanced for CO on graphite because of the stronger quadrupolar coupling between the molecules and the smaller reduced crystal field, which also leads to the preference of pinwheel phases to accommodate the increased pressure in the compressed monolayer regime. The crucial point is that pinwheel defects can have different chiralities depending on whether the direction of the wheel pattern is either clockwise or counterclockwise. Pairs of pinwheel defects with different chiralities resemble phenomenologically the vortex pairs of the Kosterlitz–Thouless theory [252] and may have lower energy than single defects. Their unbinding may lead to a continuous transition from an orientationally ordered phase to an "orientationally fluid phase," which in the case of CO on graphite is of course still translationally ordered. Above this first melting transition of the orientationally ordered solid there are hindered rotations of the molecules with sixfold symmetry. The molecules are subject to the strong orienting fields between neighboring quadrupoles, which might lead to short-range herringbone-like correlations over a broad temperature range as observed in the diffraction experiments [108, 242, 389]. In addition, there is the sixfold symmetry of the graphite substrate [140]. Finally, at a second melting transition both the residual orientational and the translational order are destroyed. This scenario, though being highly speculative, might explain the gradual loss of orientational order of CO on graphite. It might, however, also be possible that the hexatic-like phase above the hypothetical pinwheel-pair unbinding transition has only a narrow stability range and undergoes the second orientational transition before the translational order of the $\sqrt{3}$ lattice is destroyed; such phase diagrams are indeed observed—for example, for clock models [63, 164, 252] and an antiferromagnetic XY model on a triangular lattice [195] where helicities can be assigned to the plaquettes of the lattice. It would be very desirable to have detailed computer simulation studies based on realistic potentials to clarify this interesting question. However, such studies are still a challenge even for grossly simplified model systems [373, 374].

E. Commensurate Head–Tail Ordering

The possibility of head–tail (i.e., end-to-end or dipolar) order at low temperatures was addressed in the LEED study [389]. Two possible fully ordered arrangements of CO on graphite within one of the six in-plane herringbone $(\sqrt{3} \times \sqrt{3})R30°$ structures are shown in Fig. 53 together with expected LEED diffraction patterns. In the fully ordered case, new periodicities (b instead of $b/2$ for the corresponding head–tail disordered phase in Fig. 53*a* and a instead of $a/2$ in Fig. 53*b*) arise from the asymmetric distances of the carbon and oxygen ends from the center of the molecular charge density contour (see Fig. 1*b*) and their different scattering powers. No LEED pattern compatible with such completely ordered structures was found down to the lowest temperature studied, which was 20 K.

Head–tail-ordered herringbone as well as pinwheel structures at higher coverages (see Fig. 1 of Ref. 17) were discovered in a theoretical energy minimization ground-state study [17] using various combinations of intermolecular and physisorption potentials (see the presentation in Section V.F for more details). The predicted head–tail ordered herringbone structure at coverage unity is of the type sketched in Fig. 53*a* or Fig. 54*b*. Several types of head–tail order have been uncovered depending on coverage, including local minima for both herringbone and pinwheel structures [17]. Based on the analogy with the situation in the bulk (see Section I.B), it is suggested

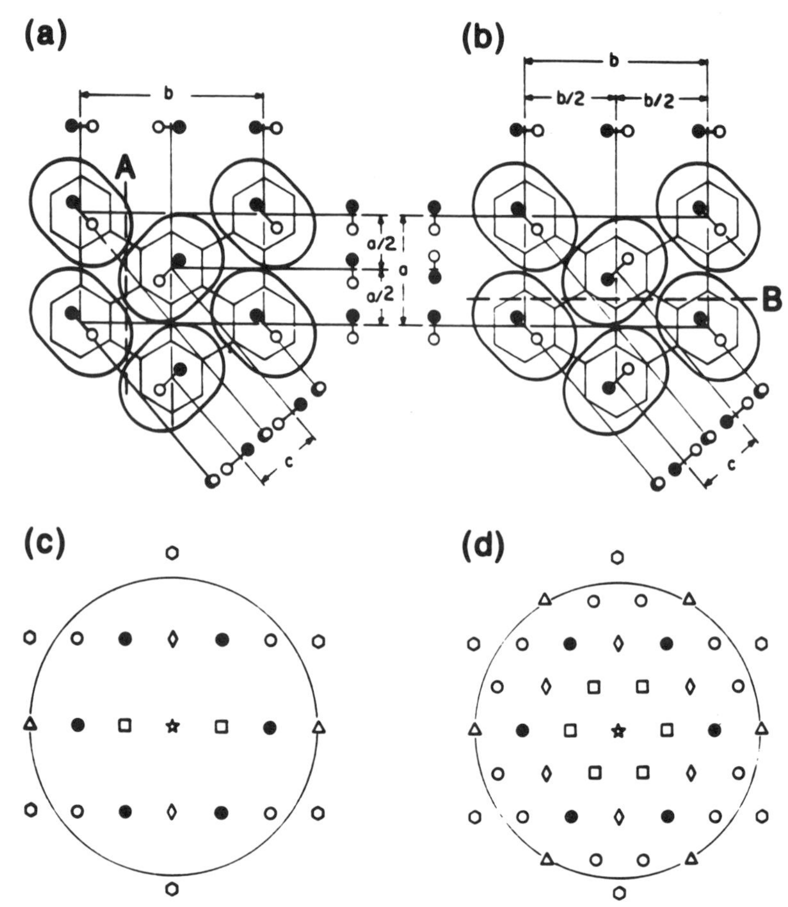

Figure 53. Possible head–tail ordered commensurate in-plane herringbone ($\sqrt{3} \times \sqrt{3})R30°$ structures and expected LEED diffraction patterns for CO on graphite. In (*a*) and (*b*), filled and unfilled circles mark the carbon and oxygen nuclei, respectively, and the centers of the outermost 95% electronic charge density contour of Fig. 1*b* are located at the centers of the graphite hexagons. The two structures are projected onto three directions. The glide lines A or B which remain preserved in the head–tail ordered structures (*a*) or (*b*), respectively, are shown as dashed lines. The LEED patterns expected from one single domain and from six domains rotated by 60° are shown in (*c*) and (*d*), respectively, for various possible structures; the large circles represent the extension of the LEED screen at 85 eV. The spots marked by triangles and squares should be observed in the case of structure (*a*), and those marked by diamonds should be observed for arrangement (*b*). Only the unfilled and filled circles corresponding to two glide lines are observed [389] at 22 K at submonolayer density. (Adapted from Fig. 4 of Ref. 389.)

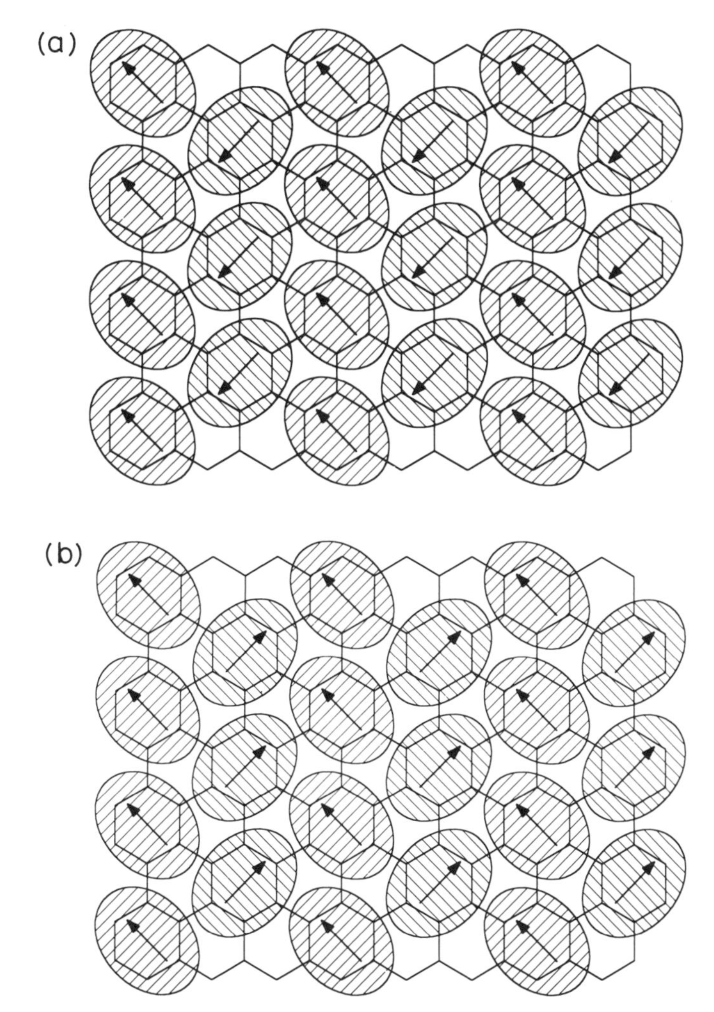

Figure 54. Possible head–tail ordered herringbone structures for a complete commensurate CO monolayer on the basal plane of graphite; only the projection on the surface plane is shown and the directions of the molecular dipole moments are indicated by arrows. (*a*) Ferrielectric structure with the net dipole perpendicular to the herringbone symmetry axis. (*b*) Ferrielectric structure with the net dipole parallel to the herringbone symmetry axis. (*c*) Antiferroelectric structure with no net dipole. The principal axes of the ellipses correspond to the 95% electronic charge density contour given in Table I. Note that each type of head–tail ordering can be combined with any of the six herringbone ground states from Fig. 6.

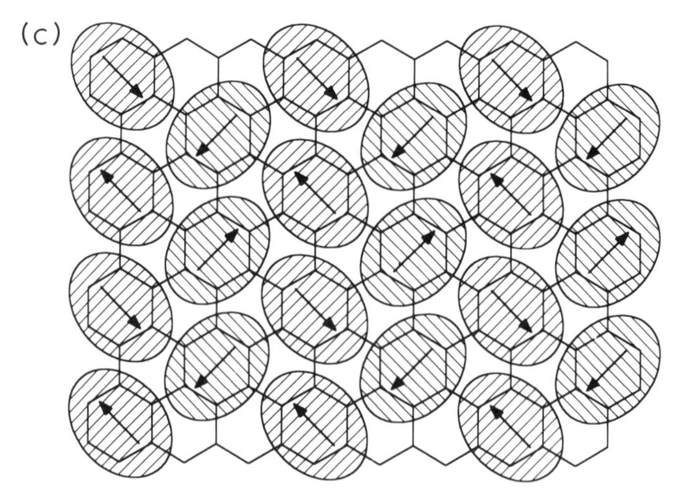

Figure 54. (*Continued*)

that the disagreement with the experiments [242, 389] available at that time is due to very long reorientation times after the physisorption process took place, which leads to a thermodynamically metastable frozen disorder.

Complete head–tail ordering was actually detected by heat capacity measurements of CO on Grafoil GTA [153] (see Fig. 55 for a direct comparison of the CO and N_2 low-temperature behavior). The small heat capacity anomaly near 5.4 K at a coverage of 0.79 monolayers was reproducible as well as independent of the previous thermal history and the cooling protocol [153]. The molar entropies were obtained by integrating the heat capacity/temperature ratio, and the difference between CO and N_2 amounts to 6.08 J/K mol at 14 K at that coverage; this reference temperature was somewhat ambiguously chosen because the head–tail disordering is already completed whereas the orientational herringbone disordering is not yet effective. This entropy difference is fairly close to the statistical value $R \ln 2 \approx 5.76$ J/K mol for complete head–tail disordering, which is clearly different from the behavior in the bulk CO and N_2 solids (see the discussion in Section I.B). The peak becomes less prominent for lower coverages, but it is not shifted below 1.050 monolayers [154, 156] and also the entropy change across the transition is not significantly affected [154]. Increasing the coverage beyond the complete monolayer smears the transition [154], but the position of the peak remains at 5.4 K. Above about 1.064 monolayers only a very broad feature remains [156]; see the change from frame (i) at 1.050 to (j) at 1.064 and (k) at 1.114 in Fig. 6 of Ref. 156. At the highest two coverages a very broad anomaly is found at 10.1 K and 11.6 K, respectively. This was in-

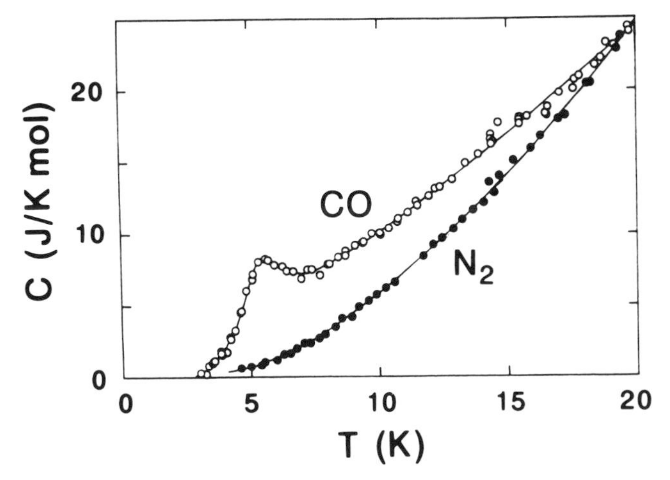

Figure 55. Molar heat capacities of commensurate CO (unfilled circles) and N_2 (filled circles) on Grafoil GTA as a function of temperature at a coverage of 0.79 monolayers. (Adapted from Fig. 1 of Ref. 153.)

terpreted as the signature of a head–tail ordering in the incommensurate pinwheel phase [156]. The total molar entropy change including the head–tail and orientational disordering amounts to about R ln 4 before melting occurs, except for the highest coverage 1.114 studied where it is speculated that the full ordering may not be completed at low temperatures [156].

The molar entropies derived from the isosteric heats of adsorption and the heat capacities are in good agreement [154, 156] at 80 K, which is additional strong evidence that the residual entropy has been removed as a result of the complete ordering below the head–tail transition. It is speculated that the 7% increase of the nearest-neighbor intermolecular distance in the two-dimensional monolayer (4.259 Å) compared to the bulk solid (3.99 Å) allows for the head–tail ordering. Concerning the interplay of head–tail and orientational disordering, it is proposed [156] that the head–tail disordering might trigger the orientational disordering at sufficiently low temperatures in the commensurate phase.

This situation can be compared [153–155] to N_2O on graphite, a molecule which has slightly larger quadrupole and dipole moments and a solid phase which is structurally very similar to the bulk α-CO solid. There is a residual entropy of about R ln 2 at low temperatures, and thus the bulk N_2O solid does not undergo a head–tail ordering transition [153, 155]. Also no heat capacity anomaly was detected for N_2O on graphite [153, 155] down to 3–4 K, which could be rationalized [153] in terms of the striking difference in the dynamical behavior of the N_2O and α-CO solids [251].

An analytical CO–CO pair potential obtained from state-of-the-art *ab initio* quantum chemical techniques was devised in Ref. 291. In order to avoid the so-called basis set superposition error the potential was not obtained based on the "supermolecule approach"—that is, by calculating the total energies of two CO molecules for various distances and orientations and fitting them to some analytical potential. Instead, the anisotropic dispersion coefficients, electrostatic, and exchange contributions are obtained separately invoking a perturbation scheme; the induction interactions are very small relative to all other energy scales and are neglected in the potential function reported in Ref. 291. The complete intermolecular potential is represented in a spherical expansion of the additive dispersion, electrostatic, and exchange contributions; a site–site fit to this expansion was derived later in Ref. 159.

The anisotropic dispersion coefficients C_6 up to C_{10} are obtained from many-body perturbation theory, and the long-range behavior of the dispersion potential is corrected with Tang–Toennies-type damping functions [291]. The electrostatic multipole moments are calculated for an isolated CO molecule up to the 32-polar contribution. The self-consistent field (Hartree–Fock) method used yields the wrong sign for the dipole moment, whereas other moments are in good agreement with previous higher-level (MP4 and SDCI) quantum chemical calculations which include correlation effects. An experimental dipole moment is used in the final parameterization of the potential; the experimental quadrupole moment is slightly overestimated by the *ab initio* calculations, but the Hartree–Fock value is still used. The exchange energy is obtained to first order from the Heitler–London formula by subtracting the electrostatic contribution already obtained from the isolated molecule; the Heitler–London contribution is calculated for three center-of-mass distances up to about 4 Å and various orientations.

The final *ab initio* potential [291] reflects the importance of the lack of inversion symmetry due to the inherent head–tail asymmetry. For the collinear dimer geometry the exchange repulsion is about 10 times larger for the OC–CO than for the CO–OC configuration at the same center-of-mass distance. This is due to a more extended charge distribution near the carbon atom compared to the oxygen atom, and to the fact that the carbon atom is further away from the center of mass than the oxygen atom [291]. For larger distances this factor actually increases, which means that the charge density decays more rapidly for the oxygen than for the carbon end of the CO molecule. Thus, the carbon atom in CO is loosely speaking "larger" than the oxygen atom. The calculated second virial coefficients agree with experimental data in the covered range from 77 K to 573 K within the experimental error bars [291].

The spherical expansion of the CO–CO potential [291] was reparameterized in terms of a site–site representation [159]. The total site–site potential as defined in Ref. 159 is written as a sum of exchange, dispersion, and electrostatic pair interactions between all sites i located at molecule A and all sites j on molecule B:

$$V_{AB} = \sum_{i \in A} \sum_{j \in B} v_{iA;jB}(R_{iA;jB}) \tag{5.2}$$

$$v_{iA;jB}(R_{iA;jB}) = v_{iA;jB}^{\text{exch}}(R_{iA;jB}) + v_{iA;jB}^{\text{disp}}(R_{iA;jB}) + v_{iA;jB}^{\text{elec}}(R_{iA;jB}) \tag{5.3}$$

$$v_{iA;jB}^{\text{exch}}(R_{iA;jB}) = A_{ij} \exp\left[-B_{ij} R_{iA;jB}\right] \tag{5.4}$$

$$v_{iA;jB}^{\text{disp}}(R_{iA;jB}) = -f_6^{ij}(R_{iA;jB}) C_{ij} R_{iA;jB}^{-6} \tag{5.5}$$

$$f_6^{ij}(R_{ij}) = 1 - \left[\sum_{k=0}^{6} \frac{(B_{ij} R_{ij})^k}{k!}\right] \exp[-B_{ij} R_{ij}] \tag{5.6}$$

where $R_{iA;jB}$ denotes the intermolecular site–site distance. The root-mean-square deviation [159] of the site–site exchange and dispersion contributions (5.4)–(5.6) from the essentially exact spherical expansion [291] is about 5%. The electrostatic part $v_{iA;jB}^{\text{elec}}(R_{iA;jB})$ is represented by pseudo-Coulomb point charges located on the carbon and oxygen atoms as well as on the center of mass (com); it reproduces the dipole and quadrupole moments. The potential parameters yielding a Kelvin scale for the energies are compiled in Tables IV and V. The differences between the spherical expansion and its site–site fit are found to be quite small for CO bulk solids [159]; for example, the lattice constants for the optimized $P2_13$ structure are 5.658 Å and 5.628 Å, respectively.

The interactions between CO molecules and a graphite surface were examined in Ref. 138 for a variety of potentials models. Total energy minimizations were performed using the nonelectrostatic contribution from Ref. 17 (i.e., the *ab initio* data [23] obtained from N_2–N_2 interactions), together with a different four-point charge representation of the electrostatics; the dispersion terms are corrected for surface-mediated effects [49, 225] as in

TABLE IV
Parameterization of CO–CO Pair Potential [159] (5.2)–(5.6): Site Positions i in Å and Pseudocharges in $(\text{K Å})^{1/2}$

i	Electrostatic	Exchange	Dispersion	Charge
O	0.4835	0.4278	0.6649	−308.79
C	−0.6447	−0.8354	−0.7134	−246.05
com	0.0	—	—	554.84

Source: From Ref. 215.

TABLE V
Parameterization of CO–CO Pair Potential [159] (5.2)–(5.6): Interaction Parameters

i, j	A_{ij} (K)	B_{ij} (Å^{-1})	C_{ij} (KÅ^6)
O, O	60,298,230	4.000	112,500
O, C	17,773,800	3.498	200,000
C, C	7,116,910	3.104	384,100

Source: From Ref. 215.

Ref. 17. In some calculations the corrugation of the periodic molecule–surface interactions obtained from isotropic atom–atom potentials [324, 326] is effectively increased by anisotropy effects [61, 165, 254, 361, 362] (see also the discussion of the N_2 case in Section III.D.1). Image charges are introduced to mimic the electrostatic response of the graphite substrate to external charges due to the physisorbed molecules. Previous instabilities [17, 282] because of the inclusion of these additional charges for CO on graphite could be cured by shifting the mirror plane by 0.25 Å toward the surface, whereas the situation for N_2 on graphite is not much affected; arguments underlying this *ad hoc* remedy are given in Appendix B of Ref. 138.

It is found [138] that the increase of the corrugation due to the inclusion of axially symmetric (experimentally determined bulk) quadrupole moments located at the carbon sites [361] which model the aspherical charge distribution in the graphite substrate [see (3.9) and (3.10) in Section III.D.1] stabilizes the commensurate herringbone structure. This structure is head–tail-ordered as in Ref. 17 (see Fig. 53*a* or Fig. 54*b*, where the molecular axes have a systematic out-of-plane tilt); the unit cell is deformed because of the displacement of the molecular centers on the two sublattices. The Brillouin-zone-center frequency gap in the phonon spectrum is estimated [138] to amount to about 10 K in the ground state.

The head–tail ordering transition as well as the random-field behavior induced by diluting the CO monolayer with N_2 and Ar impurities was studied by adiabatic heat capacity measurements [380, 381] on graphite foam. A heat capacity scan [381] at a submonolayer coverage of 0.757 including the head–tail (dipolar) and orientational (quadrupolar) transitions as well as the final melting transition of CO on graphite is shown in Fig. 56. In good agreement with Refs. 113 and 156, only a small heat capacity bump is found at the orientational disordering transition at $T_Q \approx 25.7$ K reflecting the fact that this transition is ill-defined. The increasing scatter of the data at temperatures above 35 K is caused by the decreasing sensitivity of the Germanium temperature sensor, which was used in these measurements.

The critical behavior underlying the low-temperature anomaly was in-

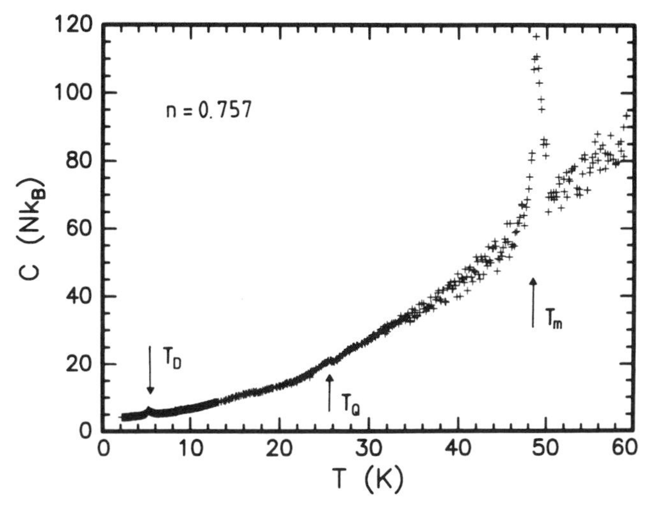

Figure 56. Total heat capacity of submonolayer CO on graphite foam at a coverage of 0.757 monolayers including the head–tail (dipolar) T_D, orientational (quadrupolar) T_Q, and melting T_m transitions. The background contribution of the substrate and the sample cell is subtracted from the data. (From Ref. 381.)

vestigated [380, 381] in detail at a coverage of 0.75 monolayers using high-purity (99.997%) CO (see Fig. 57). A linear regular background contribution approximating the lattice part of the heat capacity of the film was determined from a fit to the measured data sufficiently far from the transition and was subtracted to yield the singular part of the heat capacity shown in Fig. 57*a*. These data follow a linear behavior in a semilogarithmic plot as a function of the reduced temperature (see Fig. 57*b*). The singular part of the heat capacity could be fitted with the relation

$$C = -A \ln|1 - T/T_c| + B \qquad (5.7)$$

above and below T_c over a range of about two decades in reduced temperature; of course, rounding occurs close to the transition for small reduced temperatures $<3 \times 10^{-3}$. The critical amplitudes A above and below the transition are 0.335 ± 0.006 and 0.344 ± 0.006, respectively, which results in a ratio of 1.03 ± 0.03 close to unity. This behavior is characteristic for the two-dimensional Ising model [272] with a vanishing critical exponent $\alpha = 0$ for the heat capacity and a universal critical amplitude ratio [294] of unity. In agreement with Ref. 153, the relative entropy vanishes below the transition (i.e., the anomaly can be attributed to complete head–tail ordering), whereas it reaches a nonzero value when small amounts of N_2 impurities are present [381].

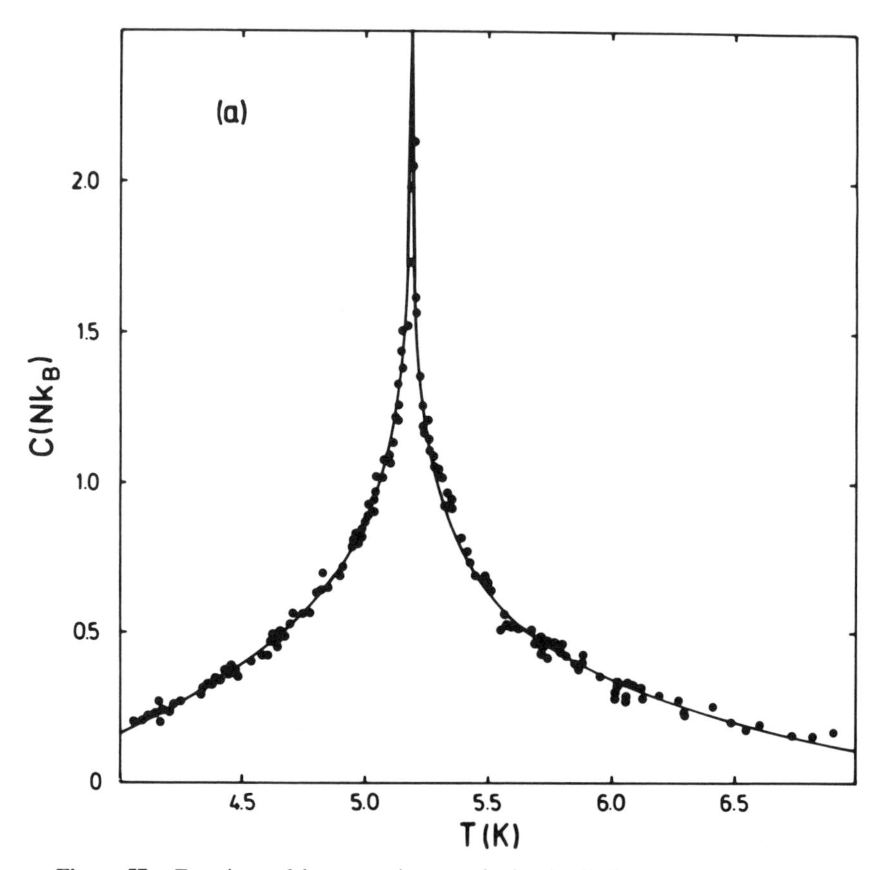

Figure 57. Experimental heat capacity near the head–tail disordering transition of CO on graphite foam at a coverage of 0.75 monolayers. (*a*) Singular contribution to the total heat capacity. (*b*) Semilogarithmic plot of the data of (*a*) as a function of the reduced temperature above (squares) and below (triangles) the transition. Solid lines are fits in terms of the two-dimensional Ising behavior (5.7). (Adapted from Fig. 1 of Ref. 380.)

The head–tail transition temperature [380] obtained by this fitting procedure is 5.18 ± 0.01 K. The slight deviation of this value from the 5.4 K determined in Ref. 153 is attributed to finite-size effects since the crystallites of graphite foam [380] are larger than those of Grafoil [153] (see Table II). The relatively sharp Ising transition to the head–tail ordered low-temperature phase implies that the orientational order of the high-temperature phase at submonolayer coverages must be almost complete. This is in contrast to the early computer simulation results of Ref. 282, where an "orientationally glassy state" with some short-range herringbone order was found at low temperatures; this is most probably an artifact of the simplified potential

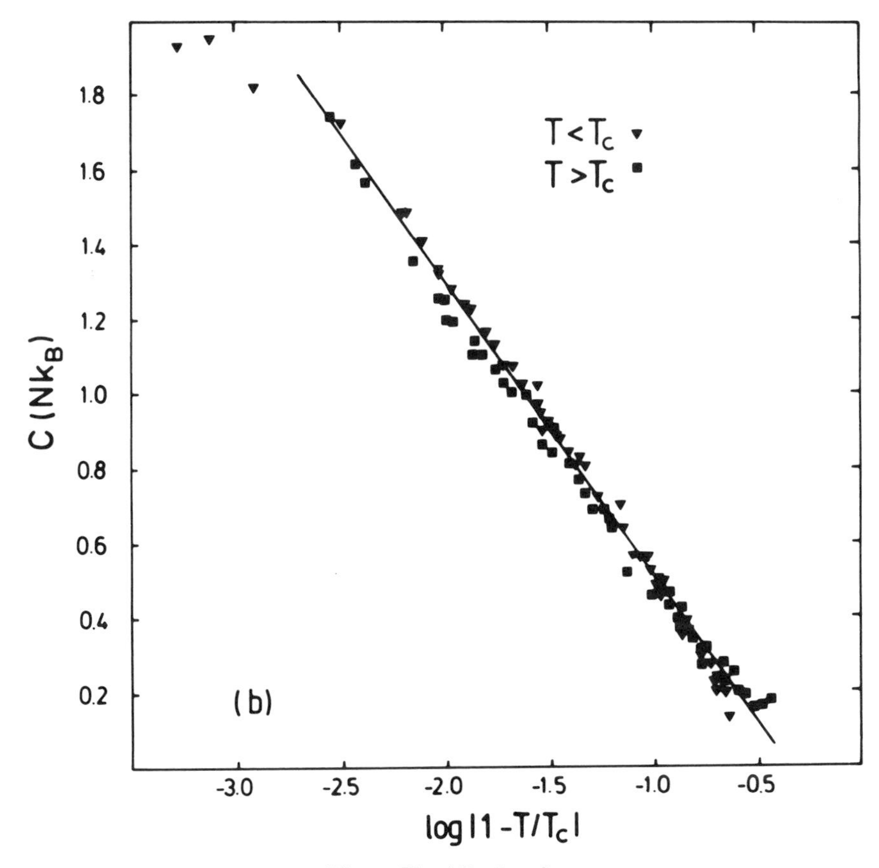

Figure 57. (*Continued*)

model used and/or insufficient sampling [282]. Heat capacity scans for various coverages show that the peak position associated to the head–tail transition remains fixed up to 1.08 monolayers where it disappears [380]. Its height, however, displays a nonmonotonic behavior (see Fig. 58). It increases from the lowest measured coverage of about 0.35 monolayers up to 0.7, reaches a broad maximum plateau in the range from 0.7–0.9, and sharply drops above a coverage of 0.9 monolayers [380, 381]. This behavior is strange, because one would expect that the heat capacity peaks are maximum at the most complete registered phase at coverage unity as was observed for the helium [46, 98] and hydrogen [120–122, 377, 379] adsorbates. The phenomenon that the heat capacity peaks are more pronounced at coverages between 0.7 and 0.9 monolayers was also observed in Ref. 156 at the orientational transition of N_2 as well as at the head–tail ordering transition of CO on graphite.

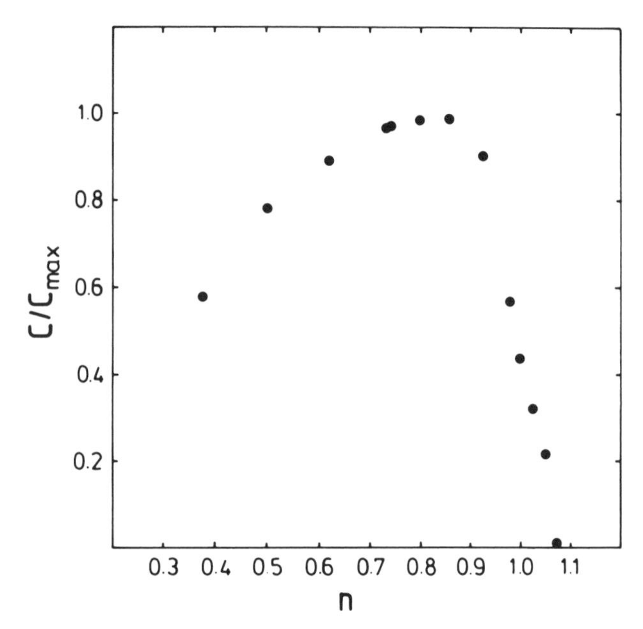

Figure 58. Normalized maximum of the heat capacity peaks of CO on graphite foam at the head–tail critical temperature $T_c = 5.18$ K as a function of the monolayer coverage. (From Ref. 381.)

One idea to explain the nonmonotonic behavior of the coverage dependence of the heat capacity maxima is stimulated by a theoretical suggestion [142, 246] derived from studying the consequences of vacancies on the phase diagram of the anisotropic–planar–rotor model discussed in Section II.B. In the case that all commensurate lattice sites are occupied by diatomic molecules confined to rotate in the plane of the surface, the herringbone structure is stabilized at low temperatures as is well known. On the other hand, when the system is diluted by 25% vacancies which are free to move, a $(2\sqrt{3} \times 2\sqrt{3})$ pinwheel structure is energetically favored. In this structure the vacancies occupy the central sites of each pinwheel and form a compositionally ordered sublattice structure (see Fig. 5*b* with the perpendicular molecules replaced by vacancies). Such a structure would of course be most completely ordered at a coverage of 0.75 monolayers, which would fit very well to the observed maximum of the heat capacity around this coverage. However, previous LEED measurements of N_2 on graphite [93] and also recent neutron diffraction measurements of CO on graphite [177] presented at the end of this section found no indication for such a phase.

Another possible explanation is guided by the behavior of the quadrupolar lattice model (2.1) + (2.3) with the possibility of out-of-plane tilts [141].

It was suggested in Ref. 113 that in the submonolayer regime, out-of-plane tilts of the axes of CO molecules may be induced by the stronger quadrupolar couplings between the molecules compared to those of N_2 molecules and the smaller reduced crystal-field strength V_c^* (see the discussion in Section V.D). In addition, it is striking (see Fig. 8 for N_2 and Fig. 49 for CO on graphite) that the melting transitions of both systems strongly shift to higher temperatures beyond coverages of about 0.8 monolayers. Based on computer simulations of N_2 on graphite, this effect has recently been attributed to different melting mechanisms (see Refs. 102, 103, 140, 301, and 302 and the discussion in Section III.B). At coverages below approximately 0.8 monolayers the vacancies are found to agglomerate in the form of large voids, which leaves patches of registered molecular networks with practically no vacancies. At melting the connected molecular networks are destroyed and the molecules start diffusing on the surface. In this coverage range the melting temperature is insensitive to changes of coverage, because the solid is stabilized by the molecular network separated by patches of vacancies. The situation changes at coverages above about 0.8 monolayers, because now network growth is nearly completed and one only has small voids and isolated mobile vacancies left. Due to the reduction of the in-plane translational degrees of freedom, one observes a rapid increase of the melting temperature in the coverage regime between 0.8 and unity [102] (see Fig. 22). At monolayer completion, melting is associated with the formation of vacancies in the adlayer by promoting molecules to the second layer. By this mechanism, sufficient free surface area is created for the molecules to translationally disorder.

By combining these ideas, one can perhaps explain the strange behavior of the coverage dependence of the heat capacity maxima of CO and N_2 at the head–tail and orientational ordering transitions. Up to coverages of about 0.8 monolayers, one has network growth and thus the maxima of the heat capacity peaks increase with coverage and the transition temperatures remain constant. Beyond this coverage, however, where network growth is complete, there may be a certain tendency of single molecules to tilt out of the surface plane, which may be more pronounced for CO than for N_2 because of the larger quadrupole moment leading to a smaller $V_c^* = V_c/\Gamma_0$ value as already discussed. This may lead to pinwheel-like defects in case of CO which reduce substantially the in-plane head–tail ordering of the molecules. This, in turn, results in a drop of the maxima of the heat capacity peaks at high coverages. Promotion of molecules to the second layer may enhance this effect near monolayer completion. Further detailed computer simulations are clearly necessary to check this speculation on a microscopic level.

A Monte Carlo finite-size-scaling study focusing on the head–tail transition of the complete commensurate CO monolayer on graphite is reported in Refs. 214 and 215. Similar to the approach used in Refs. 211 and 212, the centers of rotation of the rigid molecules were fixed on the sites of the regular $(\sqrt{3} \times \sqrt{3})R30°$ triangular superlattice on the graphite honeycomb plane, and the molecules were confined to rotate only in the surface plane. The centers of rotation were displaced uniformly for all molecules by δ (which is on the order of a few tenths of an angstrom) from the molecular centers of mass in order to take into account the consequences of the intrinsic asymmetry of the heteronuclear molecules [17, 117, 118, 138, 159] within a simple lattice model. The site–site representation [159] [see (5.2)–(5.6) together with Tables IV and V] of the *ab initio* pair potential [291] was used to describe the CO–CO interactions. Only the intermolecular interactions between a given molecule and its six nearest neighbors are taken into account as in Ref. 71, because the energetic effects of including up to the third neighbor shell are quite small (see Fig. 1 of Ref. 215 for a comparison). The molecule–surface interaction was restricted to be the first-order Fourier expansion of the periodic potential resulting from Lennard-Jones interactions between the molecular centers and the carbon atoms of the substrate [324, 326, 327], without taking into account any surface-mediated interactions [49, 138]. The system sizes ranged from $L \times L = 18 \times 18$ up to $60 \times 60 = 3600$ molecules, and typically more than 1,000,000 Monte Carlo steps were evaluated for the averages.

The order of the head–tail transition was determined based on systematic finite-size scaling of various data [214, 215]. A head–tail order parameter is defined in closest analogy to the herringbone order parameter (3.11)–(3.12) as

$$\Psi = \frac{1}{N} \sum_{i=1}^{N} \sin[\varphi(\mathbf{R}_i) - \eta_\alpha] \exp[i\mathbf{Q}_\alpha \mathbf{R}_i] \tag{5.8}$$

with the same definition of the reciprocal lattice vectors $\{\mathbf{Q}_\alpha\}$ and phases $\{\eta_\alpha\}$ as before. Note that (5.8) is a one-component order parameter for the herringbone structure $\alpha = 1, 2$, or 3 because the head–tail ordering occurs in a structure that already exhibits herringbone order; the index α is omitted because all simulations were carried out in one specific orientationally ordered phase. The behavior of the head–tail order parameter cumulant (3.22) is shown in Fig. 59 for five different system sizes, and the known universal intersection value [47, 53] characteristic for the Ising class in two dimensions $U^* \approx 0.61$ is marked by a dotted line. The cumulants intersect at a value of approximately 0.6, which is within the statistical uncertainty identical to the universal value for the Ising class. Satisfactory data collapsing in a finite-size scaling plot of the head–tail order parameter itself can be achieved using the critical exponents $\beta = 1/8$ and $\nu = 1$ for the two-dimensional Ising class as demonstrated in Fig. 3*a* of Ref. 215.

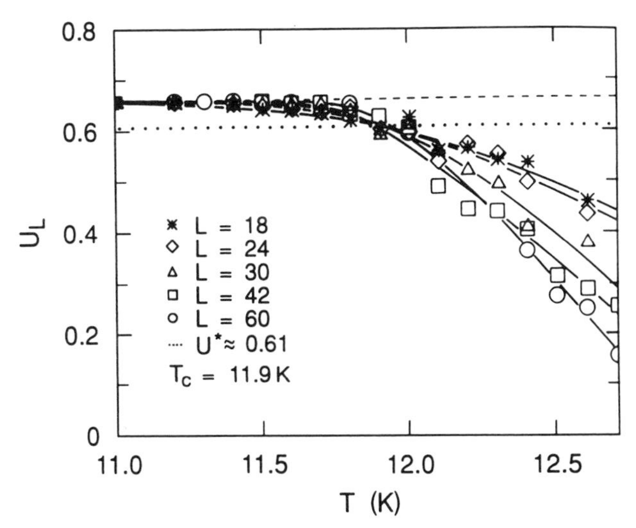

Figure 59. Fourth-order cumulants (3.22) for the head–tail order parameter (5.8) obtained from Monte Carlo simulations of complete monolayer $(\sqrt{3} \times \sqrt{3})R30°$ CO on graphite as a function of temperature ($\delta = 0.13$ Å). The dashed line marks the trivial value in the ordered phase, the dotted line marks the universal value U^* for the two-dimensional Ising universality class [47, 53], and $T_c = 11.9$ K is determined from the intersection point. Symbols for the linear dimension of the $L \times L$ system: $L = 18$ (asterisks), 24 (diamonds), 30 (triangles), 42 (squares), and 60 (circles). (Adapted from Fig. 2 of Ref. 215.)

Similarly, the isothermal susceptibility χ_T associated to the head–tail order parameter can be obtained from its fluctuation relation and scaled according to the scaling relation [11]

$$L^{-\gamma/\nu} T \chi_T^{(L)} = f_\chi(|1 - T/T_c| L^{1/\nu}) \tag{5.9}$$

assuming two-dimensional Ising behavior with $\gamma = 7/4$ and $\nu = 1$ (see Fig. 60). Using

$$\chi_T = \hat{\chi}_T^{\pm} |1 - T/T_c|^{-\gamma} \tag{5.10}$$

its critical amplitudes above $\hat{\chi}_T^+$ and below $\hat{\chi}_T^-$ the transition can be obtained from

$$f_\chi(z) = \hat{\chi}_T^{\pm} z^{-\gamma} \tag{5.11}$$

in the limit $z = |1 - T/T_c| L^{1/\nu} \to \infty$ (for more details see Section II.C of Ref. 215). The ratio $\hat{\chi}_T^+ / \hat{\chi}_T^- \approx 39.1$ obtained from the double-logarithmic plot of Fig. 60 turns out to be very close to the exact value [12] 37.69 . . .

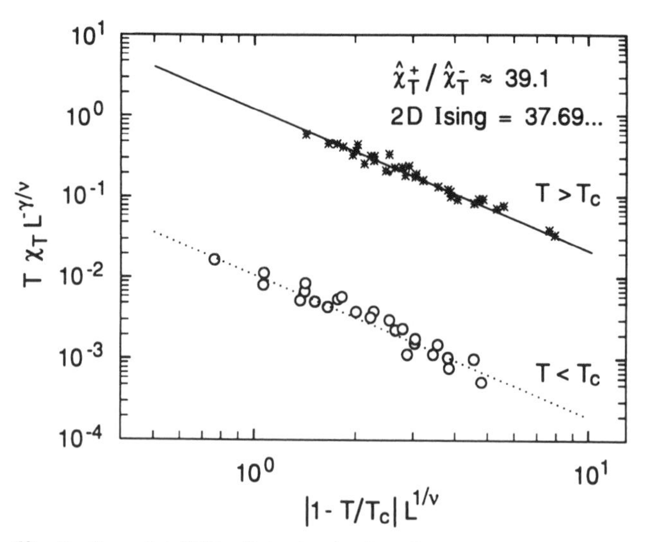

Figure 60. Scaling plot (5.9) of the head–tail order parameter susceptibility obtained from Monte Carlo simulations of complete monolayer $(\sqrt{3} \times \sqrt{3})R30°$ CO on graphite (δ = 0.13 Å) with two-dimensional Ising exponents $\gamma = 7/4$ and $\nu = 1$, and $T_c = 11.9$ K from the cumulant intersection in Fig. 59. Only the scaling regime$|1 - T/T_c|L^{1/\nu} \gg 0$ is shown, and the data above (asterisks) and below (circles) the transition are superimposed for all system sizes $L = 18 \ldots 60$; solid and dotted lines are the amplitude fits (5.11) of the data above and below the transition, respectively. (Adapted from Fig. 4*c* of Ref. 215.)

expected for this universal number in the case of two-dimensional Ising behavior.

The heat capacity does not show a systematic size dependence within the limit of the statistical resolution of the data [215]. The maxima of the system-size-dependent heat capacities $C^{(L)}$ can be fitted according to

$$\max_T C^{(L)}(T) \sim \ln L \tag{5.12}$$

the logarithmic dependence expected for the two-dimensional Ising class (5.7) (see Fig. 5 in Ref. 215). Finally, the fourth-order energy cumulant (3.13) displays a dip which approaches its limiting value as required by the scaling analysis of the two-dimensional Ising model (see (2.28)–(2.30) and Fig. 6 in Ref. 215).

The critical temperature $T_c = 11.9$ K was obtained from the intersection point of the order parameter cumulant (3.22) in Fig. 59 for the parameter choice $\delta = 0.13$ Å. This displacement δ of the center of mass of the CO molecules from the lattice sites was taken from simulations of the bulk $P2_13$

CO solid phase [159], where it was already observed that this value depends sensitively on potential and lattice structure. It is found [214, 215] that the head–tail transition temperature T_c in the monolayer is very sensitive to the choice of the displacement. Increasing it slightly to $\delta = 0.165$ Å yields a head–tail transition temperature in excellent agreement with experiments [153, 156, 380]; changing δ in reasonable bounds did not affect the qualitative behavior reported, but only shifted T_c. It would be more satisfactory, however, to determine δ from independent calculations allowing for translational degrees of freedom of the molecules. This displacement 0.165 Å is also closer to the difference of 0.2 Å between the center of mass and the geometric center of the 95% electronic charge density contour of an isolated CO molecule (see Section I.B) and is in line with the notion of the "center of interaction" introduced in Ref. 159 as well as the significant translation–rotation coupling observed in the bulk phases [117, 118, 159]. Using fully quantum-mechanical path-integral Monte Carlo simulations with 900 molecules (see Section III.D.1 for some details and further references concerning this technique), it has been shown [214, 215] that quantum fluctuations do not destroy the fully ordered ground state, but merely shift the heat capacity peak near 5 K by about 10% to lower temperatures. At variance with the herringbone order parameter, however, the head–tail order parameter reaches essentially its classical value well below the head–tail transition; the amplitude of the zero-point librations amounts to about 10.2° in the ground state.

The fully ordered ground-state structure found with the particular modeling of Ref. [214, 215] is presented in Fig. 61*b*: It clearly displays herringbone order below (Fig. 61*b*) and above (Fig. 61*a*) the head–tail transition and is of the particular ferrielectric type with a net dipole moment perpendicular to the herringbone axis as sketched in Fig. 54*a* or Fig. 53*b*. It should be noted, however, that the relative energies and thus stabilities of the various head–tail ordered structures are expected to depend very sensitively on the potential model used as demonstrated for semiempirical potentials in Ref. 17. In addition, no out-of-plane tilt was allowed by construction, but such a tilt is probable because of the slightly different interactions of the carbon and oxygen ends of the molecules with the graphite surface and because of the strong quadrupolar interactions between the molecules; note that out-of-plane degrees of freedom would result in a systematic, but probably small, out-of-plane tilt in the fully ordered ground state as a direct consequence of the head–tail order. The setting angle of 45° in the herringbone structure is not altered by the head–tail ordering of the molecular dipoles above and below this transition [214, 215] (see also the discussion of the mixed quadrupolar–dipolar model (2.8) in Section II.B). It has been found [214, 215] that the qualitative picture of the head–tail transition remains unaffected when the electrostatic dipole contribution to the CO–CO

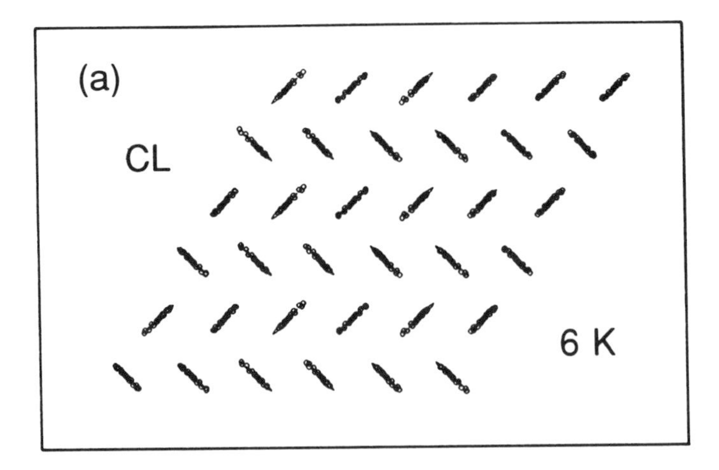

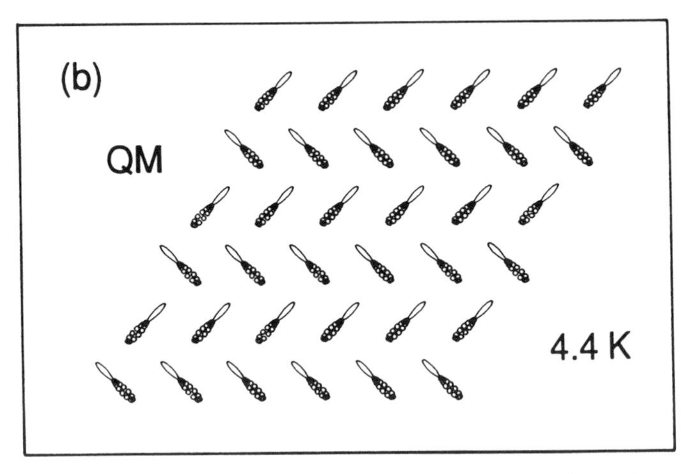

Figure 61. Angle distributions from Monte Carlo simulations of complete monolayer $(\sqrt{3} \times \sqrt{3})R30°$ CO on graphite ($L = 30$, $\delta = 0.165$ Å; for clarity only a patch of 36 molecules is shown). (*a*) Head–tail disordered phase at 6 K, classical simulation. (*b*) Fully ordered phase at 4.4 K, quantum simulation. Plotted in the direction of a given angle is the probability of finding this angle in the simulation. The oxygen ends of the molecules are marked by circles. (Adapted from Figs. 8*a* and 11*b* of Ref. 215.)

potential is switched off (see Fig. 13 of Ref. 215), the only difference being that T_c is roughly 5% higher than with the full potential; the crossover phenomenon from Ising and dipolar critical behavior close to T_c is discussed in some detail in Ref. 215 but is not investigated. Thus, it is concluded that the head–tail ordering is caused by the molecular shape asymmetry [i.e., by the anisotropy of the exchange–dispersion part (5.4)–(5.6)] rather than by the anisotropic electrostatic dipole forces, which even lower T_c and destabilize the ordered structure slightly.

Neutron diffraction experiments covering the range from the fully ordered structure to the orientationally disordered structure were performed for submonolayer CO on graphite [177, 381]. Several speculations concerning the nature of the fully ordered phase have been made. Other structures besides the herringbone structure are conceivable for this phase, such as, for example, (i) an incommensurate phase as theoretically conjectured for Kr on graphite at low temperatures [130] or (ii) a pinwheel structure with a sublattice of ordered vacancies [142, 246, 255] as already discussed in this section. A representative diffraction pattern after subtraction of the background is shown in Fig. 62 for CO on graphite (Papyex) at a temperature of 1.58 K and a coverage of 0.78 monolayers. The residual scattering intensity around $Q = 1.873$ Å^{-1} is due to some interference effect between the adlayer and the substrate in combination with imperfect subtraction of

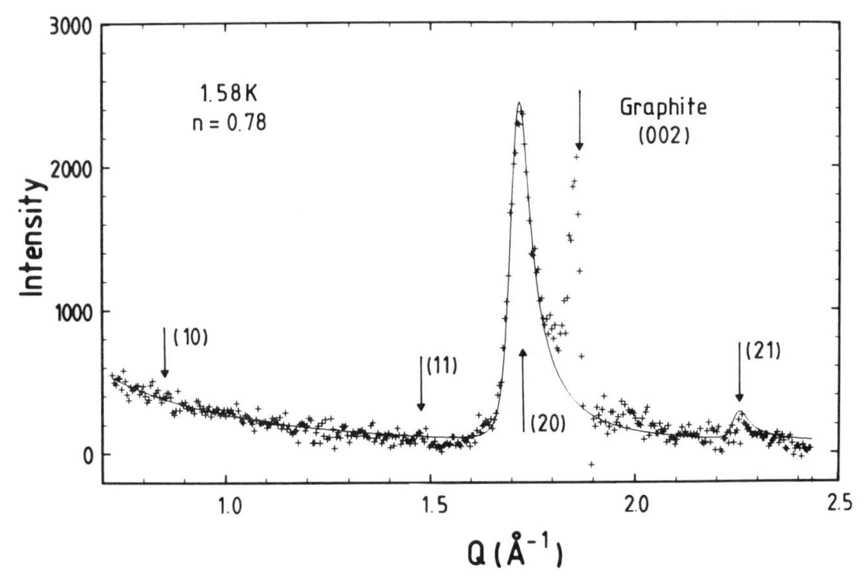

Figure 62. Neutron diffraction intensity after subtraction of the background from CO on graphite (Papyex) at 1.58 K and at a coverage of 0.78 monolayers (see also Fig. 25 for comparison). Note the presence of the (20) and (21) reflections (at $Q = 1.703$ Å^{-1} and 2.253 Å^{-1}, respectively) and the absence of the (10) and (11) reflections (at $Q = 0.852$ Å^{-1} and 1.475 Å^{-1}, respectively) as marked by the arrows; the solid line is a two-dimensional lineshape fit [309]. The diffraction pattern reveals that CO on graphite remains in the commensurate herringbone structure down to very low temperatures. (From Refs. 177 and 381.)

the strong (002) graphite reflection. The data clearly reveal (20) and (21) Bragg reflections at momentum transfers of $Q = 1.703$ Å^{-1} and $Q = 2.253$ Å^{-1}, respectively. They are indexed in terms of a $(2\sqrt{3} \times \sqrt{3})R30°$ unit cell hosting two molecules. The data were fitted by a powder-averaged Lorentzian-squared line shape [309] convoluted with the resolution function of the neutron diffractometer used (solid line). From the fit a coherence length of about 250 Å was found, which corresponds to the natural size of the crystallites of the Papyex graphite substrate sample employed for this study (see Table II). At temperatures above the orientational ordering transition of about 26 K the tiny (21) superlattice reflection at $Q = 2.253$ Å^{-1} gradually disappeared. The diffraction pattern and the intensities of the reflections prove that the monolayer remains in the commensurate orientationally ordered herringbone structure also in the head–tail-ordered low-temperature phase. No further superlattice reflections as well as no shifts of the reflections could be detected. This rules out (i) the envisaged incommensurate structure and (ii) the pinwheel structure with a sublattice of ordered vacancies. In the latter case, one would have expected (10) and (11) reflections at $Q = 0.852$ Å^{-1} and $Q = 1.475$ Å^{-1}, respectively, as indicated by the arrows in Fig. 62. Structure factor calculations yield intensity ratios of $I_{10}/I_{20} = 0.17$, $I_{11}/I_{20} = 0.145$, and $I_{21}/I_{20} = 0.24$ of the superstructure reflections (10), (11), and (21) with respect to the main (20) reflection for a pinwheel structure with an ordered sublattice of vacancies. The absence of the (10) and (11) reflections and the low intensity of the (21) reflection strongly suggests that CO on graphite retains its herringbone structure down to very low temperatures. Thus, a pinwheel-like phase with ordered vacancies can be excluded as a possible candidate for the head–tail ordered low-temperature structure [177, 381]. This result is in accord with preliminary Monte Carlo simulations reported in Ref. 215 that include the presence of mobile vacancies. It was found that even at a coverage of 0.75 the system phase separated. The vacancies left their initial positions in the ideal pinwheel-like phase where vacancies replaced the pin molecules and clustered to form voids. Unfortunately, the low neutron beam intensity and the corresponding high statistical error of the (21) superlattice reflection intensity did not allow us to fully clarify the arrangement of the CO molecules in the head–tail ordered herringbone low-temperature phase.

F. Compressed Monolayers

LEED experiments [389] revealed beyond monolayer completion at a coverage of about 1.13 diffraction patterns which are indicative of a triangular incommensurate phase with (2×2) molecular axes ordering below 35 K, which is to be contrasted with N_2 on graphite where a uniaxial incommensurate phase occurs beyond commensurate coverages. It is suggested that this structure is pinwheel-like (see Fig. 63), in close analogy to the close-packed (111) face of bulk α-CO (see Section I.B). In this orientationally ordered bulk structure the angle between the molecular axes of the wheel molecules and the (111) plane is about 20°, but the LEED data [389] did not provide detailed information concerning the tilt angle. EELS data [161] obtained at a coverage of about 1.17 monolayers and 28 K are consistent with one out of every four CO molecules being aligned perpendicular to the surface.

The commensurate and incommensurate phases are observed [389] to coexist in a narrow coverage interval but wide temperature range (see the region III in the proposed phase diagram Fig. 1 of Ref. 389). Desorption prevented the investigation of the orientational ordering transition in the high-coverage regime, but the transition to a triangular incommensurate orientationally disordered phase was estimated to occur at about 35 K. Further compression beyond the first layer at 30 K is compatible with multilayer formation.

The x-ray data [242] indicate that the orientational transition within the compressed monolayer at a coverage of about $1.17 \pm 10\%$ occurs near 35 K and is completed at about 48 K (see Fig. 37); that is, it occurs at a higher temperature than in the submonolayer commensurate regime. This transition

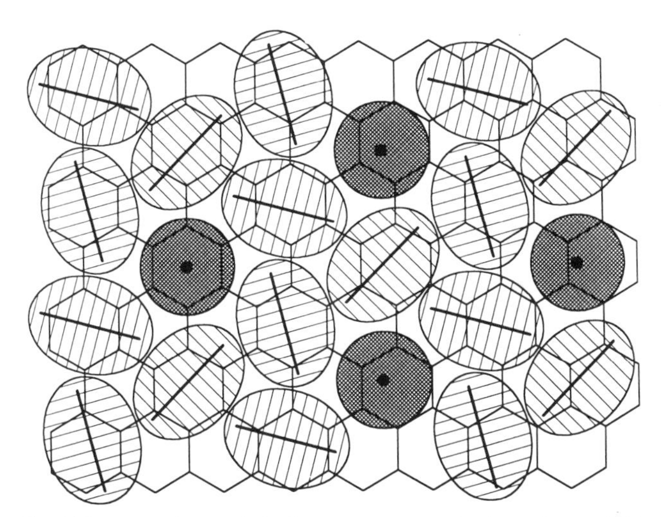

Figure 63. Schematic picture of the incommensurate pinwheel phase of CO on graphite found in the LEED experiments below about 35 K at 1.13 monolayers; this plot is very similar to the projection of the molecules onto the (111) plane of bulk α-CO. The actual orientation of the molecules is not known and the leftmost pin molecule is arbitrarily centered in a graphite hexagon. (Adapted from Fig. 7 of Ref. 389.)

is definitively accompanied by a slight expansion of the center-of-mass lattice in the orientationally disordered phase by about 0.75% (see Fig. 5*b* and Fig. 6*b* of Ref. 242 for more details). The orientational transition in the incommensurate phase is much sharper for CO than for N_2 on graphite (see Fig. 37), which is the opposite situation encountered in the commensurate phase (see Fig. 28). These observations are qualitatively confirmed in heat capacity measurements [113, 156, 381]. The incommensurate CO low-temperature phase is essentially completely orientationally ordered below 30 K, whereas the corresponding intensity still increases below 10 K for N_2 on Papyex [242] (see Fig. 37 for a direct comparison). These dissimilarities give evidence that the orientational ordering transitions of compressed (as well as commensurate) CO and N_2 layers on graphite might be significantly different [108]. A more detailed structure determination was not possible for the compressed monolayers, but there were signs that the center-of-mass lattice was distorted in the incommensurate orientationally ordered phase [242]. The (20) peak of the latter phase has a satellite at lower angles, and also shows hysteresis on heating and cooling the sample [242]. Because of the close proximity of the strong (002) graphite peak it is difficult to resolve these features, and a regular defect lattice [242] and a small distortion from an ideal triangular structure [108] were favored among several possible explanations.

LEED measurements [160] of the compressed low-temperature phase of CO on graphite at a coverage of 1.18 monolayers and a temperature of 26 K found evidence for high-order periodicity of the overlayer. The LEED spectra revealed besides the principal peaks of the pinwheel structure tiny satellite reflections, which were attributed to long-range high-order periodicity induced by interaction with the substrate corrugation. Due to the restricted resolution of the employed electron energy-loss spectrometer, however, it was not possible to decide whether this effect was caused by mass-density waves or by a periodically distorted high-order commensurate structure. In addition, multiple-scattering calculations would be required to settle this issue.

The first Monte Carlo study of CO on graphite [282] (see Section V.D for some more details) obtained at a coverage of 1.09 a herringbone ground state which disordered at a relatively sharp transition around 30 K. However, it is suggested that compression favors the generation of pinwheels. Energy minimizations [282] with four molecules per unit cell showed that within the utilized model, the herringbone phase is stable up to 1.13 monolayers with a mean tilt angle of 12° due to the fact that the carbon atom is larger than the oxygen atom. Further compression to 1.25 results in a pinwheel-like structure, and in view of the too small molecule within the applied model it is estimated [282] that this coverage might correspond to 1.15

monolayers in reality. Again, the hybrid model using the N_2 atom–atom interactions combined with the CO electrostatics suggests [282] that the intermolecular electrostatic interaction is not responsible for the observed differences between N_2 and CO. Thus, it is the intrinsic asymmetry of the CO intermolecular potential causing the dissimilarities [282].

The theoretical study [17] focused on the ground-state properties of CO monolayers on graphite. Again, the largest uncertainty is attributed to the potential models. Similarly to the hybrid model of Ref. 282 a semiempirical potential was constructed using the nonelectrostatic interactions from *ab initio* computations [23] for the isoelectronic N_2 molecule in combination with the experimental CO bond length. The electrostatic contribution was represented by three point charges reproducing some values for the set of the four lowest nonvanishing multipole moments. Another potential was obtained based on empirical fitting [238] in conjunction with the same electrostatic model. The molecule–surface interaction was represented in terms of the Fourier decomposition [324, 326]. In an alternative model treated in Ref. 17 the amplitude of the first-order corrugation term was arbitrarily increased by a factor of two in view of previous theoretical suggestions [61, 362]. The dispersion interaction is modified according to Refs. 49 and 225 to take into account the surface influence on the intermolecular interactions. Interactions between the point charges modeling the charge distribution and their images in the graphite substrate are included; however, they are evaluated with the asymptotic expression in order to avoid divergences observed otherwise [17]. Using these potentials in various combinations, the energy of two, four, and six sublattice unit cells with deformable boundary conditions was minimized truncating the lattice sums at 15 Å.

Up to 1.05 monolayers, minimization without including the corrugation term, which allows for unregistered structures, yields within the *ab initio* potential stable two-sublattice herringbone phases with out-of-plane tilt angles of about 2.5° where the carbon atom is closer to the surface [17]. This structure transforms to a four-sublattice triangular pinwheel phase at 1.05 monolayers. The central pin molecule has a small angle of about 1.8° relative to the surface normal, its center of mass is approximately 0.5 Å higher than that of the wheel molecules, and the carbon atom is again closest to the surface. The wheel molecules have an out-of-plane tilt of about 4° with the oxygen end down. The pinwheel phase remains stable up to at least 1.235 monolayers, but the tilt angle of the wheel molecules increases.

When the corrugation term is included for the registered structure, it is found that the oxygen end tilts out of the surface plane by 3.4° within an herringbone arrangement of the molecular axes [17]. At complete monolayer coverage the registered structure is lower in energy than unregistered structures at any density, a finding which does not change when the corrugation amplitude is doubled [17]. This, however, stabilized the herringbone phase relative to the pinwheel phase. The stability of the registered versus unregistered herringbone phase is found to depend on the treatment of the image and mediated dispersion interactions, but the pinwheel structure is found to be the stable phase beyond about 1.12 monolayers. Using the empirical CO–CO potential, however, yielded no stable registered herringbone structure [17]. Based on various combinations of the contributions to the potential, it is concluded that the herringbone phase with the molecules slightly tilting out of the substrate plane is stable at low coverages, whereas the pinwheel phase is thermodynamically stable above about 1.12 monolayers.

Energy minimizations of clusters with three to six molecules show that $(\sqrt{3} \times \sqrt{3})R30°$ registered structures are preferred [17]. A single molecule is not located with its center of mass on top of the center of a graphite hexagon. Instead it is displaced in the direction of the saddle point so that the overall charge density of the molecule illustrated in Fig. 1*b* is more centered.

High-resolution Fourier transform infrared spectra of CO on graphite-like films [354] and highly oriented pyrolytic graphite [144] were obtained in the frequency range corresponding to the intramolecular CO stretching mode. The low-coverage spectra [144] are characterized by a single absorption band (which is only slightly shifted to the red relative to the gas-phase frequency) and a low integrated absorption value. The isotherms [144] at 30.5 K and 35.0 K show a sharp step upon increasing the pressure, and the single band shifts and splits into several components. The step is interpreted in terms of a phase transition from the complete monolayer at lower pressure to a dense monolayer. Because the experiment is mainly sensitive to intramolecular vibrations perpendicular to the surface, the large plateau value of the integrated absorption beyond the sharp increase is interpreted in terms of a dense phase with some molecules being perpendicular on the surface; it is suggested that this phase might be a pinwheel phase. A further increase of the pressure leads to a dramatic rise of the isotherm and the appearance of a new band in the infrared spectra. These phenomena are supposed [144] to be caused by multilayer adsorption and the growth of solid CO. The step in the isotherm is smeared out considerably at 38.5 K, vanishes at 45.0 K, and reappears at 56.5 K at a higher CO partial pressure.

The combined ac heat capacity and vapor pressure isotherm study [112] resulted in a detailed mapping of the phase diagram at higher temperatures and coverages. Several of the heat capacity traces in Fig. 52 (see the arrows) are consistent [112] with the presence of a reentrant fluid phase being squeezed in between the incommensurate and commensurate solids in analogy to Kr on graphite, where the incommensurate solid to reentrant fluid melting is believed to be a Kosterlitz–Thouless transition [252, 332] related to the unbinding of the dislocation pairs in a domain wall lattice [77, 78, 151, 260]. This interpretation is supported by the coverage dependence of the compressibility deduced from the vapor pressure isotherms [112]. The unobservable essential singularity of the heat capacity [252] at a Kosterlitz–Thouless transition is identified with the ID → RF feature, which appears more like a broad shoulder at the low-temperature wing of the sharper cusp-like RF → CD anomaly (see Fig. 52).

The freezing transition from the reentrant fluid to the commensurate solid monolayer seems to be of first order at low temperatures and changes to a continuous transition at a tricritical point near 85 K. This interpretation of the heat capacity and compressibility data [112], however, needs further experimental confirmation, and the corresponding tricritical point shown in the phase diagram Fig. 51 as an open triangle is only tentative. There are also conflicting theoretical predictions concerning the order of the RF → CD transition [77, 78, 151, 260], and the present resolution of the calorimetric data [112] does not allow us to draw a firm conclusion in this respect.

Adiabatic heat capacity measurements of CO on Grafoil GTA and MAT reveal a new vertical phase boundary at 49 K between the coverages 0.95 and 1.04, which flattens out at higher temperatures [156]. It is speculated [156] (see Fig. 64) that between 1.04 and 1.05 a uniaxial incommensurate phase exists, which transforms to the triangular incommensurate pinwheel

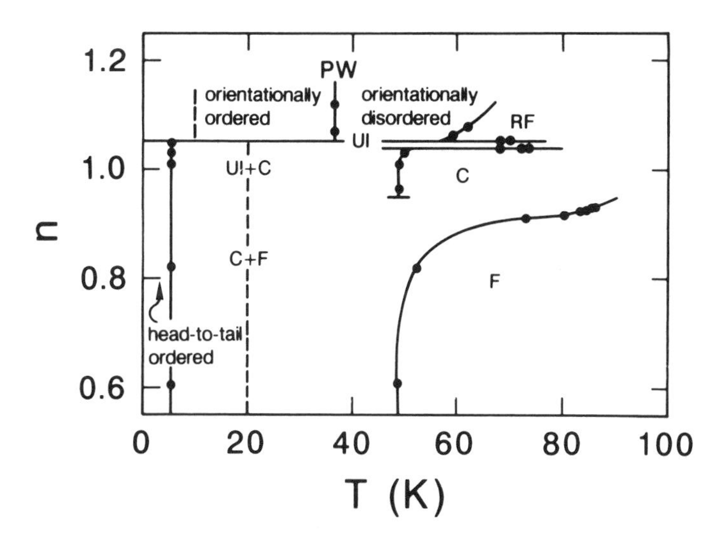

Figure 64. Partly speculative phase diagram of CO on graphite obtained from adiabatic heat capacity measurements [156]. The phases are commensurate (C), uniaxial incommensurate (UI), triangular incommensurate pinwheel (PW), fluid (F), and reentrant fluid (RF). (Adapted from Fig. 11 of Ref. 156.)

phase at higher coverages beyond 1.05 monolayers. The orientational disordering transition is still broad in this speculative uniaxial incommensurate phase because it is in the commensurate monolayer, whereas it is much sharper in the pinwheel phase. In view of the speculative uniaxial phase it is believed that the uniaxial incommensurate and commensurate phases coexist below 49 K in the coverage range from 0.95 up to 1.04 monolayers [156]. The pinwheel phase is believed [156] to undergo a head–tail ordering transition centered around 10 K.

Independent of this investigation, recent neutron diffraction experiments [176] of the slightly compressed monolayer observed a splitting of the principal Bragg reflection into a double peak in the temperature range between 50 K and 62 K. This behavior was interpreted as a possible indication for the existence of a uniaxially compressed incommensurate phase. The formation of such a phase is conceivable because of the nonspherical character of the CO molecule. However, further measurements are necessary to gain more detailed information on the possible existence of such a uniaxial phase.

The high-coverage phase diagram was analyzed in detail by ac heat capacity measurements on graphite foam [113] (see this part in the complete phase diagram, Fig. 49). It is found that the commensurate phase extends up to 1.15 monolayers because of the promotion of molecules to the second layer [113]. The transition from the commensurate to the incommensurate monolayer is assigned to be due to pinwheel ordering at low temperatures. Along the lines of the arguments given to explain the weak signatures of the orientational disordering for submonolayer CO on graphite (see the discussion in Section V.D), it is argued that the stability of the CO pinwheel incommensurate phase instead of a uniaxial incommensurate phase at higher coverages might also be explained by the smaller V_c/Γ_0 ratio allowing easier out-of-plane tilts (see Fig. 3). The orientational disordering of the pinwheel phase is studied between 1.15 and 1.36 monolayers [113] (see below). The heat capacity traces are indicative of a narrow coexistence region sandwiched between the commensurate and incommensurate solids where commensurate herringbone and incommensurate pinwheel phase coexist; note that in the corresponding phase diagram of Ref. 112 the reentrant fluid phase extended into the range where the commensurate and incommensurate solids are assigned in the low-temperature phase diagram of Ref. 113 (see also the corresponding remarks in Refs. 1 and 229). Above about 1.19 monolayers only the pinwheel phase was found. This phase transforms to an incommensurate orientationally disordered phase roughly in the range from 30 K to 35 K.

Beyond a compression to about 1.4 monolayers, evidence for a second-layer fluid on top of the incommensurate monolayer solid is found in the calorimetric study [112] (see the phase diagram, Fig. 49). A second-layer two-dimensional Ising-like critical point [112], which corresponds to the highest temperature where the second-layer vapor and second-layer liquid phases can still coexist on top of the incommensurate first-layer orientationally disordered solid, is located near 60 K. This region extends at lower temperatures down to about 53 K, where it is separated by a triple line from the coexistence regime of second-layer vapor and second-layer orientationally disordered solid on the incommensurate monolayer. Thus, the phase diagram of the second CO layer on the incommensurate first monolayer is topologically very similar to a standard three-dimensional phase diagram with (1) a liquid–vapor coexistence region ending at a critical point and (2) a liquid–vapor–solid triple line at lower temperatures. Cooling the system furthermore results at about 39 K in an orientational ordering of the second-layer orientationally disordered solid to a bilayer structure in coexistence with the second-layer vapor and the incommensurate solid in the first layer. At 36 K the orientationally disordered solid in the first layer transforms to the pinwheel phase, which coexists with a vapor phase in the second layer and the orientationally ordered bilayer at low temperatures.

The critical behavior of the peak in the specific heat at 1.29 monolayers associated with the orientational disordering transition of the pure pinwheel phase (see Fig. 65*a*) was examined in great detail [113] following the procedure of Ref. 196. To this end, the experimental specific heat data above and below the transition were described separately by the form

$$C^{\pm} = \frac{A^{\pm}}{\alpha^{\pm}}\left(|1 - T/T_c|^{-\alpha\pm} - 1\right) + B^{\pm} + G(1 - T/T_c) \qquad (5.13)$$

where a linear regular background was assumed. For the critical exponents above and below T_c it is found that $\alpha^{+} \approx \alpha^{-}$ holds within the experimental resolution in the reduced temperature range $1 \times 10^{-1} - 4 \times 10^{-3}$ yielding a critical temperature $T_c = 35.205$ K at the chosen coverage. The combined data from above and below T_c (see Ref. 113 for more technical details) could not be fitted directly to a power law of the form (5.13) because of a vanishingly small critical exponent α. However, a logarithmic divergence of the form (5.7) which results from (5.13) in the $\alpha \to 0$ limit without the linear regular background term describes the singular part of the pinwheel specific heat peak very well in the reduced temperature interval $1 \times 10^{-1} - 4 \times 10^{-3}$ (see Fig. 65*b* for the corresponding scaling plot and Fig. 65*a* for the resulting fit of the peak). Hence, this analysis shows that the orientational disordering transition of the pinwheel incommensurate solid (i.e., the PW $\to$ ID transition in Fig. 49) is continuous and belongs to the two-dimensional Ising universality class.

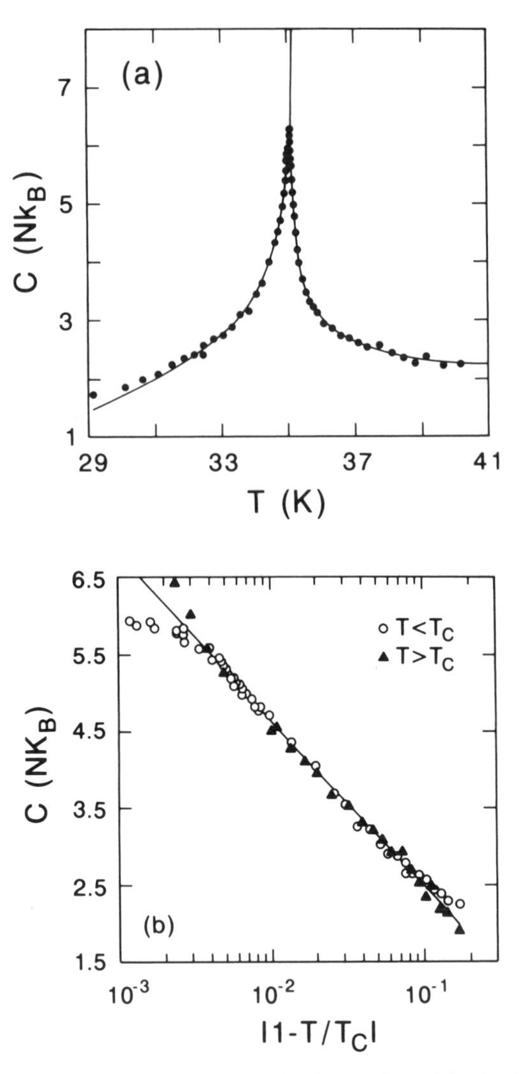

Figure 65. Experimental heat capacity near the pinwheel transition in the incommensurate solid phase of CO on graphite at a coverage of 1.29 monolayers. (*a*) Singular contribution to the total heat capacity. (*b*) Semilogarithmic plot of the data of (*a*) as a function of the reduced temperature above (triangles) and below (circles) the transition. Solid lines result from an analysis in terms of the two-dimensional Ising behavior (5.7) and (5.13). (Adapted from Figs. 5 and 4*b* of Ref. 113.)

Increasing further the coverage results in multilayer growth and wetting transitions of CO which were investigated, for example, in Refs. 190 and 397.

VI. CO ON BORON NITRIDE

A tentative phase diagram of monolayer CO on boron nitride was obtained from a vapor pressure isotherm study [229] and dielectric measurements [382] (see Fig. 66; the solid dots and diamonds indicate the adsorption isotherm and dielectric data, respectively). The isotherms and compressibilities [229] are featureless above 70 K, which signals that no monolayer phase transition takes place above this temperature. A single substep is found for 69 K and 70 K, whereas an additional kink is present in the range from 62 K to 68 K. In analogy to very similar isotherms for Kr [336], N_2 [237], and CO [112] on graphite, the substep and kink features are assigned to stem from the fluid to commensurate solid coexistence region and the melting transition from the pure commensurate solid to the reentrant fluid, respectively. According to the interpretation along the lines of Kr on graphite [323] the reentrant fluid is suggested to be a domain wall fluid—that is, a phase where commensurate patches are separated by mobile domain walls of a different density. In contrast, the domain walls in the incommensurate phase are fixed. The tricritical point is located between 70 K and 71 K because the substep vanishes above 71 K. This special point for CO on boron nitride is approximately 25 K lower than for the same adsorbate on graphite (see Section V.C). As in the case of N_2 monolayers [1] (see Section IV), this downshift of the highest temperatures at which the commensurate monolayer solid can be stabilized on boron nitride is attributed to the shallower adsorption and corrugation potentials on boron nitride relative to graphite [234].

The compressibility data are indicative [229] of a change in the nature of the melting transition from the pure commensurate solid to the reentrant fluid. Another tricritical point between 68 K and 69 K is suggested to separate a continuous melting line above this temperature from first-order melting below it, which is approximately 17 K lower than on graphite (see Section V.F). Three peaks are found in the region between 57 K and 59 K and are attributed in the order of increasing coverage to the coexistence of fluid and commensurate solid phases, the melting from the pure commensurate solid to the reentrant fluid, and the solidification of the reentrant fluid to a pure incommensurate solid phase (see Fig. 66). The resulting steep phase boundary between the incommensurate solid and reentrant fluid phase is also found for Kr on graphite [323]. The compressibility cusp becomes more symmetric and broader and evolves into a substep characteristic for regions of phase coexistence for temperatures lower than 56 K. This is

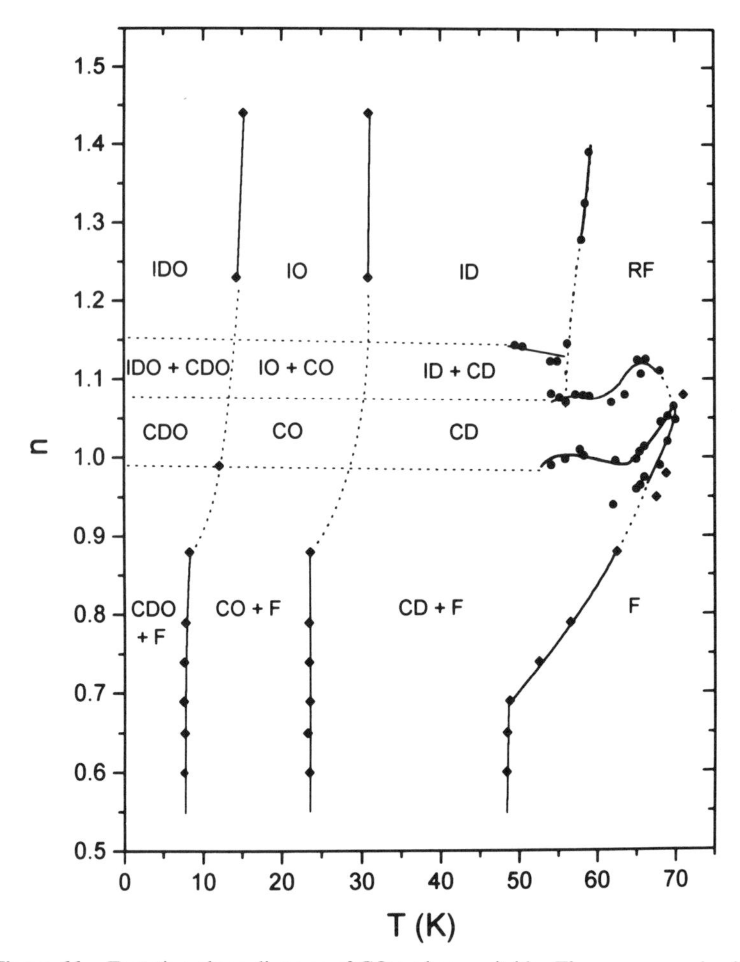

Figure 66. Tentative phase diagram of CO on boron nitride. The coverage n is given in units of the complete $\sqrt{3}$ monolayer obtained from the top of the fluid to commensurate solid isotherm substep at low temperatures less than 62 K, (see Ref. 229). The phase diagram is based on adsorption isotherm data [229] indicated by filled dots at high temperatures and coverages, as well as on dielectric data [382] marked by filled diamonds. The various phases are denoted by the following abbreviations: commensurate orientationally disordered solid, CD; incommensurate orientationally disordered solid, ID; fluid, F; reentrant fluid, RF; commensurate orientationally (quadrupolar) ordered solid, CO; incommensurate orientationally (quadrupolar) ordered solid, IO; commensurate orientationally and dipolar ordered solid, CDO; incommensurate orientationally and dipolar ordered solid, IDO. Several coexistence regimes are indicated. The solid lines correspond to phase boundaries based on measured features, whereas the dotted lines are expected phase boundaries. (Adapted from Fig. 6 of Ref. 229 and from Ref. 382.)

interpreted [229] as a change from the commensurate solid to reentrant fluid transition above 56 K to the coexistence regime of commensurate and incommensurate solids below this temperature. Thus, the reentrant fluid phase extends at most down to about 56 K (see the corresponding ID + CD–RF dotted line in Fig. 66). In summary, the topology of the phase diagrams of N_2 on boron nitride and graphite, as well as of CO on these substrates, are quite similar, and thus different from Kr on boron nitride [229]. Guided by the comparison of the behavior of identical adsorbates on substrates of different ''polarity'' (such as graphite, boron nitride, and magnesium oxide), it is argued [229] that these differences are caused by electrostatic effects because Kr has no permanent multipole moments, whereas the diatomic molecules do (see Ref. 229 for a more complete discussion).

Recently some support for the volumetric results [229] and additional information on the phase diagram were obtained from dielectric measurements [382] of CO adsorbed on boron nitride. They were carried out by employing a capacitor formed by two coaxial cylinders. The annular space between the cylinders was closely packed with boron nitride as substrate. For dielectric experiments, boron nitride (unlike graphite) has the great advantage of being an insulator (see Section I.C). An ac bridge operating between 100 Hz and 100 kHz served to measure the real and imaginary parts of the dielectric response function. Dielectric measurements have already been used previously [186] for recording dielectric isotherms of Kr and Xe on boron nitride and were also applied for studies of other polar adsorbates [223, 376].

A series of capacitance scans at various coverages between 0.6 and 0.88 monolayers and at a frequency of 1 kHz is shown in Fig. 67. As in case of CO adsorbed on graphite (see Figs. 49 and 56), three phase transitions can be clearly observed: the melting transition between 48 K and 58 K, the transition from the orientationally disordered commensurate $\sqrt{3}$ structure to the orientationally (quadrupolar) ordered herringbone structure at $T_c = 23.35$ K, and the dipolar or head–tail ordering transition at $T_c = 7.15$ K. However, it has to be stressed here that due to the lack of any structural information, the identification of the various phases was made in strong analogy to CO on graphite and is therefore highly speculative. The steplike feature at the freezing transition may be caused by a loss of translational mobility of the molecules. Surprisingly, in contrast to the heat capacity results for CO on graphite (see Section V.D and Fig. 56), very-well-defined capacitance peaks were found at the orientational (quadrupolar) ordering transition. In principle, dielectric measurements are not sensitive to the ordering of quadrupoles. However, the loss of orientational degrees of freedom of dipole moments from the orientationally disordered commensurate $\sqrt{3}$ phase, where the molecules are assumed to be more or less freely rotating, to the few discrete directions of the quadrupolar ordered (herringbone) structure may result in

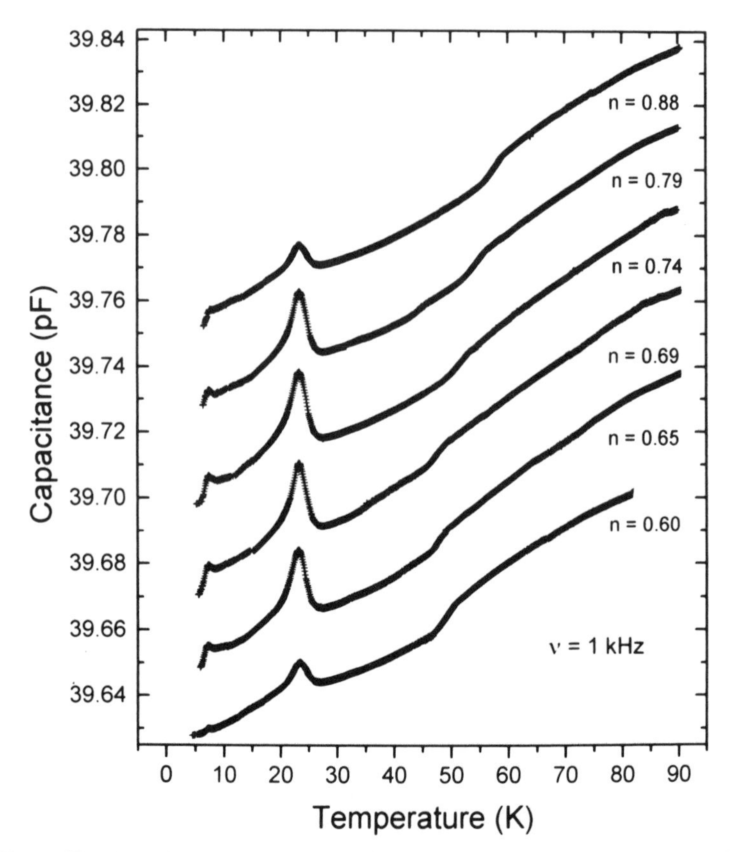

Figure 67. Capacitance as a function of temperature for various coverages n between 0.6 and 0.88 monolayers of CO on boron nitride at a measuring frequency of 1 kHz. Except for the data set at the lowest coverage, for compactness and clearness in the presentation all others have been shifted along the ordinate by adding in each case an increment of 0.024 pF to the data set at the lower coverage. The background of the sample cell and the boron nitride substrate has not been subtracted from the data, and thus the anomalies are appearing on a slightly sloped background. The steplike features between 48 K and 58 K are attributed to the melting transition, and the peaks at 23.35 K and 7.15 K are attributed to the orientational (quadrupolar) and dipolar (head-tail) ordering transitions, respectively. (From Ref. 382.)

the observed behavior. Finally, head–tail ordering of the molecules at a temperature of 7.15 K leads to peaks in the real and imaginary parts of the dielectric response function; note that the imaginary part is not displayed in Fig. 67.

The transition temperatures of CO on boron nitride differ by a few Kelvin from those on graphite. The reason might be that the substrate potential is weaker and the corrugation lower for CO on boron nitride than on graphite as was concluded from the volumetric measurements [229, 234]. In addition, the larger lattice spacing of boron nitride (see Section IC) reduces the intermolecular coupling strength; the quadrupolar interaction (2.2), e.g., decreases with the inverse fifth power of the intermolecular distance. On the other hand, the same features as for the specific heat results of CO on graphite (see Sections V.D and V.E) were found: The temperatures of the orientational and head–tail (dipolar) ordering transitions remain constant with coverage, and the peak heights rise up to a coverage of about 0.75 monolayers and then drop off again at higher coverages (see Fig. 58 for a comparison with the calorimetric results for CO on graphite and see Section V.E for a possible interpretation of this coverage dependence of the peak heights). Near monolayer completion the anomalies of the orientational and head–tail ordering transitions are totally suppressed, but shoulder-like anomalies reappear at coverages beyond 1.2 monolayers in the compressed phases not included in Fig. 67. They are located at temperatures between 12 K and 15 K for the dipolar (head–tail) ordering transition and at around 30 K for the quadrupolar (orientational) ordering transition. The melting temperature clearly shifts to higher temperatures with increasing coverage (see Fig. 67).

The phase boundaries determined from the dielectric measurements are plotted together with the adsorption isotherm data in the phase diagram (Fig. 66). The close similarity to the phase diagram of CO on graphite (see Fig. 49) is apparent and leads to the assignment of identical phases and coexistence regimes. In reasonable agreement with adsorption isotherm results [229] the melting temperature in the vicinity of the complete monolayer is about 25 K lower than that for CO on graphite. The coverage range between 0.9 and 1.2 monolayers at low temperatures is still for the most part unexplored because of the absence of well-defined dielectric signals; the dotted lines in Fig. 66 refer only to expected phase boundaries.

One important advantage of dielectric measurements compared to static methods, as, for example, heat capacity measurements, is that the former allow to probe dynamic properties of the adsorbed film by studying the frequency dependence of the real and imaginary parts of the dielectric response function. In Ref. 382 the frequency dependence of the capacitance and dissipation peak temperatures at the quadrupolar and dipolar ordering transitions occurring at 23.35 K and 7.15 K, respectively, have been investigated in the frequency range between 500 Hz and 50 kHz. In order to analyze the data, a simple Debye behavior with reorientations of individual molecules with a single time constant has been assumed as a first approximation. Such a behavior can be described by the Arrhenius law. The analysis of the data in terms of this law yielded activation energies of about 171 K for the dipolar transition and of about 2740 K for the quadrupolar transition. The observed energy barrier to reorientations, or probably 180° flips, at the dipolar transition turned out to be by a factor of 4.3 smaller than the corresponding value found for bulk CO, which was determined to be 737 K [251]. This may be a consequence of the larger lattice constant of CO in the registered phase compared to the bulk solid (i.e., 4.337 Å versus 3.99 Å, respectively) and the reduced number of nearest neighbors. Both effects may favor the attainment of a head–tail ordered ground state in physisorbed CO in contrast to the bulk system.

At present, it is not clear why the orientational ordering transition of CO on boron nitride from these measurements appears to be so much sharper than for the same system adsorbed on graphite. It is expected that structural investigations of CO on boron nitride will shed some light on this behavior. Such experiments would also be very helpful to clarify the natures of the various phases indicated in the phase diagram (Fig. 66). They may lead, in combination with results gained by other techniques, to a comprehensive understanding of the observed phenomena.

VII. CONCLUDING REMARKS

At the end of this review it should be evident that much is now known about the simple molecular physisorbates N_2 and CO on graphite as can be concluded from the very detailed knowledge of their phase diagrams up to the multilayer regime. Many puzzles could be resolved in the case of N_2 on graphite thanks to a fruitful interplay of experiment, theory, and computer simulation. Especially the early theoretical predictions of possible phases and their transitions provided many useful stimulations and also guidelines for experiments, while the subsequent simulations produced a lot of insight into the behavior on a microscopic scale. Detailed experimental data, in turn, helped to refine considerably the statistical mechanical models on which the computer simulations are based. This is clearly not the case for CO monolayers. Here, realistic simulations are largely lacking, but they could help a great deal in the microscopic understanding as they did in the case of the N_2 overlayers. On the other hand, much more precise and detailed experimental data have been published (for historical reason) for CO than for N_2 on graphite, so that especially the phase boundaries are known more reliably for CO on graphite. The microscopic structures of several phases, in particular in the compressed regime, are not yet known because of a lack of conclusive diffraction experiments in this range. One of the remaining puzzles is the herringbone-like orientational ordering transition of CO on graphite. To date, it is not fully clear why this ordering is that gradual and smeared out over a broad temperature range. Is it a Kosterlitz–Thouless-like transition as speculated in Section V.D, or is it a phase transition at all? Furthermore, the question concerning the coverage dependence of the heat capacity maxima above 0.75 monolayers for both N_2 and CO is unsettled (see Fig. 58 and the discussion for the case of the head–tail transition in Section V.E). Is it a sign for some molecules tilting out of the surface plane due to the tendency of CO in particular to form a pinwheel phase in the compressed regime? Therefore a determination of the exact orientations of the molecules would be extremely helpful for the interpretation of the results.

There is a long list of further open challenges for experiments. This is evident from the still remaining many "white spots" in the maps of the phase diagrams (see Figs. 8, 47, 49, and 66, where the microscopic structures of the phases, the nature of various phase transitions, or even the phase boundaries are unknown). For all systems the commensurate–incommensurate phase transition taking place near monolayer completion is largely unexplored. The comparison of results obtained for spherical atoms like Kr or for the ground-state spherical molecules like the hydrogens to those of the aspherical molecules N_2 and CO could result in new insights into this interesting phase transition. There are some recent indications for CO on graphite that there might exist a uniaxially compressed incommensurate phase just above monolayer completion as clearly found in the case of N_2 on graphite, but conclusive experiments still have to be performed (see the presentation in Section V.F).

In addition, there are only a few investigations which give some information on the phase diagrams of the second layer. So far it is not clear whether N_2 and CO form an independent second layer of lower density than the first one, or whether both layers adopt a common bilayer crystal structure on the substrate. There are indications from neutron diffraction measurements that in the case of N_2 on graphite, two amorphous or highly disordered layers may be adsorbed above the fully compressed monolayer prior to bilayer crystallization, which is in contrast to present interpretations of heat capacity results. The orientational disordering transitions of the bilayer or the second layer of both systems have not been investigated in detail, and the same is true for the melting transition of the second layer of N_2. Are there any differences between N_2 and CO on graphite, or are the speculative phase boundaries included in Fig. 8 correct? In addition, a lot of work remains to be done for both systems to explore (a) the growth of multilayers and (b) the transition from two-dimensional to three-dimensional behavior. This touches, however, problems of surface layering, wetting, roughening, and melting, which have not been treated in this review article.

New aspects may evolve in future experiments from studies of these diatomic molecules on various high-quality substrates. Investigations on boron nitride surfaces have just been started. Here, one of the most interesting questions is: How do the different electrical properties of this substrate compared to those of graphite influence the behavior of the adsorbed film? The phase diagram of N_2 on boron nitride is largely unexplored (see Fig. 47). In the case of CO on boron nitride, some more details are known due to recent dielectric measurements in addition to adsorption isotherm studies (see Fig. 66). The differences observed for CO on graphite or on boron nitride may be attributed, at least in part, to the different electrical properties of both substrates. Computer simulations would be highly desirable to elucidate these effects, especially since various interactions can be changed artificially and the resulting effects studied. It is also hoped that future experimental work with complementary techniques, such as x-ray diffraction and heat capacity measurements, will lead to a more profound understanding of the observed phenomena and pave new ways. In the reviewed field of surface physics of physisorbed films the potential of STM and AFM approaches to investigate the microscopic structure of adlayers is still fully unexplored, although some attempts are planned.

A vast field which is mostly unexplored by both experiment and simulation is the determination of the nature of the phase transitions in these quasi-two-dimensional systems. This is true in particular in the compressed regime with incommensurate layers. Here, relatively simple theoretical models aiming to describe commensurate–incommensurate transitions have to compete with quite sophisticated experimental techniques. But even experimentally it is often difficult, if not impossible, to distinguish between tiny coexistence regions and smeared continuous transitions. Such questions can nowadays be tackled with large-scale computer simulations, but a thorough finite-size scaling analysis of data for more than one system size has to be recognized as an indispensable part of such studies. There are a few such studies available which are devoted to orientational transitions (see the presentations in Sections III.D.2 and V.E), but nothing comparable has yet been done for positional ordering in realistic models of adsorbates. This is certainly a challenge because very large systems have to be used in order to accommodate the large-scale superstructures such as domain wall patterns. In addition, the large length scales go hand in hand with long equilibration, relaxation, and thus simulation times. Addressing such questions might also be an opportunity to apply recently developed advanced simulation and analysis techniques which have up to now only been used together with relatively simple lattice models. At this point, we wish to point out that the simple adsorbates treated in this review are exceptional at least in one respect. They serve on the one hand as well-controlled experimental realizations of statistical–mechanical model systems, which are at the roots of statistical physics in the context of studying phase transitions. On the other hand, these systems are simple enough to be simulated with great microscopic detail using quite realistic potential models, so that the very crude and often oversimplified parameterizations underlying the archetypal statistical–mechanical models are avoided.

This touches on another aspect. A delicate balance between the intermolecular interactions and the holding potential governs the relative stability of the various orientationally ordered commensurate and incommensurate structures. Usually the complicated quantum-mechanical many-body interactions of the total system comprising molecular adlayer and substrate surface are (more or less artificially) subdivided and parameterized in terms of molecule–molecule and various molecule–substrate contributions. The predictive power of the conventional simulation approaches is certainly limited by the accuracy and reliability of these underlying potential models. The orientational ordering of N_2 in the commensurate phase is fairly robust against variations of the anisotropic potential, and very simple caricatures such as the anisotropic–planar–rotor model already contain the essential physics. However, the situation changes as soon as the calculation of the melting behavior, especially the melting temperature as a function of coverage, is attempted, and it gets worse in compressed and incommensurate phases. Here, the details of how the molecule–surface interactions are approximated are much more important. The slightly more complicated physics in the case of CO on graphite leads to severe problems in the theoretical description, so that the orientational transition even in the commensurate phase is not yet fully understood. Evidently, these problems become more serious the more complex the systems are.

Recently, however, the *ab initio* molecular dynamics technique [58] was applied to adsorbates such as water and its dissociation on an oxide surface [188] or the chemisorption of a big organic molecule [119]. In the Car–Parrinello molecular dynamics approach [58] the forces acting on the particles are obtained in each timestep from *ab initio* electronic structure calculations—for example, in the framework of density functional theory (for reviews see Refs. 126, 280, and 295). Thus, the interactions of the molecules with the surface are described with the accuracy of state-of-the-art electronic structure theory without involving any pair-interaction approximation, partitioning in different contributions, and subsequent fitting of potential functions. It has already been shown that the structural parameters of bulk graphite are satisfactorily described in the local density approximation [66]. The *ab initio* molecular dynamics approach [58] is, of course, computationally demanding, but in the future we will certainly witness *ab initio* investigations of phase transitions in adsorbates.

Acknowledgments

It is a great pleasure for us to thank the following colleagues for their various contributions to our own research activities in this field (including helpful discussions) and/or to this particular review: St.-A. Arlt, D. Arndt, K. Binder, E. Britten, S. C. Fain, Jr., M. L. Klein, K. Knorr, K.-D. Kortmann, B. Krömker, J. Lee, B. Leinböck, E. Maus, P. Nielaba, O. Opitz, M. Parrinello, M. Schick, S. Sengupta, O. E. Vilches, H. Weber, and M. Willenbacher. We are grateful to many colleagues, in particular to L. W. Bruch and S. C. Fain, Jr., for their useful remarks and suggestions on the preprint version of this review, and to F. Y. Hansen and W. A. Steele for sending us preprints. Our research was supported in part by Deutsche Forschungsgemeinschaft, Sonderforschungsbereich 262 Mainz, Materialwissenschaftliches Forschungszentrum Mainz, BMBF Projekt 03-WI3MAI-4, IBM Zurich Research Laboratory, HRLZ Jülich, RHRK Kaiserslautern, and ZDV Mainz. We gratefully acknowledge these grants.

NOTE ADDED IN PROOF

We became aware that the following references were not known to us during the writing of this chapter: D. Schmeisser, F. Greuter, E. W. Plummer, and H.-J. Freund, "Photoemission from Ordered Physisorbed Adsorbate Phases: N_2 on Graphite and CO on Ag(111)," *Phys. Rev. Lett.* **54,** 2095 (1985); M. Seel and H.-J. Freund, "Ab initio Bandstructure of a Layer of Nitrogen Molecules Oriented in a Herringbone Structure," *Solid State Commun.* **55,** 895 (1985); V. L. Eden, F. J. Giessibl, D. P. E. Smith, and K. Aggarwal, "Low Temperature UHV STM Observations of Weakly Physisorbed Atoms and Molecules on Graphite," *Bull. Am. Phys. Soc.* **37,** 327 (1992); F. Y. Hansen, L. W. Bruch, and H. Taub, "Mechanism of Melting in Submonolayer Films of Nitrogen Molecules Adsorbed on the Basal Planes of Graphite," *Phys. Rev. B* **52,** 8515 (1995); W. A. Steele, "Monolayers of Linear Molecules Adsorbed on the Graphite Basal Plane: Structures and Intermolecular Interactions," *Langmuir* **12,** 145 (1996); F. Y. Hansen, L. W. Bruch, and H. Taub, "Molecular Dynamics of a Solid with Vacancies: Submonolayer Nitrogen on Graphite," submitted to *Phys. Rev. B* (1996); L. W. Bruch, M. W. Cole, and E. Zaremba, "Physical Adsorption: Forces and Phenomena," to be published by Oxford Press.

Finally, we would like to direct the reader's attention to the most recent developments: P. Laitenberger and R. E. Palmer, "Electron Stimulated Desorption from Selected Phases of Physisorbed CO/Graphite: Evidence for Structure Sensitive Desorption Dynamics," *J. Phys.: Condens. Matter* **8,** L71 (1996); M. Pilla and N. S. Sullivan, "First Order Phase Transitions in the Orientational Ordering of Monolayers of N_2 on Graphite Substrate Studied by the Monte Carlo Technique," *Bull. Am. Phys. Soc.* **41,** 192 (1996); H. Wu and G. B. Hess, "Physisorption of Multilayer Carbon Monoxide on Graphite," *Bull. Am. Phys. Soc.* **41,** 192 (1996); P. Shrestha and A. D. Migone, "Multilayer N_2 Films Adsorbed on BN," *Bull. Am. Phys. Soc.* **41,** 193 (1996); R. A. Wolfson, L. Arnold, P. Shrestha, and A. D. Migone, "Quality of Four Grades of BN as Substrate for Physisorption Studies," *Bull. Am. Phys. Soc.* **41,** 193 (1996).

The closing date of the literature search as covered in this chapter including this "Note Added in Proof" is March 1996.

References

1. M. T. Alkhafaji, P. Shrestha, and A. D. Migone, *Phys. Rev. B* **50,** 11088 (1994).
2. M. P. Allen and S. F. O'Shea, *Mol. Simulation* **1,** 47 (1987).
3. M. P. Allen and D. J. Tildesley, *Computer Simulations of Liquids* (Clarendon Press, Oxford, 1987).
4. K. L. D'Amico, J. Bohr, D. E. Moncton, and D. Gibbs, *Phys. Rev. B* **41,** 4368 (1990).
5. T. Atake, H. Suga, and H. Chihara, *Chem. Lett. (Japan)* 567 (1976).
6. A. van der Avoird, P. E. S. Wormer, F. Mulder, and R. Berns, *Topics Curr. Chem.* **93,** 1 (1980).
7. L. V. Azároff, *Elements of X-Ray Crystallography* (McGraw-Hill, New York, 1968).
8. G. E. Bacon, *Neutron Diffraction* (Clarendon Press, Oxford, 1975).
9. R. F. W. Bader, W. H. Henneker, and P. E. Cade, *J. Chem. Phys.* **46,** 3341 (1967).
10. P. Bak, *Rep. Prog. Phys.* **45,** 587 (1982).
11. M. N. Barber, in *Phase Transitions and Critical Phenomena,* Vol. 8, C. Domb and J. L. Lebowitz (eds.) (Academic Press, London, 1983).
12. E. Barouch, B. M. McCoy, and T. T. Wu, *Phys. Rev. Lett.* **31,** 1409 (1973).
13. Y. Baskin and L. Meyer, *Phys. Rev.* **100,** 544 (1955).
14. M. Bée, *Quasielastic Neutron Scattering* (Hilger, Bristol, 1988).
15. R. A. Beebe, J. Biscoe, W. R. Smith, and C. B. Wendell, *J. Am. Chem. Soc.* **69,** 95 (1947).
16. R. A. Beebe, R. L. Gale, and T. C. W. Kleinsteuber, *J. Phys. Chem.* **70,** 4010 (1966).
17. J. Belak, K. Kobashi, and R. D. Etters, *Surf. Sci.* **161,** 390 (1985).
18. T. H. M. van den Berg and A. van der Avoird, *Phys. Rev. B* **40,** 1932 (1989).
19. T. H. M. van den Berg and A. van der Avoird, *Phys. Rev. B* **43,** 13926 (1991).
20. A. N. Berker, S. Ostlund, and F. A. Putnam, *Phys. Rev. B* **17,** 3650 (1978).
21. A. J. Berlinsky and A. B. Harris, *Phys. Rev. Lett.* **40,** 1579 (1978).
22. B. J. Berne and D. Thirumalai, *Ann. Rev. Phys. Chem.* **37,** 401 (1986).
23. R. M. Berns and A. van der Avoird, *J. Chem. Phys.* **72,** 6107 (1980).
24. V. R. Bhethanabotla and W. A. Steele, *J. Chem. Phys.* **91,** 4346 (1989).
25. V. R. Bhethanabotla and W. A. Steele, *Phys. Rev. B* **41,** 9480 (1990).
26. M. Bienfait, in *Dynamics of Molecular Crystals,* J. Lascombe (ed.), *Studies in Physical and Theoretical Chemistry*, Vol. 46 (Elsevier, Amsterdam, 1987), p. 353.
27. M. Bienfait and J.-M. Gay, in *Phase Transitions in Surface Films 2,* H. Taub, G. Torzo, H. J. Lauter, and S. C. Fain, Jr. (eds.) (Plenum, New York, 1991), p. 307.
28. K. Binder, in *Phase Transitions and Critical Phenomena,* Vol. 5b, C. Domb and M. S. Green (eds.) (Academic Press, London, 1976), p. 1.
29. K. Binder (ed.), *Monte Carlo Methods in Statistical Physics* (Springer, Berlin, 1979).
30. K. Binder, *J. Stat. Phys.* **24,** 69 (1981).
31. K. Binder, *Phys. Rev. Lett.* **47,** 693 (1981).
32. K. Binder, *Z. Phys. B* **43,** 119 (1981).
33. K. Binder (ed.), *Applications of the Monte Carlo Method in Statistical Physics* (Springer, Berlin, 1984).

34. K. Binder, *Ferroelectrics* **73,** 43 (1987).
35. K. Binder and D. W. Heermann, *Monte Carlo Simulation in Statistical Physics—An Introduction* (Springer, Berlin, 1988).
36. K. Binder and D. P. Landau, *Adv. Chem. Phys.* **76,** 91 (1989).
37. K. Binder (ed.), *The Monte Carlo Method in Condensed Matter Physics* (Springer, Berlin, 1992).
38. K. Binder, K. Vollmayr, H.-P. Deutsch, J. D. Reger, M. Scheucher, and D. P. Landau, *Int. J. Mod. Phys. C* **3,** 1025 (1992).
39. R. J. Birgeneau, P. A. Heiney, and J. P. Pelz, *Physica B* **109/110,** 1785 (1982).
40. R. J. Birgeneau and P. M. Horn, *Science* **232,** 329 (1986).
41. M. J. Bojan and W. A. Steele, *Langmuir* **3,** 116 (1987).
42. M. J. Bojan and W. A. Steele, *Langmuir* **3,** 1123 (1987).
43. C. Borgs and W. Janke, *J. Phys. I (France)* **2,** 2011 (1992).
44. C. J. F. Böttcher, *Theory of Electric Polarization,* Vol. I and II (Elsevier Science Publishers, Amsterdam, 1993).
45. A. Bourdon, C. Marti, and P. Thorel, Phys. Rev. Lett. **35,** 544 (1975).
46. M. Bretz, J. G. Dash, D. C. Hickernell, E. O. McLean, and O. E. Vilches, *Phys. Rev. A* **8,** 1589 (1973).
47. A. D. Bruce, *J. Phys. A* **18,** L873 (1985).
48. L. W. Bruch, *J. Chem. Phys.* **79,** 3148 (1983).
49. L. W. Bruch, *Surf. Sci.* **125,** 194 (1983).
50. L. W. Bruch, in *Phase Transitions in Surface Films 2,* H. Taub, G. Torzo, H. J. Lauter, and S. C. Fain, Jr. (eds.) (Plenum, New York, 1991), p. 67.
51. S. Brunauer, P. H. Emmett, and E. Teller, *J. Am. Chem. Soc.* **60,** 309 (1938).
52. J. C. Burford and G. M. Graham, *Can. J. Phys.* **47,** 23 (1969).
53. T. W. Burkhard and B. Derrida, *Phys. Rev. B* **32,** 7273 (1985).
54. D. M. Butler, G. B. Huff, R. W. Toth, and G. A. Stewart, *Phys. Rev. Lett.* **35,** 1718 (1975).
55. D. M. Butler, J. A. Litzinger, G. A. Stewart, and R. B. Griffiths, *Phys. Rev. Lett.* **42,** 1289 (1979).
56. Z.-X. Cai, *Phys. Rev. B* **43,** 6163 (1991).
57. J. H. Campbell and M. Bretz, *Phys. Rev. B* **32,** 2861 (1985).
58. R. Car and M. Parrinello, *Phys. Rev. Lett.* **55,** 2471 (1985).
59. G. Cardini and S. F. O'Shea, *Surf. Sci.* **154,** 231 (1985).
60. J. L. Cardy, M. P. M. den Nijs, and M. Schick, *Phys. Rev. B* **27,** 4251 (1983).
61. W. E. Carlos and M. W. Cole, *Surf. Sci.* **91,** 339 (1980).
62. E. Chacón and P. Tarazona, *Phys. Rev. B* **39,** 7111 (1989).
63. M. S. S. Challa and D. P. Landau, *Phys. Rev. B* **33,** 437 (1986).
64. M. S. S. Challa, D. P. Landau, and K. Binder, *Phys. Rev. B* **34,** 1841 (1986).
65. M. H. W. Chan, A. D. Migone, K. D. Miner, and Z. R. Li, *Phys. Rev. B* **30,** 2681 (1984).
66. J.-C. Charlier, X. Gonze, and J.-P. Michenaud, *Europhys. Lett.* **28,** 403 (1994).
67. R. Chen, P. Trucano, and R. F. Stewart, *Acta Cryst. A* **33,** 823 (1977).
68. E. Cheng, M. W. Cole, J. Dupont-Roc, W. F. Saam, and J. Treiner, *Rev. Mod. Phys.* **65,** 557 (1993).

69. P. S. Y. Cheung and J. G. Powles, *Mol. Phys.* **32,** 1383 (1976).
70. H.-Y. Choi and E. J. Mele, *Phys. Rev. B* **40,** 3439 (1989).
71. H.-Y. Choi, A. B. Harris, and E. J. Mele, *Phys. Rev. B* **40,** 3766 (1989).
72. T. T. Chung and J. G. Dash, *J. Chem. Phys.* **64,** 1855 (1976).
73. T. T. Chung and J. G. Dash, *Surf. Sci.* **66,** 559 (1977).
74. G. Ciccotti, D. Frenkel, and I. R. McDonald (eds.), *Simulation of Liquids and Solids—Molecular Dynamics and Monte Carlo Methods in Statistical Mechanics* (North-Holland, Amsterdam, 1987).
75. L. J. Clarke, *Surface Crystallography* (Wiley, Chichester, 1985).
76. J. O. Clayton and W. F. Giauque, *J. Am. Chem. Soc.* **54,** 2610 (1932).
77. S. N. Coppersmith, D. S. Fisher, B. I. Halperin, P. A. Lee, and W. F. Brinkman, *Phys. Rev. Lett.* **46,** 549 (1981). Erratum: *ibid.* **46,** 869(E) (1981).
78. S. N. Coppersmith, D. S. Fisher, B. I. Halperin, P. A. Lee, and W. F. Brinkman, *Phys. Rev. B* **25,** 349 (1982).
79. J.-P. Coulomb, T. S. Sullivan, and O. E. Vilches, *Phys. Rev. B* **30,** 4753 (1984).
80. J. P. Coulomb and O. E. Vilches, *J. Phys. (Paris)* **45,** 1381 (1984).
81. J. P. Coulomb, in *Phase Transitions in Surface Films 2,* H. Taub, G. Torzo, H. J. Lauter, and S. C. Fain, Jr. (eds.) (Plenum, New York, 1991), p. 113.
82. A. D. Crowell and J. S. Brown, *Surf. Sci.* **123,** 296 (1982).
83. J. Cui, S. C. Fain, Jr., H. Freimuth, H. Wiechert, H. P. Schildberg, and H. J. Lauter, *Phys. Rev. Lett.* **60,** 1848 (1988); Erratum: *Phys. Rev. Lett.* **60,** 2704(E) (1988).
84. J. Cui and S. C. Fain, Jr., *Phys. Rev. B* **39,** 8628 (1989).
85. B. D. Cullity, *Elements of X-Ray Diffraction* (Addison-Wesley, Reading, 1978).
86. R. F. Curl, Jr., H. P. Hopkins, Jr., and K. S. Pitzer, *J. Chem. Phys.* **48,** 4064 (1968).
87. J. G. Dash, *Films on Solid Surfaces* (Academic Press, New York, 1975).
88. J. G. Dash and J. Ruvalds (eds.), *Phase Transitions in Surface Films* (Plenum, New York, 1980).
89. J. G. Dash and R. D. Puff, *Phys. Rev. B* **24,** 295 (1981).
90. R. D. Diehl and S. C. Fain, Jr., *Phys. Rev. B* **26,** 4785 (1982).
91. R. D. Diehl and S. C. Fain, Jr., *J. Chem. Phys.* **77,** 5065 (1982).
92. R. D. Diehl, M. F. Toney, and S. C. Fain, Jr., *Phys. Rev. Lett.* **48,** 177 (1982).
93. R. D. Diehl and S. C. Fain, Jr., *Surf. Sci.* **125,** 116 (1983).
94. L. D. Dillon, R. E. Rapp, and O. E. Vilches, *J. Low Temp. Phys.* **59,** 35 (1985).
95. E. Domany, M. Schick, J. S. Walker, and R. B. Griffiths, *Phys. Rev. B* **18,** 2209 (1978).
96. O. M. B. Duparc and R. D. Etters, *J. Chem. Phys.* **86,** 1020 (1987).
97. R. E. Ecke, J. G. Dash, and R. D. Puff, *Phys. Rev. B* **26,** 1288 (1982).
98. R. E. Ecke and J. G. Dash, *Phys. Rev. B* **28,** 3738 (1983).
99. J. Eckert, W. D. Ellenson, J. B. Hastings, and L. Passell, *Phys. Rev. Lett.* **43,** 1329 (1979).
100. C. A. English and J. A. Venables, *Proc. Roy. Soc. (London) A* **340,** 57 (1974).
101. R. D. Etters, V. Chandrasekharan, E. Uzan, and K. Kobashi, *Phys. Rev. B* **33,** 8615 (1986).
102. R. D. Etters, M. W. Roth, and B. Kuchta, *Phys Rev. Lett.* **65,** 3140 (1990).
103. R. D. Etters, B. Kuchta, and J. Belak, *Phys. Rev. Lett.* **70,** 826 (1993).

104. H. Evans, D. J. Tildesley, and T. J. Sluckin, *J. Phys. C: Solid State Phys.* **17,** 4907 (1984).
105. M. D. Evans, N. Patel, and N. S. Sullivan, *J. Low Temp. Phys.* **89,** 653 (1992).
106. S. C. Fain, Jr., M. F. Toney, and R. D. Diehl, in *Proceedings of the Ninth International Vacuum Congress and Fifth International Conference on Solid Surfaces,* J. L. Segovia (ed.) (Imprenta Moderna, Madrid, 1983), p. 129.
107. S. C. Fain, Jr., and H. You, in *The Structure of Surfaces I,* M. A. van Hove and S. Y. Tong (eds.), *Springer Series in Surface Sciences,* Vol. 2 (Springer, Berlin, 1985), p. 413.
108. S. C. Fain, Jr., *Ber. Bunsenges. Phys. Chem.* **90,** 211 (1986).
109. S. C. Fain, Jr., *Carbon* **25,** 19 (1987).
110. D. K. Fairobent, W. F. Saam, and L. M. Sander, *Phys. Rev. B* **26,** 179 (1982).
111. J. W. O. Faul, U. G. Volkmann, and K. Knorr, *Surf. Sci.* **227,** 390 (1990).
112. Y. P. Feng and M. H. W. Chan, *Phys. Rev. Lett.* **64,** 2148 (1990).
113. Y. P. Feng and M. H. W. Chan, *Phys. Rev. Lett.* **71,** 3822 (1993).
114. R. P. Feynman and A. R. Hibbs, *Quantum Mechanics and Path Integrals* (McGraw-Hill, New York, 1965).
115. M. E. Fisher, *Rev. Mod. Phys.* **46,** 597 (1974).
116. M. E. Fisher and A. N. Berker, *Phys. Rev. B* **26,** 2507 (1982).
117. P. F. Fracassi and M. L. Klein, *Chem. Phys. Lett.* **108,** 359 (1984).
118. P. F. Fracassi, R. Righini, R. G. Della Valle, and M. L. Klein, *Chem. Phys.* **96,** 361 (1985).
119. I. Frank, D. Marx, and M. Parrinello, *J. Am. Chem. Soc.* **117,** 8037 (1995).
120. H. Freimuth and H. Wiechert, *Surf. Sci.* **162,** 432 (1985).
121. H. Freimuth and H. Wiechert, *Surf. Sci.* **178,** 716 (1986).
122. H. Freimuth, H. Wiechert, and H. J. Lauter, *Surf. Sci.* **189/190,** 548 (1987).
123. H. Freimuth, H. Wiechert, H. P. Schildberg, and H. J. Lauter, *Phys. Rev. B* **42,** 587 (1990).
124. H. Fröhlich, *Theory of Dielectrics: Dielectric Constant and Dielectric Loss* (Clarendon, Oxford, 1949).
125. C. R. Fuselier, N. S. Gillis, and J. C. Raich, *Solid State Commun.* **25,** 747 (1978).
126. G. Galli and M. Parrinello, in *Computer Simulations in Materials Science,* M. Meyer and V. Pontikis (eds.) (Kluwer, Dordrecht, 1991).
127. E. K. Gill and J. A. Morrison, *J. Chem. Phys.* **45,** 1585 (1966).
128. M. J. Gillan, in *Computer Modelling of Fluids, Polymers, and Solids,* C. R. A. Catlow, S. C. Parker, and M. P. Allen (eds.) (Kluwer, Dordrecht, 1990).
129. M. A. Glaser and N. A. Clark, *Adv. Chem. Phys.* **83,** 543 (1993).
130. M. B. Gordon and J. Villain, *J. Phys. C* **18,** 3919 (1985).
131. J. M. Gottlieb and L. W. Bruch, *Phys. Rev. B* **40,** 148 (1989); see also Ref. 132.
132. J. M. Gottlieb and L. W. Bruch, *Phys. Rev. B* **41,** 7195 (1990).
133. C. G. Gray and K. E. Gubbins, *Theory of Molecular Fluids,* Vol. 1 (Clarendon, Oxford, 1984).
134. R. B. Griffiths, *Phys. Rev. B* **7,** 545 (1973).
135. Y. Grillet, F. Rouquerol, and J. Rouquerol, *J. Phys. (Paris) Colloq.* **38,** C4–57 (1977).

136. E. M. Hammonds, P. Heiney, P. W. Stephens, R. J. Birgeneau, and P. M. Horn, *J. Phys. C* **13,** L301 (1980).
137. F. Y. Hansen, V. L. P. Frank, H. Taub, L. W. Bruch, H. J. Lauter, and J. R. Dennison, *Phys. Rev. Lett.* **64,** 764 (1990).
138. F. Y. Hansen, L. W. Bruch, and S. E. Roosevelt, *Phys. Rev. B* **45,** 11238 (1992).
139. F. Y. Hansen, J. C. Newton, and H. Taub, *J. Chem. Phys.* **98,** 4128 (1993).
140. F. Y. Hansen and L. W. Bruch, *Phys. Rev. B* **51,** 2515 (1995).
141. A. B. Harris and A. J. Berlinsky, *Can. J. Phys.* **57,** 1852 (1979).
142. A. B. Harris, O. G. Mouritsen, and A. J. Berlinsky, *Can. J. Phys.* **62,** 915 (1984).
143. D. W. Heermann, *Computer Simulation Methods in Theoretical Physics* (Springer, Berlin, 1990).
144. J. Heidberg, M. Warskulat, and M. Folman, *J. Electron Spectrosc. Related Phenomena* **54/55,** 961 (1990).
145. P. A. Heiney, P. W. Stephens, S. G. J. Mochrie, J. Akimitsu, R. J. Birgeneau, and P. M. Horn, *Surf. Sci.* **125,** 539 (1983).
146. K. Heinz, *Rep. Prog. Phys.* **58,** 637 (1995).
147. G. B. Hess, in *Phase Transitions in Surface Films 2,* H. Taub, G. Torzo, H. J. Lauter, and S. C. Fain, Jr. (eds.) (Plenum, New York, 1991), p. 357.
148. H. Hoinkes, *Rev. Mod. Phys.* **52,** 933 (1980).
149. H. Hong, C. J. Peters, A. Mak, R. J. Birgeneau, P. M. Horn, and H. Suematsu, *Phys. Rev. B* **40,** 4797 (1989).
150. M. A. van Hove, W. H. Weinberg, and C. M. Chan, *Low-Energy Electron Diffraction, Springer Series in Surface Sciences,* Vol. 6 (Springer, Berlin, 1986).
151. D. A. Huse and M. E. Fisher, *Phys. Rev. B* **29,** 239 (1984).
152. A. Inaba and H. Chihara, *Can. J. Chem.* **66,** 703 (1988).
153. A. Inaba, T. Shirakami, and H. Chihara, *Chem. Phys. Lett.* **146,** 63 (1988).
154. A. Inaba and H. Chihara, *Order/Disorder in Solids* **7,** 33 (1989).
155. A. Inaba, T. Shirakami, and H. Chihara, *Surf. Sci.* **242,** 202 (1991).
156. A. Inaba, T. Shirakami, and H. Chihara, *J. Chem. Thermodynamics* **23,** 461 (1991).
157. W. B. J. M. Janssen, T. H. M. van den Berg, and A. van der Avoird, *Surf. Sci.* **259,** 389 (1991).
158. W. B. J. M. Janssen, T. H. M. van den Berg, and A. van der Avoird, *Phys. Rev. B* **43,** 5329 (1991).
159. W. B. J. M. Janssen, J. Michiels, and A. van der Avoird, *J. Chem. Phys.* **94,** 8402 (1991).
160. E. T. Jensen and R. E. Palmer, *J. Phys.: Condens. Matter* **1,** SB7 (1989).
161. E. T. Jensen and R. E. Palmer, *Surf. Sci.* **233,** 269 (1990).
162. W. Jin, S. D. Mahanti, and S. Tang, *Solid State Commun.* **66,** 877 (1988).
163. F. Jona, J. A. Strozier, Jr., and W. S. Yang, *Rep. Prog. Phys.* **45,** 527 (1982).
164. J. V. José, L. P. Kadanoff, S. Kirkpatrick, and D. R. Nelson, *Phys. Rev. B* **16,** 1217 (1977).
165. Y. P. Joshi and D. J. Tildesley, *Mol. Phys.* **55,** 999 (1985).
166. L. G. Joyner and P. H. Emmett, *J. Am. Chem. Soc.* **70,** 2353 (1948).
167. K. Kern and G. Comsa, in *Chemistry and Physics of Solid Surfaces VII, Springer Series*

in Surface Sciences, Vol. 10, R. Vanselow and R. Howe (eds.) (Springer, Berlin, 1988), p. 65.
168. K. Kern and G. Comsa, in *Phase Transitions in Surface Films 2,* H. Taub, G. Torzo, H. J. Lauter, and S. C. Fain, Jr. (eds.) (Plenum, New York, 1991), p. 41.
169. R. Kikuchi, *Phys. Rev.* **81,** 988 (1951).
170. D. Kirin, B. Kuchta, and R. D. Etters, *J. Chem. Phys.* **87,** 2332 (1987).
171. J. K. Kjems, L. Passell, H. Taub, and J. G. Dash, *Phys. Rev. Lett.* **32,** 724 (1974).
172. J. K. Kjems, L. Passell, H. Taub, J. G. Dash, and A. D. Novaco, *Phys. Rev. B* **13,** 1446 (1976).
173. H. Kleinert, *Gauge Fields in Condensed Matter,* Vol. II (World Scientific, Singapore, 1989), Part III.
174. M. A. Klenin and S. F. Pate, *Phys. Rev. B* **26,** 3969 (1982).
175. B. C. Kohin, *J. Chem. Phys.* **33,** 882 (1960).
176. K.-D. Kortmann, M. Maurer, H. Wiechert, N. Stüßer, and X. Hu, in *BENSC Experimental Reports 1993,* Y. Kirschbaum and R. Michaelsen (eds.) (Berlin, 1994), p. 71.
177. K.-D. Kortmann, H. Wiechert, and N. Stüßer, in *BENSC Experimental Reports 1994,* Y. Kirschbaum, H. Gast, and R. Michaelsen (eds.) (Berlin, 1995), p. 144.
178. J. van Kranendonk, *Solid Hydrogen* (Plenum, New York, 1983).
179. P. R. Kubik and W. N. Hardy, *Phys. Rev. Lett.* **41,** 257 (1978).
180. P. R. Kubik, W. N. Hardy, and H. Glattli, *Can. J. Phys.* **63,** 605 (1985); see also Ref. 179.
181. B. Kuchta and R. D. Etters, *Phys. Rev. B* **36,** 3400 (1987).
182. B. Kuchta and R. D. Etters, *Phys. Rev. B* **36,** 3407 (1987).
183. B. Kuchta and R. D. Etters, *J. Chem. Phys.* **88,** 2793 (1988).
184. B. Kuchta and R. D. Etters, *J. Comput. Phys.* **108,** 353 (1993).
185. S. Kumar, M. Roth, B. Kuchta, and R. D. Etters, *J. Chem. Phys.* **97,** 3744 (1992).
186. J. P. Laheurte, J. C. Noiray, M. Obadia, and J. P. Romagnan, *Surf. Sci.* **122,** 330 (1982).
187. D. P. Landau, *Phys. Rev. B* **27,** 5604 (1983).
188. W. Langel and M. Parrinello, *Phys. Rev. Lett.* **73,** 504 (1994).
189. Y. Larher, *J. Chem. Phys.* **68,** 2257 (1978).
190. Y. Larher, F. Angerand, and Y. Maurice, *J. Chem. Soc. Faraday Trans. 1* **83,** 3355 (1987).
191. Y. Larher, F. Millot, and C. Tessier, *J. Chem. Phys.* **88,** 1474 (1988).
192. H. J. Lauter, V. L. P. Frank, H. Taub, and P. Leiderer, *Physica B* **165/166,** 611 (1990).
193. H. J. Lauter, H. Godfrin, V. L. P. Frank, and P. Leiderer, in *Phase Transitions in Surface Films 2,* H. Taub, G. Torzo, H. J. Lauter, and S. C. Fain, Jr. (eds.) (Plenum, New York, 1991), p. 135.
194. I. D. Lawrie and S. Sarbach, in *Phase Transitions and Critical Phenomena,* Vol. 9, C. Domb and J. L. Lebowitz (eds.) (Academic Press, London, 1984). See Section V.F; note that β_2 in this review is identical to β_u of Ref. 134.
195. D. H. Lee, J. D. Joannopoulos, J. W. Negele, and D. P. Landau, *Phys. Rev. B* **33,** 450 (1986).
196. F. L. Lederman, M. B. Salamon, and L. W. Shacklette, *Phys. Rev. B* **9,** 2981 (1974).

197. J. Lee and J. M. Kosterlitz, *Phys. Rev. B* **43,** 3265 (1991).
198. R. LeSar and M. S. Shaw, *J. Chem. Phys.* **84,** 5479 (1986).
199. F. Li, J. R. Brookeman, A. Rigamonti, and T. A. Scott, *J. Chem. Phys.* **74,** 3120 (1981).
200. M. S. H. Ling and M. Rigby, *Mol. Phys.* **51,** 855 (1984).
201. S. W. Lovesey, *Theory of Neutron Scattering from Condensed Matter,* Vol. 1 and 2 (Oxford University Press, Oxford, 1986).
202. H. Lüth, *Surfaces and Interfaces of Solids* (Springer, Berlin, 1993).
203. R. M. Lynden-Bell, J. Talbot, D. J. Tildesley, and W. A. Steele, *Mol. Phys.* **54,** 183 (1985).
204. M. Lysek, P. Day, M. LaMadrid, and D. Goodstein, *Rev. Sci. Instrum.* **63,** 5750 (1992).
205. I. Lyuksyutov, A. G. Naumovets, and V. Pokrovsky, *Two-Dimensional Crystals* (Academic Press, Boston, 1992).
206. S. D. Mahanti and S. Tang, *Superlattices Microstruct.* **1,** 517 (1985).
207. D. Marx, P. Nielaba, and K. Binder, *Phys. Rev. Lett.* **67,** 3124 (1991).
208. D. Marx, *Surf. Sci.* **272,** 198 (1992).
209. D. Marx and P. Nielaba, *Phys. Rev. A* **45,** 8968 (1992).
210. D. Marx, P. Nielaba, and K. Binder, *Phys. Rev. B* **47,** 7788 (1993).
211. D. Marx, O. Opitz, P. Nielaba, and K. Binder, *Phys. Rev. Lett.* **70,** 2908 (1993).
212. D. Marx, S. Sengupta, and P. Nielaba, *J. Chem. Phys.* **99,** 6031 (1993).
213. D. Marx, S. Sengupta, and P. Nielaba, *Ber. Bunsenges. Phys. Chem.* **98,** 525 (1994).
214. D. Marx, S. Sengupta, P. Nielaba, and K. Binder, *Phys. Rev. Lett.* **72,** 262 (1994).
215. D. Marx, S. Sengupta, P. Nielaba, and K. Binder, *Surf. Sci.* **321,** 195 (1994).
216. D. Marx, S. Sengupta, P. Nielaba, and K. Binder, *J. Phys.: Condens. Matter* **6,** A175 (1994).
217. D. Marx, S. Sengupta, O. Opitz, P. Nielaba, and K. Binder, *Mol. Phys.* **83,** 31 (1994).
218. D. Marx and P. Nielaba, *J. Chem. Phys.* **102,** 4538 (1995).
219. R. Marx, B. Vennemann, and E. Uffelmann, *Phys. Rev. B* **29,** 5063 (1984).
220. R. Marx, *Phys. Rep.* **125,** 1 (1985).
221. R. Marx and B. Christoffer, *Phys. Rev. B* **37,** 9518 (1988).
222. L. Mattera, F. Rosatelli, C. Salvo, F. Tommasini, U. Valbusa, and G. Vidali, *Surf. Sci.* **93,** 515 (1980).
223. E. Maus, W. Weimer, H. Wiechert, and K. Knorr, *Ferroelectrics* **108,** 77 (1990).
224. R. L. McIntosh, *Dielectric Behavior of Physically Adsorbed Gases* (Marcel Dekker, New York, 1966).
225. A. D. McLachlan, *Mol. Phys.* **7,** 381 (1964).
226. J. P. McTague and M. Nielsen, *Phys. Rev. Lett.* **37,** 596 (1976).
227. J. P. McTague and A. D. Novaco, *Phys. Rev. B* **19,** 5299 (1979).
228. L. Mederos, E. Chacón, and P. Tarazona, *Phys. Rev. B* **42,** 8571 (1990).
229. J. M. Meldrim and A. D. Migone, *Phys. Rev. B* **51,** 4435 (1995).
230. M. W. Melhuish and R. L. Scott, *J. Phys. Chem.* **68,** 2301 (1964).
231. A. D. Migone, M. H. W. Chan, K. J. Niskanen, and R. B. Griffiths, *J. Phys. C: Solid State Phys.* **16,** L1115 (1983).

232. A. D. Migone, H. K. Kim, M. H. W. Chan, J. Talbot, D. J. Tildesley, and W. A. Steele, *Phys. Rev. Lett.* **51,** 192 (1983).

233. A. D. Migone, Z. R. Li, and M. H. W. Chan, *Phys. Rev. Lett.* **53,** 810 (1984).

234. A. D. Migone, M. T. Alkhafaji, G. Vidali, and M. Karimi, *Phys. Rev. B* **47,** 6685 (1993).

235. A. Milchev, D. W. Heermann, and K. Binder, *J. Stat. Phys.* **44,** 749 (1986).

236. F. Millot, Y. Larher, and C. Tessier, *J. Chem. Phys.* **76,** 3327 (1982).

237. K. D. Miner, M. H. W. Chan, and A. D. Migone, *Phys. Rev. Lett.* **51,** 1465 (1983).

238. K. Mirsky, *Chem. Phys.* **46,** 445 (1980).

239. A. C. Mitus, D. Marx, S. Sengupta, P. Nielaba, A. Z. Patashinskii, and H. Hahn, *J. Phys.: Condens. Matter* **5,** 8509 (1993).

240. R. Moreh and O. Shahal, *Phys. Rev. Lett.* **43,** 1947 (1979).

241. R. Moreh and O. Shahal, *Surf. Sci. Lett.* **177,** L963 (1986).

242. K. Morishige, C. Mowforth, and R. K. Thomas, *Surf. Sci.* **151,** 289 (1985).

243. F. C. Motteler and J. G. Dash, *Phys. Rev. B* **31,** 346 (1985).

244. O. G. Mouritsen and A. J. Berlinsky, *Phys. Rev. Lett.* **48,** 181 (1982).

245. O. G. Mouritsen, *Phys. Rev. B* **32,** 1632 (1985).

246. O. G. Mouritsen, *Computer Studies of Phase Transitions and Critical Phenomena* (Springer, Berlin, 1984), Chapters 5.3 and 5.4.

247. Y. Murakami and H. Suematsu, *Surf. Sci.* **242,** 211 (1991).

248. C. S. Murthy, K. Singer, M. L. Klein, and I. R. McDonald, *Mol. Phys.* **41,** 1387 (1980).

249. C. S. Murthy, S. F. O'Shea, and I. R. McDonald, *Mol. Phys.* **50,** 531 (1983).

250. K. J. Naidoo, J. Schnitker, and J. D. Weeks, *Mol. Phys.* **80,** 1 (1993).

251. K. R. Nary, P. L. Kuhns, and M. S. Conradi, *Phys. Rev. B* **26,** 3370 (1982).

252. D. R. Nelson, in *Phase Transitions and Critical Phenomena,* Vol. 7, C. Domb and J. L. Lebowitz (eds.) (Academic Press, London, 1983).

253. X.-Z. Ni and L. W. Bruch, *Phys. Rev. B* **33,** 4584 (1986).

254. D. Nicholson, R. F. Cracknell, and N. G. Parsonage, *Mol. Simulation* **5,** 307 (1990).

255. E. J. Nicol, C. Kallin, and A. J. Berlinsky, *Phys. Rev. B* **38,** 556 (1988).

256. M. Nielsen and J. P. McTague, *Phys. Rev. B* **19,** 3096 (1979).

257. M. Nielsen, K. Kjaer, J. Bohr, and J. P. McTague, *J. Electron Spectrosc. Related Phenomena* **30,** 111 (1983).

258. B. Nienhuis, *J. Phys. A* **15,** 199 (1982).

259. B. Nienhuis, E. K. Riedel, and M. Schick, *Phys. Rev. B* **27,** 5625 (1983).

260. M. den Nijs, in *Phase Transitions and Critical Phenomena,* Vol. 12, C. Domb and J. L. Lebowitz (eds.) (Academic Press, London, 1988).

261. K. J. Niskanen and R. B. Griffiths, *Phys. Rev. B* **32,** 5858 (1985).

262. K. J. Niskanen, *Phys. Rev. B* **33,** 1830 (1986); see also the comment in Ref. 191.

263. A. D. Novaco and J. P. McTague, *Phys. Rev. Lett.* **38,** 1286 (1977).

264. A. D. Novaco and J. P. McTague, *J. Phys. (Paris) Colloq.* **38,** C4–116 (1977).

265. A. D. Novaco, *Phys. Rev. B* **19,** 6493 (1979).

266. A. D. Novaco and J. P. McTague, *Phys. Rev. B* **20,** 2469 (1979).

267. A. D. Novaco, *Phys. Rev. B* **22,** 1645 (1980).
268. A. D. Novaco, *Phys. Rev. B* **35,** 8621 (1987).
269. A. D. Novaco, *Phys. Rev. Lett.* **60,** 2058 (1988).
270. A. D. Novaco and J. P. Wroblewski, *Phys. Rev. B* **39,** 11364 (1989).
271. A. D. Novaco, *Phys. Rev. B* **46,** 8178 (1992).
272. L. Onsager, *Phys. Rev.* **65,** 117 (1944).
273. O. Opitz, D. Marx, S. Sengupta, P. Nielaba, and K. Binder, *Surf. Sci. Lett.* **297,** L122 (1993).
274. S. F. O'Shea and M. L. Klein, *Chem. Phys. Lett.* **66,** 381 (1979).
275. S. F. O'Shea and M. L. Klein, *Phys. Rev. B* **25,** 5882 (1982).
276. S. Ostlund and A. N. Berker, *Phys. Rev. Lett.* **42,** 843 (1979).
277. S. Ostlund and A. N. Berker, *Phys. Rev. B* **21,** 5410 (1980).
278. R. E. Palmer and R. F. Willis, *Surf. Sci. Lett.* **179,** L1 (1987).
279. M. Parrinello and A. Rahman, *Phys. Rev. Lett.* **45,** 1196 (1980).
280. M. C. Payne, M. P. Teter, D. C. Allen, T. A. Arias, and J. D. Joannopoulos, *Rev. Mod. Phys.* **64,** 1045 (1992).
281. J. B. Pendry, *Low Energy Electron Diffraction* (Academic Press, London, 1974).
282. C. Peters and M. L. Klein, *Mol. Phys.* **54,** 895 (1985).
283. C. Peters and M. L. Klein, *Phys. Rev. B* **32,** 6077 (1985).
284. C. Peters and M. L. Klein, *Faraday Discuss. Chem. Soc.* **80,** 199 (1985).
285. C. Pierce and B. Ewing, *J. Phys. Chem.* **68,** 2562 (1964).
286. J. Piper, J. A. Morrison, C. Peters, and Y. Ozaki, *J. Chem. Soc. Faraday Trans. 1* **79,** 2863 (1983).
287. J. Piper, J. A. Morrison, and C. Peters, *Mol. Phys.* **53,** 1463 (1984).
288. V. Pereyra, P. Nielaba, and K. Binder, *J. Phys.: Condens. Matter* **5,** 6631 (1993).
289. V. Pereyra, P. Nielaba, and K. Binder, *Z. Phys. B* **97,** 179 (1995).
290. F. Pobell, *Matter and Methods at Low Temperatures* (Springer, Berlin, 1992).
291. A. van der Pol, A. van der Avoird, and P. E. S. Wormer, *J. Chem. Phys.* **92,** 7498 (1990).
292. V. L. Pokrovsky and A. L. Talapov, *Phys. Rev. Lett.* **42,** 65 (1979).
293. V. Privman (ed.), *Finite Size Scaling and Numerical Simulation of Statistical Systems* (World Scientific, Singapore, 1990).
294. V. Privman, P. C. Hohenberg, and A. Aharony, in *Phase Transitions and Critical Phenomena,* Vol. 14, C. Domb and J. L. Lebowitz, (eds.) (Academic Press, London, 1991).
295. D. K. Remler and P. A. Madden, *Mol. Phys.* **70,** 921 (1990).
296. R. C. Richardson and E. N. Smith, *Experimental Techniques in Condensed Matter Physics at Low Temperatures* (Addison-Wesley, Redwood City, 1988).
297. E. K. Riedel, *Physica A* **106,** 110 (1981).
298. S. E. Roosevelt and L. W. Bruch, *Phys. Rev. B* **41,** 12236 (1990).
299. B. Rosenblum, A. H. Nethercot, Jr., and C. H. Townes, *Phys. Rev.* **109,** 400 (1958).
300. S. Ross and W. Winkler, *J. Colloid. Sci.* **10,** 319 (1955).
301. M. Roth and R. D. Etters, *Phys. Rev. B* **44,** 6581 (1991).

302. M. W. Roth, *Phys. Rev. B* **51,** 7778 (1995).
303. J. Rouquerol, S. Partyka, and F. Rouquerol, *J. Chem. Soc. Faraday Trans. 1* **73,** 306 (1977).
304. L. Sanche and M. Michaud, *Phys. Rev. B* **30,** 6078 (1984).
305. L. M. Sander and J. Hautman, *Phys. Rev. B* **29,** 2171 (1984).
306. R. H. Savage and C. Brown *J. Am. Chem. Soc.* **70,** 2362 (1948).
307. M. Schick, *Prog. Surf. Sci.* **11,** 245 (1981).
308. M. Schick, *Surf. Sci.* **125,** 94 (1983).
309. H. P. Schildberg and H. J. Lauter, *Surf. Sci.* **208,** 507 (1989).
310. D. Schmeißer, I. W. Lyo, F. Greuter, E. W. Plummer, H.-J. Freund, and M. Seel, *Ber. Bunsenges. Phys. Chem.* **90,** 228 (1986).
311. K. E. Schmidt and D. M. Ceperley, in *The Monte Carlo Method in Condensed Matter Physics,* K. Binder (ed.) (Springer, Berlin, 1992).
312. T. A. Scott, *Phys. Rep.* **C27,** 89 (1976).
313. J. L. Seguin, J. Suzanne, M. Bienfait, J. G. Dash, and J. A. Venables, *Phys. Rev. Lett.* **51,** 122 (1983).
314. W. Selke, *Physica A* **177,** 460 (1991).
315. S. Sengupta, D. Marx, and P. Nielaba, *Europhys. Lett.* **20,** 383 (1992).
316. S. Sengupta, O. Opitz, D. Marx, and P. Nielaba, *Europhys. Lett.* **24,** 13 (1993).
317. T. Shinoda, T. Atake, H. Chihara, Y. Mashiko, and S. Seki, *Kogyo Kagaku Zasshi* **69,** 1619 (1966).
318. P. Shrestha, M. T. Alkhafaji, M. M. Lukowitz, G. Yang, and A. D. Migone, *Langmuir* **10,** 3244 (1994).
319. N. D. Shrimpton, M. W. Cole, W. A. Steele, and M. H. W. Chan, in *Surface Properties of Layered Structures,* G. Benedek (ed.) (Kluwer Academic Publisher, Dordrecht, 1992), p. 219.
320. I. F. Silvera, *Rev. Mod. Phys.* **52,** 393 (1980).
321. S. K. Sinha (ed.), *Ordering in Two Dimensions* (North Holland, New York, 1980).
322. S. K. Sinha, in *Methods of Experimental Physics,* Vol. 23, *Neutron Scattering, Part B,* D. L. Price and K. Sköld (eds.) (Academic Press, San Diego, 1987), p. 1.
323. E. D. Specht, A. Mak, C. Peters, M. Sutton, R. J. Birgeneau, K. L. D'Amico, D. E. Moncton, S. E. Nagler, and P. M. Horn, *Z. Phys. B* **69,** 347 (1987).
324. W. A. Steele, *Surf. Sci.* **36,** 317 (1973).
325. W. A. Steele, *The Interaction of Gases with Solid Surfaces* (Pergamon Press, Oxford, 1974).
326. W. A. Steele, *J. Phys. (Paris) Colloq.* **38,** C4-61 (1977).
327. W. A. Steele, *J. Phys. Chem.* **82,** 817 (1978).
328. W. A. Steele, A. V. Vernov, and D. J. Tildesley, *Carbon* **25,** 7 (1987).
329. P. W. Stephens, P. A. Heiney, R. J. Birgeneau, P. M. Horn, D. E. Moncton, and G. S. Brown, *Phys. Rev. B* **29,** 3512 (1984).
330. D. E. Stogryn and A. P. Stogryn, *Mol. Phys.* **11,** 371 (1966).
331. J. Stoltenberg and O. E. Vilches, *Phys. Rev. B* **22,** 2920 (1980).
332. K. J. Strandburg, *Rev. Mod. Phys.* **60,** 161 (1988); Erratum: *Rev. Mod. Phys.* **61,** 747(E) (1989).

333. K. J. Strandburg (ed.), *Bond-Orientational Order in Condensed Matter Systems* (Springer, New York, 1992).
334. P. F. Sullivan and G. Seidel, *Phys. Rev.* **173,** 679 (1968).
335. N. S. Sullivan and J. M. Vaissiere, *Phys. Rev. Lett.* **51,** 658 (1983).
336. R. M. Suter, N. J. Colella, and R. Gangwar, *Phys. Rev. B* **31,** 627 (1985).
337. R. H. Swendsen, J.-S. Wang, and A. M. Ferrenberg, in Ref. 37.
338. W. C. Swope and H. C. Andersen, *Phys. Rev. A* **46,** 4539 (1992).
339. W. C. Swope and H. C. Andersen, *J. Chem. Phys.* **102,** 2851 (1995).
340. J. Talbot, D. J. Tildesley, and W. A. Steele, *Mol. Phys.* **51,** 1331 (1984). Note that the exponents 5 and 2 in Eq. (6) are misplaced; see Eq. (2.35) in Ref. 324.
341. J. Talbot, D. J. Tildesley, and W. A. Steele, *Faraday Discuss. Chem. Soc.* **80,** 91 (1985).
342. J. Talbot, D. J. Tildesley, and W. A. Steele, *Surf. Sci.* **169,** 71 (1986).
343. S. Tang, S. D. Mahanti, and R. K. Kalia, *Phys. Rev. B* **32,** 3148 (1985).
344. S. Tang, S. D. Mahanti, and R. K. Kalia, *Phys. Rev. Lett.* **56,** 484 (1986).
345. P. Tarazona and E. Chacón, *Phys. Rev. B* **39,** 7157 (1989).
346. H. Taub, K. Carneiro, J. K. Kjems, L. Passell, and J. P. McTague, *Phys. Rev. B* **16,** 4551 (1977).
347. H. Taub, G. Torzo, H. J. Lauter, and S. C. Fain, Jr. (eds.), *Phase Transitions in Surface Films 2* (Plenum, New York, 1991).
348. A. Terlain and Y. Larher, *Surf. Sci.* **93,** 64 (1980).
349. D. Thirumalai, R. W. Hall, and B. J. Berne, *J. Chem. Phys.* **81,** 2523 (1984).
350. R. K. Thomas, *Prog. Solid State Chem.* **14,** 1 (1982).
351. A. Thomy, X. Duval, and J. Regnier, *Surf. Sci. Rep.* **1,** 1 (1981).
352. D. J. Tildesley and R. M. Lynden-Bell, *J. Chem. Soc. Faraday Trans. 2* **82,** 1605 (1986).
353. M. F. Toney and S. C. Fain, Jr., *Phys. Rev. B* **36,** 1248 (1987).
354. E. Tsidoni, Y. Kozirovski, M. Folman, and J. Heidberg, *J. Electron Spectrosc. Related Phenomena* **44,** 89 (1987).
355. C. Uher, *Cryogenics* **20,** 445 (1980).
356. C. Uher, *Phys. Rev. B* **25,** 4167 (1982).
357. C. Uher and L. M. Sander, *Phys. Rev. B* **27,** 1326 (1983).
358. A. Vernov and W. Steele, *Surf. Sci.* **171,** 83 (1986).
359. A. V. Vernov and W. A. Steele, *Langmuir* **2,** 219 (1986).
360. A. V. Vernov and W. A. Steele, *Langmuir* **2,** 606 (1986).
361. A. Vernov and W. A. Steele, *Langmuir* **8,** 155 (1992).
362. G. Vidali and M. W. Cole, *Phys. Rev. B* **29,** 6736 (1984).
363. G. Vidali, G. Ihm, H.-Y. Kim, and M. W. Cole, *Surf. Sci. Rep.* **12,** 133 (1991).
364. J. Villain and M. B. Gordon, *Surf. Sci.* **125,** 1 (1983).
365. E. Vives and P. A. Lindgård, *Phys. Rev. B* **47,** 7431 (1993).
366. U. G. Volkmann and K. Knorr, *Phys. Rev. Lett.* **66,** 473 (1991).
367. H. Vollmayr, *Phys. Rev. B* **46,** 733 (1992).
368. K. Vollmayr, J. D. Reger, M. Scheucher, and K. Binder, *Z. Phys. B* **91,** 113 (1993).

369. J. Walton, J. Brookeman, and A. Rigamonti, *Phys. Rev. B* **28,** 4050 (1983).
370. R. Wang, S.-K. Wang, H. Taub, J. C. Newton, and H. Shechter, *Phys. Rev. B* **35,** 5841 (1987).
371. S.-K. Wang, J. C. Newton, R. Wang, H. Taub, J. R. Dennison, and H. Shechter, *Phys. Rev. B* **39,** 10331 (1989).
372. B. E. Warren, *X-Ray Diffraction* (Dover, New York, 1990).
373. H. Weber and D. Marx, *Europhys. Lett.* **27,** 593 (1994).
374. H. Weber, D. Marx, and K. Binder, *Phys. Rev. B* **51,** 14636 (1995).
375. G. K. White, *Experimental Techniques in Low Temperature Physics* (Clarendon Press, Oxford, 1979).
376. H. Wiechert, E. Maus, and K. Knorr, *Jpn. J. Appl. Phys.* **26,** Suppl. 26-3, 889 (1987).
377. H. Wiechert, *Physica B* **169,** 144 (1991).
378. H. Wiechert, in *Excitations in Two-Dimensional and Three-Dimensional Quantum Fluids,* A. F. G. Wyatt and H. J. Lauter (eds.) (Plenum Press, New York, 1991).
379. H. Wiechert, H. Freimuth, and H. J. Lauter, *Surf. Sci.* **269/270,** 452 (1992).
380. H. Wiechert and St.-A. Arlt, *Phys. Rev. Lett.* **71,** 2090 (1993).
381. H. Wiechert and coworkers, to be published.
382. M. Willenbacher, E. Britten, E. Maus, and H. Wiechert, to be published.
383. D. P. Woodruff and T. A. Delchar, *Modern Techniques of Surface Science* (Cambridge University Press, Cambridge, 1986).
384. C. J. Wright and C. M. Sayers, *Rep. Prog. Phys.* **46,** 773 (1983).
385. F. Y. Wu, *Rev. Mod. Phys.* **54,** 235 (1982).
386. R. W. G. Wyckoff, *Crystal Structures,* Vol. I (Wiley, New York, 1963).
387. G. Yang, A. D. Migone, and K. W. Johnson, *Rev. Sci. Instrum.* **62,** 1836 (1991).
388. H. You and S. C. Fain, Jr., *Faraday Discuss. Chem. Soc.* **80,** 159 (1985).
389. H. You and S. C. Fain, Jr., *Surf. Sci.* **151,** 361 (1985).
390. H. You and S. C. Fain, Jr., *Phys. Rev. B* **33,** 5886 (1986).
391. H. You and S. C. Fain, Jr., *Phys. Rev. B* **34,** 2840 (1986).
392. H. You, S. C. Fain, Jr., S. Satija, and L. Passell, *Phys. Rev. Lett.* **56,** 244 (1986).
393. A. Zangwill, *Physics at Surfaces* (Cambridge University Press, Cambridge, 1988).
394. Q. M. Zhang, H. K. Kim, and M. H. W. Chan, *Phys. Rev. B* **32,** 1820 (1985).
395. Q. M. Zhang, H. K. Kim, and M. H. W. Chan, *Phys. Rev. B* **33,** 413 (1986).
396. Q. M. Zhang and J. Z. Larese, *Phys. Rev. B* **43,** 938 (1991).
397. G. Zimmerli and M. H. W. Chan, *Phys. Rev. B* **45,** 9347 (1992).

AUTHOR INDEX

Numbers in parentheses are reference numbers and indicate that the author's work is referred to although his name is not mentioned in the text. Numbers in *italic* show the pages on which the complete references are listed.

SUBJECT INDEX

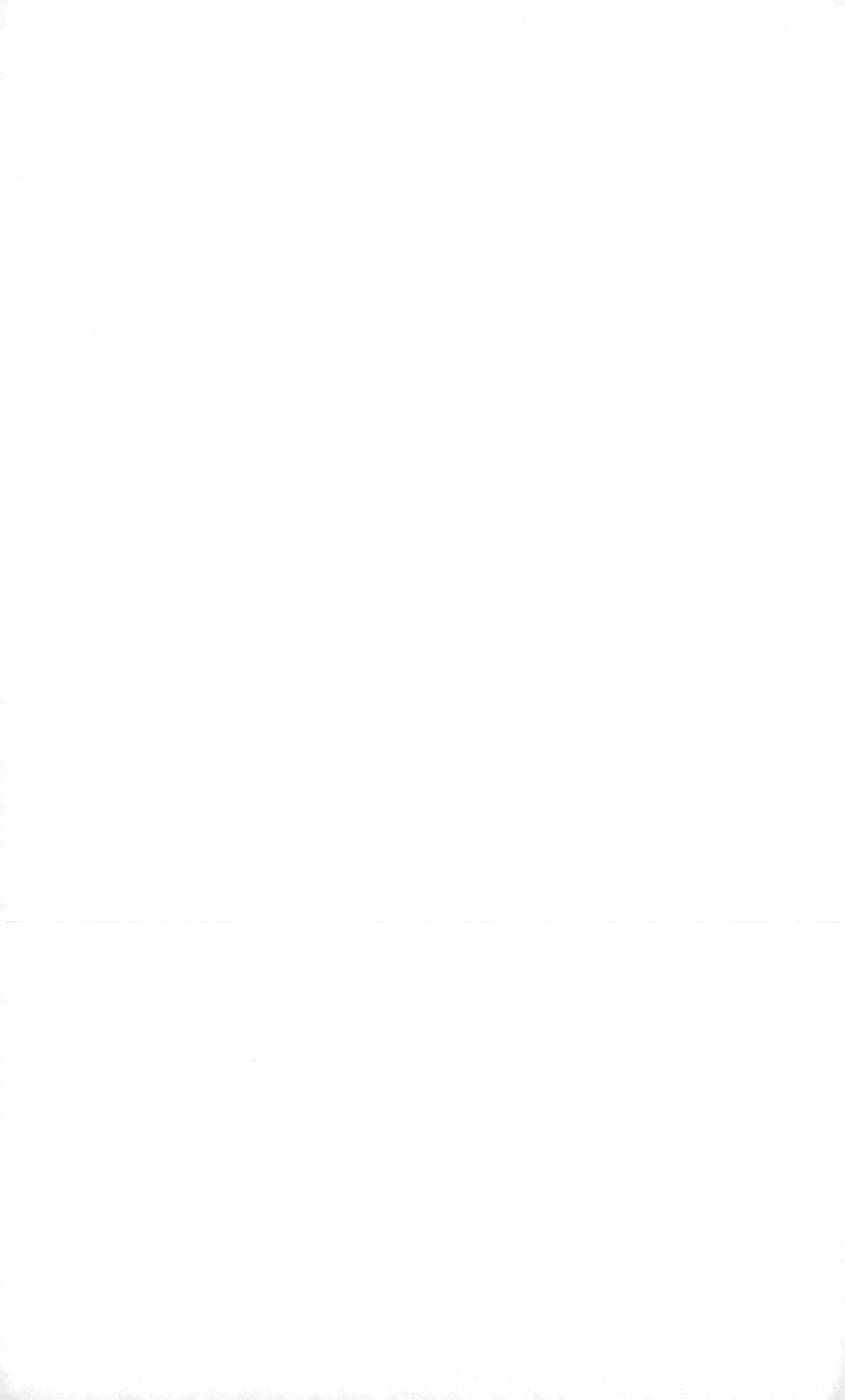